THE ANGULAR MOMENTUM OF LIGHT

Recent developments in the angular momentum of light present fresh challenges to long-established concepts and pave the way for new and wide-ranging applications. The scope for structured light such as optical vortices, in particular, now extends from microfluidics to quantum information.

This comprehensive edited collection deals with light carrying spin and orbital angular momentum, covering both fundamental and applied aspects. Written by internationally leading specialists, the chapters have been compiled to reflect the latest scientific progress and to address the multitude of theoretical, experimental and technical issues associated with this vibrant and exciting field.

The volume is an authoritative reference for academic researchers and graduate students engaged in theoretical or experimental study of optical angular momentum and its applications. It will also benefit professionals in physics, optics and optical engineering, chemistry and biology.

DAVID L. ANDREWS is Professor of Chemical Physics at the University of East Anglia where he leads theoretical research on fundamental photonics and molecular energy transport. He is a Fellow of the SPIE, the Institute of Physics and the Royal Society of Chemistry.

MOHAMED BABIKER is Professor of Physics at The University of York and has previously held this role at the University of Essex. Professor Babiker has been researching light beams with orbital angular momentum for the past 20 years. He is a Fellow of the Institute of Physics.

THE ANGULAR MOMENTUM OF LIGHT

Edited by

DAVID L. ANDREWS
University of East Anglia
MOHAMED BABIKER
The University of York

CAMBRIDGE
UNIVERSITY PRESS

CAMBRIDGE
UNIVERSITY PRESS

32 Avenue of the Americas, New York NY 10013-2473, USA

Cambridge University Press is part of the University of Cambridge.

It furthers the University's mission by disseminating knowledge in the pursuit of education, learning and research at the highest international levels of excellence.

www.cambridge.org
Information on this title: www.cambridge.org/9781107006348

First published 2013

A catalogue record for this publication is available from the British Library

Library of Congress Cataloguing in Publication data

The angular momentum of light / edited by David L. Andrews and Mohamed Babiker.
pages cm
Includes bibliographical references and index.
ISBN 978-1-107-00634-8 (Hardback)
1. Angular momentum. 2. Quantum optics. I. Andrews, David L., 1952– II. Babiker, Mohamed.
QC446.3.A54A54 2012
535′.15–dc23

2011052383

ISBN 978-1-107-00634-8 Hardback

Contents

Contributors

A. Aiello
Universität Erlangen-Nürnberg, Institut für Optik, Information und Photonik, Staudtstrasse 7/B2, Erlangen, D-91058, Germany

M. A. Alonso
University of Rochester, The Institute of Optics, Rochester, New York, NY 14627, USA

D. L. Andrews
University of East Anglia, School of Chemistry, Norwich Research Park, Norwich, NR4 7TJ, UK

M. Babiker
University of York, Department of Physics, York, YO10 5DD, UK

S. M. Barnett
Scottish Universities Physics Allliance (SUPA), Department of Physics, University of Strathclyde, John Anderson Building, Glasgow, G4 0NG, UK

Z. Bialynicka-Birula
Institute of Physics, Lotnikow 32/46, Warszawa, 2668, Poland

I. Bialynicki-Birula
Center for Theoretical Physics, Lotnikow 32/46, Warszawa, 2668, Poland

K. Y. Bliokh
National University of Ireland, Applied Optics Group, School of Physics, Galway, Ireland

J. Courtial
University of Glasgow, School of Physics and Astronomy, Kelvin Building, Glasgow, G12 8QQ, UK

V. R. Daria
The Australian National University, Research School of Physics and Engineering, and John Curtin School of Medical Research, Canberra, ACT 0200, Australia

N. Davidson
Weizmann Institute of Science, Department of Physics of Complex Systems, Rehovot, 76100, Israel

A. S. Desyatnikov
The Australian National University, Nonlinear Physics Centre, Research School of Physics and Engineering, Canberra, ACT 0200, Australia

N. Drozd
Center for Theoretical Physics, Lotnikow 32/46, Warszawa 2668, Poland

E. R. Eliel
Universiteit Leiden, Huygens Laboratory, Niels Bohrweg 2, Leiden, 2333 CA, The Netherlands

O. Firstenberg
Technion – Israel Institute of Technology, Department of Physics, Haifa, 32000, Israel

S. Franke-Arnold
University of Glasgow, School of Physics and Astronomy, Kelvin Building, Glasgow, G12 8QQ, UK

E. J. Galvez
Colgate University, Department of Physics and Astronomy, 13 Oak Drive, Hamilton, New York, NY 13345, USA

D. Giovannini
University of Glasgow, School of Physics and Astronomy, Kelvin Building, Glasgow, G12 8QQ, UK

J. Glückstad
Technical University of Denmark, DTU Fotonik, Dept of Photonics and Engineering, Anker Engelundsvej 1, 101A, Lyngby, DK-2800 Kgs., Denmark

J. B. Götte
University of Bristol, HH Wills Physics Laboratory, Tyndall Avenue, Bristol, BS8 1TL, UK

Y. S. Kivshar
The Australian National University, Nonlinear Physics Centre, Research School of Physics and Engineering, Canberra, ACT 0200, Australia

M. P. J. Lavery
University of Glasgow, School of Physics and Astronomy, Kelvin Building, Glasgow, G12 8QQ, UK

V. E. Lembessis
King Saud University, Department of Physics and Astronomy, P.O. Box 2455, Riyadh 11451, Saudi Arabia

G. Molina-Terriza
Macquarie University, Faculty of Science, Dept of Physics and Astronomy, C5C-350 Level 3 West, Sydney, NSW 2109, Australia

G. Nienhuis
Universiteit Leiden, Huygens Laboratorium, Postbus 9504, Leiden 2300, The Netherlands

M. J. Padgett
University of Glasgow, School of Physics and Astronomy, Kelvin Building, Glasgow, G12 8QQ, UK

D. Palima
Technical University of Denmark, DTU Fotonik, Dept of Photonics and Engineering, Anker Engelundsvej 1, 101A, Lyngby, DK-2800 Kgs., Denmark

R. Pugatch
Weizmann Institute of Science, Department of Physics of Complex Systems, Rehovot, 76100, Israel

J. Romero
University of Glasgow, School of Physics and Astronomy, Kelvin Building, Glasgow, G12 8QQ, UK

M. Shuker
Technion – Israel Institute of Technology, Department of Physics, Haifa, 32000, Israel

M. P. van Exter
Universiteit Leiden, Huygens Laboratory, Niels Bohrweg 2, Leiden, 2333 CA, The Netherlands

J. P. Woerdman
Universiteit Leiden, Huygens Laboratory, Niels Bohrweg 2, Leiden, 2333 CA, The Netherlands

Preface

For any sceptic of the continued capacity of science to uncover new truth, to pave the way for previously unimagined applications, there is hardly a better corrective than to invite reflection on recent discoveries in the science of light. It may be unscientific to say that light is unfathomable, but it certainly is a characteristic of the subject that there is always more to be learned, just when the utmost depths seem within grasp. There is no better illustration than the specific subject of the volume before you.

It has long been known that light conveys energy, and the associated linear momentum has also been understood since the days of Maxwell and Bartoli.[1] With angular momentum the history is more recent, and the property a little less straightforward. What we quickly learned is that light has a propensity to convey angular momentum, depending on its state. The pioneering work in which Beth[2] established a link with circular polarisation is nonetheless already three-quarters of a century old. Once the quantum theory of light was developed, many would have surmised that the science was complete, the concept of angular momentum so beautifully related to the unit spin of the photon–the hallmark of a boson. But what has been discovered in the past quarter century has shown that the spin angular momentum is only half the story – and the other half has no ending yet in sight.

Recent developments in the angular momentum of light are leading to new and wide-ranging applications, even as the subject presents fresh challenges to long established and cherished concepts. The present volume of contributed chapters has been compiled to reflect how far the science has now progressed, and to address the multitude of theoretical, experimental and technical issues associated with this vibrant and exciting field. The chapters are as follows.

In Chapter 1, Götte and Barnett formally introduce orbital angular momentum, the topic that has opened the door to a broader understanding of angular momentum in light. The twisted wave-front structures that deliver such features have a characteristic form whose vortex properties are then discussed by Molina-Terriza in Chapter 2. Chapter 3 by Galvez describes the detailed relationships between polarization and wave-front morphology, revealing the connectivity between spin and orbital angular momentum. Desyatnikov

[1] J. C. Maxwell, *A Treatise on Electricity and Magnetism*, Vol. 2 (Macmillan, London, 1873); A. Bartoli, Il calorico raggiante e il secondo principio di termodynamica, *Il Nuovo Cimento* **15**, 196–202 (1884).

[2] R. A. Beth, Mechanical detection and measurement of the angular momentum of light, *Phys. Rev.* **50**, 115–25 (1936).

and Kivshar address the additional complexities that arise when twisted light propagates through an optically nonlinear medium, in Chapter 4. In Chapter 5, Nienhuis provides a study of the connections between ray optics, wave optics and quantum mechanics, associated with optical angular momentum. In a second contribution by Götte and Barnett, Chapter 6 continues the theme with an analysis of the interplay between discrete and quantum mechanically uncertain angular properties. Chapter 7 by Bialynicki-Birula and Bialynicka-Birula then provides the theory of rotational frequency shift, a dynamically controllable feature that affords another tier of opportunities for application.

In Chapter 8, Bliokh *et al.* give a detailed account of the interactions between the two forms of optical angular momentum, spin and orbital. Theory continues with Chapter 9 by Andrews and Babiker, detailing the quantum electrodynamical basis for the interactions of optical vortices, lending insights into the connected issues of chirality and propagated angular momentum. In a second contribution by Bialynicki-Birula *et al.*, Chapter 10 focuses attention on the trapping of charged particles by Bessel beams, a theme continued in a second contribution by Babiker *et al.*, addressing charged and uncharged atoms in Laguerre–Gaussian light, in Chapter 11. As an introduction to specifically quantum optical applications, Romero *et al.* describe in Chapter 12 the practical aspects of producing and deploying light with orbital angular momentum. In Chapter 13, also from Padgett's group, the means of experimentally determining and separating beams with such angular momentum are described. More experimental detail is provided in Chapter 14 by Daria *et al.*, concerning the efficient generation and deployment of twisted optical beams. In Chapter 15, Firstenberg *et al.* address the issues of coherent diffusion, and finally in Chapter 16 van Exter *et al.* return to a distinctly quantum mechanical topic, developing and exploiting the orbital angular momentum of light to achieve azimuthal entanglement.

For the editors it has been a particular pleasure to undertake the responsibility of collating contributions from an outstanding set of authors, who include many of the key players in recent developments. On behalf of all our readers, we express our indebtedness to them for bringing a new level of clarity and accessibility to this fascinating subject.

1

Light beams carrying orbital angular momentum

J. B. GÖTTE AND S. M. BARNETT

For optical fields the notion of a total angular momentum has long been known. The concept of a light beam carrying orbital angular momentum, however, was unfamiliar until it was discovered that Laguerre–Gaussian beams, within the paraxial approximation, carry a well-defined orbital angular momentum [1, 2]. This discovery started the modern interest in orbital angular momentum of light. In this chapter we discuss the theoretical framework of orbital angular momentum of light in terms of fields and light beams and how to generate these. The material in this chapter is based in parts on the PhD thesis of Götte [3].

1.1 Introduction

A quantitative treatment of the mechanical effects of light became possible only after light had been integrated into Maxwell's dynamical theory of electromagnetic waves. With this theory Poynting [4] derived a continuity equation for the energy in the electromagnetic field. After Heaviside [5, 6] introduced the vectorial notation for the Maxwell equations this continuity equation could be written in its modern form using the Poynting vector. Interestingly, the linear momentum density in the electromagnetic field is also given by the Poynting vector apart from constant factors depending on the chosen system of units. Poynting [7] also derived an expression for the angular momentum of circularly polarised light by means of a mechanical analogue in the form of a rotating shaft. Later, Poynting's expression was verified by measuring the torque on a quarter wave-plate due to circularly polarised light [8]. By this time quantum mechanics had been firmly established and Beth [8] showed that the same quantitative result could be obtained whether the torque was calculated classically, or from the assumption that each quantum of circularly polarised light carries an angular momentum of $\pm\hbar$.

While spin angular momentum in light beams was thus established in terms of circularly polarised light, the orbital angular momentum of light beams remained a less familiar concept. This does not mean, however, that the orbital angular momentum of light fields

---

The Angular Momentum of Light, ed. D. L. Andrews and M. Babiker. Published by Cambridge University Press.

in general has not been studied [9]. The total angular momentum density is given by the cross product of the radial vector and the linear momentum density, which is a form typical for orbital angular momentum [10]. This intriguing fact shows the difficulties in splitting the total angular momentum into spin and orbital parts for a general electromagnetic field. It was discovered at Leiden University that Laguerre–Gaussian light beams, familiar from paraxial optics or laser theory and realisable in a laboratory, have a well-defined separation of the total angular momentum into spin and orbital parts [1, 2]. Both can be transferred separately to a trapped particle causing a rotation around the centre of the particle, owing to spin angular momentum, or around the centre of the beam owing to orbital angular momentum [11]. This has led to a wide use of the angular momentum of light in the field of optical micromanipulation and trapping [12, 13].

The separation into spin and orbital angular momentum is also crucial for the quantum mechanical properties of light. The two orthogonal polarisations, for example left and right circularly polarised light, and thus the spin angular momentum have long been used as an optical implementation of a qubit in quantum information [14]. The possibility to produce pairs of photons entangled in both spin and orbital angular momentum [15, 16] has created a strong interest in optical angular momentum as a resource in quantum information. In contrast to spin angular momentum, orbital angular momentum is not restricted to two orthogonal states, which is why it is used as a high-dimensional quantum system (see Chapters 12 and 16).

In this chapter we review the concept of orbital angular momentum of light beams. Starting from the Maxwell equations we look at the continuity equation for linear and angular momentum of the electromagnetic field. We derive the exact Helmholtz equation and then the paraxial approximation to it in order to study light beams with orbital angular momentum in the paraxial and non-paraxial regime. This allows us to establish that light beams with a particular azimuthal phase structure have orbital angular momentum. At the end of this chapter we will describe some methods to generate these beams in a laboratory.

1.2 Mechanical properties of optical fields

The set of Maxwell equations describes the electromagnetic field including its mechanical properties. The conservation of mechanical properties such as energy, linear and angular momentum are expressed by continuity equations. These equations describe the change of quantities in a continuum and are therefore stated in terms of densities. The rate of change for a density ρ of a locally conserved quantity is given by the divergence of the respective flux density $\mathbf{j}$ and the source density q:

$$\frac{\partial \rho}{\partial t} + \nabla \cdot \mathbf{j} = q. \tag{1.1}$$

Here, ρ could stand for the density of electromagnetic charge or energy as well as the density of linear or angular momentum, while $\mathbf{j}$ and q are the corresponding flux and source densities respectively. In case a quantity is locally conserved the source density vanishes identically.

The relevant continuity equation for the study of optical angular momentum governs the local conversation of the total angular momentum. However, for the electromagnetic field the continuity equations for the mechanical quantities are intricately linked; the flux density of the energy is also the density for the linear momentum and the angular momentum densities are directly related to the linear momentum densities. In the following we therefore review the continuity equations for energy, and the linear and angular momentum of the electromagnetic field.

1.2.1 Energy of the electromagnetic field

Electromagnetic energy can be converted into mechanical or thermal energy. This conversion of energy will be mediated through work exerted on a charge q through the Lorentz force [17, 18]

$$\mathbf{F} = q\left(\mathbf{E} + \mathbf{v} \times \mathbf{B}\right), \tag{1.2}$$

which is stated here for a point charge q with velocity $\mathbf{v}$. In this chapter we will use vectorial notation along with notation in components. The Cartesian component of a vector $\mathbf{a}$ will be denoted by a_i, where $i = 1, 2, 3$ represents the x, y or z component respectively. The Cartesian components of the Lorentz force may thus be we written as

$$F_i = q\left(E_i + \epsilon_{ijk} v_j B_k\right), \tag{1.3}$$

where we have used the Levi–Civita symbol ϵ_{ijk} to write the cross product $\mathbf{v} \times \mathbf{B}$ in components. The Levi–Civita symbol is equal to 1 if the combination of the indices ijk is an even permutation of 123 and it is equal to -1 for an odd permutation. Otherwise it is zero. We also use the convention of implicitly summing over doubly occurring indices.

The work exerted on a point charge between two arbitrary times t_0 and t_1 is thus given by

$$W(t_0, t_1) = \int_{t_0}^{t_1} \mathrm{d}t\ v_i F_i = q \int_{t_0}^{t_1} \mathrm{d}t\ v_i E_i. \tag{1.4}$$

The magnetic field does not contribute to the work as the cross product $\mathbf{v} \times \mathbf{B}$ is perpendicular to $\mathbf{v}$. For a continuous distribution of charges ρ in a velocity field $\mathbf{v}$ the energy source density of the electromagnetic fields is expressed as [10]

$$q^E = -j_i E_i = -\mathbf{j} \cdot \mathbf{E}, \tag{1.5}$$

where the charge flux density $\mathbf{j}$ is given by $\rho\mathbf{v}$. The work exerted reduces the energy of the electric field, which explains the minus sign in the expression above. The expression in Eq. (1.5) has the dimension of power which is lost through work done on the charges. The loss in power is balanced by the rate of change of the energy of the electromagnetic field and the energy flux. In the absence of any charged particles, however, the continuity equation has no source density term and the rate of change of the energy density is balanced by the divergence of the energy density flux only.

To find the expressions for the energy density and the energy flux density we can use the Maxwell equations for external charge and flux densities [18, 19]

$$\nabla \cdot \mathbf{E} = \frac{\rho}{\epsilon_0}, \tag{1.6a}$$

$$\nabla \times \mathbf{E} = -\frac{\partial \mathbf{B}}{\partial t}, \tag{1.6b}$$

$$\nabla \cdot \mathbf{B} = 0, \tag{1.6c}$$

$$\nabla \times \mathbf{B} = \mu_0 \left(\mathbf{j} + \epsilon_0 \frac{\partial \mathbf{E}}{\partial t} \right). \tag{1.6d}$$

On substituting Eq. (1.6d) into Eq. (1.5) and using Eq. (1.6b) to substitute $\nabla \times \mathbf{E}$ we find that

$$-\mathbf{j} \cdot \mathbf{E} = \frac{\partial}{\partial t} \left(\frac{\epsilon_0}{2} E^2 + \frac{1}{2\mu_0} B^2 \right) + \nabla \cdot \left(\frac{1}{\mu_0} \mathbf{E} \times \mathbf{B} \right). \tag{1.7}$$

This has the required form of a continuity equation for the energy of the electromagnetic field if we identify

$$\rho^E \equiv w = \frac{\epsilon_0}{2} E^2 + \frac{1}{2\mu_0} B^2 \tag{1.8}$$

with the energy density of the electromagnetic field and

$$\mathbf{j}^E \equiv \mathbf{S} = \frac{1}{\mu_0} \mathbf{E} \times \mathbf{B} \tag{1.9}$$

with the energy flux density. The quantity $\mathbf{S}$ is often called the Poynting vector as it was Poynting [4] who discovered the law for the transfer of energy in the electromagnetic field. The continuity equation is therefore also known as Poynting's theorem. A continuity equation does not determine the density and flux density completely; the continuity equation Eq. (1.1) is also fulfilled for a density ρ^E augmented by the divergence $\nabla \cdot \mathbf{V}$ of an arbitrary vector field $\mathbf{V}$ if its time derivative $\partial \mathbf{V}/\partial t$ is subtracted from the flux density $\mathbf{j}^E$. If $\mathbf{V}$ falls off sufficiently quickly, both expressions for the energy density give rise to the same total energy. In addition it is always possible to add the curl $\nabla \times \mathbf{W}$ of an arbitrary vector field $\mathbf{W}$ to the energy flux density $\mathbf{j}^E$ as only the divergence of the flux density enters the continuity equation [20]. It is, however, possible to determine the energy density uniquely from relativistic considerations [18, 21] and these suggest that we should set $\mathbf{V}$ and $\mathbf{W}$ equal to zero. Finally, it should be noted that the Poynting vector in Eq. (1.9) is not a generally valid expression for the energy flux density, but rather only in the special case of external sources in vacuum, where the electric displacement $\mathbf{D}$ is related to the electric field $\mathbf{E}$ by $\mathbf{D} = \epsilon_0 \mathbf{E}$ and ϵ_0 is the free space electric permittivity. Also, the magnetic induction $\mathbf{B}$ has to be related to the magnetic field $\mathbf{H}$ by $\mathbf{B} = \mu_0 \mathbf{H}$ and μ_0 is the free space magnetic permittivity. For the propagation of light beams we will treat the electromagnetic wave as a closed system without any sources. In this case the expression for the energy flux density is rigorous [22].

1.2.2 Linear momentum of the electromagnetic field

The linear momentum is a vector and we will use the notation in components to express the continuity equation for a vector quantity:

$$\frac{\partial \rho_k^P}{\partial t} + \frac{\partial j_{ik}^P}{\partial x_i} = q_k^P, \tag{1.10}$$

where the superscript P denotes the locally preserved quantity, in this case the momentum. For the linear momentum the rate of change of the momentum density and the divergence of the momentum flux density are balanced by the momentum source density. It is not surprising that the source density is given by minus the Lorentz force density:

$$q_k^P \equiv f_k = -\rho E_k - \epsilon_{klm} j_l B_m. \tag{1.11}$$

On using the Maxwell equations (1.6) to manipulate the expression for the source density the continuity equation for the linear momentum can be derived if we identify

$$\rho_k^P \equiv \frac{1}{c^2}\mu_0 S_k = \epsilon_0 \epsilon_{klm} E_l B_m \tag{1.12}$$

with the linear momentum density. This is identical to the energy flux density apart from the factor μ_0/c^2. The momentum flux density is given by

$$j_{ik}^P \equiv -\mathcal{T}_{ik} = \epsilon_0 \left(\frac{1}{2}E^2\delta_{ik} - E_i E_k\right) + \frac{1}{\mu_0}\left(\frac{1}{2}B^2\delta_{ik} - B_i B_k\right), \tag{1.13}$$

where $\mathcal{T}$ is the symmetric energy-stress tensor [18].

For a symmetric momentum flux density the angular momentum continuity equation may be fulfilled with a density, flux density and source density which are given by a vector product of the radial position and the respective linear density. This is similar to a rigid body where the angular momentum is solely due to orbital angular momentum which is given by the cross product $\mathbf{L} = \mathbf{r} \times \mathbf{P}$ and where the torque $\boldsymbol{\tau} = \mathbf{r} \times \mathbf{F}$ is completely determined by the force $\mathbf{F}$. In a continuum theory this indicates that there is no intrinsic angular momentum and no intrinsic torque, and that for sufficiently small volumes all angular quantities are already given by the corresponding linear quantity. For classical fluids this a well justified assumption, but circularly polarised light carries spin angular momentum [10]. This indicates the difficulties in assigning the terms 'spin' and 'orbital' to specific parts of the total angular momentum. For light beams this difficulty may be resolved, at least in part, by using the angular momentum flux rather than the angular momentum density itself [23].

1.2.3 Angular momentum of the electromagnetic field

The density, flux density and source density for the angular momentum are directly related to the respective linear quantities [10]:

$$\rho_k^J = \epsilon_{klm} x_l \rho_m^P, \tag{1.14a}$$

$$j_{ik}^J = \epsilon_{ijl} x_j j_{lk}^P, \tag{1.14b}$$

$$q_k^J = \epsilon_{klm} x_l q_m^P. \tag{1.14c}$$

The given angular quantities fulfil the angular momentum continuity equation owing to the symmetry of the linear momentum flux density j_{lk}^P. This is because

$$\frac{\partial}{\partial x_k} j_{ik}^J = \epsilon_{ijl} x_j \frac{\partial}{\partial x_k} j_{lk}^P, \tag{1.15}$$

as $\epsilon_{ijl} j_{lj}^P = 0$ by virtue of the symmetry in l and j of j_{lj}^P. The whole continuity equation for the angular momentum can thus be written as

$$\epsilon_{ijk} x_j \left(\frac{\partial}{\partial t} \rho_k^P + \frac{\partial}{\partial x_l} j_{kl}^P - q_k^P \right) = 0, \tag{1.16}$$

which is always fulfilled because of the validity of the continuity equation for linear momentum.

For the purpose of this chapter we will be concerned with the Maxwell equations in free space, so that all the source densities are equal to zero. The density of the optical angular momentum is

$$\boldsymbol{\rho}^J = \epsilon_0 \mathbf{r} \times (\mathbf{E} \times \mathbf{B}) \quad \text{or} \quad \rho_i^J = \epsilon_0 \left(E_i x_j B_j - B_i x_j E_j \right), \tag{1.17}$$

and we introduce $\mathcal{M}$ for the angular momentum flux density which has the components

$$\mathcal{M}_{ik} \equiv j_{ik}^J = \epsilon_{ilk} x_l w - \epsilon_{ilm} x_l \left(\epsilon_0 E_m E_k + \frac{1}{\mu_0} B_m B_k \right), \tag{1.18}$$

where w is the energy density from Eq. (1.8). For a light beam with an optical axis along the z direction the angular momentum flux through a plane of constant z is given by the integral

$$M_{zz} \equiv M_{33} = \iint \mathrm{d}x\, \mathrm{d}y\, \mathcal{M}_{33}. \tag{1.19}$$

With the help of these expressions, we are in the position to calculate the angular momentum of light beams and its separation into spin and orbital parts.

1.2.4 Spin and orbital angular momentum

It is natural, and also essential for the purpose of this book, to try to separate the angular momentum into spin and orbital parts. The former we shall associate with optical polarization and the latter with angular momentum derived from the phase structure of our light beams. In terms of the field, we require decomposition of the total angular momentum of the form

$$\mathbf{J} = \int \mathrm{d}V \epsilon_0 \, \mathbf{r} \times (\mathbf{E} \times \mathbf{B}) = \mathbf{L} + \mathbf{S}, \tag{1.20}$$

where **L** is the orbital part and **S** is the spin part. Perhaps the most obvious way we might make this separation is to write **B** in terms of the vector potential **A** and then after a little manipulation find [24]

$$\begin{aligned} \mathbf{J} &= \epsilon_0 \int \mathrm{d}V \, [E_i(\mathbf{r} \times \nabla)A_i + \mathbf{E} \times \mathbf{A} - \nabla_i(E_i \mathbf{r} \times \mathbf{A})] \\ &= \epsilon_0 \int \mathrm{d}V \, [E_i(\mathbf{r} \times \nabla)A_i + \mathbf{E} \times \mathbf{A}] - \epsilon_0 \int (\mathbf{r} \times \mathbf{A})\mathbf{E} \cdot \mathrm{d}\mathbf{S}, \end{aligned} \tag{1.21}$$

where we have used Gauss' theorem. If the fields fall off sufficiently quickly then the surface integral will be zero and we can identify the orbital and spin parts as

$$\mathbf{L} = \epsilon_0 \int \mathrm{d}V E_i (\mathbf{r} \times \nabla) A_i, \tag{1.22}$$

$$\mathbf{S} = \epsilon_0 \int \mathrm{d}V \, \mathbf{E} \times \mathbf{A}. \tag{1.23}$$

The analogy with the operator for angular momentum in quantum theory suggests the association of the first part with the orbital angular momentum. The second part, however, clearly depends on the vectorial nature of the field and hence its polarization, which makes it natural to associate it with the spin.

The appearance of the vector potential in **L** and **S** has caused much confusion and it has been suggested, for this reason, that the separation might not be physically meaningful [24–28]. It is now known, however, that the separation is meaningful although neither part, **L** nor **S**, is itself an angular momentum, as defined by the commutation relations [29, 30]. The reason for this can be traced to the impossibility of separately rotating either the direction of the fields or their spatial distribution without violating transversality [31].

It is perhaps interesting to mention one further subtlety associated with the orbital and spin parts of the angular momentum. In the absence of charges and currents, Maxwell's equations are invariant under the Heaviside–Lamor or duplex transformation [18, 32, 33].

$$\mathbf{E} \longrightarrow \cos\theta \, \mathbf{E} + \sin\theta \, c\mathbf{B}, \tag{1.24}$$

$$\mathbf{B} \longrightarrow \cos\theta \, \mathbf{B} - \sin\theta \, \frac{1}{c}\mathbf{E}. \tag{1.25}$$

All physically significant quantities for the free field, such as ρ^E, ρ^P and ρ^J, are necessarily also invariant under this transformation. We can write **L** and **S** in a form which manifestly obeys this symmetry by introducing a second vector potential, **C**, for which [34]

$$\mathbf{E} = -c\,\nabla \times \mathbf{C}, \tag{1.26}$$

$$\mathbf{B} = -\frac{1}{c}\frac{\partial}{\partial t}\mathbf{C}. \tag{1.27}$$

The Heaviside–Larmor symmetry then requires that quantities of physical interest should be invariant under

$$\mathbf{A} \longrightarrow \cos\theta\,\mathbf{A} + \sin\theta\,\mathbf{C}, \tag{1.28}$$

$$\mathbf{C} \longrightarrow \cos\theta\,\mathbf{C} - \sin\theta\,\mathbf{A}. \tag{1.29}$$

This allows us to write **L** and **S** in the form [31]

$$\mathbf{L} = \frac{\epsilon_0}{2}\int \mathrm{d}V \left(E_i(\mathbf{r}\times\nabla)A_i + \frac{1}{c}B_i(\mathbf{r}\times\nabla)C_i\right), \tag{1.30}$$

$$\mathbf{S} = \frac{\epsilon_0}{2}\int \mathrm{d}V \left(\mathbf{E}\times\mathbf{A} + \frac{1}{c}\mathbf{B}\times\mathbf{C}\right). \tag{1.31}$$

It is straightforward to show that these have the same value as Eqs. (1.22) and (1.23). They do suggest, however, different values for the *densities* of the orbital and spin parts of the angular momentum.

1.3 Wave equations

In this section we derive the Helmholtz equation and its paraxial approximation for electromagnetic waves in free space. For a consistent application of the paraxial approximation the wave equation has to be derived for the vector potential rather than the electric field or the magnetic induction.

1.3.1 Helmholtz equation

We are mostly concerned with the propagation of waves carrying orbital angular momentum in free space and so we do not consider the presence of charge densities ρ or flux densities **j** in the Maxwell equations. In this case the Maxwell equations are a set of homogenous differential equations for the electric field **E** and the magnetic induction **B**:

$$\nabla\cdot\mathbf{E} = 0, \tag{1.32a}$$

$$\nabla\times\mathbf{E} = -\frac{\partial\mathbf{B}}{\partial t}, \tag{1.32b}$$

$$\nabla\cdot\mathbf{B} = 0, \tag{1.32c}$$

$$\nabla\times\mathbf{B} = \epsilon_0\mu_0\frac{\partial\mathbf{E}}{\partial t}. \tag{1.32d}$$

On taking the curl of either Eq. (1.32b) or Eq. (1.32d) and substituting the other equation it is possible to derive two wave equations for **E** and **B**. For the electric field the double curl $\nabla \times (\nabla \times \mathbf{E})$ may be replaced by $\nabla(\nabla \cdot \mathbf{E}) - \nabla^2\mathbf{E}$ and analogously for the magnetic induction **B**. In both cases the first term is equal to zero because of the Maxwell equations (1.32a) and (1.32c). We are then left with the wave equations

$$\nabla^2\mathbf{E} - \frac{1}{c^2}\frac{\partial^2}{\partial t^2}\mathbf{E} = 0, \tag{1.33a}$$

$$\nabla^2\mathbf{B} - \frac{1}{c^2}\frac{\partial^2}{\partial t^2}\mathbf{B} = 0. \tag{1.33b}$$

In the following we introduce complex electric and magnetic fields and we limit our attention to monochromatic beams with angular frequency ω. This allows us to separate the electric and magnetic field into a part with the spatial dependence and a time dependent factor $\exp(-i\omega t)$:

$$\mathbf{E} = \mathrm{Re}[\boldsymbol{\mathcal{E}} \exp(-i\omega t)] \quad \text{and} \quad \mathbf{B} = \mathrm{Re}[\boldsymbol{\mathcal{B}} \exp(-i\omega t)]. \tag{1.34}$$

By substituting this separated ansatz into the wave equations (1.33) we obtain the Helmholtz equations for $\boldsymbol{\mathcal{E}}$ and $\boldsymbol{\mathcal{B}}$:

$$\nabla^2\boldsymbol{\mathcal{E}} + k^2\boldsymbol{\mathcal{E}} = 0, \tag{1.35a}$$

$$\nabla^2\boldsymbol{\mathcal{B}} + k^2\boldsymbol{\mathcal{B}} = 0, \tag{1.35b}$$

where $k = \omega/c$ is the overall wavenumber.

As detailed by Lax *et al.* [35], the paraxial approximation to the Helmholtz equation (1.35a), as applied to a single polarization, is inconsistent with the Maxwell equation (1.32a) and needs to be treated with care. To avoid inconsistencies it is possible to use a Helmholtz equation for the vector potential **A** [36] which is not required to have a vanishing divergence in the Lorentz gauge. The vector potential is related to the magnetic induction by

$$\mathbf{B} = \nabla \times \mathbf{A}. \tag{1.36}$$

On substituting the definition of the vector potential into Faraday's law in Eq. (1.32b) we find that the curl of the combined vector field $\mathbf{E} + (\partial\mathbf{A}/\partial t)$ is equal to zero:

$$\nabla \times \left(\mathbf{E} + \frac{\partial}{\partial t}\mathbf{A}\right) = 0. \tag{1.37}$$

This is an indication for the existence of a scalar potential Φ defined as

$$\nabla\Phi = -\left(\mathbf{E} + \frac{\partial}{\partial t}\mathbf{A}\right). \tag{1.38}$$

The vector potential and the scalar potential are not completely determined by the electric field and the magnetic induction. The derivative of a scalar function ζ may be added to the vector potential if the time-derivative of ζ is subtracted from the scalar potential [18]:

$$\mathbf{A} \longrightarrow \mathbf{A}' = \mathbf{A} + \nabla\zeta, \quad \Phi \longrightarrow \Phi' = \Phi - \frac{\partial}{\partial t}\zeta, \tag{1.39}$$

where ζ is an arbitrary function of $\mathbf{r}$ and t. The most convenient gauge for our purposes is the Lorentz gauge, as the Coulomb gauge requires that the divergence of the vector potential is equal to zero, which leads to unnecessary complications when making the paraxial approximation [37]. The Lorentz gauge is given by the condition

$$\nabla \cdot \mathbf{A} + \frac{1}{c^2}\frac{\partial}{\partial t}\Phi = 0. \tag{1.40}$$

This also requires that the scalar function ζ obeys a wave equation:

$$\nabla^2\zeta - \frac{1}{c^2}\frac{\partial^2}{\partial t^2}\zeta = 0. \tag{1.41}$$

With the help of the Eqs. (1.36) and (1.38) it is possible to substitute the electric field and the magnetic induction in the Maxwell equations with the scalar and vector potential. The substitution satisfies the Maxwell equations (1.32b) and (1.32c) identically. The dynamical behaviour of $\mathbf{A}$ and Φ is determined by the two remaining Maxwell equations (1.32a) and (1.32d). Together with the Lorentz condition in Eq. (1.40) the two remaining Maxwell equations yield two wave equations for the scalar and the vector potential:

$$\nabla^2\Phi - \frac{1}{c^2}\frac{\partial^2}{\partial t^2}\Phi = 0, \tag{1.42a}$$

$$\nabla^2\mathbf{A} - \frac{1}{c^2}\frac{\partial^2}{\partial t^2}\mathbf{A} = 0. \tag{1.42b}$$

From these, the Helmholtz equation for the vector potential is obtained with the ansatz $\mathbf{A} = \mathrm{Re}[\boldsymbol{\mathcal{A}}\exp(-\mathrm{i}\omega t)]$ for a monochromatic wave. With $k = \omega/c$ this yields the following equation:

$$\nabla^2\boldsymbol{\mathcal{A}} + k^2\boldsymbol{\mathcal{A}} = 0. \tag{1.43}$$

Once a solution for the vector potential has been found, the electric field can be calculated from the vector potential via

$$\boldsymbol{\mathcal{E}} = \mathrm{i}\omega\left(\boldsymbol{\mathcal{A}} + \frac{\nabla(\nabla\cdot\boldsymbol{\mathcal{A}})}{k^2}\right). \tag{1.44}$$

For this relation we have assumed the monochromatic form for the scalar and vector potential. The given equation follows then from the definition of the scalar potential in Eq. (1.38) and the Lorentz condition in Eq. (1.40).

The vectorial form of the Helmholtz equation can be turned into a scalar wave equation by choosing a constant polarisation vector of unit modulus. We are considering the case where the vector fields have a polarisation transverse to the optical axis, which is taken along the z direction. We write, therefore

$$\mathcal{V} = \mathbf{t}\xi(\mathbf{r}), \tag{1.45}$$

where $\mathcal{V}$ stands for the electric field, the magnetic induction or the vector potential and $\mathbf{t}$ is the polarisation vector of unit modulus. With this ansatz the vector form of the Helmholtz equation (1.35) and (1.43) can be turned into a scalar form:

$$\nabla^2 \xi + k^2 \xi = 0. \tag{1.46}$$

This equation is the starting point for the analysis of the exact form of light beams with orbital angular momentum. In the next section we find an approximation of the Helmholtz equation for beams forming a small angle with the optical axis.

1.3.2 Paraxial approximation

In geometric optics a ray is called 'paraxial' if its inclination to the optical axis is small. In wave optics the angular spectrum of a paraxial light beam consists predominantly of plane waves which are paraxial to the propagation direction of the beam, which we chose to be the z direction. The largest component of a paraxial wavevector $\mathbf{k}$ is therefore the z component with

$$k_z = \sqrt{k^2 - \kappa^2} \approx k - \frac{\kappa^2}{2k}, \tag{1.47}$$

where $\kappa = \sqrt{k_x^2 + k_y^2}$ is the transverse wave number. As an ansatz for a solution to the scalar Helmholtz equation (1.46) we consider a function which explicitly depends on the phase factor $\exp(ikz)$:

$$\xi(\mathbf{r}) = u(\mathbf{r}) \exp(ikz). \tag{1.48}$$

Here, $u(\mathbf{r})$ is an amplitude distribution. The profile may change with the distance z owing to diffraction or propagation effects. Compared with the primary variation $\exp(ikz)$ these changes will be small for a well-collimated beam. This is the reason why the primary factor $\exp(ikz)$ has been written explicitly. On substituting the ansatz of Eq. (1.48) into the scalar Helmholtz equation (1.46) we find a partial differential equation for the amplitude distribution u:

$$\nabla_t^2 u + \frac{\partial^2}{\partial z^2} u + 2ik \frac{\partial}{\partial z} u = 0. \tag{1.49}$$

This differential equation can be approximated by neglecting the second derivative of u with respect to z in comparison with the other two terms. If the profile changes only slowly with z the transverse variation of the profile can become dominant so that the following relation holds [38]:

$$\left|\frac{\partial^2 u}{\partial z^2}\right| \ll \left|\nabla_t^2 u\right|. \tag{1.50}$$

If the paraxial approximation is applied to the Helmholtz equation for the electric field the relation above fails because a free electric field the Maxwell equation is necessarily divergenceless. For the transversely polarised electric field of Eq. (1.45) this requires in turn that the directional derivative in the transverse direction $\nabla_{\mathbf{t}} E$ is equal to zero everywhere:

$$\nabla \cdot \mathbf{E} = \mathbf{t} \cdot \nabla E = 0. \tag{1.51}$$

This also requires, however, that the transverse Laplacian is equal to zero, which contradicts the relation in Eq. (1.50) [35]. Using the Helmholtz equation for the vector potential in the Lorentz gauge avoids these inconsistencies and Davis [37] showed that the results are equivalent in the lowest orders. The second estimate for the paraxial approximation is less problematic. If the profile is slowly varying in z the second derivative is much smaller than the first derivative multiplied by the overall wavenumber:

$$\left|\frac{\partial^2 u}{\partial z^2}\right| \ll k\left|\frac{\partial u}{\partial z}\right|. \tag{1.52}$$

Neglecting the second-order derivative with respect to z constitutes the paraxial approximation and results in the familiar paraxial wave equation:

$$\nabla_t^2 u + 2ik\frac{\partial}{\partial z}u = 0. \tag{1.53}$$

The paraxial approximation can be seen as a small angle approximation, where the ratio κ/k is small [38].

1.4 Paraxial and non-paraxial optics

In this section we present light beams as a solution to the paraxial and non-paraxial wave equation, which is the full Helmholtz equation derived earlier in this chapter. The orbital angular momentum of light was first studied in the paraxial approximation and, for a while it seemed to be a well-defined quantity only within this approximation. In this way it was the Laguerre–Gaussian beams [38] that were first associated with modes of well-defined orbital angular momentum [1]. The angular momentum flux as the defining quantity for angular momenta in electromagnetic fields, however, makes it possible to extend the analysis into the non-paraxial regime.

1.4.1 Paraxial light beams

Solutions to the paraxial wave equation are well known from laser optics [38]. In particular, Hermite–Gaussian beams are commonly employed for physical systems with a rectangular symmetry. Hermite–Gaussian beams do not carry orbital angular momentum but they form a complete set of solutions. It is possible, therefore, to transform between these beams and Laguerre–Gaussian beams – another set of complete solutions. Laguerre–Gaussian beams are also solutions to the paraxial wave equation but arise naturally when cylindrical polar coordinates are employed. In many respects the most simple solutions are the Bessel beams for which transverse parts and longitudinal parts are completely factorised. These also form a complete set of solutions, but Bessel beams are solutions to both paraxial and full Helmholtz equations which makes them useful for an analysis of the propagation of beams with orbital angular momentum in the non-paraxial regime.

Hermite–Gaussian beams

In Cartesian coordinates it is possible to separate the amplitude function $u(x, y, z)$ into two functions, either of which depends on one transverse variable, x or y, and the longitudinal variable z in the following way:

$$u_{nl}^{\mathrm{HG}}(x, y, z) = u_n^{\mathrm{HG}}(x, z) u_l^{\mathrm{HG}}(y, z), \tag{1.54}$$

where n and l give the order of the Hermite polynomial in u_n or u_l. Each of these functions obeys a paraxial wave equation in one transverse dimension. In the following we consider only one function and choose x as the single transverse variable. The paraxial equation for $u_n(x, z)$ has the form

$$\left(\frac{\partial^2}{\partial x^2} + 2\mathrm{i}k\frac{\partial}{\partial z}\right) u_n^{\mathrm{HG}}(x, z) = 0. \tag{1.55}$$

A normalised solution to this equation is given by the product of a Gaussian with a Hermite polynomial in the following form:

$$u_n^{\mathrm{HG}}(x, z) = \frac{C_n^{\mathrm{HG}}}{\sqrt{w(z)}} \exp\left[\mathrm{i}k\frac{x^2 z}{2(z_R^2 + z^2)}\right] \exp\left(-\frac{x^2}{w^2(z)}\right) \times \exp\left[-\mathrm{i}\left(n + \frac{1}{2}\right)\chi(z)\right] H_n\left(\frac{\sqrt{2}x}{w(z)}\right). \tag{1.56}$$

Here, $C_n^{\mathrm{HG}} = \sqrt{1/(2^n n!)}(2/\pi)^{(1/4)}$ is the normalisation constant and H_n is the Hermite polynomial of order n. Owing to the orthogonality of the Hermite polynomials [39] the Hermite–Gaussian beams are orthonormal when integrated over the transverse variable x:

$$\int_{-\infty}^{\infty} u_n^{\mathrm{HG}}(x, z) \left[u_m^{\mathrm{HG}}(x, z)\right]^* \mathrm{d}x = \delta_{nm}. \tag{1.57}$$

The width of the beam changes on propagation and is determined by the Gaussian spot size $w(z)$, which gives the radial distance at which the Gaussian term falls off to $1/e$ of its value on axis:

$$w(z)^2 = \frac{2(z_R^2 + z^2)}{kz_R} = w_0^2\left[1 + \left(\frac{z}{z_R}\right)^2\right]. \tag{1.58}$$

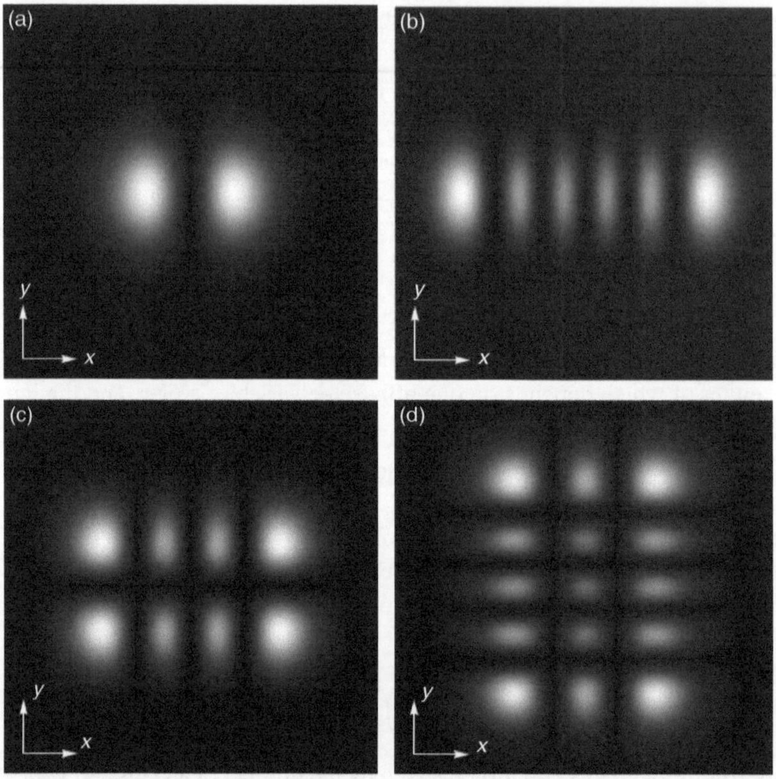

Figure 1.1 Intensity profiles of Hermite–Gaussian beams. (a) $n = 1, l = 0$. (b) $n = 5, l = 0$. (c) $n = 3, l = 1$. (d) $n = 2, l = 4$.

The Rayleigh range z_R is a measure for the focal region of the beam and is given by

$$z_R = \frac{\pi w_0^2}{\lambda}, \tag{1.59}$$

where λ is the wavelength and w_0 is the beam waist at the focal point, which is also the origin of the z axis. The term $(n + \frac{1}{2})\chi(z)$ is the Gouy phase which describes the phase jump of π that occurs over the focal region of any spherical converging wave [38, 40]. This behaviour is apparent from the form of the function $\chi(z)$:

$$\tan\chi = \frac{z}{z_R}. \tag{1.60}$$

In Fig. 1.1 the intensity profiles for a number of different Hermite–Gaussian beams are shown. The profiles are plotted for the full Hermite–Gaussian mode $u_{nl}^{\mathrm{HG}}(x, y, z)$. The indices n and l indicate the number of nodal lines in the respective x and y direction.

Laguerre–Gaussian beams

Another complete set of beams are the Laguerre–Gaussian modes, but instead of the rectangular symmetry of the Hermite–Gaussian modes these beams show a rotational symmetry. Consequently, the Laguerre–Gaussian beams are solutions to the paraxial wave equation in cylindrical coordinates:

$$\left(\frac{1}{\rho}\frac{\partial}{\partial\rho} + \frac{\partial^2}{\partial\rho^2} + \frac{1}{\rho^2}\frac{\partial^2}{\partial\varphi^2} + 2\mathrm{i}k\frac{\partial}{\partial z}\right) u_{mp}^{\mathrm{LG}} = 0. \tag{1.61}$$

Here, ρ is the radial coordinate and φ is the azimuthal angle. The normalised form of the Laguerre–Gaussian beams is given by

$$u_{mp}^{\mathrm{LG}}(\rho, \varphi, z) = \frac{C_{mp}^{\mathrm{LG}}}{\sqrt{w(z)}} \left(\frac{\rho\sqrt{2}}{w(z)}\right)^{|m|} \exp\left(-\frac{\rho^2}{w^2(z)}\right) L_p^{|m|}\left(\frac{2\rho^2}{w^2(z)}\right)$$
$$\times \exp\left[-ik\frac{\rho^2 z}{2(z_R^2 + z^2)}\right] \exp(im\varphi) \exp\left[-\mathrm{i}(2p + |m| + 1)\chi(z)\right]. \tag{1.62}$$

Here, $L_p^{|m|}$ are the generalised Laguerre polynomials and the appropriate normalisation constants are given by $C_{mp}^{\mathrm{LG}} = \sqrt{2^{|m|+1} p!/[\pi(p + |m|)!]}$. The Laguerre–Gaussian beams form an orthonormal set in the mode index p when integrated over the radial coordinate:

$$\int_0^\infty u_{mp}^{\mathrm{LG}}(\rho, \varphi, z) \left[u_{mq}^{\mathrm{LG}}(\rho, \varphi, z)\right]^* \rho\, d\rho = \delta_{pq}. \tag{1.63}$$

The orthogonality is due to the properties of the generalised Laguerre polynomials [41]. The profiles of the Laguerre–Gaussian modes show concentric rings, the number of which is determined by the mode index p (see Fig. 1.2). The other mode index is contained in the azimuthal phase term $\exp(im\varphi)$, which gives rise to $|m|$ intertwined helical wavefronts, that is surfaces of equal phase. The handedness of these helixes is determined by the sign of m. This azimuthal phase structure is the cause for the OAM of Laguerre–Gaussian modes. The Laguerre–Gaussian modes are also orthogonal in the azimuthal mode index m when integrated over the azimuthal variable φ. They are orthonormal in both indices under integration over the full ρ, φ plane:

$$\int_0^{2\pi} \mathrm{d}\phi \int_0^\infty \rho\, \mathrm{d}\rho\, u_{np}^{\mathrm{LG}}(\rho, \varphi, z) \left[u_{mq}^{\mathrm{LG}}(\rho, \varphi, z)\right]^* = \delta_{nm}\delta_{pq}. \tag{1.64}$$

In all generality the choice of the 2π radians interval for the angle φ is arbitrary and it is more general to integrate from θ_0 to $\theta_0 + 2\pi$. This will be discussed in more detail in Chapter 6.

Hermite–Gaussian modes and Laguerre–Gaussian modes both form complete sets and one family of beams can therefore be represented as a superposition of the other.

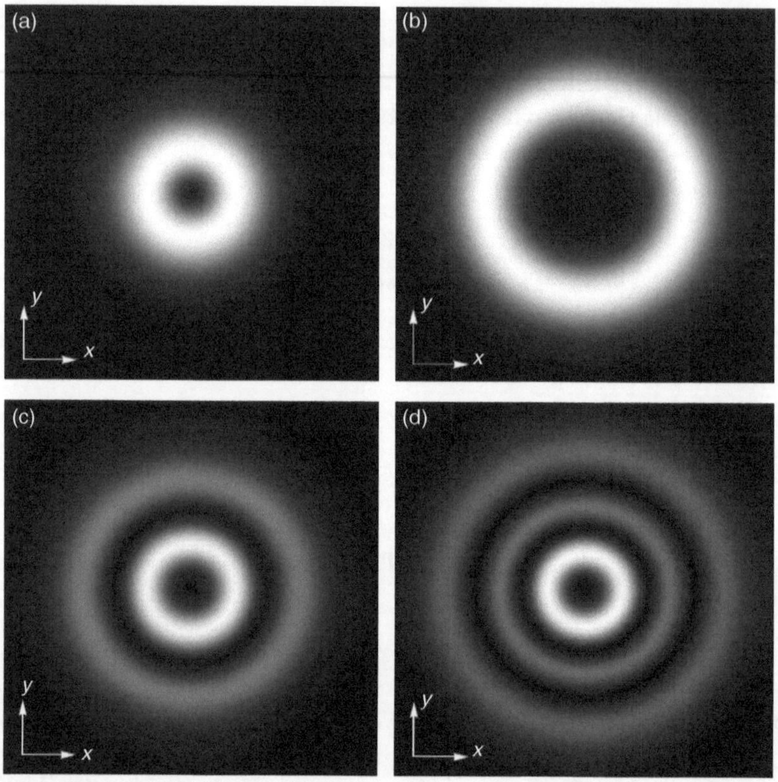

Figure 1.2 Intensity profiles of Laguerre–Gaussian beams. (a) $m = 1, p = 0$. (b) $m = 5, p = 0$. (c) $m = 2, p = 1$. (d) $m = 2, p = 2$.

Beijersbergen *et al.* [2] published a very clear analysis of the mode conversion, but instead of using the indices m and p to describe the Laguerre–Gaussian modes, they considered indices n and l analogously to the Hermite–Gaussian beams. The two sets of indices are related by

$$m = n - l \quad \text{and} \quad p = \min(n, l). \tag{1.65}$$

Any Laguerre–Gaussian mode can now be expressed in terms of Hermite–Gaussian modes as follows:

$$u_{nl}^{\mathrm{LG}}(x, y, z) = \sum_{t=0}^{n+l} \mathrm{i}^t b(n, l, t) u_{n+l-t,t}^{\mathrm{HG}}(x, y, z), \tag{1.66}$$

with the coefficients, $b(n, l, t)$, given by

$$b(n, l, t) = \sqrt{\frac{(n+l)!t!}{2^{n+l}n!l!}} \frac{1}{k!} \frac{d^t}{\mathrm{d}s^t} \left[(1-s)^n (1-s)^m\right]\Big|_{s=0}. \tag{1.67}$$

This transformation is based on earlier results by Abramochkin and Volostnikov [42]. Hermite–Gaussian and Laguerre–Gaussian beams can also be generated from a Gaussian seed function by acting on the fundamental Gaussian function with differential operators [43], or as limiting cases of general polynomial solutions [44].

Complex argument form

Both the Hermite–Gaussian modes and the Laguerre–Gaussian modes can also be written with complex arguments. This is due to a freedom in the scaling parameter for the Hermite or Laguerre polynomials in the respective rectangular or cylindrical paraxial equations. In the standard form, where the polynomials have real arguments, this scaling parameter is chosen differently from the scaling parameter of the Gaussian term. This results in the known form of the Hermite–Gaussian and Laguerre–Gaussian modes in Eqs. (1.56) and (1.62), where the Gaussian term is split into a real and imaginary part. The real part contains the Gaussian spot size $w(z)$ and the imaginary part contains the radius of curvature $R(z) = (z^2 + z_R^2)/z$.

This 'inelegant' difference in the scaling parameter is overcome in the complex argument form, which is therefore also called the 'elegant' form [38, 45]. In this form the scaling parameter for the polynomial is chosen to be identical to the scaling parameter of the Gaussian term and is basically determined by a function $\tilde{q}$ which is given in terms of the radius of curvature $R(z)$ and and Gaussian spot size $w(z)$:

$$\frac{1}{q(z)} \equiv \frac{1}{R(z)} + \mathrm{i}\frac{\lambda}{\pi w^2(z)} = \frac{\mathrm{i}}{z_R + \mathrm{i}z}. \tag{1.68}$$

If we use this scaling parameter the expression for the Hermite–Gaussian mode changes to

$$u_n^{\mathrm{eHG}}(x, z) = C_n^{\mathrm{eHG}} \left(\frac{k}{z_R}\right)^{\frac{1}{4}} \left(\frac{-\mathrm{i}z_R}{q(z)}\right)^{\frac{n+1}{2}} H_n\left(\sqrt{\frac{-\mathrm{i}kx^2}{2q(z)}}\right) \exp\left(\frac{\mathrm{i}kx^2}{2q(z)}\right), \tag{1.69}$$

which is of a simpler form than the standard Hermite–Gaussian modes. The Gouy phase of the standard form is related to the term $[-\mathrm{i}z_R/q(z)]^{(n+1)/2}$ in the elegant form which also changes the scaling parameter. The elegant variant represents an equally valid and complete set of solutions to the paraxial wave equations. They are normalisable with normalisation constant $C_n^{\mathrm{eHG}} = \sqrt{(2^n n!/(2n)!}(1/\pi)^{(1/4)}$, but they are not orthogonal as the Hermite–Gaussian modes in the standard form. Instead, they are biorthogonal to a set of adjoint functions v_n^{HG} given by

$$v_n^{\mathrm{eHG}}(x, z) = H_n\left(\sqrt{\frac{\mathrm{i}kx^2}{2q^*(z)}}\right). \tag{1.70}$$

The orthogonality relation is now given by

$$\int_{-\infty}^{\infty} u_n^{\mathrm{eHG}}(x,z)\left[v_m^{\mathrm{eHG}}(x,z)\right]^* \mathrm{d}x = c_n \delta_{nm}, \tag{1.71}$$

where c_n is an appropriate normalisation constant.[1]

The Laguerre–Gaussian beams can also be given in an elegant form. By replacing the standard argument $\sqrt{2}\rho/w(z)$ with the complex argument $\sqrt{-\mathrm{i}k\rho^2/[2q(z)]}$ the Laguerre polynomial has the same argument as the Gaussian term. However, this also affects the Gouy phase and the term $[\sqrt{2}\rho/w(z)]^{|m|}$. The final result is given by

$$\begin{aligned} u_{mp}^{\mathrm{eLG}}(\rho,\varphi,z) = C_{mp}^{\mathrm{eLG}} \left(\frac{k}{2}\right)^{|m|+p+1} \rho^{|m|} \left(\frac{-\mathrm{i}}{q(z)}\right)^{|m|+p+1} \exp(\mathrm{i}m\varphi) \\ \times L_p^{|m|}\left(\frac{-\mathrm{i}k\rho^2}{2q(z)}\right) \exp\left(\frac{\mathrm{i}k\rho^2}{2q(z)}\right), \end{aligned} \tag{1.73}$$

which is identical to the paraxial limit of a non-paraxial extension to Laguerre–Gaussian beams discussed by Barnett and Allen [46]. The appropriate normalisation constant is given by $C_{mp}^{\mathrm{eLG}} = p!/\sqrt{\pi(2p+m)!}$.

Bessel beams

The full Helmholtz equation supports a class of solutions, the transverse profile of which is invariant on propagation. Consequently, these beams have been called 'diffraction-free beams' by Durnin *et al.* [47]. They are exact solutions to the full Helmholtz equation, but they have an infinitely extended transverse profile. Under laboratory conditions the realisation of such beams will be limited by the physical requirement of a finite aperture [48, 49]. The radial, azimuthal and axial parts are all separated in the Bessel beams, which have an amplitude distribution of the form

$$\xi_m^{\mathrm{B}}(\rho,\varphi,z) = J_m(\kappa\rho)\exp(\mathrm{i}m\varphi)\exp(\mathrm{i}k_z z), \tag{1.74}$$

Bessel beams have the same azimuthal phase factor $\exp(\mathrm{i}m\varphi)$ as the Laguerre–Gaussian modes, and hence carry also orbital angular momentum. In the transition from the full Helmholtz equation to the paraxial wave equation the transverse part of the differential equation is left unchanged, but the relation between the different wavenumbers k, κ and k_z changes. The ansatz for the paraxial wave equation Eq. (1.53) is given by

$$u_m^B(\rho,\varphi,z) = J_m(\kappa\rho)\exp(\mathrm{i}m\varphi)\exp(\mathrm{i}\tilde{k}_z z), \tag{1.75}$$

[1] For completeness we give the normalisation constant c_n as

$$c_n = \frac{(2m)!!\pi^{\frac{1}{4}}}{\sqrt{(2m-1)!!}} \left(\frac{k}{z_R}\right)^{\frac{1}{4}} \left(\frac{-\mathrm{i}z_r}{q(z)}\right)^{\frac{n+1}{2}} \left(\frac{2\mathrm{i}q(z)}{k}\right)^{\frac{1}{2}}, \tag{1.72}$$

where !! denotes the double factorial.

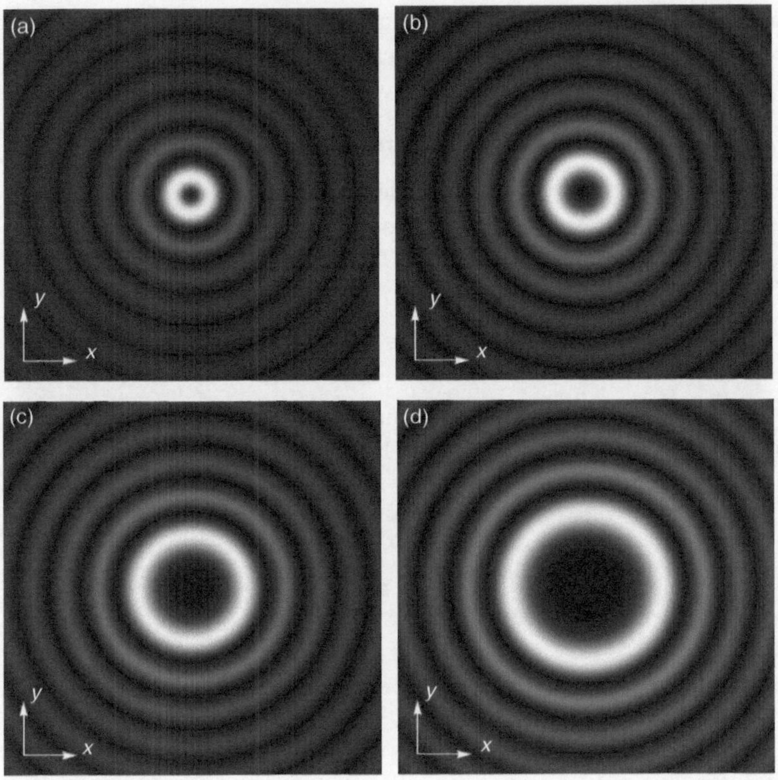

Figure 1.3 Intensity profiles of Bessel beams. (a) $m = 1$. (b) $m = 2$. (c) $m = 4$. (d) $m = 6$.

where $\tilde{k}_z$ is a reduced longitudinal wave number with $\tilde{k}_z = k_z - k$. Its presence is due to the separated factor $\exp(ikz)$ in the paraxial ansatz (see Eq. (1.49)). On comparing the paraxial wave equation (1.53) with the defining differential equation for Bessel functions [see 39]

$$(\kappa\rho)^2 J_m''(\kappa\rho) + \kappa\rho J_m'(\kappa\rho) + \left((\kappa\rho)^2 - m^2\right) J_m(\kappa\rho) = 0, \tag{1.76}$$

one can see that u_m^B will be a solution of the paraxial wave equation if

$$\kappa^2 = -2k\tilde{k}_z \quad \longrightarrow \quad k_z = k - 2\frac{\kappa^2}{2k}. \tag{1.77}$$

This form of the longitudinal wave number is identical to the approximate form for small κ/k. This is not surprising as u_m^{B} is the solution to the paraxial wave equation. In the next section we will see that the Bessel beams are also solutions to the full Helmholtz equation but with a different expression for k_z.

The transverse profiles of the Bessel beams are shown in Fig. 1.3. Bessel beams do not change on propagation and there is no diffraction or dilation of the profile. The difference between paraxial and non-paraxial Bessel beams is only apparent in the longitudinal factor. The transverse profiles are the same for both cases.

1.4.2 Non-paraxial light beams

Bessel beams are also solutions to the full Helmholtz equation, which can be seen if we substitute the ansatz ξ_m^B from Eq. (1.74) into the full Helmholtz equation in cylindrical coordinates (see Eq. (1.35)):

$$\left(\frac{1}{\rho}\frac{\partial}{\partial\rho} + \frac{\partial^2}{\partial\rho^2} + \frac{1}{\rho^2}\frac{\partial}{\partial\varphi^2} + \frac{\partial^2}{\partial z^2} + k^2\right)\xi_m^{\mathrm{B}} = 0. \tag{1.78}$$

To satisfy Bessel's equation (1.76) the expression for k_z has to be different from the paraxial case and is given by exact expression in Eq. (1.47):

$$\kappa^2 = k^2 - k_z^2 \quad \longrightarrow \quad k_z = \sqrt{k^2 - \kappa^2}. \tag{1.79}$$

Bessel modes can thus be transformed from solving the Helmholtz equation to solve the paraxial equation, by the same change in the expression for k_z that characterises the paraxial approximation.

By virtue of the completeness of Bessel beams it is possible to expand the Laguerre–Gaussian modes in the following way:

$$u_{mp}^{\mathrm{LG}}(\rho,\varphi,z) = \int_0^\infty d_{mp}(\kappa) J_m(\kappa\rho)\exp(\mathrm{i}m\varphi)\exp(-\mathrm{i}\kappa^2 z/(2k))\mathrm{d}\kappa. \tag{1.80}$$

The exact form of the function $d_{mp}(\kappa)$ can be determined from the expression for the paraxial Laguerre–Gaussian beam. To transform the superposition into a solution of the full Helmholtz equation we have to change the longitudinal phase factor. The radial and the azimuthal part are left unchanged. In this way we can construct a non-paraxial form of the Laguerre–Gaussian beam:

$$\xi_{mp}^{\mathrm{LG}}(\rho,\varphi,z) = \int_0^k d_{mp}(\kappa) J_m(\kappa\rho)\exp(\mathrm{i}m\varphi)\exp\left(\mathrm{i}\sqrt{k^2-\kappa^2}z\right)\mathrm{d}\kappa. \tag{1.81}$$

The two expansions differ not only in the longitudinal part, but also in the boundaries of the integration. While the paraxial expansion includes all transverse wavevectors $0 \le \kappa < \infty$, the non-paraxial expansion is limited to the range $0 \le \kappa \le k$. For values of $\kappa > k$ the contributions to the expansion become evanescent. To determine the function $d_{mp}(\kappa)$ we can multiply both sides of Eq. (1.80) by $\rho J_m(\kappa'\rho)$ and integrate both sides over ρ. On applying the Fourier–Bessel theorem in the form [50]

$$\int_0^\infty J_m(\kappa\rho)J_m(\kappa'\rho)\,\rho\,\mathrm{d}\rho = \frac{1}{\kappa'}\delta(\kappa-\kappa'), \tag{1.82}$$

we can determine the amplitudes $d_{mp}(\kappa)$ as

$$d_{mp}(\kappa) = \kappa \exp(-im\varphi) \exp\left(\mathrm{i}\frac{\kappa^2}{2k}z\right) \int_0^\infty u_{mp}^{\mathrm{LG}}(\rho, \varphi, z) J_m(\kappa\rho)\, \rho\, \mathrm{d}\rho,$$
$$= \exp\left(-\frac{\kappa^2}{2k}z_R\right)\left(\frac{\kappa}{k}\right)^{2p+|m|+1}. \tag{1.83}$$

On substituting this result into the expansion in Eq. (1.81) we obtain a non-paraxial extension of the Laguerre–Gaussian modes.

1.5 Angular momentum of light beams

It is necessary to make a link between the various modes described in the previous section and the angular momentum carried by the associated field. We are interested in the component of the angular momentum parallel to the direction of propagation. The natural way to make this link is via the flux of angular momentum [23], which represents the rate of flow of angular momentum. It also has the benefit that it may be applied directly to both paraxial and non-paraxial light.

1.5.1 Angular momentum flux

In Section 1.2.3 we defined the angular momentum flux density and the total angular momentum flux. Here, we are discussing how this quantity provides a meaningful separation into spin and orbital angular momentum parts. It is interesting to note that it is not possible to distinguish the spin part from the orbital part by their behaviour under a change of axis. As long as the total linear momentum flux through a surface is perpendicular to this surface, the total angular momentum flux is the same for every axis parallel to the beam [23]. A similar result has also been derived for the angular momentum by Berry [51].

Although the expression in Eq. (1.18) is a valid expression for the angular momentum flux density, it is helpful to substitute the real electric field and the real magnetic induction with the complex field amplitudes of a monochromatic beam with frequency ω (see Eq. (1.34)). For the complex field amplitudes the Maxwell equations (1.32) are given by

$$\mathcal{E}_j = \mathrm{i}\frac{c^2}{\omega}\varepsilon_{jkl}\frac{\partial}{\partial x_k}\mathcal{B}_l, \tag{1.84a}$$

$$\mathcal{B}_j = -\mathrm{i}\frac{1}{\omega}\varepsilon_{jkl}\frac{\partial}{\partial x_k}\mathcal{E}_l. \tag{1.84b}$$

With help of the complex field amplitudes it is possible to calculate the cycle–averaged angular momentum flux density M_{zz}. The real field components in Eq. (1.18) are expressed as sums of the complex field amplitudes and their conjugates. On integration over the cycle time $T = 2\pi/\omega$ only terms without the time-varying factor $\exp(-\mathrm{i}\omega t)$ are left. This allows us to write the cycle average of the angular momentum flux as [23]

$$\mathcal{M}_{zz} = \frac{1}{2}\,\mathrm{Re}\left[y\left(\epsilon_0 \mathcal{E}_x \mathcal{E}_z^* + \frac{1}{\mu_0}\mathcal{B}_x^* \mathcal{B}_z\right) - x\left(\epsilon_0 \mathcal{E}_y \mathcal{E}_z^* + \frac{1}{\mu_0}\mathcal{B}_y^* \mathcal{B}_z\right)\right]. \tag{1.85}$$

The z components of the electric field and the magnetic induction can be eliminated by use of the Maxwell equations (1.84). The total, cycle-averaged angular momentum flux is obtained by integrating this quantity over the whole transverse plane. By partial integration and subsequent manipulations this integral can be brought into the following form:

$$M_{zz} = \frac{\epsilon_0 c^2}{2\omega}\,\mathrm{Re}\left\{-\mathrm{i}\iint \rho\,\mathrm{d}\rho\,\mathrm{d}\varphi\left[(\mathcal{E}_x \mathcal{B}_x^* + \mathcal{E}_y \mathcal{B}_y^*) + \frac{1}{2}\left(-\mathcal{B}_x^* \frac{\partial}{\partial\varphi}\mathcal{E}_y + \mathcal{E}_y \frac{\partial}{\partial\varphi}\mathcal{B}_x^* - \mathcal{E}_x \frac{\partial}{\partial\varphi}\mathcal{B}_y^* + \mathcal{B}_y^* \frac{\partial}{\partial\varphi}\mathcal{E}_x\right)\right]\right\}, \tag{1.86}$$

which is given in cylindrical polar coordinates. In this form the separation of the total angular momentum flux into a spin and orbital part becomes most apparent. The spin part is given by

$$M_{zz}^{\mathrm{spin}} = \frac{\epsilon_0 c^2}{2\omega}\,\mathrm{Re}\left\{-\mathrm{i}\iint \rho\,\mathrm{d}\rho\,\mathrm{d}\varphi\,(\mathcal{E}_x \mathcal{B}_x^* + \mathcal{E}_y \mathcal{B}_y^*)\right\}, \tag{1.87}$$

and depends on the polarisation of the light beam, whereas the orbital part is connected to changes in the azimuthal variable:

$$M_{zz}^{\mathrm{orbital}} = \frac{\epsilon_0 c^2}{4\omega}\,\mathrm{Re}\left\{-\mathrm{i}\iint \rho\,\mathrm{d}\rho\,\mathrm{d}\varphi\left(-\mathcal{B}_x^* \frac{\partial}{\partial\varphi}\mathcal{E}_y + \mathcal{E}_y \frac{\partial}{\partial\varphi}\mathcal{B}_x^* - \mathcal{E}_x \frac{\partial}{\partial\varphi}\mathcal{B}_y^* + \mathcal{B}_y^* \frac{\partial}{\partial\varphi}\mathcal{E}_x\right)\right\}. \tag{1.88}$$

The physical significance of this separation follows on considering the effects of birefringence on the beam. The imparted phase shift will be different for the x and y components but it will not incur an azimuthal phase shift. The orbital part is therefore left unchanged. However, if an optical element imprints an azimuthal phase shift all the components will change in the same way, which leaves the spin part unchanged.

We are now able to calculate the spin and orbital parts of the total angular momentum flux M of a light beam beyond the paraxial approximation. To this end we consider a class of general non-paraxial beams introduced by Barnett and Allen [46]. The components of the electric field for this beam are given by the Hankel transform

$$\mathcal{E}_x = \alpha \int_0^k \mathrm{d}\kappa\, d_{mp}(\kappa) \exp(\mathrm{i}m\varphi) \exp\left(\mathrm{i}\sqrt{k^2-\kappa^2}z\right) J_m(\kappa\rho), \tag{1.89a}$$

$$\mathcal{E}_y = \beta \int_0^k \mathrm{d}\kappa\, d_{mp}(\kappa) \exp(\mathrm{i}m\varphi) \exp\left(\mathrm{i}\sqrt{k^2-\kappa^2}z\right) J_m(\kappa\rho), \tag{1.89b}$$

$$\mathcal{E}_z = \int_0^k \mathrm{d}\kappa\, d_{mp}(\kappa) \exp(\mathrm{i}m\varphi) \exp\left(\mathrm{i}\sqrt{k^2-\kappa^2}z\right) \frac{\kappa}{2\sqrt{k^2-\kappa^2}}$$
$$\times \left[(\mathrm{i}\alpha-\beta)\exp(-\mathrm{i}\varphi)J_{m-1}(\kappa\rho) - (\mathrm{i}\alpha+\beta)\exp(-\mathrm{i}\varphi)J_{m+1}(\kappa\rho)\right]. \tag{1.89c}$$

This beam is a solution to the vector form of the Helmholtz equation (1.35a) and also satisfies the transversality condition $\nabla \cdot \mathbf{E} = 0$. The complex numbers α and β determine the polarisation and are chosen such that their squared moduli sum to unity: $|\alpha|^2 + |\beta|^2 = 1$. The components of the magnetic induction are given by Eq. (1.84) [see 23]:

$$\mathcal{B}_x = \frac{1}{\omega}\int_0^k \mathrm{d}\kappa \frac{d_{mp}(\kappa)\exp\left(\mathrm{i}\sqrt{k^2-\kappa^2}z\right)}{\sqrt{k^2-\kappa^2}} \left[-\beta\frac{2k^2-\kappa^2}{2}\mathrm{e}^{\mathrm{i}m\varphi}J_m(\kappa\rho)\right.$$
$$\left. + \frac{\kappa^2}{4}\left(\mathrm{e}^{\mathrm{i}(m-2)\varphi}(\mathrm{i}\alpha-\beta)J_{m-2}(\kappa\rho) - \mathrm{e}^{\mathrm{i}(m+2)\varphi}(\mathrm{i}\alpha+\beta)J_{m+2}(\kappa\rho)\right)\right], \tag{1.90a}$$

$$\mathcal{B}_y = \frac{1}{\omega}\int_0^k \mathrm{d}\kappa \frac{d_{mp}(\kappa)\exp\left(\mathrm{i}\sqrt{k^2-\kappa^2}z\right)}{\sqrt{k^2-\kappa^2}} \left[\alpha\frac{2k^2-\kappa^2}{2}\mathrm{e}^{\mathrm{i}m\varphi}J_m(\kappa\rho)\right.$$
$$\left. + \frac{\kappa^2}{4}\left(\mathrm{e}^{\mathrm{i}(m-2)\varphi}(-\alpha-\mathrm{i}\beta)J_{m-2}(\kappa\rho) - \mathrm{e}^{\mathrm{i}(m+2)\varphi}(\alpha-\mathrm{i}\beta)J_{m+2}(\kappa\rho)\right)\right], \tag{1.90b}$$

$$\mathcal{B}_z = \frac{1}{\omega}\int_0^k \mathrm{d}\kappa \frac{\kappa d_{mp}(\kappa)\exp\left(\mathrm{i}\sqrt{k^2-\kappa^2}z\right)}{2}$$
$$\times \left[(-\alpha+\mathrm{i}\beta)\mathrm{e}^{\mathrm{i}(m+1)\varphi}J_{m+1}(\kappa\rho) - (\alpha+\mathrm{i}\beta)\mathrm{e}^{\mathrm{i}(m-1)\varphi}J_{m-1}(\kappa\rho)\right]. \tag{1.90c}$$

For this particular beam the total linear momentum flux has no component in the transverse direction. The total angular momentum flux is therefore the same for every axis parallel to the z direction.

Originally, Barnett and Allen [46] considered a specific form of the function $d_{mp}(\kappa)$ which is different from the result in Eq. (1.83), but there is some degree of freedom in choosing this function, so that it is equally possible to use the function $d_{mp}(\kappa)$ given in Eq. (1.83). This particular function falls off sufficiently quickly as κ tends to zero, which ensures a finite energy, momentum and angular momentum per unit length. If we now calculate the spin and orbital part of the total angular momentum flux according to Eqs. (1.87) and (1.88) we find

$$M_{zz}^{\text{spin}} = \mathrm{i}(\alpha\beta^* - \alpha^*\beta)\frac{\pi\epsilon_0 c^2}{2\omega^2}\int_0^k \mathrm{d}\kappa \, |d_{mp}(\kappa)|^2 \frac{2k^2 - \kappa^2}{\kappa\sqrt{k^2 - \kappa^2}}, \tag{1.91}$$

$$M_{zz}^{\text{orbital}} = m\frac{\pi\epsilon_0 c^2}{2\omega^2}\int_0^k \mathrm{d}\kappa \, |d_{mp}(\kappa)|^2 \frac{2k^2 - \kappa^2}{\kappa\sqrt{k^2 - \kappa^2}}. \tag{1.92}$$

Only terms for which the azimuthal factor $\exp(im\varphi)$ is multiplied by its conjugate contribute to the total flux, this also ensures that only products of Bessel functions with equal order have to be considered. These terms can be calculated using the Fourier–Bessel theorem (see Eq. (1.82)). In the original paper on angular momentum of light Allen *et al.* [1] examined the ratio between angular momentum density and energy flux both per unit length. This established the fact that in the paraxial approximation Laguerre–Gaussian beams, or in fact any beam with the particular azimuthal phase factor $\exp(im\varphi)$, has well-defined orbital angular momentum. On considering the ratio between either the spin or orbital part of the total angular momentum flux through the transverse plane and the total energy flux though the same plane we can recover the original results beyond the paraxial approximation. The cycle-averaged energy flux through the transverse plane is given by the integral (see Eq. (1.9))

$$\begin{aligned} J_z^E &= \frac{1}{2\mu_0\omega}\,\mathrm{Re}\iint \rho\,\mathrm{d}\rho\,\mathrm{d}\phi(\mathcal{E}_x\mathcal{B}_y^* - \mathcal{E}_y\mathcal{B}_x^*), \\ &= \frac{\pi}{2\mu_0\omega}\int_0^k \mathrm{d}\kappa\,|d_{mp}(\kappa)|^2 \frac{2k^2 - \kappa^2}{\kappa\sqrt{k^2 - \kappa^2}}. \end{aligned} \tag{1.93}$$

This expression includes the same integral as the spin and orbital parts of the total angular momentum flux. The ratio of the respective quantities is thus given by

$$\frac{M_{zz}^{\text{spin}}}{J_z^E} = \frac{\sigma_z}{\omega}, \qquad \frac{M_{zz}^{\text{orbital}}}{J_z^E} = \frac{m}{\omega}, \qquad \frac{M_{zz}}{J_z^E} = \frac{\sigma_z + m}{\omega}. \tag{1.94}$$

Here, we have set $\sigma_z = \mathrm{i}(\alpha\beta^* - \alpha^*\beta)$, which determines the polarisation. For circular polarised light $\beta = \alpha\exp(\pm\mathrm{i}\pi/2)$, so that σ_z takes the values ± 1. This analysis shows that it is the azimuthal phase factor $\exp(im\varphi)$ that is responsible for the well-defined orbital angular momentum.

1.5.2 Azimuthal phase structure

The previous analysis showed that beams with a phase factor $\exp(im\varphi)$ have a well-defined orbital angular momentum. This is pleasing as quantum wavefunctions of this form are eigenstates of the z-component of the angular momentum with eigenvalue $m\hbar$. This analogy was a key idea in the early development of our topic [1]. As the ratio of the orbital angular momentum flux to the energy flux is given by m/ω and the energy of a single photon is given by $\hbar\omega$, this seems to suggest that each photon in a beam with the phase factor $\exp(im\varphi)$ carries an orbital angular momentum of $\hbar m$.

This angular momentum may be linked with the the associated phase factor for an optical vortex [52, 53]. Optical vortices are phase singularities of the optical field and hence have zero amplitude. Light beams with a phase factor of $\exp(im\varphi)$ have an optical vortex on axis with a topological charge of m. This can be seen most clearly from the definition of the vortex charge, which is calculated by integrating the phase along a closed path around the singularity [54]

$$Q = \frac{1}{2\pi}\oint \mathrm{d}\chi = \frac{1}{2\pi}\oint \mathbf{ds}\cdot\nabla\chi, \tag{1.95}$$

where $\mathbf{ds}$ is the line element. For a azimuthal phase of $\chi = m\varphi$ this integral results in a vortex charge of m. To complete this section we show the phase structures for a number of different values of m in Fig. 1.4. We can see that for a phase factor of $\exp(im\varphi)$ there are $|m|$ phase ramps in a complete circle. Whether the phase ramps clockwise or anti-clockwise depends on the sign of m. On propagation we can identify $|m|$ intertwined helices as surfaces of equal phase, which are shown on the right panel in Fig. 1.4. The handedness of these helices is determined by the sign of m.

1.6 Generating light beams with orbital angular momentum

In most experimental setups the transverse field distribution of a laser beam is best described by a superposition of Hermite–Gaussian modes. If the laser cavity has rectangular symmetry this follows naturally from the symmetry of the Hermite–Gaussian modes, but even for a cavity with circular symmetry the output mode is hardly ever a pure Laguerre–Gaussian mode. As discussed earlier in this chapter (see Section 1.4.1) Laguerre–Gaussian beams can be represented as superpositions of Hermite–Gaussian beams. In a circular cavity astigmatism breaks the frequency degeneracy between the constituent Hermite–Gaussian modes, so that instead of the pure Laguerre–Gaussian mode a so-called 'doughnut'-mode is observed. One way to generate a Laguerre–Gaussian mode is to remove the astigmatism, either by specially designing the laser cavity [55] or by using cylindrical lenses [2].

There are two other commonly employed methods of generating Laguerre–Gaussian modes. Rather than transforming Hermite–Gaussian modes into Laguerre–Gaussian modes these two methods change the orbital angular momentum content within the family of Laguerre–Gaussian modes. In this way a Laguerre–Gaussian mode with non-zero orbital angular momentum can be generated from a pure Gaussian mode. The first way of manipulating light beams in this way is to use spiral phase plates [56]. These are optical elements with an optical thickness that increases with the azimuthal angle, until it drops sharply after completing the circle. As a result, an incident Gaussian beam emerges with a helical phase front. Spiral phase plates require a very small mechanical tolerance to ensure that the step height corresponds exactly to an integer multiple of the optical wavelength. Spiral phase

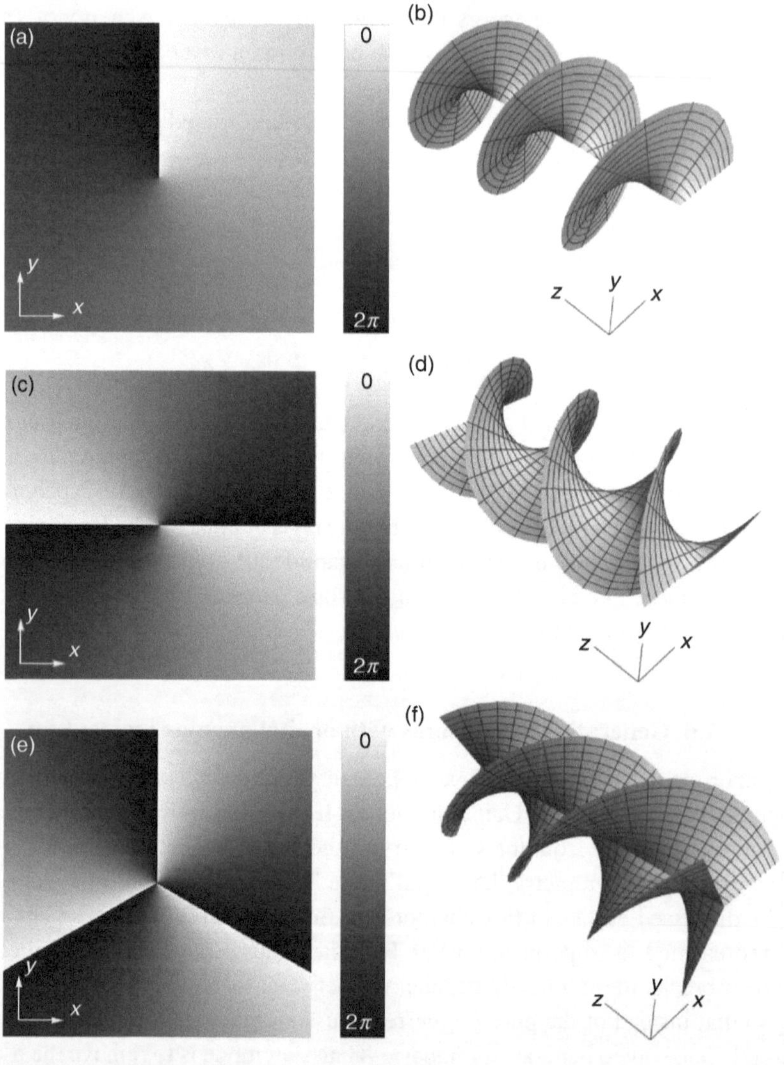

Figure 1.4 The left panel shows plots of the phase structure associated with a phase factor $\exp(im\varphi)$ for different values of m. When propagated light beams with the depicted phase structures have surfaces of equal phase in the form of intertwined helices. (a,b) $m = 1$. (c,d) $m = -2$. (d,e) $m = 3$.

plates can also be deliberately designed such that the step height does not correspond to an integer multiple. This gives rise to the phenomenon of fractional orbital angular momentum which will be covered in more detail in Chapter 6.

Holograms give results which are very similar to spiral phase plates. They resemble diffraction gratings, but modified such that the first order diffracted beam has a specific phase and amplitude structure. The required diffraction grating can be produced experimentally by interfering the desired light beam with a reference beam, say a pure Gaussian

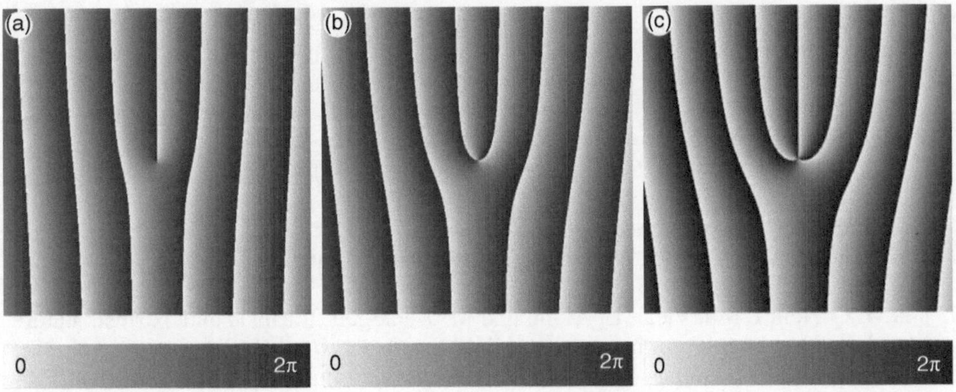

Figure 1.5 Plot of the phase profile of holographic patterns for generating light carrying orbital angular momentum. The patterns are designed to give the emerging light an azimuthal phase factor of $\exp(im\varphi)$ in the first order diffracted beam. (a) $m = 1$. (b) $m = 2$. (c) $m = 3$.

beam. The interference pattern can be put on a holographic film which, once photographically developed, generates the desired beam when placed in the reference beam. It is also possible to calculate the required diffraction grating. This has the advantage that the resulting pattern can be created on a spatial light modulator, a commercially available optical element which manipulates the intensity or phase of the light beam. With the help of spatial light modulators the pattern of the diffraction grating can be modified in real-time. This makes computer-generated holograms a very versatile experimental tool, which is used widely to manipulate light beams carrying orbital angular momentum [57]. In Fig. 1.5 we show the holographic patterns designed to create light beams with an azimuthal phase factor $\exp(im\varphi)$ for $m = 1, 2, 3$. The patterns show a characteristic fork dislocation, where the number of prongs is given by $|m|$.

References

[1] L. Allen, M. W. Beijersbergen, R. J. C. Spreeuw, and J. P. Woerdman, Orbital angular momentum of light and the transformation of Laguerre–Gaussian modes, *Physical Review A*, vol. 45, no. 11, pp. 8185–90, June 1992, reprinted in [57, Paper 2.1]. [Online]. Available: http://link.aps.org/abstract/PRA/v45/p8185.

[2] M. W. Beijersbergen, L. Allen, H. E. L. O. van der Veen, and J. P. Woerdman, Astigmatic laser mode converters and transfer of orbital angular momentum, *Optics Communication*, vol. 96, no. 1-3, pp. 123–32, 1993.

[3] J. B. Götte, Integral and fractional orbital angular momentum of light, Ph.D. dissertation, University of Strathclyde, 2006.

[4] J. H. Poynting, On the transfer of energy in the electromagnetic field, *Philosophical Transactions of the Royal Society*, vol. 175, pp. 343–61, 1884.

[5] O. Heaviside, *Electromagnetic Theory*. London: 'The Electrician' Printing and Publishing Company, 1894, vol. 1.

[6] O. Heaviside, *Electromagnetic Theory*. London: 'The Electrician' Printing and Publishing Company, 1894, vol. 2.
[7] J. H. Poynting, The wave motion of a revolving shaft, and a suggestion as to the angular momentum in a beam of circularly polarised light, *Proceedings of the Royal Society of London, Series A*, vol. 82, pp. 560–7, 1909.
[8] R. A. Beth, Mechanical detection and measurement of the angular momentum of light, *Physical Review*, vol. 50, pp. 115–25, 1936.
[9] M. E. Rose, *Multipole Fields*. New York: John Wiley & Sons, 1955.
[10] H. Römer and M. Forger, *Klassische Feldtheorie*. Weinheim: VCH Verlagsgesellschaft, 1993.
[11] A. T. O'Neil, I. MacVicar, L. Allen, and M. J. Padgett, Intrinsic and extrinsic nature of the orbital angular momentum of a light beam, *Physical Review Letters*, vol. 88, no. 5, p. 053601, January 2002. [Online]. Available: http://link.aps.org/abstract/PRL/v88/e053601.
[12] A. Ashkin, *Optical Trapping and Manipulation of Neutral Particles Using Lasers*. Singapore: World Scientific, 2006.
[13] M. J. Padgett, J. Molloy, and D. McGloin, *Optical Tweezers: Methods and Applications*. (eds) Philadelpha, PA: Taylor and Francis, 2009.
[14] S. M. Barnett, *Quantum Information*. Oxford: Oxford University Press, 2009.
[15] A. E. Mair, A. Vaziri, G. Weihs, and A. Zeilinger, Entanglement of the orbital angular momentum states of photons, *Nature*, vol. 412, pp. 313–16, 2001.
[16] S. Franke-Arnold, S. M. Barnett, M. J. Padgett, and L. Allen, Two-photon entanglement of orbital angular momentum states, *Physical Review A*, vol. 65, p. 033823, 2003.
[17] H. A. Lorentz, *Versuch einer Theorie der electrischen und optischen Erscheinungen in bewegten Körpern*. Leiden: Brill, 1895.
[18] J. D. Jackson, *Classical Electrodynamics*, 3rd edn. New York: John Wiley & Sons, 1998.
[19] J. C. Maxwell, *A Treatise on Electricity and Magnetism*. Oxford: Oxford at the Clarendon Press, 1873.
[20] J. A. Stratton, *Electromagnetic Theory*. New York: McGraw-Hill, 1941.
[21] L. D. Landau and E. M. Lifshitz, *The Classical Theory of Fields*, 4th edn. Burlington,VT: Butterworth-Heinemann, 1975, vol. 2.
[22] F. Rohrlich, Electromagnetic momentum, energy, and mass, *American Journal of Physics*, vol. 38, no. 11, pp. 1310–6, 1970. [Online]. Available: http://link.aip.org/link/?AJP/38/1310/1.
[23] S. M. Barnett, Optical angular-momentum flux, *Journal of Optics B*, vol. 4, no. 2, pp. S7–16, 2002. [Online]. Available: http://www.iop.org/EJ/abstract/1464-4266/4/2/361.
[24] C. Cohen-Tannoudji, J. Dupont-Roc, and G. Grynberg, *Photons and Atoms*. New York: John Wiley & Sons, 1989.
[25] J. M. Jauch and F. Rohrlich, *The Theory of Photons and Electrons*. Cambridge, MA: Addison-Wesley, 1955.
[26] H. Yilmaz, *Introduction to the Theory of Relativity and the Principles of Modern Physics*. New York: Blaisdell, 1965.
[27] J. W. Simmons and M. J. Guttmann, *States, Waves and Photons*. Reading, MA: AddisonWesley, 1970.
[28] A. O. Barut, *Electrodynamics and Classical Theory of Fields and Particles*. New York: Dover Publications, 1980.

[29] S. J. van Enk and G. Nienhuis, Commutation rules and eigenvalues of spin and orbital angular momentum of radiation fields, *Journal of Modern Optics*, vol. 41, no. 5, pp. 963–77, 1994. [Online]. Available: http://dx.doi.org/10.1080/09500349414550911.

[30] S. J. Van Enk and G. Nienhuis, Spin and orbital angular momentum of photons, *Europhysics Letters*, vol. 25, no. 7, pp. 497–502, March 1994. [Online]. Available: http://dx.doi.org/10.1209/0295-5075/25/7/004.

[31] S. M. Barnett, Rotation of the electromagnetic field and the nature of optical angular momentum, *Journal of Modern Optics*, vol. 57, no. 14–15, pp. 1339–1343, September 2010.

[32] O. Heaviside, On the forces, stresses, and fluxes of energy in the electromagnetic field, *Philosophical Transactions of the Royal Society*, vol. 183, pp. 423–80, 1892.

[33] J. Larmor, A dynamical theory of the electric and luminiferous medium. part iii. relations with material media, *Philosophical Transactions of the Royal Society*, vol. 190, pp. 205–493, 1897.

[34] H. Bateman, *The Mathematical Analysis of Electrical and Optical Wave-Motion*. Cambridge: Cambridge University Press, 1915, reprinted New York: Dover Publications, 1955.

[35] M. Lax, W. H. Louisell, and B. McKnight, From Maxwell to paraxial wave optics, *Physical Review A*, vol. 11, no. 4, pp. 1365–70, April 1975.

[36] H. A. Haus, *Waves and Fields in Optoelectronics* Englewood Cliffs, NJ: Prentice-Hall, 1984, no. 2.

[37] L. W. Davis, Theory of electromagnetic beams, *Physical Review A*, vol. 19, no. 3, pp. 1177–9, March 1979. [Online]. Available: http://link.aps.org/abstract/PRA/v19/p1177.

[38] A. E. Siegman, *Lasers*. Sausalito, CA: University Science Books, 1986.

[39] G. Stephenson and P. M. Radmore, *Advanced Mathematical Methods for Engineerging and Science Students*. Cambridge: Cambridge University Press, 1993.

[40] L. G. Gouy, Sur une propriete nouvelle des ondes lumineuses, *Compted Rendus de l'Academie des Sciences*, vol. 110, p. 1251, 1890.

[41] M. Abramowitz and I. S. Stegun, *Handbook of Mathematical Functions*. Mineola, NY: Dover Publications, Inc., 1974, reprint. Originally published: National Bureau of Standards, corrected edition, 1964.

[42] E. Abramochkin and V. Volostnikov, Beam transformation and nontransformed beams, *Optics Communications*, vol. 83, no. 1,2, pp. 123–35, 1991.

[43] J. Enderlein and F. Pampaloni, Unified operator approach for deriving Hermite–Gaussian and Laguerre–Gaussian laser modes, *Journal of the Optical Society of America A*, vol. 21, no. 8, pp. 1553–8, 2004. [Online]. Available: www.opticsinfobase.org/josaa/abstract.cfm?URI=josaa-21-8-1553.

[44] M. R. Dennis, J. B. Götte, R. P. King, M. A. Morgan, and M. A. Alonso, Paraxial and nonparaxial polynomial beams and the analytic approach to propagation, *Optics Letters*, vol. 36, no. 22, pp. 4452–4, 2011.

[45] A. E. Siegman, Hermite–Gaussian functions of complex argument as optical-beam eigenfunctions, *Journal of the Optical Society of America*, vol. 63, no. 9, pp. 1093–4, 1973. [Online]. Available: www.opticsinfobase.org/josa/abstract.cfm?URI=josa-63-9-1093.

[46] S. M. Barnett and L. Allen, Orbital angular momentum and nonparaxial light beams, *Optics Communication*, vol. 110, no. 5–6, pp. 670–8, 1994.

[47] J. Durnin, J. J. Micelli, and J. H. Eberly, Diffraction-free beams, *Physial Review Letters*, vol. 58, no. 15, pp. 1499–501, 1987. [Online]. Available: http://link.aps.org/abstract/PRL/v58/p1499.
[48] J. Durnin, Exact solutions for nondiffracting beams, I: the scalar theory, *Journal of the Optical Society of America A*, vol. 4, no. 4, p. 651, 1987. [Online]. Available: www.opticsinfobase.org/josaa/abstract.cfm?URI=josaa-4-4-651.
[49] D. McGloin and K. Dholakia, Bessel beams: diffraction in a new light, *Contemporary Physics*, vol. 46, no. 1, pp. 15–28, 2005. [Online]. Available: http://dx.doi.org/10.1080/0010751042000275259.
[50] A. G. Gray and G. B. Mathews, *A Treatise on Bessel functions and their Application to Physics*. London: Macmillan and Co., 1895. (The authors do not use the notation of the Dirac delta distribution, which had not yet been introduced at the time the book was published).
[51] M. V. Berry, Paraxial beams of spinning light, in M. S. Soskin and M. V. Vastnetsov eds, Singular Optics. SPIE, 1998, pp. 6–11.
[52] J. F. Nye and M. V. Berry, Dislocations in wave trains, *Proceedings of the Royal Society of London, Series A*, vol. 336, no. 1605, pp. 165–90, 1974. [Online]. Available: www.journals.royalsoc.ac.uk/link.asp?id=308238272p258w1t.
[53] M. R. Dennis, K. O'Holleran, and M. J. Padgett, Optical vortices and polarization singularities, *Progress in Optics*, vol. 53, pp. 293–363, 2009.
[54] J. F. Nye, *Natural Focussing and the Fine Structure of Light*. Bristol: Institute of Physics Publishing, 1999.
[55] C. Tamm and C. O. Weis, Bistability and optical switching of spatial patterns in a laser, *Journal of the Optical Society of America B*, vol. 7, no. 6, pp. 1034–8, 1990. [Online]. Available: www.opticsinfobase.org/josab/abstract.cfm?URI=josab-7-6-1034.
[56] M. W. Beijersbergen, R. P. C. Coerwinkel, M. Kristensen, and J. P. Woerdman, Helical-wavefront laser beams produced with a spiral phaseplate, *Optics Communication*, vol. 112, no. 5–6, pp. 321–7, 1994.
[57] L. Allen, S. M. Barnett and M. J. Padgett, *Optical Angular Momentum*. Bristol: Institute of Physics Publishing, 2003.

2

Vortex transformations and vortex dynamics in optical fields

G. MOLINA-TERRIZA

2.1 Introduction

Vortices and vorticity are common objects of study in the field of fluid dynamics [1]. In a fluid, vortices appear as spinning, often turbulent, flows, giving rise to common phenomena such as tornadoes and whirlpools. In their more discreet manifestations, fluid vortices are thought to be responsible for the destruction of bridges due to the formation of vortex streets and structural resonance [2] and for the flight of insects [3], ending a long debate in the community about the reversibility of the movement of insect wings. Because of the importance of all those natural phenomena, the study of vortex dynamics and the creation and annihilation of hydrodynamical vortices is a matter of intense study.

When the first properties of superfluids were found, the question of whether vortices could appear in such irrotational fluids was addressed by prominent scientists [4, 5]. It was found that vortices could also appear in such systems, but they could only appear as point vortices, i.e. singular locations where the vorticity of the fluid was infinite, whereas everywhere else the vorticity was strictly zero. This property was also found later on in Bose–Einstein condensates (BECs), where the quest for exciting and controlling the point vortices of quantum systems is still a hot topic. The dynamics of and processes related to the birth and destruction of vortex–antivortex pairs play an important role in the physics of these exotic systems.

Optical fields also present vortices [6, 7]. They are present in most of the cylindrical modes of the optical fields, thus appearing naturally as cavity modes. They can also be observed in speckle fields, where the random interference of different waves gives rise to singular points [8]. The technology to control optical vortices is very mature and has allowed a multitude of applications, as the readers of this book can easily check, in fields as diverse as nanophotonics or astronomy. In the case of optical fields the "velocity" which swirls around the vortex core is related to the gradient of the phase, which in turn is related to the transversal Poynting vector (or electromagnetic momentum) of the field. This momentum can be transferred to material particles, which has found enormous

The Angular Momentum of Light, ed. D. L. Andrews and M. Babiker. Published by Cambridge University Press.

applications in the fields of optical trapping, developing into a new tool: the optical spanner [9, 10]. Also, the interesting property that the field vanishes at the position of the optical vortex is used in lithographic applications [11].

As we will show in the next section, optical vortices share many properties with point vortices in superfluids and BECs. In particular, they present an interesting dynamics. For most of the applications already mentioned, the dynamics of optical vortices plays a very important role and has to be carefully controlled. The creation and annihilation of optical vortices and the exact control over their position in the transversal plane are factors that need to be considered in any application.

A very important feature seldom explored in the literature is that optical vortices present very rich dynamics even in linear optics, where the fields can be computed analytically or easily analyzed numerically. Exotic effects that can be easily understood in the linear optics regime, such as vortex transformations (not too dissimilar to alchemical reactions), have since found counterparts their in nonlinear systems. These effects can also be mapped to other systems, such as the aforementioned BECs. As we will see below, when we try to understand vortex dynamics solely following vortex lines, the dynamical equations can become nonlinear, which is part of the richness of such dynamics. The fact that this complicated dynamical system of equations can be easily solved using the complete wave function of the optical field deserves further study.

Finally, the properties of optical vortices only start in the linear optics regime. A wealth of interesting effects can be found when we propagate optical vortices in a nonlinear optical material. From Kerr-like nonlinearities, to the frequency mixing of different waves, to nonlocal systems, the dynamics of optical vortices keeps giving us interesting physical effects to explore and understand [12].

This chapter will review some of the concepts mentioned above. It will not be exhaustive as the literature is vast and the interested reader can find review articles on current topics of vortex dynamics, particularly in nonlinear systems [12]. The intention is to present a tutorial where some of the more interesting aspects of the linear optics dynamics of vortices is overviewed. The focus will be on clarity and understandability for the reader unfamiliar with such concepts.

2.2 Optical vortices

Let us start by defining our system of study. We will mainly be interested in vortices embedded in nearly monochromatic optical beams, i.e. electromagnetic fields in the optical frequency range with a well-characterized direction of propagation, which we take to be the z axis. Then, we only need to work within the paraxial approximation, where the polarization $\mathbf{u}$ of the beam is well defined and perpendicular to z. In this case, the electric field will be $\mathbf{E} \propto E(x, y, z) \exp(i\omega t - ikz)\mathbf{u} + c.c.$. The complex scalar field $E(x, y, z)$ then follows the paraxial propagation equation, which in normalized units reads:

$$\mathrm{i}\frac{\partial E}{\partial z} + \frac{1}{2}\nabla^2 E + g(E, E^*; x, y, z) = 0. \tag{2.1}$$

We have added a function g for use later on. This function can be nonlinear, but we will usually require that it at least satisfies $g(E=0)=0$. Then the above equation can accommodate several interesting situations, such as propagation in graded index fibers, Kerr nonlinearities, or even wave mixing in the undepleted regime.

In this context, optical vortices are defined as those points in the transversal plane (x, y) where the amplitude of the scalar field vanishes and its phase varies continuously around the singular point. The most commonly studied optical vortices have a field that near the core of the vortex is proportional to $E \propto r^{|m|}\exp(im\phi)$ in cylindrical coordinates, or $E \propto (x + \mathrm{sign}(m)\mathrm{i}y)^{|m|}$ in Cartesian coordinates. The parameter m is the topological charge of the optical vortex, i.e. 2π windings of the phase of the field around the vortex, and can be defined even if the vortex is not of the shape given above as

$$m = \frac{1}{2\pi}\oint_C \vec{\nabla} S \cdot \mathrm{d}\vec{l}, \tag{2.2}$$

where C is a closed contour surrounding the vortex position, and S is the phase of the scalar field: $E=|E|\exp(\mathrm{i}S)$. In the particular case presented, m is also related to the local content of orbital angular momentum of the field. In general, an optical vortex appears at the crossings of the lines where the real and imaginary parts of E are zero. In the example above these lines cross perpendicularly, which gave this kind of vortices the adjective "canonical." In general this is not the case, as we will show below.

Another important characteristic of vortices is that when $|m|=1$, the vortex is only displaced under small fluctuations. Higher order vortices, however, usually split into vortices of unit charge (positive or negative).

2.2.1 Morphology of noncanonical vortices

We can now describe in more detail a class of vortices that we call "noncanonical" vortices. In the vicinity of a vortex of charge one, located at $(x, y)=(x_1, y_1)$, the scalar field can be approximated as $E(x \simeq x_1, y \simeq y_1) = a_x(x-x_1) + a_y(y-y_1) + O(x^2, y^2)$, where a_x, a_y are complex quantities. To simplify the expressions we will assume that $a_x \neq 0$. Then, we can re-express the field as

$$E(x \simeq x_1, y \simeq y_1) = C((x-x_1) + \mathrm{i}A(y-y_1)) + O(x^2, y^2), \tag{2.3}$$

where $C=a_x$ and $A=-\mathrm{i}a_y/a_x$. The parameter C does not affect the morphology of the optical vortex and then we have reduced the problem to a single complex variable A, which we call the noncanonical parameter. When $A=\pm 1$ the field is of the form $(x \pm \mathrm{i}y)$, which is the canonical expression we had previously shown. The sign of the real part of A defines the topological charge of the vortex. Other ways of parametrizing noncanonical vortices have been used in the study of *speckle-fields* [13–15] and are mainly based on the use of a_x and a_y or their combinations. In Fig. 2.1 the amplitude and phase of a typical noncanonical vortex are shown and in Fig. 2.2, we show a few examples of noncanonical vortices.

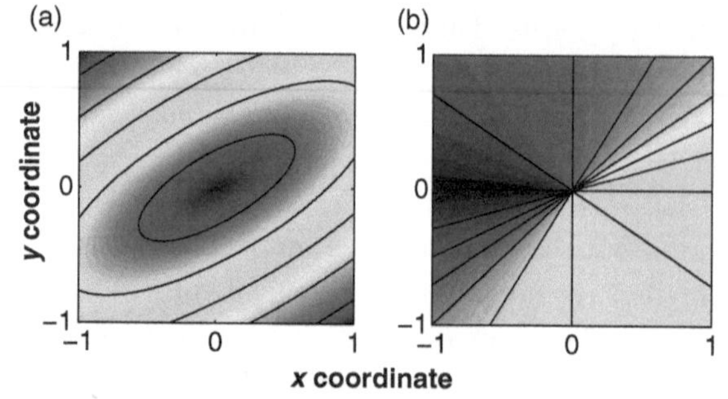

Figure 2.1 Amplitude (a) and phase (b) of the core of a noncanonical vortex with $A = 1 + \mathrm{i}$. Contour lines are spaced by 0.2 units in (a) and $\pi/8$ units in (b).

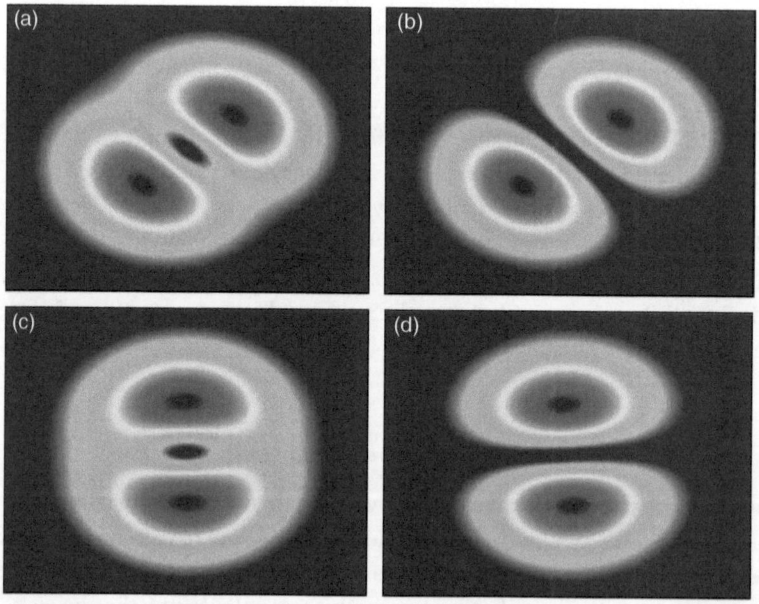

Figure 2.2 Examples of Gaussian beams carrying noncanonical vortices. The noncanonical parameter A varies in each panel: (a) $A = 0.7 + 0.7\mathrm{i}$; (b) $A = 0.1 + 0.9\mathrm{i}$; (c) $A = 2 - 0.1\mathrm{i}$; (d) $A = 7 + 7\mathrm{i}$.

Given a certain noncanonical parameter A we can easily calculate what the field looks like. Experimentally, the inverse question is of relevance, i.e. given a certain field, what is the associated noncanonical parameter? We can partially answer this question by determining the intensity field of a noncanonical vortex:

$$I = | C(x + \mathrm{i}Ay) |^2 + O(x^3, y^3) \simeq | C |^2 \left((x - A_{\mathrm{i}}y)^2 + A_r^2 y^2\right). \tag{2.4}$$

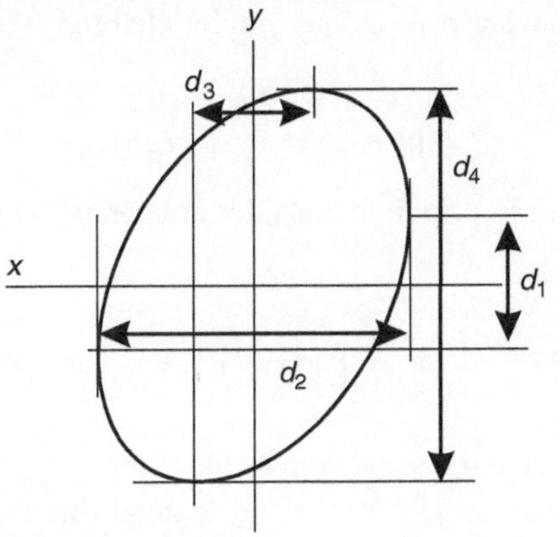

Figure 2.3 Locus of constant intensity for a typical noncanonical vortex; d_1, d_2, d_3, and d_4 are different distances which allow A to be estimated.

The contour curves of intensity are given by the equation $(x - A_i y)^2 + A_r^2 y^2 = T^2$, where T parametrizes each different curve. Note that we can in principle find the sign of A_i with only intensity measurements, but not the sign of A_r. Then, if we want to know the topological charge of the vortex, we have to resort to interferometric means. In Fig. 2.3 we review the most important points of such a curve.

Following the notation in Fig. 2.3 and from the intensity contour level equations, we can find that:

$$\begin{aligned}
d_1 &= 2T/|A_r|, \\
d_2 &= 2T \mid A \mid /|A_r|, \\
d_3 &= 2T A_i/|A_r|, \\
d_4 &= 2T A_i/|A_r| \mid A \mid . \qquad (2.5)
\end{aligned}$$

Experimentally then, we should get as close as possible to the vortex and take different measurements of d_1 and d_2 (or d_3 ar d_4) for different contour levels to then find $|A_r|$ and $|A|$ (or $A_i/|A_r|$ and $|A|$).

We will end this subsection with the remark that noncanonical vortices are the most general kind of stable isolated vortices. As the field at the vortex position has to vanish, it means that we can always linearize the real and imaginary parts around the vortex position and obtain the equations described above.

2.2.2 *Multiple vortices*

After considering the most general form of isolated vortices, let's turn to the case where several vortices appear in the same beam. We will consider that in the region of interest the complex scalar field is described by a function V. Canonical vortices with topological charges $|m| > 1$ can be described as m single charged vortices coincident at the same point:

$$V = r^m \exp(\mathrm{i}m\phi) = (x + sign(m)\mathrm{i}y)^m. \tag{2.6}$$

Following this idea, one can in principle consider noncanonical vortices of order m in the same way:

$$V = \prod_{k=1}^{|m|} (x + \mathrm{i}A_k y) \tag{2.7}$$

where the real parts of A_k have the same sign as m.

There are some complications when we try to extend this kind of structure to several displaced vortices. The easiest way of constructing functions of several vortices is by multiplying the independent vortex functions, i.e.

$$V = \prod_{k=1}^{N} (x - x_k + \mathrm{i}A_k(y - y_k)), \tag{2.8}$$

where N is the number of nested vortices and x_k, y_k, A_k are the positions and noncanonical parameters of each vortex. This clearly is not the only way of obtaining a given vortex structure. One can build nonfactorizable expressions where, locally, each vortex can be expressed as in Eq. (2.3). In order to do that, one can design the real and imaginary zero lines of the field so that their crossings give the desired topological content. As an example, we can construct a symmetric vortex quadrupole (two pairs of positive and negative canonical vortices in opposite vertices of a square) with $V = (x - x_0 + \mathrm{i}y)\,(x + x_0 + \mathrm{i}y)\,(x - \mathrm{i}(y - x_0))\,(x - \mathrm{i}(y + x_0))$, which is a fourth-order polynomial, or we could use $V = x^2 + y^2 - x_0^2 + \mathrm{i}2xy$. In this latter case, V is a second-order polynomial. The vortex dynamics in either case can be very different and we have to choose the expression that better reflects our experimental needs.

2.3 Equations of evolution for the isolated vortices

We will now study fields that contain at least one vortex. The complex scalar function can then be expressed as

$$E(x, y, z) = V(x, y, z)F(x, y, z), \tag{2.9}$$

where $V = \{x - x_1(z) + \mathrm{i}A(z)[y - y_1(z)]\}$ is the field corresponding to the noncanonical vortex located at $(x_1, y_1) = \vec{r}_1$ in the transversal plane and $F(x, y, z)$ is the shape of the beam hosting the vortex. The function $F(x, y, z)$ can in principle include contributions from other vortices.

We will start by assuming that in our case the vortices are nondegenerate, i.e. two vortices cannot be in the same place, and with unit charge. In this case, $F(x_1(z), y_1(z), z) \neq 0$. Then, we can insert Eq. (2.9) in Eq. (2.1), in order to obtain

$$F\vec{\nabla}V \cdot \left[-\mathrm{i}\frac{\mathrm{d}\vec{r}_1}{\mathrm{d}z} + \vec{\nabla}\ln F\right] + V\left[\mathrm{i}\frac{\mathrm{d}F}{\mathrm{d}z} + \frac{1}{2}\nabla^2 \ln F\right]$$
$$+ g(E, E^*; x, y, z) - \frac{\mathrm{d}A}{\mathrm{d}z}(y - y_0)F = 0. \tag{2.10}$$

We can now develop this equation around the vortex position and expand it in terms of $(x - x_1, y - y_1)$. We obtain the following equation [16]:

$$\frac{\mathrm{d}x_1}{\mathrm{d}z} + \mathrm{i}\,A\,\frac{\mathrm{d}y_1}{\mathrm{d}z} = -\mathrm{i}L_1\{\ln[F(x = x_1, y = y_1, z)]\}, \tag{2.11}$$

where L_1 is the differential operator $L_1 = \partial/\partial x + \mathrm{i}A\partial/\partial y$. We also find that the noncanonical parameter A changes dynamically as

$$\frac{\mathrm{d}A}{\mathrm{d}z} - \left[L_2L_1\{\ln[F(x, y, z)]\} - \frac{L_2\{g(E, E^*; x, y, z)\}}{F(x, y, z)}\right]_{x=x_1, y=y_1} = 0, \tag{2.12}$$

where $L_2 = \mathrm{i}A\partial/\partial x - \partial/\partial y$. Our set of equations would be complete if we knew the evolution of the host beam $F(x, y, z)$. This equation can also be found from Eq. (2.10). Nevertheless, solving this set of equations is usually at least as complex as solving directly the paraxial equation, Eq. (2.1). Equations (2.11) and (2.12) are very useful when the host beam F can be determined or at least known approximately.

2.3.1 Dynamics of a single vortex in linear homogeneous media

One situation where the host beam can be determined independently of the vortex is the case of propagation in linear homogeneous media, i.e. $g = 0$. In this case, if the host beam is Gaussian, it will follow the equation $F = C(z)\exp\left[\mathrm{i}x^2/q_x(z) + \mathrm{i}y^2/q_y(z)\right]$, where $C(z)$ is the complex amplitude of the beam and $q_x(z)$ and $q_y(z)$ are the complex beam widths in the corresponding directions [17]. This example is particularly interesting because it reproduces the conditions one usually has when experimentally propagating a Gaussian beam in free space. Then, the equations of propagation for the vortex are

$$\frac{\mathrm{d}x_1}{\mathrm{d}z} + \mathrm{i}\,A\,\frac{\mathrm{d}y_1}{\mathrm{d}z} = \frac{2x_1}{q_x} + \mathrm{i}A\frac{2y_1}{q_y}, \tag{2.13}$$

$$\frac{\mathrm{d}A}{\mathrm{d}z} = 2A\left[\frac{1}{q_x} - \frac{1}{q_y}\right]. \tag{2.14}$$

We will start with the case of a symmetric host beam. In this case, $q_x = q_y = 2z - \mathrm{i}w_0^2$, and the solution of Eq. (2.14) is simply $A(z) = A_0$. The noncanonical parameter is then conserved through the propagation. This simple example allows us to show the effects of the noncanonical parameter on the vortex dynamics. Without loss of generality we can

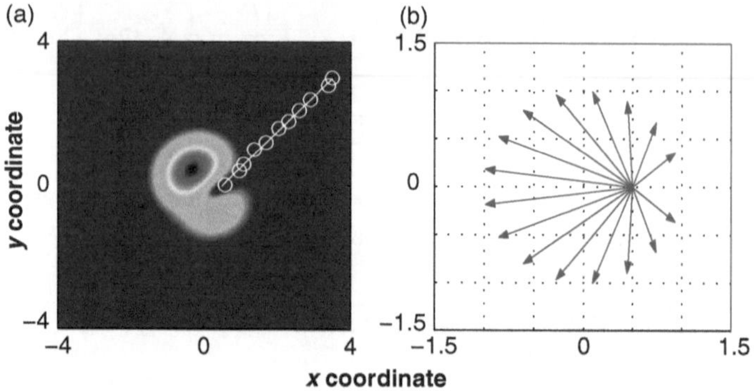

Figure 2.4 Evolution of different noncanonical vortex positions in free-space diffracting Gaussian beams from $z = 0$ up to $z = 3$. (a) $A_0 = 1 + \mathrm{i}$, we also show the intensity pattern at $z = 0$. The white solid line shows the vortex positions according to our model, circles represent the exact positions calculated numerically. (b) Vortex trajectories for different noncanonical parameters. The values of A_0 have been calculated so that at $z = 3$ all vortices are located along a circumference of unit radius. Arrows indicate the vortex trajectory orientation. All upwards (downwards) trajectories correspond to vortices of charge -1 $(+1)$. In all cases $w_0 = 1$, $x_{1,0} = 0.5$.

set $(x_1(0), y_1(0)) = (x_{1,0}, 0)$, due to the symmetry of the host beam. With this condition Eq. (2.13) can be solved:

$$x_1(z) = x_{1,0}\left(1 - \frac{2A_{i,0}}{w_0^2 A_{r,0}}\, z\right), \tag{2.15}$$

$$y_1(z) = -\frac{2x_{1,0}}{w_0^2 A_{r,0}}\, z. \tag{2.16}$$

These are the parametric equations of a straight line, which has an angle with respect to the x axis given by A. When $A_0 = \pm 1$ the trajectory is a vertical line, which is the result found by Indebetouw [18, 19] for canonical vortices. In Fig. 2.4 a series of different trajectories for different values of A_0 is presented. It can be seen that the noncanonical parameter strongly affects the trajectory of the vortex and its "velocity" dx_1/dz.

2.3.2 *Inversion of the topological charge of a vortex*

When the host beam is not symmetric the noncanonical parameter changes during the propagation. An elliptical beam where the two foci are in the same plane $z = 0$ is characterized by $q_x = 2z - \mathrm{i}w_{0x}^2$ and $q_y = 2z - \mathrm{i}w_{0y}^2$. In this case, Eqs. (2.13) and (2.14) can be integrated to obtain the following expression for A:

$$A(z) = A_0 \frac{w_{0y}^2}{w_{0x}^2}\left(\frac{\mathrm{i}w_{0x}^2 - 2z}{\mathrm{i}w_{0y}^2 - 2z}\right) \tag{2.17}$$

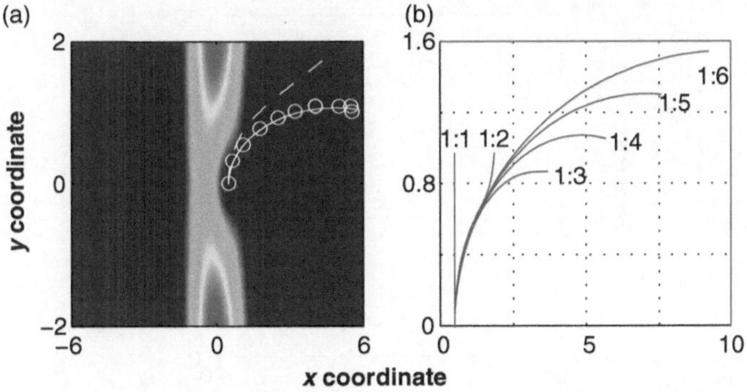

Figure 2.5 Evolution of the positions of optical vortices in elliptical beams. Vortices at $z = 0$ are canonical $A_{0,\mathrm{r}} = 1$, $A_{0,\mathrm{i}} = 0$. Labels in (b) show the host beam initial ellipticity $w_{0x} : w_{0y}$. In (a) $w_{0x} = 1$, $w_{0y} = 4$; the light pattern at $z = 0$ is also shown. Solid lines and circles show the trajectory predicted by the model and the numerically calculated vortex positions respectively; the dashed line shows the resulting trajectory if the evolution of the noncanonical parameter is not taken into account. All cases with $x_{1,0} = 0.5$.

and the vortex position:

$$\frac{x_1(z)}{q_x} + \mathrm{i}A_0 \frac{w_{0y}^2}{w_{0x}^2} \frac{y_1(z)}{q_y} = \frac{x_{1,0}}{\mathrm{i}w_{0x}^2}. \tag{2.18}$$

Representative examples are shown in Fig. 2.5. In the same figure the reader can find the prediction of a model where the noncanonical parameter is not considered. This example shows that in order to ascertain the dynamics of optical vortices in the general case it is essential to include the internal structure of the vortex.

From a dynamical point of view the fact that the noncanonical parameter may change during propagation is surprising. Careful examination of Eq. (2.17) shows that the real part of A can change during propagation. In particular, if $\frac{A_{\mathrm{r},0}}{A_{\mathrm{i},0}} \leq \frac{|w_x^2 - w_y^2|}{2w_x w_y}$ the real part of A can even change its sign, which means that the vortex can dynamically suffer a transformation from having positive topological charge to negative charge.

As briefly mentioned in the introduction, the topological content of an optical beam can typically change when vortex–antivortex pairs are created or annihilated, or when optical vortices appear or disappear from the reservoir at infinity, $(x, y) \to \infty$. The total topological charge of an optical beam is only affected by this latter mechanism, meanwhile the former keeps the total topological charge invariant. The possibility presented above is a new mechanism of vortex transformation. In this case, the vortex transforms itself via continuous deformations of its noncanonical parameter. From Eq. (2.18) it can be shown that if the vortex is at the center of the beam, i.e. $x_{1,0} = 0$, then the position of the vortex remains always the same, independently of its noncanonical parameter. Then, the vortex will proceed with the charge inversion without any displacement. This new transformation mechanism was first investigated in [20] and has found applications in other

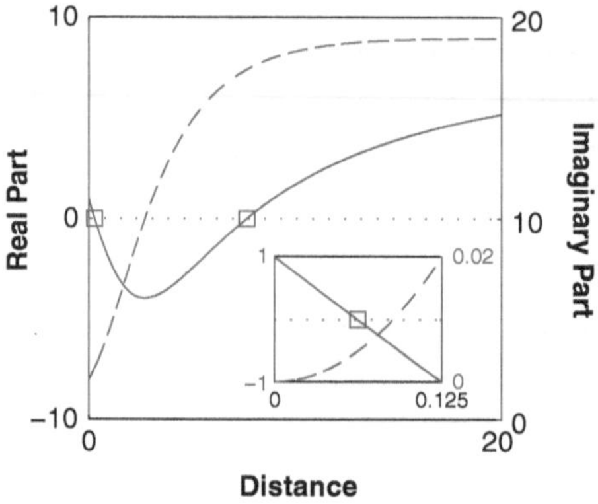

Figure 2.6 Evolution of the noncanonical parameter of a vortex. Solid and dashed lines represent respectively the evolution of the real and imaginary parts of A. Axes are in arbitrary units. The squares show the distances where the inversion of the topological charge occurs. Initial conditions: vortex with $A(0) = 1 + 2i$ nested in an elliptical Gaussian beam with $w_{0x} = 1$, $w_{0y} = 3$. Inset shows the evolution of a canonical vortex after a cylindrical lens.

fields, as for example in BECs [21] or in acoustic vortices [22]. Also, this vortex transformation is the hidden mechanism by which some mode converters work, such as the *π-converter* [23, 24].

Let us study this mechanism in more detail. From Eq. (2.17), and if the noncanonical parameter and host beam widths are in the ranges of the inequality shown in the previous paragraphs, there are two charge inversions at distances:

$$z_{\text{inv}} = -4(w_x^2 - w_y^2)\frac{A_i}{A_r} - \left(1 \pm \sqrt{1 - 4\frac{A_r^2}{A_i^2}\frac{w_x^2 w_y^2}{(w_x^2 - w_y^2)^2}}\right). \tag{2.19}$$

In Fig. 2.6 we have represented the evolution of A for a typical case. It can be observed that, as expected, the real part of A changes sign on passing through zero. Morphologically, this means that the vortex core becomes more and more elliptical until it becomes a line traversing the whole beam. This is another type of phase singularity, called *edge dislocation*. In Fig. 2.7 there is an example of such a transformation.

A case of interest is when $w_x \gg w_y$ and $A_r = 0$ at $z = 0$. This situation models very well the effect of a canonical vortex embedded in a collimated circular Gaussian beam after traversing an astigmatic lens, with focal plane at $z = 0$. In this case, it can be shown that A_r changes sign at $z = 0$ as expected and is approximately linear around $z = 0$. In the inset of Fig. 2.6 we show a similar case which models the experiment in [20].

When the vortex is not exactly located at the center of the beam, the dynamics of the vortex become coupled with A, as in Fig. 2.5. Near the location where the vortex suffers this transformation process, it typically wanders away from the beam center. At the exact position where $\Re e(A) = 0$ the vortex disappears to infinity, and immediately after this

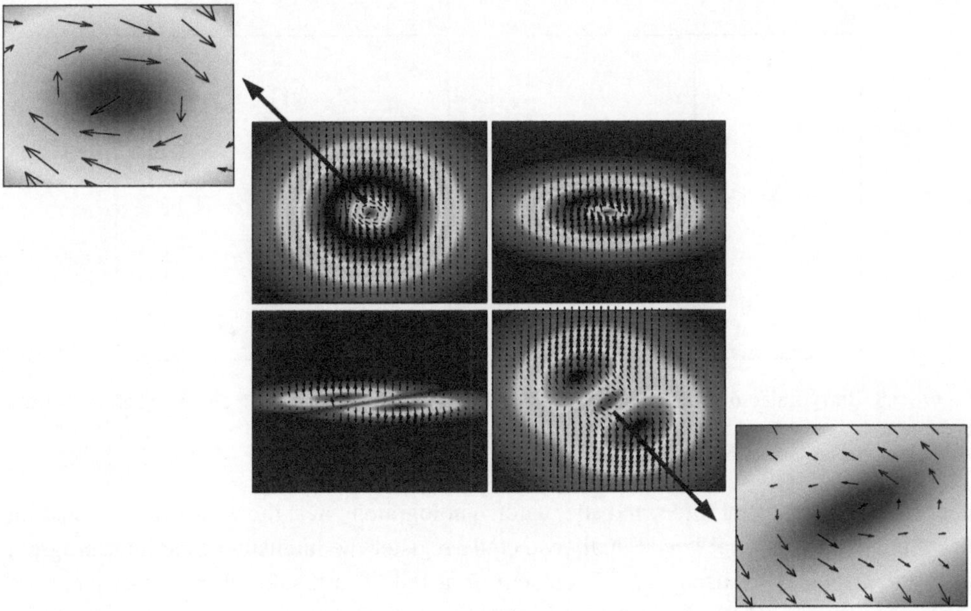

Figure 2.7 Evolution of the amplitude and phase of an optical vortex after a cylindrical lens. Figures read from top-left to bottom-right. The panels correspond to distances from the lens of $z = 0$, $z = f/2$, $z = f$, and $z = 2f$ respectively. Arrows indicate the direction of the gradient of the phase. Enlargements of the vortex cores in panels 1 and 4 show that the direction of the optical flow has changed.

z position another inverted vortex reappears from the opposite point of the beam, traveling towards the center of the beam. The closer the initial beam is to the center, the more abrupt is the evolution of the vortex. Practically, when the initial conditions are near the stationary point, it is not possible to distinguish between the *edge dislocation* and the deformed vortex.

All these effects are easier to understand by means of Berry's vortex trajectories [25]. In this way, an edge dislocation simply appears as a three-dimensional vortex line which lies in the transversal plane. Then, one can understand the case where the vortex does not move as a degenerate case, as represented in Fig. 2.8.

2.3.3 Experimental demonstration of the vortex inversion

In [20] the dynamical inversion of the topological charge of a vortex was experimentally demonstrated. The linearly polarized light from a Helium–Neon laser at 633 nm was divided with a beam-splitter, in order to allow for interferograms to detect the inversion of the charge. The reference beam included a spatial filter, neutral density filters to control the amplitude of the reference beam, and a lens to compensate for the curvature of the beam and obtain straight lines in the interference pattern. In the main beam path we

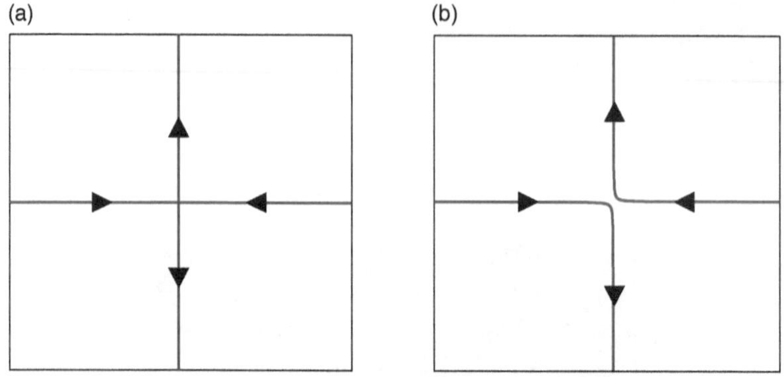

Figure 2.8 Berry trajectories for the vortex lines during the inversion of the charge when the vortex is (a) exactly located on the axis and (b) slightly off–axis.

used, besides the spatial filters and attenuator, a hologram to nest the vortex in the Gaussian beam and the cylindrical lens. Finally, we could register the intensity or the interferogram at different planes by using a CCD camera on a rail. The results of the experiment are shown in Fig. 2.9, where the inversion of the charge can be clearly seen after crossing the focal plane.

2.3.4 Vortices and angular momentum

The orbital angular momentum in a linear homogeneous medium has to be conserved along the propagation direction. Then, the process of the transformation of the topological content of an optical beam in these media occurs without altering the total angular momentum of the field. In particular, the transformation discussed in the previous subsection, where a vortex inverts its charge, leaves the angular momentum of the field unchanged.

It can be shown that a cylindrical lens does not change the total angular momentum of a beam with cylindrical symmetry [26]. It nevertheless mixes the eigenfunctions of angular momentum. The input beam, being cylindrically symmetric, has to be an eigenmode of the angular momentum. The output beam, however, no longer has this cylindrical symmetry and, thus, must be a superposition of several angular momentum eigenmodes.

Generally, if the input beam before the lens is an eigenvector of the angular momentum with eigenvalue m, i.e. $E_{(-)}(x, y) = u_m(x, y)$, the output field is

$$E_{(+)}(x, y; z = 0) = v_m(\rho) \exp\left(-i\frac{k\rho^2}{4f}\right) \sum_{n=-\infty}^{\infty} (-i)^n J_n\left(\frac{k\rho^2}{4f}\right) \exp\left[i(m + 2n)\varphi\right], \quad (2.20)$$

where (ρ, φ) are cylindrical coordinates, v_m is the input radial distribution, and J_n is the Bessel function of order n. Remarkably, if the input beam has an even (odd) topological

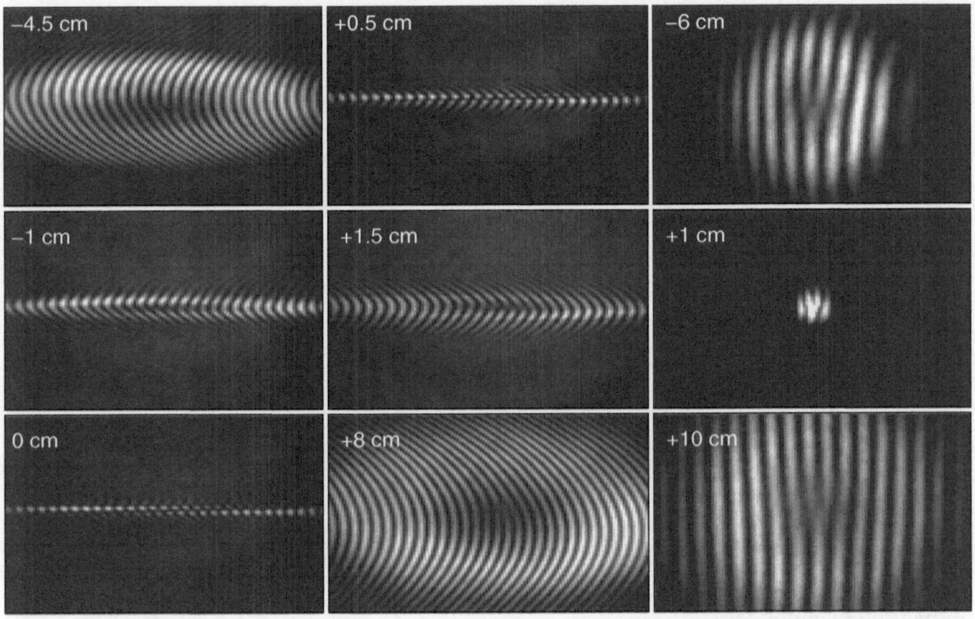

Figure 2.9 Experimental results for the inversion of the topological charge. Images were obtained with a CCD camera at different positions after the lens. The labels in each image show the actual position after the lens. Left and central columns show the noncanonical evolution of an initially canonical vortex nested in a 5 mm width Gaussian beam, after passing through a cylindrical lens with 10 cm focal length. Right column shows the light pattern under identical conditions but using a spherical lens.

charge m, the output field only contains even (odd) modes. The evolution of each of the modes composing $E_{(+)}$ is independent, owing to the linearity of the system. Then, the total intensity of each separate mode is conserved along the propagation. The radial distribution of the modes, however, may change along the propagation.

Then, returning to the unit charge vortex described in the previous sections, it can be shown that near the vortex (at $x = y \simeq 0$) the only important contribution comes from the modes of order $+1$ and -1. From this point of view the inversion of the topological charge is only due to the competition of the local amplitude of these two modes in the vicinity of the vortex, as shown in Fig. 2.10. Also, from this point of view, although the total angular momentum should be conserved, one should also observe that the local angular momentum around the vortex should indeed change sign. This is clearly seen in Fig. 2.7.

2.3.5 Another interpretation of the charge inversion

We will end this section with probably the easiest explanation of the vortex transformation described in earlier sections. We have used the equations of vortex dynamics. We have also

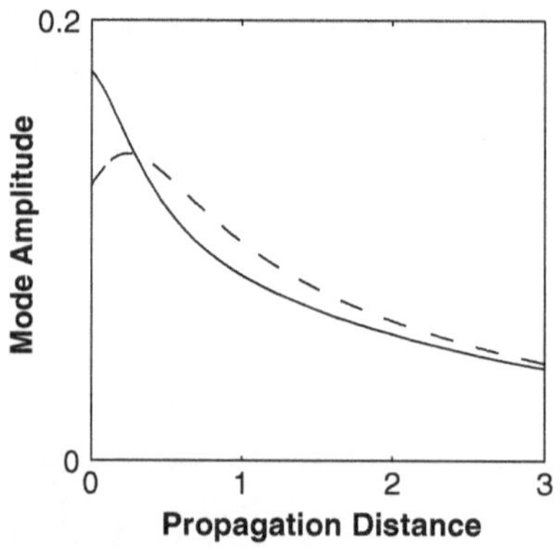

Figure 2.10 Evolution of the amplitude of the modes $m = 1$ (solid line) and $m = -1$ (dashed line) at radial position $\rho = 0.01$. Parameters are as in Fig. 2.6. The amplitudes of all other contributing modes are much smaller than those shown.

shown that this effect is compatible with the conservation of angular momentum and that the inversion of the topological charge can also be described as a competition of angular momentum eigenmodes.

The inversion of the topological charge could also have been predicted from simple geometrical optics arguments. When the optical beam goes through the cylindrical lens and propagates beyond the focal plane, the resulting field is a mirror image in one of the transverse directions, while remaining the same in the perpendicular one, i.e. $x \to x$ and $y \to -y$. It is clear that this transformation will invert the charge of the vortex.

The simplicity of this interpretation gives us a very easy way of visualizing the net effect. However, it doesn't provide any insight into the exact mechanism of the vortex transformation. We lose all the information about the vortex position or that the transformation occurs through an edge dislocation.

2.4 Interactions between vortices

The main message intended in the previous section is that, in principle, the evolution of a vortex can be understood using dynamical equations such as the ones presented in Eqs. (2.11) and (2.12). We have shown the importance of the noncanonical parameter to understand the trajectories of a single vortex. In principle, the same equations could be used to understand the behavior of point vortices in Bose–Einstein condensates, nonlinear optical vortices, etc. If one overlooks the noncanonical parameter and leaves it constant, imposing, for example that the vortex will always be canonical, Eq. (2.11) becomes the same as in the literature of point vortices, when using the appropriate host beam. In general, if we set[1] $A(z) = 1$, and we write the host beam in terms of its amplitude and phase:

[1] As we have seen in the previous section, fixing the noncanonical parameter is not physical and it only works in a few special cases. Then, any dynamical equation, in any system following Eq. (2.1) which does not consider A is unphysical. (We include this remark so that the interested reader can compare with the different equations found in the literature.)

$F(x, y, z) = \Psi(x, y, z)\exp(i\Phi(x, y, z))$, where, of course, Ψ and Φ are real variables, the dynamical equation takes the following more recognizable form:

$$\frac{\mathrm{d}\mathbf{r_1}}{\mathrm{d}z} = \frac{1}{2}\left(\hat{\mathbf{z}} \times \nabla(\ln \Psi) - \nabla\Phi\right), \tag{2.21}$$

where $\mathbf{r_1} = (x_1, y_1)$, $\hat{\mathbf{z}}$ is the unit vector pointing in the z direction, the operator $\nabla = (\partial/\partial x, \partial/\partial y)$, and the left part of the equation has to be evaluated at (x_1, y_1). Written in this form, the different contributions from the phase and amplitude of the host beam to the motion of a canonical vortex become clear. The internal parameter A mixes these contributions and affects the whole motion as explained earlier.

The simplicity of Eq. (2.21) allows us to understand some properties which could have been hidden from the set of Eqs. (2.11)–(2.12). For example, we can show that if we have several vortices embedded in a Gaussian beam, the motion of one of the vortices is affected separately from, on one side, the Gaussian beam and, on the other side, from the motion of all the other vortices. Another interesting feature which can be recovered from Eq. (2.21) is that in the case of multiple canonical vortices of the same sign, embedded in a Gaussian beam, the effect of the "field" due to the other vortices cancels out. Then, as we already knew, the motion of the vortices is independent of each other and only affected by the Gaussian host beam. Note that in this situation it is consistent to use Eq. (2.21), because all the vortices are canonical and keep their noncanonical parameters constant through propagation.

2.4.1 GRaded INdex media as a laboratory for vortex interactions

In general, the motion of vortices is affected by the host beam and other vortices. In our study we would like to be able to focus just on the effect between vortices, although still keeping physically realizable fields. For that reason we are going to resort to a set of inhomogeneous media called GRaded INdex (GRIN) media, which can be modeled using

$$g(E, E^*; x, y, z) = -(\tilde{n}_{2,x}z^2 + \tilde{n}_{2,y}y^2)E. \tag{2.22}$$

This type of inhomogeneity can be found in certain kinds of multimode fibers. It is also the first approximation that can be made of fields with a low cubic nonlinearity and models the trapping potential for a BEC. The effect of the quadratic index of refraction is to produce a waveguiding effect, thus keeping Gaussian beams with the appropriate width invariant under propagation.

In this way, the evolution of a single vortex using Eqs. (2.13) and (2.14) becomes trivial. In order to solve them, we just need the expressions for $q_x(z)$ and $q_y(z)$ for a GRIN medium, which are obtained from

$$\dot{Q}_x + 2Q_x^2 + n_{2,x} = 0, \ \dot{Q}_y + 2Q_y^2 + n_{2,y} = 0, \tag{2.23}$$

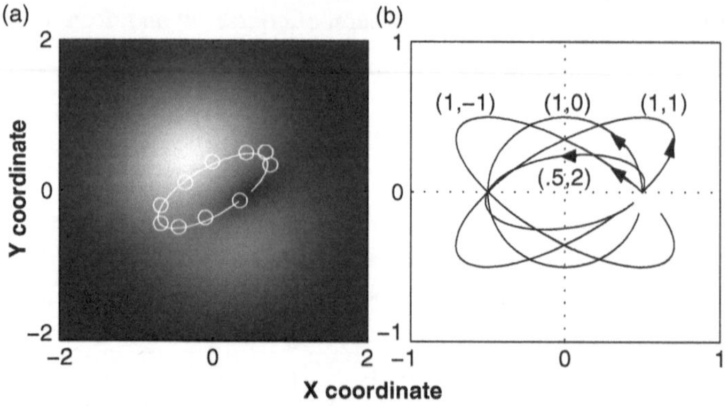

Figure 2.11 Evolution of the positions for different noncanonical vortices nested in Gaussian beams propagating in GRIN media, from $z=0$ to $z=3$. (a) The noncanonical parameter of the vortex is $A_0=1+\mathrm{i}$ and we also show the intensity pattern of the input beam. White lines show the positions of the vortices as predicted by the model and the circles show the corresponding numerical calculated ones from the paraxial equation. (b) Vortex trajectories for different noncanonical parameters. Labels in (b) show the value of the noncanonical parameter $(A_{0,\mathrm{r}}, A_{0,\mathrm{i}})$. In all cases $n_2=2$, $w_0=1$, $x_{1,0}=0.5$.

where $Q_\nu = 1/q_\nu$ with $\nu = x, y$ and $\dot{Q}_\nu = \mathrm{d}Q_\nu/\mathrm{d}z$. If $Q_\nu(0) = \mathrm{i}\sqrt{n_\nu/2}$, $\dot{Q}_\nu = 0$ and, as mentioned earlier, the complex beam width remains constant throughout the propagation. This value corresponds to a beam waist $w_\nu^2 = \sqrt{2/n_2}$.

Then, we can normalize all our problem to $w_x = w_y = 1$. Solving for the evolution of the optical vortex under these conditions shows that the vortex line forms a spiral around the propagation axis. In the transversal plane, the trajectories form closed elliptical curves with period $T=\pi$. The ellipticity and orientation depend on the noncanonical parameter of the vortex [16]. In Fig. 2.11 we show typical trajectories of different noncanonical vortices

2.4.2 Vortex interaction in GRIN media

We can now follow our program to try to understand the interaction of optical vortices. We will assume that our field contains N topological defects with parameters A_i. We will start by assuming that the initial field can be expressed as

$$E(x, y, z=0) = C\prod_{i=1}^{N}(x - x_\mathrm{i} + \mathrm{i}A_i(y - y_i))\exp\left(-\frac{x^2}{w_x^2} - \frac{y^2}{w_y^2}\right). \qquad (2.24)$$

In order to proceed as in the previous sections, the equation for the trajectory of the j vortex depends on the function $F(x, y, z)$, which we expect to be of the form

$$F(x, y, z) = C\prod_{i\neq j}(x - x_i(z) + \mathrm{i}A_i(z)(y - y_i(z)))\exp\left(-\frac{x^2}{w_x^2} - \frac{y^2}{w_y^2}\right). \qquad (2.25)$$

where $x_i(z)$, $y_i(z)$, $A_i(z)$ are unknown functions which can be found by solving the corresponding system of equations for each vortex i. We will then have a system of $3N$ equations and $3N$ unknowns. This system can even be solved analytically for certain situations. Unfortunately, the results obtained by this method do not correspond with the solution obtained by directly solving the paraxial equation [27].

The problem is that our ansatz for the field F was not properly chosen. In general, even if the initial field $E(x, y, 0)$ is a product of vortices, the evolution of the field entangles the different vortices. The resulting field at any other z cannot be factorized again as a product of vortices. Of course, there are particular cases where the ansatz Eq. (2.25) represents the actual field. One such case is when all vortices are canonical and have the same topological charge. This is a trivial case, because, as shown earlier, the vortices do not interact at all. In the case of two vortices, we can exactly calculate which fields allow for a description according to the above ansatz, even when $z \neq 0$. The condition is that $A_1(0)A_2(0) = 1$. When this condition is fulfilled, $A_1(z) = A_1(0)$ and $A_2(z) = A_2(0)$, and once again both vortices evolve independently of each other.

2.4.3 *Interaction of two vortices*

We can turn now to the simplest nontrivial case, i.e. the interesting case where vortices fulfill $A_1(0)A_2(0) \neq 1$. In this situation our ansatz fails. For example, the initial factorizable "vortex function" $V(x, y, z=0) = ((x - x_1(0)) + \mathrm{i}A_1(0)y)((x - x_2(0)) + \mathrm{i}A_2(0)y)$ becomes an unfactorizable polynomial at $z \neq 0$, as we will immediately see. Solving the paraxial equation with this initial condition, the resulting field is

$$\begin{aligned} E(x, y; z) &= \Big(x^2 \mathrm{e}^{\mathrm{i}2\Omega_x z} + \mathrm{i}(A_1 + A_2) xy \mathrm{e}^{\mathrm{i}(\Omega_y - \Omega_x) z} - A_1 A_2 y^2 \mathrm{e}^{\mathrm{i}2\Omega_y z} \\ &\quad - \mathrm{i}(A_1 x_2 + A_2 x_1) y \mathrm{e}^{\mathrm{i}\Omega_y z} - (x_1 + x_2) x \mathrm{e}^{\mathrm{i}\Omega_x z} \\ &\quad + \left[x_1 x_2 - \frac{1 - \mathrm{e}^{\mathrm{i}2\Omega_x z}}{2\Omega_x} + A_1 A_2 \frac{1 - \mathrm{e}^{\mathrm{i}2\Omega_y z}}{2\Omega_y} \right] \Big) \exp\left(-\frac{x^2}{w_x^2} - \frac{y^2}{w_y^2} \right). \end{aligned} \tag{2.26}$$

Note that we have decided for simplicity to restrict the initial position of the vortices to lie on the x axis. Ω_x and Ω_y are the "frequencies of rotation" related to the Gaussian beam widths in the corresponding directions.

One interesting case, which allows us to understand the richness of the problem, is the symmetrical dipole $A_1 = -A_2 = 1$, $x_1 = -x_2 = x_s$ in a symmetric Gaussian beam $w_x = w_y = 1$. The trajectories in this case are given by [27]:

$$y(z) = \frac{2x_s^2 - 1}{2x_s} \sin(2z), \tag{2.27}$$

$$x(z) = \pm \left[x_s^2 - y^2(z) \right]^{1/2}. \tag{2.28}$$

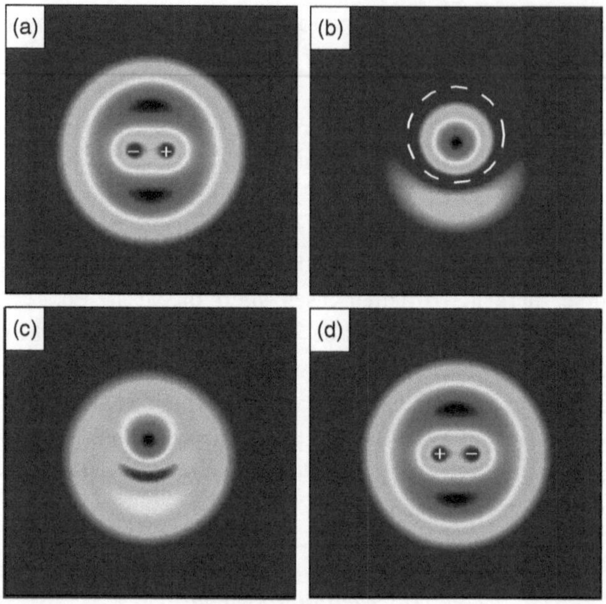

Figure 2.12 Creation and annihilation processes for a vortex dipole. Panels show the intensity patterns for different propagation distances: $z/\pi = 0, 0.25, 0.4, 0.5$. Signs + and − in (a) and (d) show the topological charge of the vortices. In (b), the dashed line corresponds to the position of the circular edge dislocation. $A_1(0) = 1$, $A_2(0) = -1$, $x_s = 0.35$,$w_x = w_y = 1$.

From these equations it can be observed that if the initial vortices are well separated (i.e. $x_s > 1/2$), their trajectories never intersect and their vortex lines are open trajectories in three-dimensional space. In this case, the optical field always contains two vortices. However, if the initial separation is $x_s \leq 1/2$, their trajectories lead to a collision and they annihilate each other. The wavefront remains vortex-less until a propagation distance Δz_{rev}, where a new pair of vortices appear. This creation and annihilation of vortices is periodic. In Fig. 2.12 we show the shape of the field in different observation planes perpendicular to the propagation direction. In Fig. 2.12(b) one can observe a beautiful pattern of light. This pattern is a circular edge dislocation, which appears in between the annihilation and the creation of the vortices at a particular z position.

As we have already seen before, this edge dislocation can be understood as a closed vortex line which lies in the plane of observation. If the symmetry of the problem does not allow it, or if we tilt the observation plane, we can observe again a pair of vortices being created and annihilated. As always, this will happen whenever the vortex line intersects the observation plane.

Finally, note that if $w_x \neq w_y$ we obtain an effect like the topological charge inversion explained in the previous section. If the field only contains one vortex, we will have periodical inversions of the charge. However, if there is more than one vortex in the field, the topological reactions are much more complicated and one can observe edge dislocations with a variety of shapes.

2.5 Conclusion and discussion

In this chapter we have shown the reader different ways of understanding the dynamics of optical vortices. The subject of vortex dynamics in optical fields is interesting from the point of view of applications, but it also allows us to better understand the dynamics of vortices in much more complicated systems. Optical fields present in this way a beautiful bench-top laboratory to explore the physics of condensed matter systems or other phenomena involving vortices and topological defects.

In particular, we have shown the enormous importance of considering the internal parameter of the vortex, i.e. the so-called noncanonical parameter, to understand its evolution. The dynamics of the vortex then becomes a coupled system between its internal morphology and its position. We have shown that this system allows for a topological reaction which is different from the usual creation and annihilation of oppositely charged vortices, namely the dynamical inversion of the topological charge through an edge dislocation.

Finally, we have discussed the effect of vortex interactions and the possibility of describing the creation and annihilation of vortices dynamically. Unfortunately, the problem becomes much more complicated because one can no longer consider each vortex independently and has to examine the whole "N-vortex" field. In principle we could consider this to be the end of the story. We could solve the equations for the polynomial coefficients of the vortex fields, but the information about where the vortices are located would be given by the crossings of the real and imaginary lines. This is, once again, as difficult as directly solving the corresponding paraxial equation and looking for the position of the vortices in the resulting field. Nevertheless, we would like to point out that Eq. (2.11) and (2.12) are exact. As we have shown in this chapter, any nondegenerate vortex can be locally described in this way. Then, we still have the hope that clever ways of interpreting the dynamics of vortices in general media can be found from these equations. This will give us the chance to understand previously unexplored new physical effects associated with the vortex line.

References

[1] H. J. Lugt. *Vortex Flow in Nature and Technology*. New York: J. Wiley and Sons, 1983.
[2] T. von Kármán. *Aerodynamics: Selected Topics in the Light of Their Historical Development*. New York; Dover Publications, 2004.
[3] C. P. Ellington, C. van der Berg, A. P. Willmott, and A. L. R. Thomas. Leading edge vortices in insect flight. *Nature*, 384: 626–30, 1996.
[4] Richard P. Feynmann. Application of quantum mechanics to liquid helium. In C. J. Gorter (ed.), *Progress in Low Temperature Physics*, chapter 2. New York: Interscience Publishers Inc., 1991.
[5] L. Onsager. Statistical hydrodinamics. *Il Nuovo Cimento*, 6(Suppl. 2):249, 1949.
[6] J. F. Nye and M. V. Berry. Dislocations in wave trains. *Proc. R. Soc. Lond. A*, 336:165–90, 1974.
[7] P. Coullet, L. Gil, and F. Rocca. Optical vortices. *Opt. Commun.*, 73:403–8, 1989.

[8] N. B. Baranova, B. Ya. Zel'dovich, A. V. Mamaev, N. F. Pilipetskii, and V. V. Schkunov. Dislocations of the wave-front of a speckle-inhomogeneous field (theory and experiment). *Sov. Phys. JETP Lett.*, 33:195–9, 1981.
[9] H. He, M. E. J. Friese, N. R. Heckenberg, and H. Rubinsztein-Dunlop. Direct obervation of transfer of angular momentum to absorptive particles from a laser beam with a phase singularity. *Phys. Rev. Lett.*, 75:826–9, 1995.
[10] L. Paterson, M. P. MacDonald, W. Sibbert, P. E. Bryant, and K. Dholakia. Controlled rotation of optically microscopic particles. *Science*, 292:912–4, 2001.
[11] M. D. Levenson, T. J. Ebihara, G. Dai, Y. Morikawa, N. Hayashi, and S. M. Tan. Optical vortex masks for via levels. *J. Microlith. Microfa. Microsyst*, 3:293–304, 2004.
[12] A. S. Desyatnikov, Y. S. Kivshar, and L. Torner. Optical vortices and vortex solitons. In E. Wolf (ed.), *Progress in Optics*, vol. XLVII, pp. 291–391. Amsterdam: Elsevier Science, 2005.
[13] Y. Y. Schechner and J. Shamir. Parameterization and orbital angular momentum of anisotropic vortices. *J. Opt. Soc. Am. A*, 13:967–73, 1996.
[14] I. Freund. Optical vortices in Gaussian random wave fields: statistical probability densities. *J. Opt. Soc. Am. A*, 11:1644–52, 1994.
[15] I. Freund and V. Freilikher. Parameterization of anisotropic vortices. *J. Opt. Soc. Am. A*, 14:1902–10, 1997.
[16] G. Molina-Terriza, E. M. Wright, and L. Torner. Propagation and control of noncanonical optical vortices. *Opt. Lett.*, 26:163–5, 2001.
[17] A. E. Siegman. *Lasers*. Mill Valley, CA: University Science Books, 1986.
[18] G. Indebetouw. Optical vortices and their propagation. *J. Mod. Opt.*, 40:73–87, 1993.
[19] D. Rozasand, C. T. Law, and G. A. Swartzlander. Propagation dynamics of optical vortices. *J. Opt. Soc. Am. B*, 14:3054–65, 1997.
[20] G. Molina-Terriza, J. Recolons, J. P. Torres, L. Torner, and E. M. Wright. Observation of the dynamical inversion of the topological charge of an optical vortex. *Phys. Rev. Lett.*, 87:023902, 2001.
[21] J. J. Garcia-Ripoll, G. Molina-Terriza, V. M. Perez-Garcia, and L. Torner. Structural instability of vortices in bose-einstein condensates. *Phys. Rev. Lett.*, 87:140403, 2001.
[22] R. Marchiano and J.-L. Thomas. Doing arithmetic with nonlinear acoustic vortices. *Phys. Rev. Lett.*, 101:064301, 2008.
[23] L. Allen, M. W. Beijersbergen, R. J. C. Spreeuw, and J. P. Woerdman. Orbital angular momentum of light and the transformation of Laguerre–Gaussian laser modes. *Phys. Rev. A*, 45:8185–9, 1992.
[24] M. W. Beijersbergen, L. Allen, H. E. L. O. van der Veen, and J. P. Woerdman. Astigmatic laser mode converters and transfer of orbital angular momentum. *Opt. Commun.*, 96:123–32, 1993.
[25] Michael Berry. Much ado about nothing: optical dislocation lines (phase singularities, zeros, vortices...). In M. S. Soskin (ed.) , *Singular Optics, Proceedings of the SPIE*, vol. 3487, pp. 1–5, 1998.
[26] S. J. van Enk and G. Nienhuis. Eigenfunction description of laser beams and orbital angular momentum of light. *Opt. Commun.*, 94:147–58, 1992.
[27] G. Molina-Terriza, L. Torner, E. M. Wright, J. J. Garcia-Ripoll, and V. M. Perez-Garcia. Vortex revivals with trapped light. *Opt. Lett.*, 26:1601–3, 2001.

3

Vector beams in free space

E. J. GALVEZ

3.1 Introduction

Higher-order modes of optical beams have been the subject of many studies over the past 20 years [1, 2]. Because laser resonators deliver in principle quite complex spatial patterns, spatial modes of beams were studied intensively when lasers were first developed [3]. Beams in higher-order spatial modes are solutions of the paraxial wave equation, with Hermite–Gauss and Laguerre–Gauss beams being the most important families of beams. These are solutions of the wave equation in Cartesian and cylindrical coordinates, respectively. The latter beams have been at the heart of a revival of research on higher-order modes due to the orbital angular momentum that they carry [4]. They have also received much attention due to the phase singularities present in their transverse amplitude [5]. Higher-order modes have stimulated much research in the application of forces and torques to objects in optical tweezers [1, 2]. Their usefulness has carried higher-order spatial modes further into new paths, in studies with non-classical sources of light and applications in quantum information [6].

The beams mentioned above are scalar solutions of the wave equation and thus independent of the polarization of the light. Vector beams are formed by the non-separable combinations of spatial and polarization modes. This enhanced modal space produces an interesting set of beams that offer new effects and applications. The origin of these beams is not recent either, as the possibility of combining higher-order spatial modes and polarization started with those early studies of modes as well. The first type of vector beam to receive attention is the "radial" vector beam, a ring-shaped beam with a linear polarization pointing radially. The recognition that a longitudinal component of the electric field appears when the radial beam is focused generated interest for application to charged particle acceleration [7]. A revival of interest in vector beams came with the realization that, when focused, radial beams produce a tighter waist [8, 9]. This came at a time of much research in technologies to manipulate microscopic objects with optical tweezers. The production of the richer variety of modes has been made easy with the advent of spatial light

The Angular Momentum of Light, ed. D. L. Andrews and M. Babiker. Published by Cambridge University Press.

modulators [10]. There is also a renewed and genuine interest in the rich combinations of spatial modes and polarizations that can be attained. Vector beams are an elementary subset of rich optical patterns that exhibit polarization singularities [11].

The purpose of this chapter is to offer a comprehensive description of vector beams. We will devote a tutorial-type treatment to the main case involving first-order singly ringed beams. Other beams can be understood by extensions of the presented description. Section 3.2 is a general description of Hermite–Gauss and Laguerre–Gauss beams. In Subsection 3.2.2 we set the framework by presenting the relations between the two families of scalar first-order beams. Section 3.3 is a necessary review of the polarization formalism. Section 3.4 treats vector beams, with subsections on first-order modes followed by higher-order modes. Section 3.5 gives a description of experimental approaches in the production of vector beams (3.5.1) and their diagnosis with a polarizer (3.5.2). Section 3.6 ventures into a modern application: the study of vector modes of non-classical light.

3.2 Spatial modes

3.2.1 *Hermite–Gauss and Laguerre–Gauss beams*

Hermite–Gauss and Laguerre–Gauss modes are solutions of the paraxial wave equation. Hermite–Gauss modes are specified by Hermite functions of the orthogonal transverse coordinates, with positive-integer indices m and n. The expression for the Hermite–Gauss mode is given by [12]

$$HG_{m,n} = \frac{b_{n,m}}{w} H_m\left(\frac{\sqrt{2}x}{w}\right) H_n\left(\frac{\sqrt{2}y}{w}\right) \mathrm{e}^{-(x^2+y^2)/w^2} \mathrm{e}^{ik(x^2+y^2)/[2R]} \mathrm{e}^{-i\varphi}, \qquad (3.1)$$

where w is the width of the mode, k the wavenumber, R the radius of curvature of the wavefront, and

$$\varphi = (N+1)\tan^{-1}(z/z_{\mathrm{R}}) \qquad (3.2)$$

the Gouy phase, with z_{R} being the Raleigh range, and $N = n + m$ the order of the mode. The amplitude constant is given by

$$b_{m,n} = \left(\frac{2^{1-N}}{\pi n! m!}\right)^{1/2}. \qquad (3.3)$$

Laguerre–Gauss modes are specified by associated Laguerre functions $L_p^{|\ell|}$, with indices p (a positive integer) and ℓ (an integer). The expression for the mode is given by

$$LG_p^{\ell} = \frac{a_{p,\ell}}{w}\left(\frac{\sqrt{2}r}{w}\right)^{|\ell|} L_p^{|\ell|}\left(\frac{2r^2}{w^2}\right) \mathrm{e}^{-r^2/w^2} \mathrm{e}^{ikr^2/2R} \mathrm{e}^{i\ell\phi} \mathrm{e}^{-i\varphi}. \qquad (3.4)$$

The order of the mode is given by $N = 2p + |\ell|$. The constant $a_{p,\ell}$ in Eq. (3.4) is given by

$$a_{p,\ell} = p! \left(\frac{2}{\pi p!(|\ell| + p)!)} \right)^{1/2}. \tag{3.5}$$

3.2.2 Relations between first-order Hermite–Gauss and Laguerre–Gauss beams

Since Hermite–Gauss and Laguerre–Gauss families of beam modes are complete sets of functions they can be expressed in terms of each other. For a detailed treatment we refer the reader to [12]. Here we will give a simplified presentation of the relations between first-order modes.

There are two first-order Laguerre–Gauss modes. Their radial index p is zero, so they are singly ringed. They have a phase that advances either with, or counter to, the azimuthal angle ϕ. They can be represented more compactly as

$$LG_0^\ell = A_1 r \mathrm{e}^{i\ell\phi} \; G \; W, \tag{3.6}$$

where one mode has $\ell = +1$ and the other $\ell = -1$. The constants plus $L_0^1(v) = 1$ can be combined into a new coefficient

$$A_1 = \frac{a_{0,\pm 1}\sqrt{2}}{w^2} = \frac{2}{\sqrt{\pi} w^2}, \tag{3.7}$$

the Gaussian envelope term is

$$G = \mathrm{e}^{-r^2/w^2} \tag{3.8}$$

and the overall phase term, which includes the Gouy phase, is

$$W = \mathrm{e}^{-i(x^2+y^2)/(2R) - i\varphi}. \tag{3.9}$$

The two first-order Hermite–Gauss modes are two-lobed patterns with opposite phase, orthogonally oriented to each other. In Cartesian coordinates the two modes are

$$HG_{10} = B_1 x \; G \; W \tag{3.10}$$

and

$$HG_{01} = B_1 y \; G \; W, \tag{3.11}$$

where we have combined the constants and $H_1(v) = 2v$ into the coefficient

$$B_1 = \frac{b_{1,0} 2\sqrt{2}}{w^2} = \frac{2\sqrt{2}}{\sqrt{\pi} w^2}. \tag{3.12}$$

Notice that $A_1 = B_1/\sqrt{2}$.

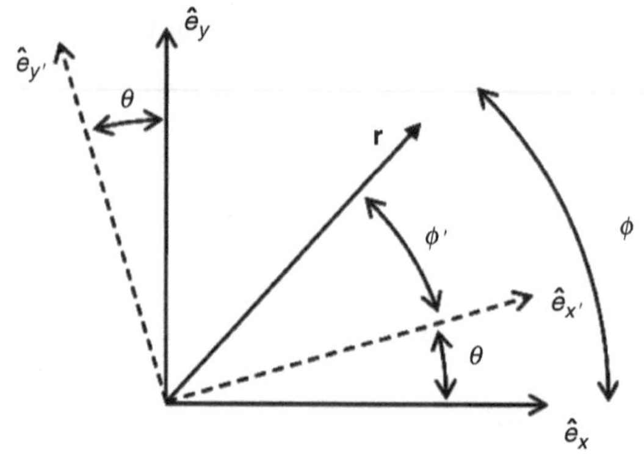

Figure 3.1 Vector representation of two reference frames rotated by θ about the origin.

Using the Euler identity it is easy to show that Eq. (3.6) is

$$\begin{aligned} LG_0^{\pm 1} &= A_1(x \pm \mathrm{i}y)\; G\; W \\ &= \frac{1}{\sqrt{2}}(HG_{10} \pm \mathrm{i}HG_{01}). \end{aligned} \tag{3.13}$$

The mode HG_{10} rotated by an angle θ, as shown in Fig. 3.1, is

$$\begin{aligned} HG_{10}(\theta) &= B_1 x'\; G\; W \\ &= B_1 r\cos(\phi - \theta)\; G\; W \\ &= HG_{10}\cos\theta + HG_{01}\sin\theta, \end{aligned} \tag{3.14}$$

where the primed coordinates refer to those in the rotated reference frame: $x' = r\cos\phi' = r\cos(\phi - \theta)$. Note also that manipulating Eq. (3.14) yields

$$\begin{aligned} HG_{10}(\theta) &= \frac{A_1}{\sqrt{2}} r\left(\mathrm{e}^{\mathrm{i}(\phi-\theta)} + \mathrm{e}^{-\mathrm{i}(\phi-\theta)}\right) G\; W \\ &= \frac{1}{\sqrt{2}}\left(LG_0^{+1}\mathrm{e}^{-\mathrm{i}\theta} + LG_0^{-1}\mathrm{e}^{\mathrm{i}\theta}\right). \end{aligned} \tag{3.15}$$

This way we get

$$HG_{10} = \frac{1}{\sqrt{2}}\left(LG_0^{+1} + LG_0^{-1}\right), \tag{3.16}$$

$$HG_{01} = \frac{-\mathrm{i}}{\sqrt{2}}\left(LG_0^{+1} - LG_0^{-1}\right). \tag{3.17}$$

Conversions between Hermite–Gauss and Laguerre–Gauss beams have been demonstrated using cylindrical lenses, which insert mode-dependent Gouy phases [4, 12].

3.3 A review of polarization modes

We will denote the state of linear polarization along the x-direction by $\hat{e}_x$, and similarly, the linear polarization along the y-direction by $\hat{e}_y$. The unit vector forming an angle ϕ with the horizontal is then defined as

$$\begin{aligned}\hat{e}_\phi &= \mathbf{r}/r \\ &= \hat{e}_x \cos\phi + \hat{e}_y \sin\phi. \end{aligned} \tag{3.18}$$

We will denote by $\hat{e}_{x'}$ and $\hat{e}_{y'}$ respectively the x and y polarizations rotated by an angle θ, as defined by

$$\hat{e}_{x'} = \hat{e}_x \cos\theta + \hat{e}_y \sin\theta \tag{3.19}$$

and

$$\hat{e}_{y'} = -\hat{e}_x \sin\theta + \hat{e}_y \cos\theta. \tag{3.20}$$

This is also shown in Fig 3.1. An important case is when $\theta = \pi/4$, where the unit vectors are also known as the "diagonal" and "anti-diagonal" states. They are given by

$$\hat{e}_{\mathrm{D}} = \frac{1}{\sqrt{2}}(\hat{e}_x + \hat{e}_y) \tag{3.21}$$

and

$$\hat{e}_{\mathrm{A}} = \frac{1}{\sqrt{2}}(-\hat{e}_x + \hat{e}_y). \tag{3.22}$$

We denote the states of right (R) and left (L) circular polarization as:

$$\hat{e}_{\mathrm{R}} = \frac{1}{\sqrt{2}}(\hat{e}_x - i\hat{e}_y) \tag{3.23}$$

and

$$\hat{e}_{\mathrm{L}} = \frac{1}{\sqrt{2}}(\hat{e}_x + i\hat{e}_y), \tag{3.24}$$

respectively.

A general elliptically polarized state is given by

$$\hat{e}_{\theta,\chi} = \hat{e}_{\mathrm{R}} e^{\mathrm{i}\theta} \cos\chi + \hat{e}_{\mathrm{L}} e^{-\mathrm{i}\theta} \sin\chi. \tag{3.25}$$

This state can be understood best with the Poincaré sphere of Fig. 3.2. Angle χ is half the polar angle. This way the latitude on the sphere is given by $\pi/2 - 2\chi$. Angle θ is half the azimuthal angle of polar coordinates, or longitude of the sphere, as shown in Fig. 3.2.

Since this is not a common representation of the general state of polarization, we will show how it is related to the more conventional state of polarization in terms of the Cartesian components. Consider the particular case of the elliptically polarized state with no rotation

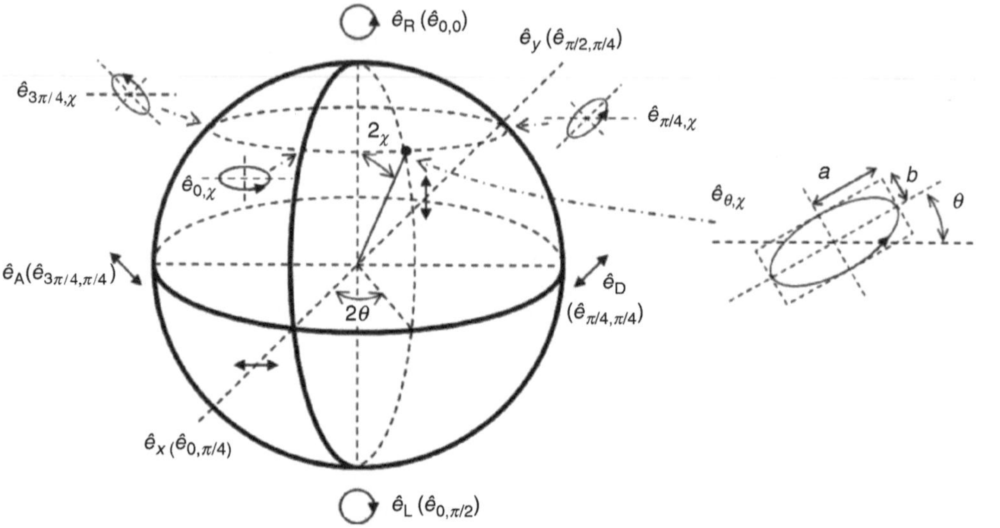

Figure 3.2 The Poincaré sphere is used to represent all the states of polarization. The latitude encodes the ellipticity of the polarization ellipse via $\epsilon = \tan(\pi/4 - \chi) = b/a$, where 2χ is the polar angle, and the longitude 2θ encodes the rotation of the ellipse by θ.

(i.e., with its axes aligned with the Cartesian axes x and y). This state may be represented by $\hat{e}_{0,\chi}$ in Fig 3.2. According to Eq. (3.25) this state is given by

$$\hat{e}_{0,\chi} = \cos\chi\ \hat{e}_R + \sin\chi\ \hat{e}_L. \tag{3.26}$$

By substituting Eqs. (3.23) and (3.24) into Eq. (3.26) we get

$$\hat{e}_{0,\chi} = \cos(\pi/4 - \chi)\ \hat{e}_x - \mathrm{i}\sin(\pi/4 - \chi)\ \hat{e}_y. \tag{3.27}$$

This is the equation of an ellipse with ellipticity

$$\epsilon = \tan(\pi/4 - \chi) = \frac{b}{a}, \tag{3.28}$$

where b and a are the semi-minor and semi-major axes of the ellipse, respectively.

The more general case of an elliptically polarized state is one where the axes of the ellipse are rotated by an angle θ. Thus, to obtain it we need to rotate the state of Eq. (3.27) by θ. Since the components relate only to the ellipticity and not the orientation of the ellipse, we can express the rotated state as the same ellipse but relative to the rotated axes:

$$\hat{e}_{\theta,\chi} = \cos(\pi/4 - \chi)\ \hat{e}_{x'} - \mathrm{i}\sin(\pi/4 - \chi)\ \hat{e}_{y'}, \tag{3.29}$$

where $\hat{e}_{x'}$ and $\hat{e}_{y'}$ are given by Eqs. (3.19) and (3.20), respectively. On the Poincaré sphere this state is represented by the point with azimuthal angle 2θ and polar angle 2χ, as shown in Fig 3.2. By putting $\hat{e}_{x'}$ and $\hat{e}_{y'}$ in terms of $\hat{e}_R$ and $\hat{e}_L$ it can be shown that Eq. (3.29) reduces to Eq. (3.25). Alternatively, it can be shown using Jones matrix algebra that a rotation operation by θ in the Cartesian basis $(\hat{e}_x, \hat{e}_y)$ is equivalent to a phase shift by 2θ in the circular basis $(\hat{e}_R, \hat{e}_L)$.

In the same notation as Eq. (3.25), a state of linear polarization rotated by θ is given by

$$\hat{e}_{\theta,\pi/4} = \frac{1}{\sqrt{2}}(\hat{e}_{\mathrm{R}}\mathrm{e}^{\mathrm{i}\theta} + \hat{e}_{\mathrm{L}}\mathrm{e}^{-\mathrm{i}\theta}). \tag{3.30}$$

Note the similarity between this equation and Eq. (3.15). Indeed it has been shown that the Poincaré sphere can also be used to represent all the first-order spatial modes [13, 14].

3.4 Vector beams

In the discussion of spatial modes in the previous section the state of polarization was not a factor because the entire beam has the same state of polarization. The situation changes when we superimpose two modes that have different polarizations. These are what we call *vector beams*. As will become clear later, vector beams can be made using either of the two families of modes, Hermite–Gauss and Laguerre–Gauss. This section has been written in a tutorial mode to better illustrate the mathematical form of the various modes; each of the important results will be presented in three forms: one in terms of Hermite–Gauss modes, a second one in terms of Laguerre–Gauss modes and a third one in a form that gives an intuitive picture of the mode.

3.4.1 Theoretical description of first-order vector beams

General form of first-order vector beams

A general type of vector mode can be expressed in terms of Hermite–Gauss modes by

$$V_{\theta,\alpha} = \frac{1}{\sqrt{2}}(HG_{10}\hat{e}_{x'} + \mathrm{e}^{\mathrm{i}\alpha}HG_{01}\hat{e}_{y'}). \tag{3.31}$$

When α is 0 or π the vector mode contains a profile that has linear polarization arranged in some form. For example, when $\alpha = 0$ the mode can be expressed as

$$\begin{aligned} V_{\theta,0} &= \frac{1}{\sqrt{2}}(HG_{10}\hat{e}_{x'} + HG_{01}\hat{e}_{y'}) \\ &= \frac{1}{\sqrt{2}}(LG_0^{+1}\mathrm{e}^{\mathrm{i}\theta}\hat{e}_{\mathrm{R}} + LG_0^{-1}\mathrm{e}^{-\mathrm{i}\theta}\hat{e}_{\mathrm{L}}) \\ &= A_1 r\ G\ W\ (\hat{e}_x\cos(\phi+\theta) + \hat{e}_y\sin(\phi+\theta)) \\ &= A_1 r\ G\ W\ \hat{e}_{\phi+\theta}. \end{aligned} \tag{3.32}$$

Equation (3.32) represents, a mode consisting of an amplitude ring with a linear polarization along points of the ring oriented an angle θ with the radial direction, as illustrated in Fig. 3.3(a).

As seen above, vector modes can also be represented in terms of Laguerre–Gauss modes. A convenient notation for the general mode in this representation is

$$W_{\theta,\chi} = LG_0^{+1}\mathrm{e}^{\mathrm{i}\theta}\hat{e}_{\mathrm{R}}\cos\chi + LG_0^{-1}\mathrm{e}^{-\mathrm{i}\theta}\hat{e}_{\mathrm{L}}\sin\chi. \tag{3.33}$$

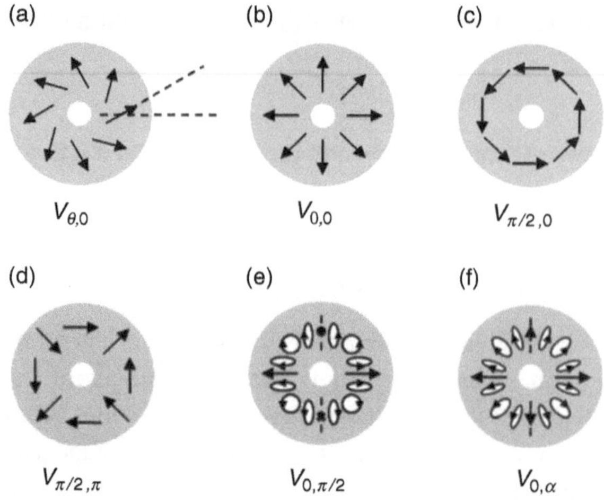

Figure 3.3 Illustration of several types of vector modes of first-order spatial modes; see text. Shades represent the regions of intensity well above zero and arrows represent the direction of the oscillation of the electric field. The vector modes $V_{\theta,\alpha}$ represent equal amplitude superpositions, with a relative phase α, of Hermite–Gauss modes HG_{10} and HG_{01} with corresponding set of orthogonal linear polarization states rotated by an angle θ.

Equations (3.31) and (3.33) convert exactly only for certain cases (e.g., $V_{\theta,0} = W_{\theta,\pi/4}$). Below we discuss several particular cases of first-order vector modes.

Radial vector beams

We begin with the most interesting type of vector beam: the radial vector beam. It corresponds to the case when $\theta = 0$ and $\alpha = 0$:

$$\begin{aligned} V_{0,0} &= \frac{1}{\sqrt{2}}(HG_{10}\hat{e}_x + HG_{01}\hat{e}_y) \\ &= \frac{1}{\sqrt{2}}\left(LG_0^{+1}\hat{e}_{\mathrm{R}} + LG_0^{-1}\hat{e}_{\mathrm{L}}\right) \\ &= A_1 r\ G\ W\ (\hat{e}_x \cos\phi + \hat{e}_y \sin\phi) \\ &= A_1 r\ G\ W\ \hat{e}_\phi. \end{aligned} \tag{3.34}$$

The product rG gives an amplitude consisting of a single ring. The mode expressed by Eq. (3.34) has a polarization vector that forms an angle ϕ with the x-axis. That is, points along the ring have polarization aligned along the radial direction, as illustrated in

Fig. 3.3(b). As mentioned earlier, radial beams are the oldest known type of vector beam, first created in 1972 by a laser with an intra-cavity element [15]. It was recognized that when focused the mode has an axial electric field at the focal point. The first application of this envisioned charged-particle acceleration [7]. More recently these modes have received renewed attention when it was realized that they focus onto a tighter waist than with uniform polarization [8, 9, 16], with potential application in the manipulation of objects with optical tweezers.

Azimuthal vector beams

Another interesting case of Eq. (3.32) arises when $\theta=\pi/2$, which results in a mode with a polarization that is always tangential to the ring of the mode, as shown in Fig. 3.3(c). It can be expressed formally as:

$$\begin{aligned} V_{\pi/2,0} &= \frac{1}{\sqrt{2}}(HG_{10}\hat{e}_y - HG_{01}\hat{e}_x) \\ &= \frac{\mathrm{i}}{\sqrt{2}}\left(LG_0^{+1}\hat{e}_\mathrm{R} - LG_0^{-1}\hat{e}_\mathrm{L}\right) \\ &= A_1 r\ G\ W\ \hat{e}_{\phi+\pi/2}. \end{aligned} \tag{3.35}$$

This type of vector beam is known as the azimuthally or tangentially polarized vector beam.

Antivortex vector beams

The next cases involve Eq. (3.31) with $\alpha=\pi$. The mode is given by

$$\begin{aligned} V_{\theta,\pi} &= \frac{1}{\sqrt{2}}(HG_{10}\hat{e}_{x'} - HG_{01}\hat{e}_{y'}) \\ &= \frac{1}{\sqrt{2}}\left(LG_0^{+1}\mathrm{e}^{-\mathrm{i}\theta}\hat{e}_\mathrm{R} + LG_0^{-1}\mathrm{e}^{\mathrm{i}\theta}\hat{e}_\mathrm{L}\right) \\ &= A_1 r\ G\ W\ \hat{e}_{-\phi+\theta}. \end{aligned} \tag{3.36}$$

Particular cases for $\theta=0$ and $\theta=\pi/2$ are respectively

$$\begin{aligned} V_{0,\pi} &= \frac{1}{\sqrt{2}}(HG_{10}\hat{e}_x - HG_{01}\hat{e}_y) \\ &= \frac{1}{\sqrt{2}}\left(LG_0^{+1}\hat{e}_\mathrm{R} + LG_0^{-1}\hat{e}_\mathrm{L}\right) \\ &= A_1 r\ G\ W\ \hat{e}_{-\phi}, \end{aligned} \tag{3.37}$$

and

$$V_{\pi/2,\pi} = \frac{1}{\sqrt{2}}(HG_{10}\hat{e}_y + HG_{01}\hat{e}_x)$$

$$= \frac{-\mathrm{i}}{\sqrt{2}}(LG_0^{+1}\hat{e}_\mathrm{L} - LG_0^{-1}\hat{e}_\mathrm{R})$$

$$= A_1 r\ G\ W\ \hat{e}_{-\phi+\pi/2}. \tag{3.38}$$

These are the antivortex modes, where the polarization is linear but rotates counter to the azimuthal angle, as shown pictorially in Fig. 3.3(d) for $V_{\pi/2,\pi}$. When $\theta \neq 0, \pi/2$ the vector has an added angular shift θ, similar to the case for $V_{\theta,0}$.

Hybrid/mixed vector beams

Next consider the case where $\alpha = \pi/2$. Let us assume $\theta = 0$ for simplicity. The expression for the mode profile is given by

$$V_{0,\pi/2} = \frac{1}{\sqrt{2}}(HG_{10}\hat{e}_x + \mathrm{i}HG_{01}\hat{e}_y)$$

$$= \frac{1}{\sqrt{2}}(LG_0^{+1}\hat{e}_\mathrm{D} - LG_0^{-1}\hat{e}_\mathrm{A})$$

$$= A_1 r\ G\ W\ (\hat{e}_x \cos\phi + \mathrm{i}\hat{e}_y \sin\phi). \tag{3.39}$$

This is a mode where the state of polarization changes from point to point along the ring. As ϕ increases the state of polarization changes from linear (along x) at $\phi = 0$ to states of polarization of increasing ellipticity, reaching the state of right circular polarization for $\phi = \pi/4$. As ϕ continues to increase the mode becomes elliptically polarized again, reaching linear polarization along y at $\phi = \pi/2$. Note that the linear polarization at this point is $\pi/2$ out of phase from the polarization at $\phi = 0$. This may be explained via a space-variant form of the Pancharatnam–Berry phase. Continuing to increase ϕ the state becomes elliptical again but with the winding of the polarization vector in the opposite direction, and reaching left circular polarization at $\phi = 3\pi/4$. This mode is illustrated in Fig. 3.3(e). At $\phi = \pi$ the polarization is linear along x but out of phase by π from the state at $\phi = 0$.

When α has a value between 0 and $\pi/2$ the vector mode is

$$V_{0,\alpha} = \frac{1}{\sqrt{2}}(HG_{10}\hat{e}_x + \mathrm{e}^{\mathrm{i}\alpha}\ HG_{01}\hat{e}_y)$$

$$= A_1 r\ G\ W\ (\hat{e}_x \cos\phi + \mathrm{e}^{\mathrm{i}\alpha}\hat{e}_y \sin\phi). \tag{3.40}$$

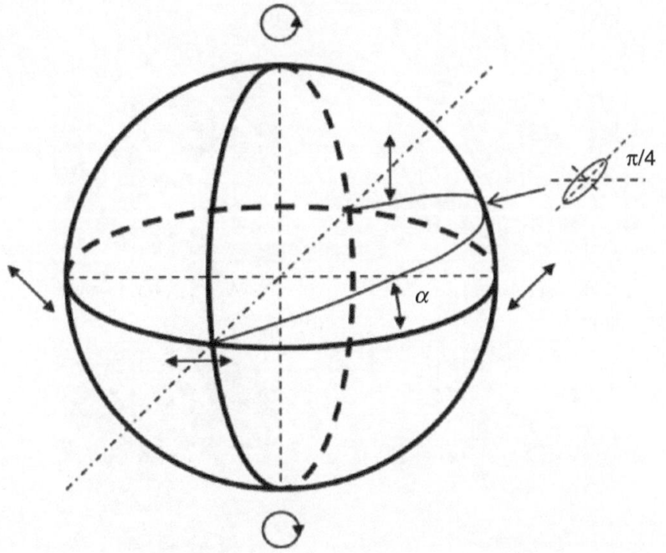

Figure 3.4 Poincaré sphere showing the geodesic path that the state of polarization follows for points along the ring of a vector mode represented by state $V_{0,\alpha}$.

The mode at $\phi = 0$ is linearly polarized along x, and at $\phi = \pi/2$ it is linearly polarized along y but phase-shifted by α. At $\phi = \pi/4$ the mode is elliptical, oriented by $\pi/4$, and with an ellipticity $\epsilon = \tan(\alpha/2)$. This is also shown in Fig. 3.3(f). The sequence of states follows a geodesic path on the Poincaré sphere, shown in Fig. 3.4. Following such a path means that the phase of the wave at a given angle ϕ can be obtained using a geometric-phase construction. Finally, should θ be increased from zero the pattern of polarization states will rotate with θ.

3.4.2 Higher-order vector beams

In principle any non-separable combination of orthogonal spatial and polarization modes will produce vector beams. Since the number of orthogonal modes increases with the order of the modes, the number of possibilities for vector beams increases as well. Here we will consider only a few simple cases. Since there are only two orthogonal polarization modes, then the possibilities for making vector beams reduces to the combination of two spatial modes with the two polarization modes.

We will consider the vector modes expressed in terms of Laguerre–Gauss modes. Our labels for this more general case will need to be different; we will label the vector modes with the same notation of Eq. (3.33) but explicitly specifying the radial and azimuthal indices of the two modes. This way, the superposition of $LG^{\ell_1}_{p_1}\hat{e}_R$ and $LG^{\ell_2}_{p_2}\hat{e}_L$ with equal amplitudes is represented by $W_{\theta,\pi/4}(p_1, \ell_1, p_2, \ell_2)$. Drawing on similarities with the

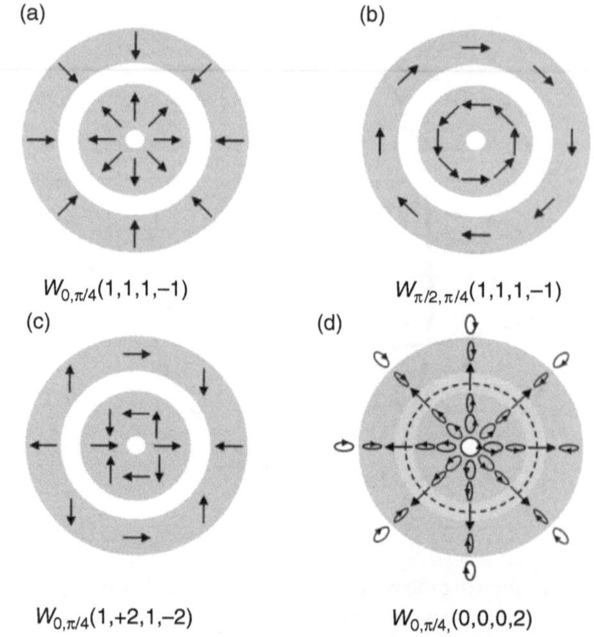

Figure 3.5 Examples of higher-order vector modes in state $W_{\theta,\chi}(p_1, \ell_1, p_2, \ell_2)$, which is a superposition of Laguerre–Gauss modes and circular polarization states $LG_{p_1}^{\ell_1}\hat{e}_{\mathrm{R}}$ and $LG_{p_2}^{\ell_2}\hat{e}_{\mathrm{L}}$, with relative phase 2θ and relative amplitude $\tan\chi$.

first-order modes, if we mix Laguerre–Gauss beams with $p \neq 0$ but $|\ell| = 1$, then we get the full assortment of cases seen with first-order modes:

$$\begin{aligned} W_{\theta,\pi/4}(p, 1, p, -1) &= \frac{1}{\sqrt{2}}\left(LG_p^{+1}\mathrm{e}^{\mathrm{i}\theta}\hat{e}_{\mathrm{R}} + LG_p^{-1}\mathrm{e}^{-\mathrm{i}\theta}\hat{e}_{\mathrm{L}}\right) \\ &= A_{p,1} r\, L_p^1\, G\, W\, (\hat{e}_x \cos(\phi + \theta) + \hat{e}_y \sin(\phi + \theta)) \\ &= A_{p,1} r\, L_p^1\, G\, W\, \hat{e}_{\phi+\theta}, \end{aligned} \tag{3.41}$$

where $A_{p,\ell}$ is a constant. The resulting mode looks almost identical to $V_{\theta,0}$. The difference lies with the term rL_p^1G, which instead of a single ring expresses a series of $p+1$ concentric rings of alternating opposite phase, as shown in the example of Figs 3.5(a) (radial) and 3.5(b) (azimuthal), for $\theta = 0$ and $\theta = \pi/2$, respectively.

If we use Laguerre–Gauss modes of increasing azimuthal index $|\ell|$ we can no longer obtain either radial or azimuthal vector modes because the phase of the spatial mode advances $|\ell|$-times faster than the circular polarization states. Possible vector modes of this type are

$$W_{\theta,\pi/4}(p,\ell,p,-\ell) = \frac{1}{\sqrt{2}}(LG_p^{+\ell}\mathrm{e}^{\mathrm{i}\theta}\hat{e}_\mathrm{R} + LG_p^{-\ell}\mathrm{e}^{-\mathrm{i}\theta}\hat{e}_\mathrm{L})$$
$$= A_{p,\ell}\, r^{|\ell|}\, L_p^\ell\, G\, W\, (\hat{e}_x \cos(\ell\phi+\theta) + \hat{e}_y \sin(\ell\phi+\theta))$$
$$= A_{p,\ell}\, r^{|\ell|}\, L_p^\ell\, G\, W\, \hat{e}_{\ell\phi+\theta}. \tag{3.42}$$

An example of this type of higher-order mode is illustrated in Fig. 3.5(c). These types of modes have been explored experimentally in [10].

In the vector modes discussed so far the polarization state varies only along the angular coordinate. If we use a mixture of higher-order modes of different order we get vector modes where the state of polarization also varies along the radial direction. An illustrative example involves mixing singly ringed modes (i.e., with $p=0$) but with different values of ℓ, say ℓ_1 and ℓ_2 with $|\ell_1| > |\ell_2|$. The new mode is then

$$W_{\theta,\pi/4}(0,\ell_1,0,\ell_2) = \frac{1}{\sqrt{2}}(LG_0^{\ell_1}\mathrm{e}^{\mathrm{i}\theta}\hat{e}_\mathrm{R} + LG_0^{\ell_2}\mathrm{e}^{-\mathrm{i}\theta}\hat{e}_\mathrm{L}). \tag{3.43}$$

The rings for each mode peak at different radii, $r_\mathrm{peak} = \sqrt{\frac{\ell}{2}}w$. At

$$r_v = \frac{w}{\sqrt{2}}\left(\frac{|\ell_2|!}{|\ell_1|!}\right)^{\frac{1}{2(|\ell_2|-|\ell_1|)}} \tag{3.44}$$

the two modes have the same amplitude [17], with $LG_0^{\ell_1}$ having greater amplitude than $LG_0^{\ell_2}$ for $r < r_v$, and the converse for $r > r_v$. The state of polarization at points along the profile is given by Eq. (3.25), with $\chi = \tan^{-1}(|LG_0^{\ell_2}|/|LG_0^{\ell_1}|)$ and $\theta = (\ell_1 - \ell_2)\phi/2$. As discussed previously, the angle 2χ is the polar angle in the Poincaré sphere of Fig. 3.2, and so $\chi = \pi/4$ corresponds to $r = r_v$ and a state of linear polarization, with the polarization rotating $(\ell_1 - \ell_2)/2$-times the transverse angle at the same radius. For $r \neq r_v$ the polarization is elliptical, with ellipticity $\epsilon = \tan(\pi/4 - \chi)$. Thus, going inward from $r = r_v$ the orientation of the ellipse is the same but with an ellipticity that decreases with r, to become left circular polarization at $r = 0$. Similarly, for $r > r_v$ the ellipticity also decreases, reaching right circular polarization in the limit $r \to \infty$. Thus, the beam contains *all* the states of polarization, a type of beam also known as "full-Poincaré beam" [18]. This vector mode is illustrated in Fig. 3.5(d) for the particular case when $\ell_1 = 0$ and $\ell_2 = 2$. The point $r = 0$ is also a C point of polarization singularity [11]. More types of rich patterns are present for other combinations of parameters. An interesting dichotomy to do with the previous mode is that if we pass that mode through a polarizer we obtain the linearly polarized superposition of Laguerre–Gauss modes that gives rise to composite phase vortices, located at $r = 0$, $r = r_v$ and $r = \infty$ [17].

3.5 Experimental approaches

In this section we discuss a selected group of methods of production and the standard detection method: polarization projection. We refer the reader to [19] for a comprehensive discussion of experimental methods.

3.5.1 Experimental methods of production

Vector beams have been produced in free space, in optical fibers and via intra-laser-cavity emission. The simplest to understand are the free-space methods, shown in Fig. 3.6. The setup of Fig. 3.6(a) uses Dove prisms for image rotation. The input mode is HG_{10} (or HG_{01}) but with linear polarization forming an angle of 45 degrees with the horizontal. A polarizing beam splitter splits the image evenly into the two linear polarization components. The horizontal component goes through a Dove prism rotated by 45 degrees. Ideally we would want the image to be rotated but not the polarization. However, the total internal reflection in the Dove prism changes the state of polarization, so an optical element (not shown) needs to be inserted to re-shift the polarization so that it returns into being horizontal. An alternative to the Dove prisms is an afocal pair of cylindrical lenses, also known as a π-mode converters [12]. A second polarizing beam splitter reassembles the beam into the vector beam. A rotator/flipper of polarization but not mode, such as a half-wave plate,

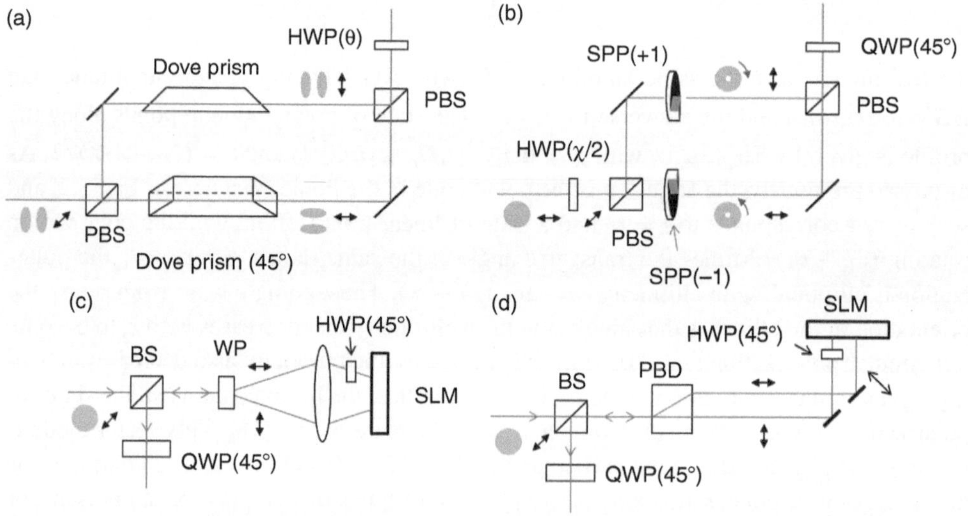

Figure 3.6 Schematic of methods to produce vector beams in free space. Cases (a) and (b) prepare vector beams according to Eqs. (3.31) and (3.33), respectively. Cases (c) and (d) can prepare any type of vector beam using SLMs. Optical elements include polarizing beam splitters (PBS), non-polarizing beam splitters (BS), half-wave plates (HWP), quarter-wave plates (QWP), spiral phase plates (SPP), Wollaston prism (WP) and polarizing beam displacer (PBD).

placed after the interferometer rotates the polarization relative to the modes. The combination of this rotation and the interferometer phase allows a full exploration of the vector modes as defined by Eq. (3.31).

The alternative to the previous method encodes the vector mode in terms of Laguerre–Gauss modes. It also uses a Mach–Zehnder interferometer that encodes Laguerre–Gauss modes in its arms via spiral phase plates placed in each of the arms of the interferometer [20], as shown in Fig. 3.6(b). Alternatively, one can use less efficient forked diffraction gratings and steer the first-order diffracted beams to the interferometer outputs. The input beam is linearly polarized but rotated from its initial state (e.g., horizontal) into forming an angle χ with the horizontal. This input beam is also in the fundamental mode. The first polarizing beam splitter selects the proper amplitudes of the two component modes. After the second polarizing beam splitter a quarter-wave plate forming an angle of 45 degrees with the horizontal puts the two orthogonal linear polarization states into circularly polarized states of opposite helicity. This method recreates the state of Eq. (3.33).

There are other more direct methods, that put the light directly into a radial vector beam. These arrangements take a linearly polarized input beam in the fundamental mode and pass them through a polarizing element that changes the polarization along the beam into a radial one. Two solutions have been proposed and implemented. One uses a wave plate divided into four sectors [21], with the axis of the wave plate in each sector such that the output beam has four quadrants each with polarization pointing outwards. A similar approach has been proposed with liquid-crystal polarization converters, which are commercially available, where the rotation of the polarization can be done gradually [22]. A third method uses a space-variant subwavelength grating [23].

Another possibility is to use spatial light modulators (SLM). Their versatility allows a much richer variety of vector modes, especially for higher-order cases [10]. Figures 3.6(c) and 3.6(d) show schemes where input light that is polarized along the same direction (e.g., diagonal) pass through a non-polarizing beam splitter, which is the output element. The linear polarization states are split by a polarizing prism: either a Wollaston prism, as shown in Fig. 3.6(c) [10]; or a polarizing beam displacer, as shown in Fig. 3.6(d). After the polarizing prisms the light is steered to the SLM, with one beam passing through a half-wave plate to have its polarization rotated. One of the shortcomings of the SLM is that they diffract more efficiently one polarization orientation. Conversely, the versatility of the SLM is that we can encode both component spatial modes with the same device. The diffracted modes then retrace their path, get recombined by the polarizing prism, and are steered by the non-polarizing beam splitter. These schemes allow any vector mode to be produced. Should it be done via Eq. (3.33), then a quarter-wave plate placed after the beam splitter converts each linearly polarized component into circular polarization, similar to the case of Fig. 3.6(b). If Hermite–Gauss modes are created instead then the final element is replaced by a half-wave plate to create the vector mode according to Eq. (3.31).

Vector beams can also be produced either directly out of a laser's output by either modification of the laser cavity [24] or by use of intra-cavity elements [25]. An interesting and rich possibility is the production of these beams in optical fibers [26–28]. Finally, recent novel approaches use the passage of optical beams through birefringent crystals [29], nematic liquid crystals [30] and glass with stress-induced birefringence [18].

3.5.2 Detection of vector beams by polarization projection

An important question regards the detection of vector beams. Polarization projection provides a simple means. If we pass the vector beam given by Eq. (3.31) through a polarizer oriented at an angle β, the transmitted beam will be

$$\begin{aligned} V'_{\theta,\alpha} &= \frac{1}{\sqrt{2}}[HG_{10}\cos(\beta-\theta) + e^{i\alpha} HG_{01}\sin(\beta-\theta)]\hat{e}_\beta \\ &= \frac{1}{\sqrt{2}} B_1 r\ G\ W\ [\cos\phi\cos(\beta-\theta) + e^{i\alpha}\sin\phi\sin(\beta-\theta)]\hat{e}_\beta. \end{aligned} \tag{3.45}$$

For $\alpha = 0$ the resulting mode is, following Eq. (3.14),

$$V'_{\theta,0} = \frac{1}{\sqrt{2}} HG_{10}(\beta-\theta)\hat{e}_\beta, \tag{3.46}$$

which is the mode HG_{10} rotated by an angle $\beta - \theta$. Should the mode be a radial mode (i.e., $\theta = 0$), then the x-axis of the mode will be aligned with the transmission axis of the polarizer. Similarly, should the vector beam be azimuthal (i.e., $\theta = \pi/2$), then the y-axis of the mode will be aligned with the polarizer. Thus, one can use the orientation of the mode and its polarization to diagnose the vector mode. When $\alpha = \pi$ a similar analysis gives a mode HG_{10} rotated by an angle $-(\beta-\theta)$. Thus, the HG_{10} mode orientation would rotate counter to the rotation of the polarizer. Alternatively we can think of the transmitted mode in terms of intensity nodes. Following Eq. (3.45) for $\alpha = 0$, the pattern has a node at angles ϕ satisfying

$$\ell\phi - (\beta-\theta) = \left(q + \frac{1}{2}\right)\frac{\pi}{2}, \tag{3.47}$$

where $\ell = 1$ in this case and q is an integer. This way, as the polarizer is rotated (i.e., changing β) the node angle, given by $\phi = (q + 1/2)\pi/2 + (\beta-\theta)$, rotates with or counter to the polarizer rotation.

If $\alpha = \pm\pi/2$ then the projected mode will be

$$V'_{\theta,\pm\pi} = \frac{1}{\sqrt{2}}[HG_{10}\cos(\beta-\theta) + iHG_{01}\sin(\beta-\theta)]\hat{e}_\beta. \tag{3.48}$$

This gives modes HG_{10} and HG_{01} for $\beta-\theta = 0$ and $\beta-\theta = \pi/2$, respectively, and $LG_0^{\pm 1}$ for $\beta-\theta = \pm\pi/4$. At other values of the angle $\beta-\theta$ the mode is not pure. For other values of α the method does not give a recognizable mode at any angle $\beta-\theta$. Demonstrations of these projections have often appeared in the literature [10, 24, 26]. Figure 3.7 shows the

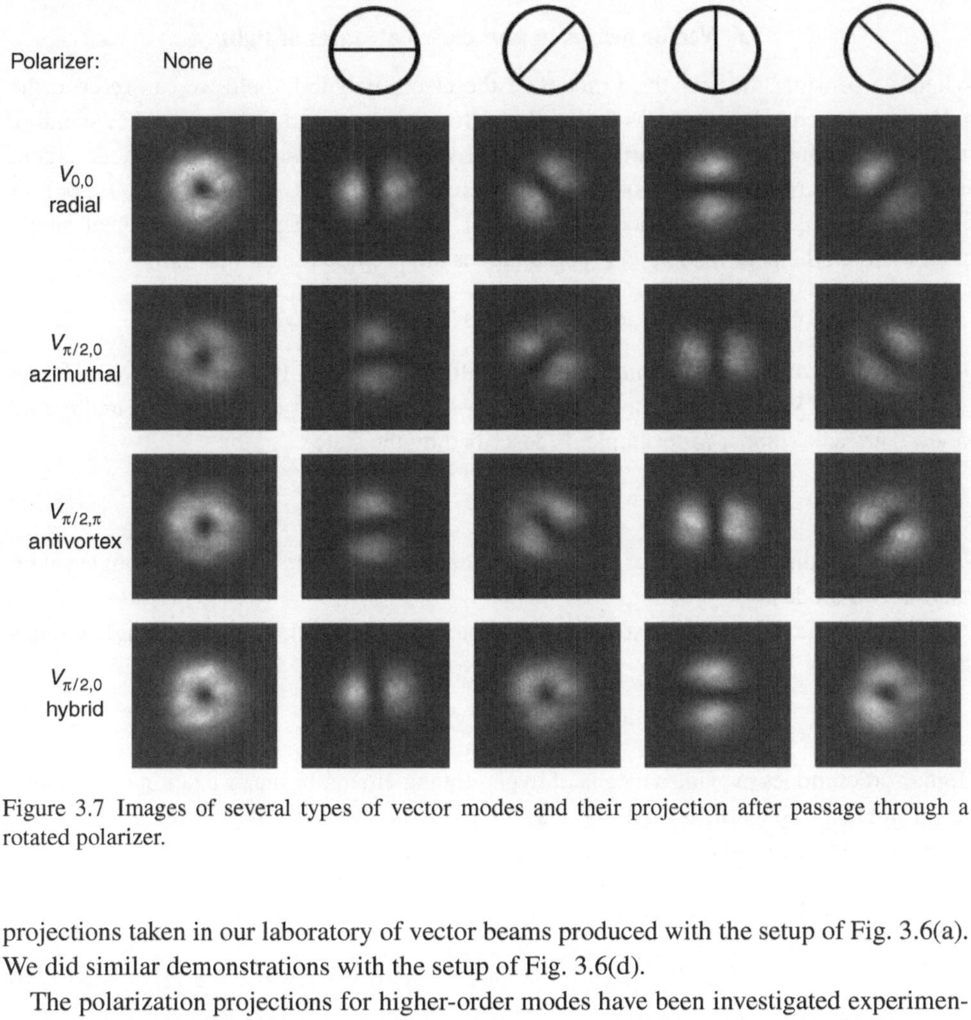

Figure 3.7 Images of several types of vector modes and their projection after passage through a rotated polarizer.

projections taken in our laboratory of vector beams produced with the setup of Fig. 3.6(a). We did similar demonstrations with the setup of Fig. 3.6(d).

The polarization projections for higher-order modes have been investigated experimentally using spatial light modulators [10]. If we pass the state $W_{\theta,\pi/4}(p, \ell, p, -\ell)$ through a polarizer we get a pattern of $2|\ell|$ symmetric nodes – a well-known result of combining two linearly polarized Laguerre–Gauss modes of opposite helicity. The angle formed by the q-th node follows Eq. (3.46):

$$\phi_q = \frac{1}{\ell}\left[\left(q + \frac{1}{2}\right)\frac{\pi}{2} + \beta - \theta\right]. \tag{3.49}$$

The interesting feature of this result is that the transmitted pattern rotates $1/\ell$ times as fast as the polarizer [10]. Should we flip the helicity of the polarization relative to the modes in Eq. (3.42) (equivalent to making $\alpha = \pi$ in Eq. (3.45) for first-order modes), a minus sign would precede the right-hand side of Eq. (3.49), and the pattern would rotate counter to the polarization rotation.

3.6 Vector beams in non-classical states of light

When we consider the quantized modes of the electromagnetic field we can refer to the polarization-mode eigenstates as $|e_x\rangle$ and $|e_y\rangle$ (often referred to as $|H\rangle$ and $|V\rangle$, standing for horizontal and vertical polarization, respectively). The first-order spatial-mode eigenstates of the Hermite–Gauss modes can be represented by $|HG_{10}\rangle$ and $|HG_{01}\rangle$. Each of these degrees of freedom embodies a qubit of information. The general state of single photons when the polarization and spatial modes are prepared separately is

$$|\psi\rangle = \frac{1}{2}(a|e_x\rangle + b|e_y\rangle)(c|HG_{10}\rangle + d|HG_{01}\rangle), \tag{3.50}$$

where (a, b) and (c, d) are normalized probability amplitudes. It represents a one-photon two-qubit state. Vector modes, however, are non-separable states of polarization and spatial mode. The radial mode, for example, is described by the state

$$|V_{0,0}\rangle = \frac{1}{\sqrt{2}}(|HG_{10}\rangle|e_x\rangle + |HG_{01}\rangle|e_y\rangle). \tag{3.51}$$

One can view this expression as referring to the state of a single photon entangled with itself in two modes.

Today it is fairly common to produce photon pairs in polarization entangled states such as

$$|\Psi\rangle = \frac{1}{\sqrt{2}}(|e_x\rangle_1|e_x\rangle_2 + |e_y\rangle_1|e_y\rangle_2). \tag{3.52}$$

Higher-order modes provide a means of hyper-entanglement, or maximization of the number of qubits per photon. It might be feasible to create the following vector mode of two photons entangled in spatial mode and polarization:

$$|\psi_c\rangle = \frac{1}{2}(|e_x, HG_{10}\rangle_1|e_x, HG_{10}\rangle_2 + |e_x, HG_{10}\rangle_1|e_y, HG_{01}\rangle_2 + |e_y, HG_{01}\rangle_1|e_x, HG_{10}\rangle_2 - |e_y, HG_{01}\rangle_1|e_y, HG_{01}\rangle_2), \tag{3.53}$$

which is a four-qubit two-photon cluster state, a model state for doing quantum computation [31, 32]. Higher-order modes with their modal space increasing with N can expand the qubits spaces even further.

Challenges still remain though, as successful implementation of quantum computing algorithms require state projection with high efficiency. Polarization states are projected efficiently with polarizing beam-splitters or prisms. However, optical devices that project spatial modes have a long way to go to rival the efficiency of their polarization counterparts. Current methods involve converting projected modes into fundamental modes via diffractive optical elements or spatial light modulators. The latter have much versatility in projecting any modes. However, they still suffer from low efficiencies, and a polarization dependence of these efficiencies. Although technological challenges remain, spatial modes and their combination with polarization modes have a promising future in quantum computing implementations.

3.7 Conclusions

Vector beams constitute an important form of complex light that combines polarization and spatial modes in non-separable analytical forms. The complex light forms that can be generated already have unique manifestations and interactions with matter. The goal of the present chapter has been to provide a tutorial to those entering the field and for students interested in understanding and implementing experiments with vector beams. The presentation also offers a perspective on promising experiments in the investigation of new vector modes of higher-order spatial modes and new applications in quantum information.

With their long history, vector beams have made important inroads into basic science and applications. The development of new technologies, in particular spatial light modulators, are expected to produce more growth in this topic, promising an interesting future ahead.

The author wishes to thank S. M. Nomoto and W. H. Schubert for contributions and support from National Science Foundation grant PHY-0923129 and US Air Force contract FA8750-11-2-0034.

References

[1] Allen, L., Barnett, S. M., and Padgett, M. J. (eds.), 2003, *Optical Angular Momentum*. Bristol: Institute of Physics.

[2] Andrews, D. L. (ed.), 2008, *Structured Light and Its Applications*. Burlington, VT: Academic Press.

[3] Siegman, A. E., 1986. *Lasers*. Mill Valley, CA: University Science Books.

[4] Allen, L., Beijersbergen, M. W., Spreeuw, R. J. C., and Woerdman J. P., 1992. Orbital angular momentum of light and the transformation of Laguerre–Gaussian laser modes. *Phys. Rev. A*, **45**, 8185–9.

[5] Soskin, M. and Vasnetsov, M. V., 2001. Singular optics. *Progress in Optics* **42**, 219–76.

[6] Molina-Terriza, G., Torres, J.P., and Torner, L., 2007, Twisted photons. *Nature Phys.*, **3**, 305–10.

[7] Fontana, J. R. and Pantell, R. H. 1983. A high-energy, laser accelerator for using the inverse Cherenkov effect. *J. Appl. Phys.*, **54**, 4285–8.

[8] Quabis, S., Dorn., R., Eberler, M., Glöckl, O., and Leuchs, G., 2000, Focusing light to a tighter spot. *Opt. Commun.*, **179**, 1–7.

[9] Dorn, R., Quabis, S., and Leuchs, G., 2003, Sharper focus for a radially polarized light beam. *Phys. Rev. Lett.*, **91**, 233901-1–4.

[10] Maurer, C., Jesacher, A., Fürhapter, S., Bernet, S., and Ritsch-Marte, M., 2007., Tailoring of arbitrary optical vector beams. *New J. Phys.*, **9**, 78-1–20.

[11] Freund, I., 2002, Polarization singularity indices in Gaussian laser beams. *Opt. Commun.*, **201**, 251–70.

[12] Beijersbergen, M. W., Allen, L., van der Veen, H. E. L. O., and Woerdman, J. P., 1993, Astigmatic laser mode converters and transfer of orbital angular momentum. *Opt. Commun.*, **96**, 123–32.

[13] Padgett, M. J. and Courtial, J., 1999, Poincaré-sphere equivalent for light beams containing orbital angular momentum *Opt. Lett.*, **24**, 430–2.

[14] Agarwal, J., 1999, SU(2) structure of the Poincaré sphere for light beams with orbital angular momentum *Opt. Soc. Am. A*, **16**, 2914–6.
[15] Mushiake, Y., Matsumura, K., and Nakajima, N., 1972, Generation of radially polarized optical beam mode by laser oscillation. *Proc. IEEE*, **60**, 1107–9.
[16] Youngsworth, K. S. and Brown, T. G., 2000, Focusing of high numerical aperture cylindrical vector beams. *Opt. Express*, **7**, 77–87.
[17] Baumann, S. M., Kalb, D. M., MacMillan, L. H., and Galvez, E. J., 2009, Propagation dynamics of optical vortices due to Gouy phase. *Opt. Express*, **17**, 9818–27
[18] Beckley, A. M., Brown, T. G. Brown, and Alonso, M. A., 2010, Full Poincaré beams. *Opt. Express*, **18**, 10777–85.
[19] Zhan, Q., 2009, Cylindrical vector beams: from mathematical concepts to applications. *Adv. Opt. Photon.*, **1**, 1–57.
[20] Tidwell, S. C., Ford, D. H., and Kimura, W. D., 1990, Generating radially polarized beams interferometrically. *Appl. Opt.*, **29**, 2234–9.
[21] Quabis, S., Dorn, R., and Leuchs, G., 2005, Generation of a radially polarized doughnut mode of high quality. *Appl. Phys. B*, **81**, 597–600.
[22] Stalder, M. and Schadt, M., 1996, Linearly polarized light with axial symmetry generated by liquid-crystal polarization converters. *Opt. Lett.*, **21**, 1948–50.
[23] Bomzon, Z., Biener, G., Kleiner, V. and Hasman, E., 2002, Radially and azimuthally polarized beams generated by space variant dielectric subwavelength gratings. *Opt. Lett.*, **27**, 285–7.
[24] Oron, R., Blit, S., Davidson, N., Friesem, A. A., Bomzon, Z., and Hasman, E., 2000, The formation of laser beams with pure azimuthal and radial polarization. *Appl. Phys. Lett.*, **77**, 3322–4.
[25] Kozawa, Y. and Sato., S., 2005, Generation of radially polarized laser beam by use of a conical Brewster prism. *Opt. Lett.*, **30**, 3063–5.
[26] Volpe, G. and Petrov, D., 2004, Generation of cylindrical vector beams with few-mode fibers excited by Laguerre–Gaussian beams. *Opt. Commun.*, **237**, 89–95.
[27] Ramachandran, S., Kristensen, P., and Yan, M.F., 2009, Generation and propagation of radially polarized beams in optical fibers. *Opt. Lett.*, **34**, 2525–7.
[28] Milione, H. I. Sztul, Alfano, R. R., and Nolan, D. A., 2009. Stokes polarimetry of a hybrid vector beam from a spun elliptical core optical fiber. *Proc. SPIE*, **7613**, 761305–10.
[29] Fadeyeva, T. A., Shvedov, V. G., Izdebskaya, Y. V. *et al.*, 2010, Spatially engineered polarization states and optical vortices in uniaxial crystal. *Opt. Express*, **18**, 10848–63.
[30] Buinyi, I. O., Denisenko, V. G., and Soskin, M. S., 2010, Topological structure in polarization resolved conoscopic patterns for nematic liquid crystal cells. *Opt. Commun.*, **282**, 143–55.
[31] Walter, P., Aspelmeyer, M., Resch, K. J., and Zeilinger, A., 2005, Experimental violation of a cluster state Bell inequality. *Phys. Rev. Lett.*, **95**, 020403-1–4.
[32] Walter, P., Resch, K.J., Rudolph, T. *et al.*, 2005, Experimental one-way quantum computing. *Nature*, **434** 169–76.

4

Optical beams with orbital angular momentum in nonlinear media

A. S. DESYATNIKOV AND Y. S. KIVSHAR

4.1 Introduction

The nonlinear response of optical media results in an additional phase shift for propagating light waves, and has a profound effect on their evolution. When the parameters of the medium, such as the refractive index n, are modified by the nonlinear light–matter interaction, the wave dynamics is affected by the medium, and it is customary to refer to the corresponding nonlinear optical processes as the "self-action" of light. For example, in nonlinear Kerr-type materials, the refractive index can be cast in the form $n = n_0 + \mathcal{N}(I)$, with linear index n_0 and a nonlinear correction, $\mathcal{N}(I)$, which depends on the light intensity $I \propto |E|^2$, where E is the electric field $c = 1$. For transversely localized laser beams the index acquires localized inhomogeneity, resembling optical waveguides and fibers, and the beam can become "self-trapped."

The propagation of paraxial laser beams in Kerr-type media is governed by the nonlinear Schrödinger equation (NLSE) [1]

$$\mathrm{i}\frac{\partial E}{\partial z} + \nabla^2 E + \mathcal{N}(I)\, E = 0, \tag{4.1}$$

where we are choosing z and (x, y) to stand for one propagation and two transverse coordinates, respectively, and $\nabla = (\partial_x, \partial_y)$. The NLSE (4.1) is non-integrable, but it has four important integrals of motion: the power, $P = \int I\, \mathrm{d}\vec{r}$; the linear momentum, $\vec{L} = \mathrm{Im} \int E^* \nabla E\, \mathrm{d}\vec{r}$; the orbital angular momentum (OAM), $M = \vec{e}_z\, \mathrm{Im} \int E^* (\vec{r} \times \nabla E)\, \mathrm{d}\vec{r}$ with a unit vector $\vec{e}_z$; and the Hamiltonian $\mathcal{H} = \int h\, \mathrm{d}\vec{r}$ with density $h = |\nabla E|^2 - \int \mathcal{N}(I) \mathrm{d}I$.

As in any nonlinear system, the superposition principle cannot be applied to Eq. (4.1), and its analysis relies on numerical integration, e.g. by using the split-step beam propagation method (BPM). As the name suggests, BPM is based on splitting the propagation variable z into small steps, Δz, and solving the diffraction and nonlinear problems on an interval $(z, z + \Delta z)$ in separate consequent sub-steps. The linear sub-problem, $\mathrm{i}\partial_z E + \nabla^2 E = 0$, is solved by efficient fast Fourier transform, and the nonlinear sub-problem, $\mathrm{i}\partial_z E + \mathcal{N}(I)\, E = 0$, is solved explicitly for each spatial location (x, y).

The Angular Momentum of Light, ed. D. L. Andrews and M. Babiker. Published by Cambridge University Press.

The absence of analytical tools for the analysis of Eq. (4.1) led to the development of semi-analytical methods, such as perturbation expansions and the variational approach (VA). The VA requires an explicit assumption on the structure of the wave-function by the choice of trial function, $E(x, y, z) \simeq E_{\text{trial}}(x, y, \{\alpha_n\})$, with structural parameters $\alpha_n(z)$ describing the z-evolution of the field E_{trial}. The method involves minimization of the Lagrangian, $\mathcal{L} = \text{Im} \int E^* \partial_z E \mathrm{d}\vec{r} - \mathcal{H}$, with derivation of the Lagrangian equations of motion, $\frac{\mathrm{d}}{\mathrm{d}z}[\partial\mathcal{L}/\partial\dot{\alpha}_n] = \partial\mathcal{L}/\partial\alpha_n$, where $\dot{\alpha}_n \equiv \mathrm{d}\alpha_n/\mathrm{d}z$. For further details on VA see the review paper by Malomed [2].

In the rest of this chapter we employ VA to illustrate the physics of spiraling beams, subsequently verified with direct numerical BPM solutions of Eq. (4.1) and compared against available experimental data. The chapter is organized as follows. In the introduction, we provide a brief outline of the pre-history of spiraling solitons, including vortex solitons and soliton clusters, as well as discussing modulational instability (MI). Azimuthons, a general class of spiraling vortex solitons, are discussed in Section 4.2, and Section 4.3 is devoted to azimuthons and other multi-solitons in nonlocal media. Section 4.4 outlines the contribution of OAM towards suppression of nonlinear wave collapse, and Section. 4.5 concludes the chapter.

4.1.1 Vortex solitons

In a self-focusing medium with $\mathrm{d}\mathcal{N}/\mathrm{d}I > 0$, the diffraction of a light beam can be compensated by the nonlinearity and the balance between these two counter-acting "forces" corresponds to a stationary state. *Spatial optical solitons* are stationary solutions of the NLSE (4.1) which do not change their intensity profile during propagation, $\partial I/\partial z = 0$, and represent the closest available analogy to modal or self-similar diffracting beams in linear media and free propagation. Solitons can be found in a generic form,

$$E(x, y, z) = U(x, y) \exp(i\phi(x, y) + ikz), \tag{4.2}$$

with real amplitude U, phase ϕ, and soliton propagation constant k. Looking for the simplest continuous-wave (cw) solution, $U(x, y) = A = \text{const}$ and $\phi = 0$, we obtain $k = \mathcal{N}(A^2)$ from Eq. (4.1). The latter result shows how the nonlinear phase kz, accumulated during propagation, depends on the wave intensity, $I = A^2$, and nonlinearity $\mathcal{N}(I)$.

The localized bell-shaped or fundamental solitons with $\phi = 0$ possess radial symmetry, $U(x, y) = R(r)$, and can be considered as a particular case of a more general class of *vortex solitons*, introduced by Kruglov and Vlasov [3]. Vortex solitons possess a twisted phase front, $\phi = m\varphi$, where the polar coordinates $re^{i\varphi} = x + iy$ and the integer m define the topological charge of phase singularity, discovered by Nye and Berry [4]. Vortex solitons are analogous to Laguerre–Gauss (LG) modal beams, as they have a "doughnut" transverse shape and represent the eigenfunctions of the OAM operator with $M = mP$ [5]. Since the amplitude vanishes at the vortex core, the nonlinearity can be neglected in its close vicinity, and the topological properties and stability of phase singularities are similar to their linear counterparts [6, 7]. An overview of vortex solitons can be found in [8].

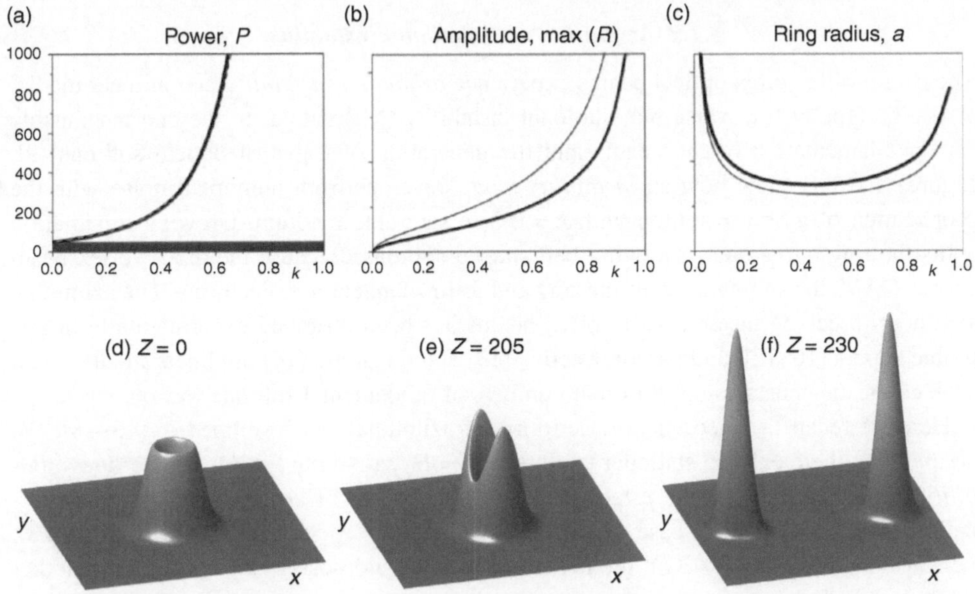

Figure 4.1 Vortex solitons with topological charge $m = 1$ in saturable medium, $\mathcal{N}(I) = I/(1 + I)$. The soliton family is parameterized by the propagation constant k versus: (a) soliton power P, (b) peak amplitude $\max(R)$, and (c) versus ring radius a. The exact numerical solutions to Eq. (4.3) (thick lines) are compared with the VA solutions (dashed and thin lines). Intensity $I = |E|^2$ of a vortex soliton with $k = 0.5$ at the input $z = 0$ in (d), and after propagation of $z = 205$ in (e) and $z = 230$ in (f).

In general, the radially symmetric soliton envelope $R(r)$ can be found only numerically,

$$\frac{\mathrm{d}^2 R}{\mathrm{d}r^2} + \frac{1}{r}\frac{\mathrm{d}R}{\mathrm{d}r} - \frac{m^2}{r^2} R - kR + \mathcal{N}\left(R^2\right) R = 0, \tag{4.3}$$

and it vanishes at the center as $R \sim r^{|m|}$ at $r \to 0$, similar to LG beams. However, the soliton tails decay exponentially, $R(r) \sim \exp\left(-r\sqrt{k}\right)/\sqrt{r}$ at $r \to \infty$, with the decay rate $\sqrt{k}$ determined by the soliton power, $P = 2\pi \int_0^\infty R^2 r \mathrm{d}r$, thus pointing at *nonlinear spatial localization*.

The VA can be applied using the LG mode as an ansatz, e.g., for a single-charged vortex with $m = 1$ we take $R_{\text{trial}} = Ar\exp(-r^2/2a^2)$, with the ring radius a and amplitude $\max(R_{\text{trial}}) = Aa$. The results for saturable nonlinearity, $\mathcal{N}(I) = I/(1 + I)$, are presented in Fig. 4.1 and compared with exact solutions to Eq. (4.3). It is important to note that the formation of vortex solitons requires the optical power to exceed some threshold, determined by TC and nonlinearity, e.g. $P > 50$ in Fig. 4.1(a). Below this threshold (the gray shaded area in Fig. 4.1(a)) the beams diffract and do not form solitons, although the diffraction rate is reduced by self-focusing.

4.1.2 Azimuthal modulational instability

The ring-profile vortex optical beams experience *azimuthal instability* in nonlinear media, a specific type of transverse modulational instability (MI) similar to the one responsible for the filamentation of the beams and the generation of trains of optical solitons [9]. Figures 4.1(d–f) show how an *instability mode* develops from numerical noise with the propagation of a vortex soliton with $k=0.5$ in saturable medium. Growing modulation splits the ring into a pair of moving bell-shaped solitons. Because the input vortex beam carries OAM, the splinters fly off the ring and follow tangential trajectories. The azimuthal symmetry-breaking instability of vortex beams has been observed experimentally in saturable vapors [10–12], and photorefractive [13] and quadratic [14] nonlinear media. In all such cases, the generation of different numbers of fundamental solitons was observed.

Here we recall the theoretical description of azimuthal MI developed in [15–18]. We assume that there exists a stationary solution $E=E_0$ satisfying Eq. (4.1). The *linear stability* of this solution can be established by the behavior of an additional small perturbation $|p| \ll |E_0|$. For the case of radially symmetric vortices determined by Eq. (4.3), i.e. $E_0 = R(r)\exp(im\varphi + ikz)$, the perturbation p should possess an azimuthal periodicity and, therefore, can be represented as Fourier series, $p(r, \varphi, z) = \sum p_s(r, z)\exp(is\varphi)$. Furthermore, as discussed in detail in [8], for any integer s only two modes p_{m+s} and p_{m-s} are coupled and build a closed eigenvalue system. Therefore, taking perturbation in the form $p(r, \varphi, z) = \mathrm{e}^{im\varphi+ikz}\{u_s(r)\mathrm{e}^{is\varphi+\gamma_s z} + v_s^*(r)\mathrm{e}^{-is\varphi+\gamma_s^* z}\}$ guarantees, owing to the completeness of the basis of azimuthal harmonics p_s, that all possible perturbations are taken into account. In this notation, if the eigenvalue γ_s has a positive real part for some s, the perturbation modes $p_{m\pm s}$ grow exponentially with the *growth rate* Re $\gamma_s > 0$, and the eigenmodes (u_s, v_s) are called *instability modes*. If p does not grow with the beam propagation, Re $\gamma_s \leq 0$, the stationary solution E_0 is linearly stable.

The number of bell-shaped splinters equals the index s of the dominating instability mode with the highest growth rate. The latter depends on the nonlinearity, the vortex topological charge m, and the soliton power. For example, a single-charged vortex (for all k and P) always has the $s=2$ mode growing faster than other modes in a saturable medium, as seen in Figs. 4.1(d–f), while this can be the $s=3$ mode in a quadratic and Kerr medium. In general, the higher-charge vortices allow for competition between different modes with close values of the growth rate, and vortex solitons are usually linearly unstable in *local media* [8]. However, recent theoretical predictions and experimental observations have revealed that *nonlocal nonlinearity* provides an effective mechanism for the stabilization of vortex and spiraling solitons against symmetry-breaking instability; we discuss nonlocal solitons in Section 4.3.

4.1.3 Two-soliton spiraling

There is an alternative view of the instability of ring-shaped vortex solitons in local media which gives further insight into the underlying physics of the beam breakup. Namely, the splinters of the vortex ring form fundamental solitons, similar to classical particles, with the interaction forces stemming from the nonlinearity. Two such solitons in Figs. 4.1 (d–f) move out along tangents to the initial ring and carry away its OAM, with the angular separation rate determined by the conservation of OAM and Hamiltonian, as shown in [18]. Thus, two solitons experience a repulsive centrifugal force, as do rotating particles in a two-body problem. In addition, the coherent interaction between solitons is determined by the overlap of their tails and thus it is phase-sensitive, as discussed in detail in [1, 19]. The splinters of the vortex ring are out-of-phase and experience an additional repulsive force which facilitates the breakup.

Therefore, to stabilize the vortex-carrying beam, it is necessary to compensate for the repulsion between solitons. One possibility is provided by a nonlocal medium, as was suggested theoretically in [20]. The strong nonlocality creates a broad trapping potential, similar to a parabolic waveguide (see Section 4.3), capable of compensating for the out-of-phase repulsion of two solitons, including dark solitons [21]. As a result, the two bright solitons, launched along skewed trajectories, can form a helicoidal structure, termed "*soliton spiraling*" [20]. The spiraling of two solitons in nonlocal liquid crystals was demonstrated in [22–24]

Another possibility to form a two-soliton system with nonzero OAM is to exploit the phase dependance of nonlinear interaction. Then, if the mutual interaction is attractive, the centrifugal repulsive force can be balanced out, and two solitons will orbit each other in a double-helix structure, as was predicted theoretically for quadratic media [25, 26]. The mechanical model, based on the potential of soliton interaction [27], has been derived, with extremal points of effective potential corresponding to spiraling bound states. Similar to any phase-sensitive soliton interaction the effective mass corresponding to the phase degree of freedom was shown to be always negative. As a result, the corresponding stationary points are of the saddle type, i.e. spiraling bound states are unstable. Depending on the initial soliton states, such as soliton velocities, relative phase, and the impact parameter, solitons can reflect, spiral, and fuse. More recently, experimental observation of related phenomena has been reported in [28].

The spiraling of *mutually incoherent* spatial solitons was observed experimentally [29] and studied theoretically [30, 31] as a possible scenario for a *dynamically stable* two-soliton bound state. Mutually incoherent solitons always attract each other in an isotropic Kerr-type medium, independently of their relative phase, and the effective potential minimum corresponds to the stable bound state. In fact, the soliton spiraling arising from an effectively vectorial beam interaction is associated with the large-amplitude oscillations of a dipole-mode vector state [32, 33] generated by the interaction of two mutually incoherent optical beams [34].

In a photorefractive medium, however, the anisotropy of the nonlinear refractive index change in the crystal results in complex trajectories which typically show partial mutual spiraling, followed by damped oscillations and the fusion of solitons, as shown in [35, 36] and summarized in [37, 38]. Therefore, the soliton spiraling is unstable in anisotropic media due to non-conservation of OAM.

4.1.4 Soliton clusters and necklaces

The concept of two-soliton spiraling above suggests that we should search for possible stable or meta-stable nonlinear states carrying OAM and optical vortices in the form of bound states of fundamental bell-shaped solitons. This idea was realized by the present authors [39] with the introduction of *rotating soliton clusters*. We considered a ring of identical weakly overlapping solitons launched in parallel, employing a Gaussian ansatz for a single beam, $G_n = A\exp\left(-\frac{|\mathbf{r}-\mathbf{r}_n|^2}{2a^2}+i\alpha_n\right)$, where $n = 1, 2, \ldots, N$, $\mathbf{r}_n = \{R\cos\varphi_n;\ R\sin\varphi_n\}$ with $\varphi_n = 2\pi n/N$ describes the soliton location in a ring-shaped array of radius R, and α_n is the phase of the nth beam. The dynamics of the slowly varying field envelope $E = \sum G_n$ is governed by the NLSE (4.1).

To allow a stationary configuration of solitons, the power flow between them should be balanced, which sets the constraint on the soliton phases, $\alpha_{i+n} - \alpha_i = \alpha_{k+n} - \alpha_k$. This condition can be satisfied provided the phase α_n has a linear dependence on n, i.e. $\alpha_n = \theta n$, where θ is the relative phase between two neighboring solitons in the ring. The periodicity on the ring, $\alpha_{n+N} = \alpha_n + 2\pi m$, requires

$$\theta = \frac{2\pi m}{N}, \tag{4.4}$$

with the integer m being the topological charge. In particular, for the case of two solitons ($N = 2$), this condition gives only two states with zero energy exchange. when m is even ($\theta = 0$, mutual attraction) and when m is odd ($\theta = \pi$, mutual repulsion).

Applying the effective-particle approach, we derived an effective interaction energy for the soliton cluster as a function of two "quantum numbers," N and m, and the cluster radius R. The effective potential has a minimum for some pair (N, m), or potential well, suggesting that such a configuration describes a stable or long-lived ring-like bound state of solitons, or a soliton cluster. The rule, generalizing the two-soliton phase-sensitive interaction, predicts the existence of a bound state of N solitons if the nonzero phase step θ is equal to or less than $\pi/2$, i.e. the interaction is attractive and can be balanced out by the net centrifugal force, stemming from nonzero OAM. Therefore, a soliton cluster with topological charge m can be quasi-stationary only for $N \geq 4m$, see Eq. (4.4), i.e. the lowest order cluster with $m = 1$ requires at least $N \geq 4$ solitons on the ring.

The semi-analytical prediction of soliton clusters was verified by a series of numerical simulations for different N-soliton rings and their propagation in a saturable medium with $\mathcal{N}(I) = I/(1+I)$. We observed the robust rotation of dynamically stable clusters as well as the "excited" states, with the radius of the cluster oscillating near the minimum of the interaction potential. An example of an excited rotating cluster in Fig. 4.2 can be identified as *multi-soliton spiraling*. Note that, several initially well-separated solitons, being launched toward the target ring in the absence of perturbations, can also excite a metastable vortex ring [40].

The concept of soliton clusters was further extended to the case of light propagation in media with other nonlinearities, such as quadratic [41], competing quadratic and

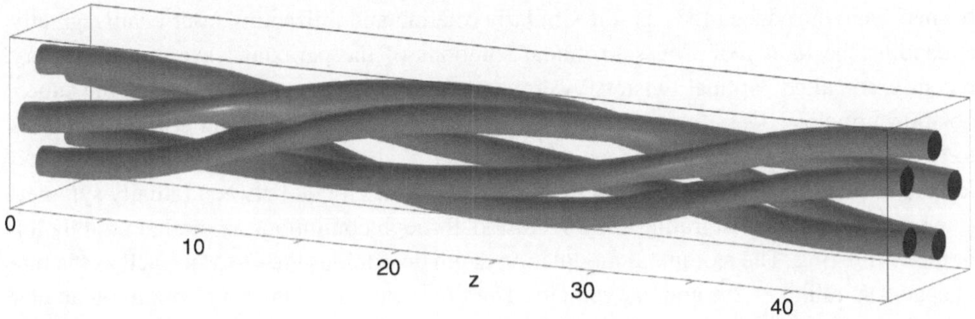

Figure 4.2 Half-peak intensity iso-surface for $N = 5$-soliton cluster with $m = 1$ in saturable medium, as in a Fig. 2(c) of [39].

self-defocusing cubic [42], and cubic-quintic [43]. The three-dimensional "soliton molecules," or robust bound states of light bullets, were introduced in [44, 45].

Similar azimuthally modulated rings, resembling "optical necklaces," were studied previously in a different physical context. Such self-trapped structures appear to be remarkably stable in pure Kerr media, $\mathcal{N}(I) = I$, where the two-dimensional beams with optical power exceeding some threshold value experience catastrophic self-focusing and collapse, see Section 4.4 and [46]. The first experimental results on the self-trapping of necklace-type beams were reported in [47, 48], with each petal of the necklace carrying sub-critical power. The diffraction of each petal is greatly suppressed within the ring, because of the collective self-trapping of the ring as a whole, while the beam transports power exceeding several times the threshold.

The concept of necklace beams was further developed in [49, 50], including necklaces with additional phase modulation and nonzero OAM [51]. The simplest way to explore necklace features [52] is to approximate a modulated vortex by $U(x, y) = R(r)(\cos m\varphi + ip \sin m\varphi)$. This ansatz describes a radially symmetric vortex with topological charge m at $p = 1$, and includes the varying modulation parameter $0 < p < 1$. The OAM of this structure is given by $M = 2mpP/(1 + p^2)$, and it vanishes for $p \to 0$. When the ring vortex is only slightly modulated, $p \approx 1$, it decays into a complex structure of filaments because of competition between different instability modes of the corresponding vortex soliton. When the modulation becomes deeper, e.g. for $p \approx 0.5$, the initial vortex transforms into a necklace-like structure, and its dynamics is modified dramatically. The modulated ring-profile structure does not decay but, instead, expands with small rotation; the rotation is much weaker because the initial OAM is reduced. Therefore, the repulsion of necklace petals prevents the appearance of dynamically stable structures.

4.2 Azimuthons

Vortex solitons and spiraling multi-hump structures such as soliton clusters and necklaces have been studied in physically different contexts. The link between such states in linear media is provided by the superposition principle. Most prominently, "spiraling light beams" were introduced [53] as self-similarly rotating and diffracting laser beams, greatly expanding the family of Gaussian modal solutions of the paraxial wave equation. Very recently, so-called "optical twisters" were suggested [54]. In nonlinear media, the superposition principle does not apply and the existence of exact spiraling soliton solutions to Eq. (4.1) is a nontrivial issue. One broad class of such states was introduced in [55] and termed "azimuthons." The family of azimuthons provides a link between radially symmetric vortex solitons and spiraling soliton clusters through continuous azimuthal modulation of the vortex ring. The azimuthal modulation depth determines the OAM as well as the rate of spatial spiraling, or the angular velocity. Therefore, instead of the usual single-parameter (power or propagation constant) soliton family, azimuthons represent a two-parameter family, with the second parameter being OAM or angular velocity. In this section we describe the basic properties and recent results on azimuthon observation.

4.2.1 Theoretical results

The first assumption to be made in order to find azimuthon solutions is to allow their self-similar rotation [53]. Looking for a general class of self-trapped beams we consider solutions in the cylindrical coordinates rotating with angular velocity ω: $E = V(r, \varphi - \omega z)\exp(ikz)$. The envelope V satisfies the stationary NLSE, derived from Eq. (4.1):

$$-kV + \frac{\partial^2 V}{\partial r^2} + \frac{1}{r}\frac{\partial V}{\partial r} + \frac{1}{r^2}\frac{\partial^2 V}{\partial \theta^2} - \mathrm{i}\omega\frac{\partial V}{\partial \theta} + \mathcal{N}(|V|^2)V = 0, \tag{4.5}$$

where k is the propagation constant in the rotating frame (r, θ) with $\theta = \varphi - \omega z$. For vortex solitons with $V(r, \theta) = R(r)\exp(im\theta)$, the correspondence between the laboratory frame (r, φ) and the rotating frame is defined by the relation $k = k_0 + m\omega$, where the angular velocity ω is an arbitrary parameter.

Since there is no analytical method of constructing exact solutions to Eq. (4.5), we employ VA with the ansatz assuming the separation of variables, $V(r, \theta) = R(r)U(\theta)$. In Kerr media with $\mathcal{N}(I) = I$ such separation results in a system of two coupled partial differential equations. The first equation for the radial envelope is similar to Eq. (4.3). The second equation for the azimuthal envelope,

$$U_{\theta\theta} - \eta U + |U|^2 U = 0, \tag{4.6}$$

is supplemented by the boundary condition $U(\theta + 2\pi) = U(\theta)\exp[\mathrm{i}2\pi(m - \Omega)]$, where the integer m is the topological charge and the angular velocity is scaled using the radial envelope $\Omega = \omega/\omega_0$, with $\omega_0 = 2k_0[R^2 r^-2]/[R^2]$ and integrals $[X] \equiv \int_0^\infty X r dr$, for details see [55]. Equation (4.6) is a completely integrable stationary one-dimensional NLSE [1] and it allows one to find exact periodic solutions, characterized by the number of peaks N on the period $0 \leq \theta \leq 2\pi$. Therefore, the number of peaks N in the intensity distribution of the modulated beam is an additional parameter generally independent of the beam topological charge m.

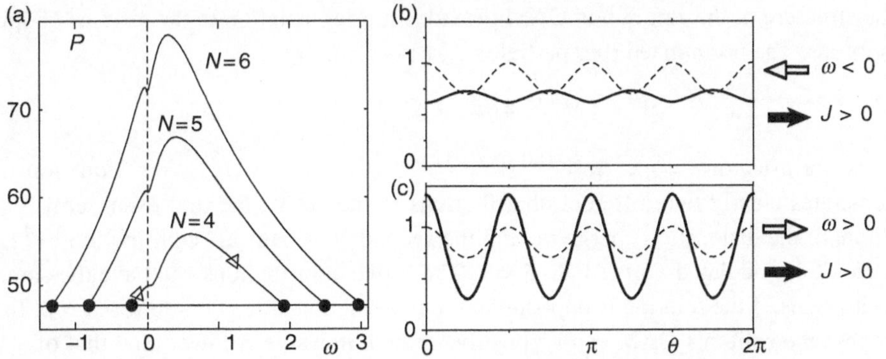

Figure 4.3 (a) The power of azimuthons with $m = 1$ in Kerr media; the horizontal line corresponds to the vortex soliton with $P = 48.4$, circles show bifurcation points, triangles indicate solutions in (b, c). Azimuthal profiles of the amplitude $|U|$ (dashed lines) and the current J (solid lines) are shown for two counter-rotating azimuthons with $N = 4$, $\omega = -0.156$ in (b), and $\omega = +1.239$ in (c). The arrows on the right show the directions of rotation and the current; in both cases the current is positive, $J > 0$, in the counter-clockwise direction. After [55].

The exact solutions of Eq. (4.6) in the reference frame rotating with the angular velocity ω generalize the solutions with $\omega = 0$ discussed in [56]. They also allow an approximate expression to be derived, which is useful in non-integrable systems,

$$|U(\theta)|^2 = 1 - n \sin^2\left(\tfrac{N\theta}{2}\right),$$
$$\operatorname{Arg} U(\theta) = \Omega\theta + \tfrac{2}{N}(m - \Omega)\tan^{-1}\left[\sqrt{1-n}\,\tan\left(\tfrac{N\theta}{2}\right)\right], \tag{4.7}$$

where $0 < n \leq 1$ is a modulation parameter.

Figure 4.3(a) shows the power vs. ω for three families of azimuthons with $m = 1$ in a Kerr medium, bifurcating from a vortex soliton at two specific cut-off values of the angular velocity obtained in the limit $n \to 0$. Note that each family includes azimuthons rotating in the direction of phase gradient, $\omega > 0$ for $m > 0$, as well as in the opposite direction, $\omega < 0$. We refer to the latter effect as *anomalous rotation*; its origin can be explained using the structure of the power flow, or optical current [57], along the ring, $J = \operatorname{Im}\, U^* U_\theta = |U|^2\, \partial_\theta \operatorname{Arg} U$. Figures 4.3(b, c) compare the azimuthal profiles of the amplitude and the current for two counter-rotating azimuthons. In both cases the current is positive, $J > 0$, which shows that the energy flow circulates in a counter-clockwise direction. However, in the anomalous case in Fig. 4.3(b), the maxima of the current along the ring correspond to the intensity minima. It means that the regions between the intensity peaks (solitons) transfer the energy faster than the solitons themselves. As a result, the solitons undergo continuous reshaping so that the visible rotation, $\omega < 0$, is against the energy flow, $J > 0$. Quite the opposite happens for the solution in Fig. 4.3(c), where the intensity peaks carry the larger energy and their motion is along the phase gradient; the angular velocity is positive for positive current and OAM.

The structure of the azimuthon OAM provides further insight into the physics of spiraling solitons. The normalized (per particle) OAM is given by

$$M/P = m\delta + \Omega\,(1-\delta), \tag{4.8}$$

with the form-factor $\delta = 4\pi^2\langle|U|^2\rangle^{-1}\langle|U|^{-2}\rangle^{-1}$, where $\langle X\rangle = \int_0^{2\pi} X\mathrm{d}\theta$. Equation (4.8) demonstrates clearly two different contributions to the OAM, the first usual term is proportional to the dislocation charge m, and the second "mechanical" contribution $\sim\Omega$, for the case of a modulated vortex with $\delta < 1$. These two contributions can have the same or opposite signs, if the rotation is opposite to the phase gradient, e.g. $\omega < 0$ for $m > 0$. In the latter case the soliton's OAM can be positive though its value is lower than that of a vortex, $M/P = m$. The non-rotating azimuthons with $\omega = 0$ appear when the modulations of amplitude and phase gradient exactly cancel each other in the expression for local current so that $\partial J/\partial\theta = 0$.

The complex azimuthon structure can be explained using an extremely simple linear model. We consider here the one-dimensional linear Schrödinger equation on the ring (cf. Eq. (4.6)),

$$\mathrm{i}u_z + u_{\varphi\varphi} = 0, \quad u(\varphi + 2\pi, z) = u(\varphi, z)\exp(\mathrm{i}2\pi m), \tag{4.9}$$

with arbitrary integer m being the topological charge or the Bloch index. The eigenmodes can be readily found, $u_m = \exp(im(\varphi - mz))$, and any solution can be represented as a superposition, $u = \sum_m A_m u_m$, with constant Fourier amplitudes A_m. Among arbitrary linear superpositions, we are interested here in a special class of solutions, which remain self-similar while rotating with constant angular velocity ω. The corresponding ansatz reads $u(\varphi, z) = U(\varphi - \omega z)\exp(ikz)$, with complex envelope $U(\theta) = \Phi(\theta)\exp(\mathrm{i}\Psi(\theta))$, where $\theta = \varphi - \omega z$. The periodicity for phase requires $\Psi(\theta + 2\pi) = \Psi(\theta) + 2\pi m$ but the amplitude can have an arbitrary number of peaks, $\Phi(\theta + 2\pi/N) = \Phi(\theta)$, where the integer $N \geq 0$ is the second azimuthal quantum number. Taking this degree of freedom into account we derive the solution (cf. Eq. (4.7)),

$$u(x, z) = \left(\cos\tfrac{N\theta}{2} \pm ip\,\sin\tfrac{N\theta}{2}\right)\exp\left[\mathrm{i}\left(m \mp \tfrac{N}{2}\right)\theta + im(m \mp N)\omega z\right], \tag{4.10}$$

where $0 < p \equiv \min\Phi \leq 1$ and the top and bottom signs correspond to two different branches with integer angular velocity, $\omega = 2m \mp N$. This seemingly complex expression has, in fact, a very simple structure. Each branch is a superposition of just two eigenmodes with amplitudes given by

$$A_m = \tfrac{1}{2}(1 + p), \quad A_{m\mp N} = \tfrac{1}{2}(1 - p). \tag{4.11}$$

In other words the, superposition of two vortex modes u_{m_1} and u_{m_2} results in a rotating "linear azimuthon" with $\omega = m_1 + m_2$ and $N = |m_1 - m_2|$, and the last expression has two branches, e.g. with $m_1 = m$ and $m_2 = m{\mp}N$ as above. The topological charge is determined by the dominating amplitude, e.g. it is $A_{m_1} > A_{m_2}$ in Eq. (4.11) and the azimuthon charge is $m_1 = m$.

The solution displayed in Eq. (4.11) cannot be directly applied to true two-dimensional problems. In the linear case different modes have different diffraction rates, and the self-similar solutions require a specific choice of mode indices, involving radial LG modes [53]. In the nonlinear case the diffraction is suppressed, but the frequencies are amplitude-dependent, i.e. the angular velocity changes with power. Nevertheless, the schematic solution displayed in Eq. (4.11) underlines some generic properties of singular states, in particular it includes solutions with *anomalous rotation*, such as the case of $m=1$, $N=3$ with two branches, $\omega=-1$ and $\omega=5$. Such anomalously rotating "linear azimuthons" were constructed with Bessel beams [58].

Furthermore, Eq. (4.11) shows how two branches of rotating azimuthon bifurcate from the vortex mode (with $p=1$ or $N=0$ in Eq. (4.11)), when the second mode is gradually added into superposition, $p<1$ (cf. Fig. 4.3(a)). A similar mechanism of azimuthon bifurcation in a nonlocal medium was recently explored in [59]. The approach includes calculation of internal modes $\{u_s, v_s\}$ (see Section 4.1.2) of a vortex soliton of charge m, and the azimuthon with N peaks can be traced from the mode with $s=\pm N$. This has strong implications in understanding azimuthon structure, as it highlights the importance of the off-center singularities, which were observed before in numerical simulations, but are not taken into account by the ansatz Eq. (4.7); see Section 4.3.2 for more details.

4.2.2 Experimental results

The experimental results reported earlier on optical vortices in local self-focusing media included the observations of vortex breakup in rubidium vapors [10, 12] and in Kerr media [60], as well as the spiraling dipole-mode solitons, or "propeller solitons," trapped by a co-propagating mutually incoherent beam in photorefractive crystals [33, 61]. Most recently, the Moiré technique was employed with a spatial light modulator (SLM) to generate robust self-trapped multi-lobe beams rotating in real time [62].

Here we discuss the first experimental observation of optical azimuthons [63]. An obstacle to observing rotating beams in anisotropic media, such as photorefractive crystals, is that the OAM is not conserved and the spiraling is suppressed [33, 61]. Therefore, the rubidium vapors are better suited as an isotropic self-focusing medium [10, 12]. In our experiments a cw Ti:sapphire laser was used, tuned to the blue-side of ^{87}Rb resonance $5S_{1/2}F=1 \rightarrow 5P_{3/2}F=0, 1, 2$ [64]. The beam propagated in an 8 cm long, hot ^{87}Rb cell with strong magnetic field applied to the cell to avoid the influence of Earth's magnetic field on the Zeeman splitting of energy levels. To match the specific azimuthon profile at the cell input a Gaussian laser beam was expanded and directed at a small angle to a reflective SLM. The SLM acts as a holographic grating that allows for precise phase and coarse intensity modulation, thus realizing phase imprinting.

The generalized phase imprinting technique, which allows us to excite azimuthons, stems from a theoretical approach unifying the complete family of optical azimuthons, including vortex solitons and soliton clusters. It relies on the azimuthal "discretization" of a vortex of charge m with imposed N-fold symmetry [55]. This discretization can be

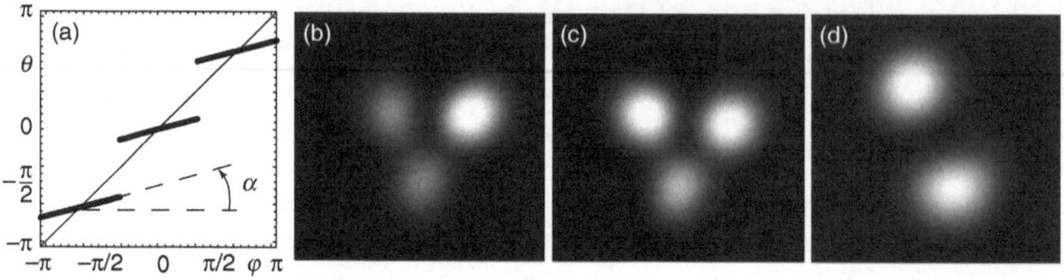

Figure 4.4 (a) Azimuthal phase profile imprinted on the input laser beam using a spatial light modulator. The thin diagonal line represents a vortex phase $\theta = m\varphi$ for $m = 1$, and the thick segments represent the piecewise Eq. (4.13) with $N = 3$. The experimental output shows intensity images for three-lobe azimuthons for input $\alpha = 1°$ in (b), 15° in (c), and 43° in (d). After [63].

described in terms of the series expansion of a vortex phase, $\theta(\varphi) = m\varphi$, near the azimuthal position, $\varphi_n = 2\pi n/N$, of the nth peak on the ring with radius R ($n = 1, .., N$):

$$\theta(\varphi) = m\varphi \simeq m\varphi_n + m(r/R)\sin(\varphi - \varphi_n) + \cdots, \tag{4.12}$$

where (r, φ) are the polar coordinates in the transverse plane.

The *zeroth-order* term in Eq. (4.12) corresponds to a *soliton cluster*, see Section 4.1.4 and [39], i.e. a ring of N beams launched in parallel to each other with a staircase-like phase distribution. The *first-order* term [55] describes the additional azimuthal tilt of all lobes in the ring (initial twist). This tilt modifies the OAM and provides parametrization to the azimuthon family.

To achieve experimental control over the first-order contribution in Eq. (4.12) required for the excitation of the complete family of azimuthons, we introduce an additional parameter of azimuthal tilt α of the plane-sector phase segments, obtained by further expansion of Eq. (4.12) near $r \simeq R$ and $\varphi \simeq \varphi_n$, see Fig. 4.4(a):

$$\theta(\varphi) = m\varphi_n + m\,(\varphi - \varphi_n)\tan\alpha, \text{ for } |\varphi - \varphi_n| < \pi/N. \tag{4.13}$$

In Eq. (4.13) $\alpha = 0$ corresponds to the zero-order soliton cluster, while for $\alpha > 0$ or $\alpha < 0$ we have the phase of each lobe tilted alongside or opposite the direction of the vortex phase twist. The vortex phase is recovered at $\alpha = \pi/4$. We also choose $\varphi_n = \pi(2n-1)/N - \pi$ to map this phase to the domain $\theta \in m(-\pi, \pi)$.

A three-fold phase pattern in Fig. 4.4(a) at the input generates the output beam with three lobes mutually trapped and strongly interacting, see Figs. 4.4(b, c). An important characteristic of azimuthons is the rate of spatial rotation, and this was estimated from the measurements of angular position of the output azimuthon lobes vs. the input tilt angle α. It is important to note that the relation between the experimentally controlled input (angular momentum $M(\alpha)$) and the measured output (angular position) lies in the dynamics of the angular velocity Ω. The latter is not directly measurable in experiment and changes during propagation through a nonlinear medium due to soliton oscillation and unavoidable loses, $\Omega = \Omega(z)$. Nevertheless, important conclusions can be drawn by comparing the output dynamics of the trapped lobes with the output trajectory of a single free lobe, i.e. a straight-propagating beam at the input angle defined by a single segment in Fig. 4.4(a). Using this approach [63], the nonlinear interaction of the azimuthons' lobes was shown to slow down their spatial twist and inversion of the direction of rotation. Thus the anomalous spiraling of three-lobe azimuthons, rotating *opposite* to the overall power flow, was observed. This nonlinear wave phenomenon is in a sharp contrast with the "mechanical" model of weakly interacting beams [51], which exhibit accelerated rotation for attracting lobes instead of slowing down and inversion.

Furthermore, when the initial vortex phase lacks an azimuthal modulation, with angles $\alpha \simeq \pi/4$ close to the vortex phase in Fig. 4.4(a), a vortex breakup with only two output beams is observed in Fig. 4.4(d). This is in a sharp contrast to a robust rotation of three-lobe azimuthon in Figs. 4.4(b, c). This observation suggests that the modulation of the power flow along the azimuthal coordinate of optical vortices in self-focusing media provides an effective mechanism for their stabilization in comparison to the unstable vortex solitons.

4.3 Solitons in nonlocal media

An important step towards stable self-trapped vortices was made [65, 66] with the theoretical predictions of the stabilization of vortex solitons against azimuthal symmetry-breaking in nonlocal media. The key mechanism of nonlocal stabilization appears to be the creation of a smooth and broad trapping potential able to suppress amplitude perturbation, rather than amplify it as in local media. This idea was later applied to higher-order and multi-hump solitons, including azimuthons. In experiments the vortex solitons, multipoles, and necklaces were observed [67, 68] using the boundary-dependent thermal nonlinearity of lead glasses. Another example of a nonlocal medium is nematic liquid crystals where spatial solitons, called "nematicons," were observed [69, 70] and the spiraling of two nematicons was demonstrated [22–24]. Here we discuss the general properties of nonlocality as a mechanism supporting stable spiraling solitons including exotic states with multiple vortices and dynamic self-transformations.

4.3.1 Stabilization of nonlocal solitons

We consider the propagation of paraxial optical beams in a nonlinear medium with nonlocal Kerr-type nonlinearity in Eq. (4.1),

$$\mathcal{N}(I) = \int \mathcal{K}(|\vec{r} - \vec{\rho}|)\, I(\vec{\rho})\, \mathrm{d}\vec{\rho} = 0, \tag{4.14}$$

described by the convolution integral with a kernel $\mathcal{K}$ of the nonlinear response, determined by a particular physical process responsible for the medium nonlinearity [71]. Here we assume the Gaussian response

$$\mathcal{K}(r) = \frac{1}{\pi\sigma^2} \exp\left(-\frac{r^2}{\sigma^2}\right), \tag{4.15}$$

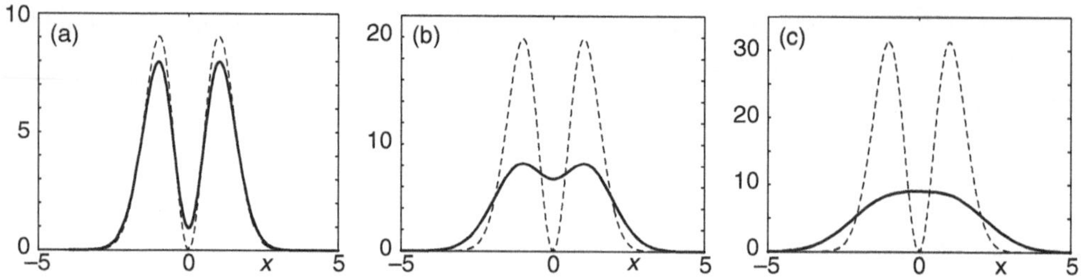

Figure 4.5 Nonlocal dipole soliton with $a=1$. Graphs show the profiles of the intensity, $I(x, y=0)$ (dashed line) and refractive index Eq. (4.18), $\mathcal{N}(x, y=0)$ (solid line), for (a) $\sigma=0.3$ and $P=38.6$, (b) $\sigma=1$ and $P=84.8$, and (c) $\sigma=1.5$ and $P=133.3$. Note the different vertical scales in (a–c).

where σ is the nonlocality parameter. When $\sigma \to 0$ we recover the local Kerr model with $\mathcal{K}(r) \to \delta(r)$ and $\mathcal{N}(I) \to I$, while in the limit of large nonlocality, $\sigma \gg a$, with characteristic transverse scale a of the localization of intensity I, the waveguide Eq. (4.14) effectively approaches the harmonic trap, $\mathcal{N}(I) \sim -Pr^2$, as discussed in [72]. We further illustrate the properties of the nonlocal potential by using a dipole soliton, approximated by the ansatz

$$E(x, y, z) = A\,x \exp\left(-\frac{x^2+y^2}{2a^2} + ikz\right), \tag{4.16}$$

with real amplitude A, half-width a, and propagation constant k. Variational solutions can be derived, see [73],

$$P \equiv \frac{\pi}{2}A^2a^4 = \frac{2\pi}{a^4}\frac{(2a^2+\sigma^2)^4}{3a^4+a^2\sigma^2+2\sigma^4}, \tag{4.17}$$

from which it follows that $P \geq 32\pi/3 \simeq 33.51$. The scaling property of Eq. (4.17) is such that, for a given power P, the solution for any σ, $A=A_1/\sigma^2$ and $a=a_1\sigma$, can be expressed using the solution A_1 and a_1 obtained for $\sigma=1$. The corresponding refractive index in Eq. (4.14) is given by

$$\mathcal{N}(I) = \frac{A^2a^6}{(a^2+\sigma^2)^3}\left(x^2+\frac{\sigma^2}{2}\left(1+\frac{\sigma^2}{a^2}\right)\right)\exp\left(-\frac{x^2+y^2}{a^2+\sigma^2}\right). \tag{4.18}$$

Figure 4.5 illustrates how the refractive index profile Eq. (4.18) changes, with the soliton power P in units normalized to the nonlocality range σ. The latter, therefore, clearly plays the role of a scaling parameter for solitons, whose shape and stability properties are fully determined by the power.

In local media the dipoles do not form stationary solutions because of the phase-sensitive repulsion between them. In contrast, the nonlocal dipole solitons exist as a stationary state even at small nonlocality σ, as in Fig. 4.5(a), although they are unstable and split into two separate beams, similar to their local counterparts. With an increase in nonlocality the refractive index profile approaches a bell-like shape, as in Fig. 4.5(c), and the dipoles became completely stable [74–77]. In fact, at high enough nonlocality the soliton in Fig. 4.5(c) can be seen as a *linear mode* of a harmonic potential [72]. Since the MI is solely a nonlinear phenomenon, it vanishes in the limit of strong nonlocality.

The nonlocal interaction then opens up the possibility of observing the formation of a variety of other higher-order solitons, similar to the families of linear modal beams, such as Laguerre–Gauss or Hermite–Gauss beams. One has to prove, however, that the solutions readily available in the linear Snyder–Mitchell model do exist as stationary solutions with nonlinear potential, as in Eq. (4.14). Such families of soliton "necklaces" and "matrices" were derived [73] in semi-analytical and exact numerical form. Surprisingly, solitons with different symmetries co-exist, i.e. they can have the same power and Hamiltonian, which leads to dynamic transformations between different states discussed in Section 4.3.3

4.3.2 Nonlocal azimuthons

Since vortex solitons can be fully stabilized by nonlocality, it is natural to search for spiraling solitons, such as the azimuthons in Section 4.2, and generalizing solitons with OAM. Surprisingly, the nonlocality allows for new azimuthon states (m, N), which are forbidden in local media. The simplest example is a rotating dipole azimuthon with $m = \pm 1$ and $N = 2$, introduced in [74]. Its envelope can be approximated by the ansatz

$$E(x, y, z) = R(r)\,(\cos\varphi + \mathrm{i}p\sin\varphi)\exp(\mathrm{i}kz), \tag{4.19}$$

with real $-1 \leq p \leq 1$ and $m = \operatorname{sign} p$. There is a continuous transformation between radially symmetric vortex solitons with $p = \pm 1$ and the dipole soliton with $p = 0$ in Eq. (4.16), and the angular velocity of spiraling grows monotonically with p. The stability of this solution was studied using direct integration of the NLSE and all such solutions were found to be unstable for small nonlocality, in full agreement with the results on unstable two-soliton spiraling in local media, see Section 4.1.3. The stability threshold is reached with an increase in soliton power (nonlocality), and this threshold strongly depends on the modulation parameter p. A rather counter-intuitive result is that the threshold power is lower for faster rotating azimuthons with smaller modulation, $p \simeq 1$. The stability, however, strongly depends on the shape of the nonlocal kernel [75] as well as the geometry of the sample [77] for realistic boundary-dependent nonlinearities.

A complete family of nonlocal azimuthons was studied in [59, 78–80]. In a sharp contrast to azimuthons with $N > 2m$ in local media, the nonlocal azimuthons with $N < 2m$ peaks were found and their stability was confirmed numerically for sufficiently large nonlocality. The dynamics of these solutions shows the development of topological structures which cannot be anticipated from the azimuthon ansatz Eq. (4.7) with a single central dislocation of charge m. Figure 4.6 shows three different azimuthons, launched initially with a single central vortex of charge $m = 6$, after a short numerical propagation. Instead of a single high-charge vortex, these solutions develop a regular pattern of single-charge phase dislocations in their dark core, depending on the symmetry order N of the intensity

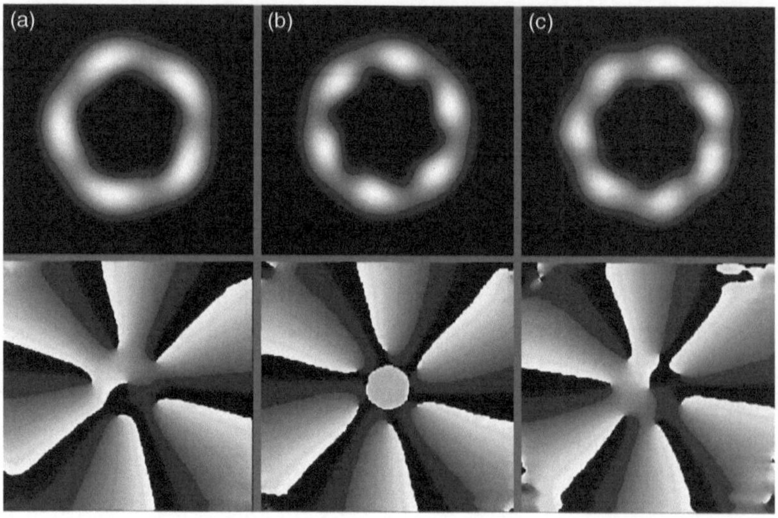

Figure 4.6 Examples of higher-order nonlocal azimuthons with $m=6$ and $n=0.5$, after numerical propagation of $z=2$, with: (a) $N=5$, $k=645.7$, $\omega=24.9$; (b) $N=6$, $k=624.3$, $\omega=21.3$; and (c) $N=7$, $k=603$, $\omega=17.8$. Top row: intensity, from black zero to white maximum. Bottom row: phase, from black $-\pi$ to white π. After [78].

ring modulation. It can be seen in Fig. 4.6 that the central dislocation is of order $m_2=1$ in (a), $m_2=0$ in (b), and $m_2=-1$ in (c). In general, the configuration of these "vortex clusters" contains a ring of N single-charged vortices (thus building a charge of $m_1=N$), and an additional vortex of charge $m_2=m-N$ at the origin, keeping the total charge $m_1+m_2=m$.

A more accurate method to obtain general azimuthon solutions beyond the limitations of the ansatz Eq. (4.7) was developed in [59]. It is based on the assumption that an azimuthon with N lobes bifurcates from an internal mode of order N of a vortex soliton of charge m, see Section 4.1.2. The superposition of a vortex soliton with its mode is similar to the "linear azimuthons" on the ring, see Eqs. (4.9)–(4.11) in Section 4.2. In two spatial dimensions it leads to the appearance of satellite vortices outside of the intensity ring, as well as inside the ring, as in Fig. 4.6. Therefore, the general notion of azimuthons as spatially modulated ring-shaped vortices [55] is expanded in nonlocal media and includes *multivortex* spiraling states.

The multivortex nonlocal solitons were introduced [81] in an attempt to find possible connections between soliton necklaces and matrices [73] using the linear generalized Hermite–Laguerre–Gaussian (HLG) modes [82]. In fact, the dipole azimuthon in Eq. (4.19) represents one of the lowest-order HLG states. A more complex *tripole soliton* carries two spatially separated vortices of the same charge [81]. It can be constructed from an HLG_{20} state, which provides a continuous single-parameter transformation ($0 \leq |\alpha| \leq \pi/4$) between a Hermite–Laguerre beam with three in-line peaks ($\alpha=0$) to a

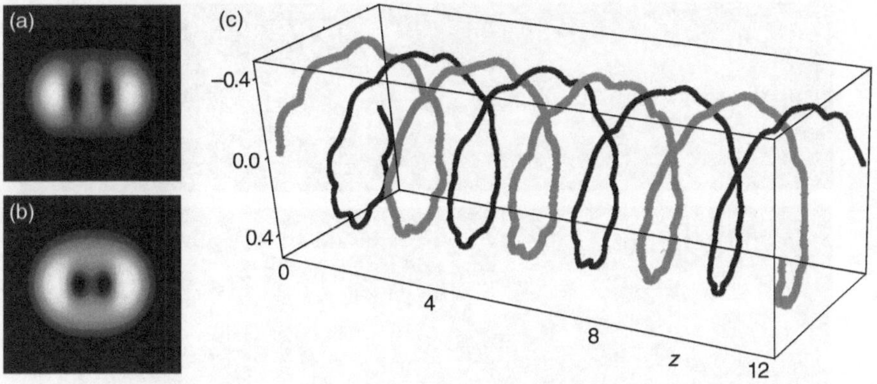

Figure 4.7 The tripole soliton intensity profiles for transformation parameter $\alpha = -\pi/8$ in (a) and $\alpha = -3\pi/16$ in (b). (c) Trajectories of two spiraling vortices of the tripole soliton in (a). After [81].

double-charge vortex beam ($\alpha = \pm\pi/4$). The transformation is accompanied with vanishing of the central intensity peak, clearly visible for $\alpha = -\pi/8$ in Fig. 4.7(a), as two vortices fuse into a single double-charge phase dislocation. Thus there appear states with two dominating intensity peaks, as with $\alpha = -3\pi/16$ in Fig. 4.7(b), which can be considered as an azimuthon with $m = 2$ and $N = 2$. That is indeed the case as a similar state bifurcates from a double-charge mode of a double-charge vortex (see Fig. 6 in [59]).

Further insight into the physics of spiraling solitons is provided by the analysis of the structure of the angular velocity [59, 81] based on the formulae derived in [83] for NLSE-type equations. The expression for the angular velocity, $\Omega = \Omega_{\mathrm{ln}} + \Omega_{\mathrm{nln}}$, has two contributions, the first is defined by the structure of the beam itself, and is thus valid in linear media and free propagation, and the second part Ω_{nln}, which depends on nonlinearity. In particular, Ω_{lin} vanishes for exact HLG modes, as expected [82], yet it is nonzero for solitons, i.e. the first source of soliton spiraling is the nonlinear deformation of the mode. Indeed, such deformed beams can be seen as a superposition of several HLG modes, and the rotation appears as a beating between the modes. Furthermore, such beams always rotate in nonlinear media because the nonlinear component Ω_{nln}, does not vanish even for the exact linear modes. Thus the self-action of a spatially modulated beam with OAM leads to its rotation. Figure 4.7(c) shows the double helix formed by two vortices of a spiraling tripole soliton.

4.3.3 Self-transforming nonlocal solitons

Nonlocal media support a great variety of spatial solitons and dramatically affects their stability. Most intriguing are periodic solitons [72, 85], including two-dimensional solitons which transform their symmetry during propagation [73, 74, 86, 87]. Quasi-periodic dynamics can be triggered by the internal modes of solitons where two different states,

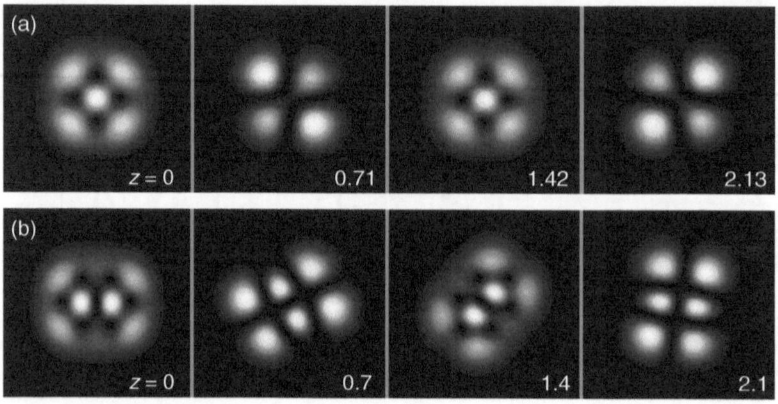

Figure 4.8 Quasi-periodic dynamics of HLG beams in a nonlocal medium. (a) HLG_{11} quadrupole beam with zero OAM. (b) HLG_{21} soliton matrix spiraling due to non-zero OAM. After [84].

such as necklaces and matrices of solitons [73], can couple without violating the conservation of power and energy. In thermal media the stability of solitons is strongly affected by the boundaries [67, 68, 77] and the transformation dynamics can be effectively controlled, e.g. by changing the aspect ratio of a rectangular sample [88].

The process of the dynamic reshaping of nonlocal solitons [73] strongly resembles astigmatic transformations in mode converters. The unifying description of the transformation between linear Hermite and Laguerre modes with a single parameter is provided by HLG beams [82]. The idea put forward in [84] is to employ an HLG ansatz in the search for self-transforming nonlocal solitons with possible dynamic transformations (during propagation) similar to the HLG family parametrization. However, the dynamic transformations of solitons are essentially more complex because of the conservation of OAM during propagation, in contrast to mode converters involving an exchange of OAM between light and optical elements. Thus OAM conservation restricts any possible conversion to states with the same OAM and also leads to soliton spiraling in nonlinear media [81]. As a result, the dipole [74] and tripole [81] spiraling solitons do not transform.

The lowest-order self-transforming soliton is a 2×2 soliton matrix, or a quadrupole-type soliton, approximated by the HLG_{11} mode with zero OAM. Figure 4.8(a) shows the dynamics of such a beam with $\alpha = \pi/8$ in nonlocal media. The reappearance of the input profile at $z = 1.42$ is evident, but the profile does not reproduce the input mode, rather the orbits in the modal space are quasi-periodic, as there are additional deformations at each period. Nevertheless, the oscillations are robust and spectral analysis [84] has shown an equidistant spectrum characteristic for discrete breathers [89]. Also noteworthy are the topological reactions accompanying transformations, e.g. the four vortices visible at $z = 0$ in Fig. 4.8(a) as dark spots disintegrate into edge-type dislocations characteristic of Hermite-type beams ($z = 0.71, 2.13$) and then reappear again ($z = 1.42$), with all topological charges flipped [90].

There are also self-transforming modes with nonzero OAM, such as the 2×3 soliton matrix in Fig. 4.8(b). This solution is generated with the ansatz of the HLG_{21} beam, which includes a Hermite mode and the double-ring single-charge Laguerre vortex. Similar to the tripole solitons, the OAM is largely defined by the modulation parameter α and practically does not change with power. However, the simulations of Eq. (4.1) reveal complex dynamics with three simultaneous processes on different scales: fast breathing oscillations, spiraling, and mode transformations. It is impossible to distinguish different frequencies from direct observations in Fig. 4.8 and the oscillation spectra include a discrete set of frequencies produced by algebraic sums of soliton frequency, rotation rate, and transformation frequency.

The transformations in Fig. 4.8 cannot be described in linearized perturbation analysis because they are not small. However, the periods of rotation and transformation are at least one order of magnitude lower than the soliton period. Therefore, the ratio of two frequencies could be used as a small adiabatic parameter to address the question of whether truly periodic two-dimensional solitons exist. There is probably also a greater variety of multi-humped solitons, stemming from linear superpositions in the highly nonlocal limit of the parabolic trapping potential, which might provide a generalization to quasi-periodic transformations between different symmetries of self-trapped optical beams.

4.4 Suppression of collapse

Another example of the nontrivial application of the OAM concept to nonlinear problems is the recent prediction of the OAM-assisted suppression of catastrophic self-focusing in nonlinear Kerr media [91]. The catastrophic self-focusing or *collapse instability* [46] in a pure Kerr medium with $\mathcal{N}(I) = I$ should not be confused with the azimuthal MI discussed in Section 4.1.2. It occurs in two (or three) dimensions when the optical power (energy) of the beam (pulse) exceeds the threshold of soliton formation, and the width of the envelope collapses while the peak intensity diverges. The collapsing filament attains the universal shape of a "Townes soliton" [92], observed in experiments by [93]. The ring-shaped vortex beams with powers exceeding the collapse threshold experience azimuthal MI before the filaments collapse [60].

Different suggestions have been made of how to arrest the collapse, in particular by modifying the medium nonlinearity [94]. The problem remains unsolved for pure Kerr media, however, although non-paraxial coupling of the polarization components was shown to provide stabilization to circularly polarized beams [95].

Here we discuss the suppression of collapse in a pure Kerr medium for fundamental bell-shaped solitons with an elliptic profile and non-zero OAM, which form a twisted profile, and thereby can also be regarded as spiraling self-trapped beams. The vortex-free beams with nonzero OAM can be generated as elliptical bell-shaped beams focused by a tilted cylindrical lens [96]. We use a Gaussian function as the VA ansatz,

$$E(x, y, z) = A(z)\, G\,(X/b(z))\; G\,(Y/c(z))\, \exp(\mathrm{i}\Psi)\,, \tag{4.20}$$

where $G(t) = \exp(-t^2/2)$ and phase $\Psi = B(z)X^2 + \Theta(z)XY + C(z)Y^2 + \varphi(z)$. In contrast to previously considered nonrotating elliptic beams [97], here the *asymmetric* phase factor $\Theta(z)$ is conjugated with the angle $\theta(z)$ of the rotating frame, $X = x\cos\theta + y\sin\theta$ and $Y = -x\sin\theta + y\cos\theta$. Since we consider beams focused at $z = 0$ by a cylindrical lens upon entry into the nonlinear media, our boundary conditions for the phase read $\{B_0, C_0, \varphi_0\} = 0$, where the subscript 0 indicates the initial value, e.g. $B_0 = B(0)$. The value $\Theta_0 \sim (\kappa/2f)\sin 2\alpha$ depends on the angle α between the major axis of the elliptically shaped beam with wavenumber κ, and the cylindrical lens of focal length f, and is independent of the other beam parameters [96].

Deriving variational equations, we obtain the integrals of motion: the power, $P = \pi A^2 bc$, and the OAM, $M = P\Theta(b^2 - c^2)/2$, as well as equations for the conjugated variables, $\dot{b} = 4bB$, $\dot{c} = 4cC$, and $\dot{\theta} = 2\Theta(b^2 + c^2)/(b^2 - c^2)$. Dots indicate derivatives with respect to the evolution variable z, so $\{\dot{P}, \dot{M}, \dot{H}\} = 0$. In what follows it is convenient to introduce the spatial scale $a(z)$, $a^2 = (b^2 + c^2)/2$, and the ellipticity parameter $\epsilon(z) = \tan^{-1}(c/b)$, then $b = \sqrt{2}\,a\cos\epsilon$ and $c = \sqrt{2}\,a\sin\epsilon$. The variational system can be cast in Hamiltonian form as follows:

$$H = \frac{P}{2}\left(\frac{\dot{a}^2}{2} + \frac{\mu}{a^2}\right), \qquad \dot{\mu} = 0, \tag{4.21}$$

$$\mu = \frac{a^4\dot{\epsilon}^2}{2} + \frac{4 - p\sin 2\epsilon}{2\sin^2 2\epsilon} + \frac{2\sigma^2}{\cos^2 2\epsilon}, \tag{4.22}$$

with $p \equiv P/\pi = A^2a^2\sin 2\epsilon$ and $\sigma \equiv M/P = a^2\Theta\cos 2\epsilon$. For a beam with its waist at $z = 0$, $\{\dot{a}_0, \dot{\epsilon}_0\} = 0$, we obtain from Eq. (4.21) $a^2(z) = a_0^2\left(1 + z^2/z_0^2\right)$, where $z_0 = a_0^2/\sqrt{2\mu}$. The latter expression shows that the critical value, which defines the collapse threshold, is the Ermakov integral of motion μ in Eq. (4.22) [98, 99]. Indeed, the elliptic spatial soliton with diverging "diffraction length", $z_0 \to \infty$, is obtained for $\mu \to 0$, while the collapse at finite length $|z_0|$ is achieved for $\mu < 0$.

The threshold power of collapse can be calculated from the condition $\mu = 0$ as

$$P_c = \frac{4\pi}{\sin 3\epsilon}(1 + \sigma^2\tan^2 2\epsilon). \tag{4.23}$$

Note how the *nonnegative* contribution of OAM to diffraction, $\sim \sigma^2$ in the expressions (4.22) and (4.23), *counteracts* the self-focusing ($\sim p$) and increases the collapse threshold power P_c.

The VA predictions of collapse suppression were confirmed [91] using extensive numerical simulations of Eq. (4.1). The analysis was based on the fact that the OAM is an additional, physically independent parameter of the elliptic beam, defined by the tilt α of a given cylindrical lens, i.e. the parameters Θ_0 and σ can be set in an experiment independent of the beam envelope p and ϵ_0. Therefore, for a given soliton solution we can introduce an additional tilt and compare beams with the same power P_s and envelope but different values of OAM. The actual value of OAM required to suppress collapse is found to be much larger than that predicted by VA. Nevertheless, as was pointed out in [96], the

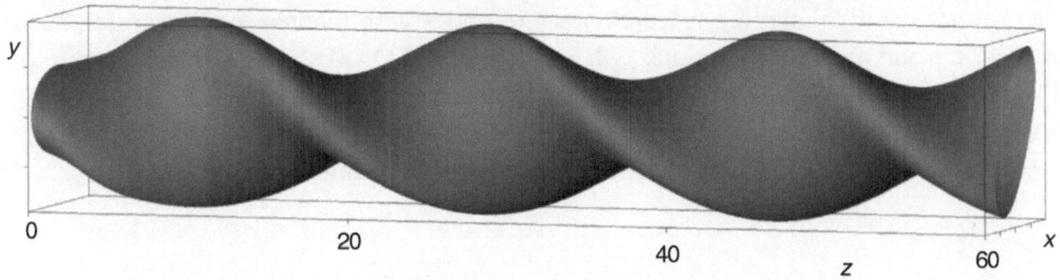

Figure 4.9 Spiraling elliptic soliton in saturable media. After [91].

very high OAM can be achieved in experiment, $\sigma \sim 10^4$, although the relatively low values of $\sigma = 6 - 12$ are required to eliminate the collapse of the beams with supercritical powers $P = (5 - 8)P_c$.

In addition to a circular soliton with $\sigma = 0$, the variational Eqs. (4.21)–(4.22) predict the existence of a novel kind of solution, a spiralling elliptic soliton with $\sigma \neq 0$. These results demonstrate that the contribution of OAM to the dynamics of elliptic beams in nonlinear self-focusing media is twofold. First, it strengthens the diffraction against self-focusing and can suppress collapse in Kerr media. Second, it preserves the elliptic profile of stably rotating solitons in optical media with collapse-free nonlinearities. An example of a stable spiraling elliptic soliton in a saturable medium is shown in Fig. 4.9.

A similar idea has been recently implemented [100] to the vortex-free elliptic clouds of Bose–Einstein condensates (BEC) with a negative scattering length trapped in a radially symmetric harmonic potential. The number of trapped ultracold atoms corresponding to the collapse threshold can be radically increased for such rotating nonlinear matter waves, by transferring OAM to the matter wave, e.g. imprinting a twisted phase front onto an elliptic BEC cloud. Owing to the formal similarity between the behavior of the coherent matter and light waves, the results could also be readily applied to the dynamics of elliptic "spiraling" modes of nonlinear optical waveguides. An interesting property of nonlinear modes of a harmonic trap is that, below the collapse threshold, the elliptic mode is stable and exhibits periodic oscillations around a stationary rotating state. The frequency of these oscillations is independent of the number of particles, and is determined solely by the trapping potential.

4.5 Conclusions

We have presented a brief account of the basic physics and recent results on spiraling self-trapped optical beams, demonstrating novel applications of the physics of orbital angular momentum, such as the suppression of collapse and self-induced mode transformations in nonlocal nonlinear media. Many closely related research problems, not covered in this chapter, are at different stages of development, such as the study of self-trapped vortices in dissipative nonlinear systems, Bose–Einstein condensates, and periodic media. There are also several rapidly emerging research areas in optics with inherent nonlinearities, such as polariton and photon condensates, where vortices and OAM should play a crucial role. Therefore, we may expect that the physical mechanisms discussed here for the applications to spatial solitons in optics will also be important in other systems with an interplay between the nonlinear self-action of waves and their orbital angular momentum.

Acknowledgements

This work has been financially supported by the Australian Research Council. We thank our students and collaborators, in particular J. Abdullaev, G. Assanto, D. Buccoliero, M. R. Dennis, Ya. V. Izdebskaya, W. Krolikowski, S. Lopez-Aguayo, A. Minovich, D. Neshev, E. A. Ostrovskaya, S. Skupin, and A. A. Sukhorukov, for useful and inspiring discussions, joint collaborative projects, and important contributions to the field of spiraling solitons.

References

[1] Kivshar, Yu. S. and Agrawal, G. P. 2003. *Optical Solitons: From Fibers to Photonic Crystals*. San Diego, CA: Academic Press.

[2] Malomed, B. A. 2002. Variational methods in nonlinear fiber optics and related fields. In Wolf, E. (ed), *Progress in Optics*, vol. 43. Amsterdam: Elsevier, pp. 71–193.

[3] Kruglov, V. I. and Vlasov, R. A. 1985. Spiral self-trapping propagation of optical beams in media with cubic nonlinearity. *Phys. Lett. A*, **111**(8–9), 401–4.

[4] Nye, J. F. and Berry, M. V. 1974. Dislocations in wave trains. *Proc. Royal Soc. London Series A Math. Phy. Sci.*, **336**(1605), 165–90.

[5] Allen, L., Barnett, S. M., and Padgett, M. J. 2003. *Optical Angular Momentum*. Institute of Physics Pub.

[6] Soskin, M. S. and Vasnetsov, M. V. 2001. Singular optics. In Wolf, E. (ed), *Progress in Optics*, vol. 42. Amsterdam: Elsevier, pp. 219–70.

[7] Dennis, M. R., O'Holleran, K., and Padgett, M. J. 2009. Singular optics: optical vortices and polarization singularities. In Wolf, E. (ed), *Progress in Optics*, vol. 53. Amsterdam: Elsevier, pp. 293–363.

[8] Desyatnikov, A. S., Kivshar, Yu. S., and Torner, L. 2005. Optical vortices and vortex solitons. In Wolf, E. (ed), *Progress in Optics*, vol. 47. Amsterdam: Elsevier, pp. 291–391.

[9] Kivshar, Yu. S. and Pelinovsky, D. E. 2000. Self-focusing and transverse instabilities of solitary waves. *Phys. Rep.*, **331**(4), 118–95.

[10] Tikhonenko, V., Christou, J., and Luther-Davies, B. 1995. Spiraling bright spatial solitons formed by the breakup of an optical vortex in a saturable self-focusing medium. *J. Opt. Soc. Am. B*, **12**(11), 2046–52.

[11] Tikhonenko, V., Christou, J., and Luther-Davies, B. 1996. Three dimensional bright spatial soliton collision and fusion in a saturable nonlinear medium. *Phys. Rev. Lett.*, **76**(15), 2698–701.

[12] Bigelow, M. S., Zerom, P., and Boyd, R. W. 2004. Breakup of ring beams carrying orbital angular momentum in sodium vapor. *Phys. Rev. Lett.*, **92**(8), 083902–4.

[13] Chen, Z., Shih, M.-F., Segev, M., Wilson, D. W., Muller, R. E., and Maker, P. D. 1997. Steady-state vortex-screening solitons formed in biased photorefractive media. *Opt. Lett.*, **22**(23), 1751–3.

[14] Petrov, D. V., Torner, L., Martorell, J., Vilaseca, R., Torres, J. P., and Cojocaru, C. 1998. Observation of azimuthal modulational instability and formation of patterns of optical solitons in a quadratic nonlinear crystal. *Opt. Lett.*, **23**(18), 1444–16.

[15] Soto-Crespo, J. M., Wright, E. M., and Akhmediev, N. N. 1992. Recurrence and azimuthal-symmetry breaking of a cylindrical Gaussian beam in a saturable self-focusing medium. *Phys. Rev. A*, **45**(5), 3168–75.

[16] Firth, W. J. and Skryabin, D. V. 1997. Optical solitons carrying orbital angular momentum. *Phys. Rev. Lett.*, **79**(13), 2450–3.

[17] Torres, J. P., Soto-Crespo, J. M., Torner, L., and Petrov, D. V. 1998. Solitary-wave vortices in quadratic nonlinear media. *J. Opt. Soc. Am. B*, **15**(2), 625–7.

[18] Skryabin, D. V. and Firth, W. J. 1998. Dynamics of self-trapped beams with phase dislocation in saturable Kerr and quadratic nonlinear media. *Phys. Rev. E*, **58**(3), 3916–30.

[19] Stegeman, G. I. and Segev, M. 1999. Optical spatial solitons and their interactions: universality and diversity. *Science*, **286**(5444), 1518–23.

[20] Poladian, L., Snyder, A. W., and Mitchell, D. J. 1991. Spiraling spatial solitons. *Opt. Commun.*, **85**(1), 59–62.

[21] Nikolov, N. N., Neshev, D., Krolikowski, W., Bang, O., Rasmussen, J. J., and Christiansen, P. L. 2004. Attraction of nonlocal dark optical solitons. *Opt. Lett.*, **29**, 286–8.

[22] Fratalocchi, A., Peccianti, M., Conti, C., and Assanto, G. 2004. Spiraling and cyclic dynamics of nematicons. *Mol. Cryst. Liq. Cryst.*, **421**, 197–207.

[23] Fratalocchi, A., Piccardi, A., Peccianti, M., and Assanto, G. 2007. Nonlinearly controlled angular momentum of soliton clusters. *Opt. Lett.*, **32**, 1447–9.

[24] Fratalocchi, A., Piccardi, A., Peccianti, M., and Assanto, G. 2007. Nonlinear management of the angular momentum of soliton clusters: theory and experiment. *Phys. Rev. A*, **75**(6), 063835.

[25] Steblina, V. V., Kivshar, Yu. S., and Buryak, A. V. 1998. Scattering and spiraling of solitons in a bulk quadratic medium. *Opt. Lett.*, **23**(3), 156–8.

[26] Buryak, A. V. and Steblina, V. V. 1999. Soliton collisions in bulk quadratic media: comprehensive analytical and numerical study. *J. Opt. Soc. Am. B*, **16**(2), 245–55.

[27] Malomed, B. A. 1998. Potential of interaction between two- and three-dimensional solitons. *Phys. Rev. E*, **58**(6), 7928–33.

[28] Simos, C., Coudrec, V., and Barthelemy, A. 2002. Experiences sur les interactions de type particulaire de solitons multicolores: fusion, répulsion, enroulement en spirale. *J. Phys. IV*, **12**(5), 211–12.

[29] Shih, M.-F., Segev, M., and Salamo, G. 1997. Three-dimensional spiraling of interacting spatial solitons. *Phys. Rev. Lett.*, **78**(13), 2551–24.

[30] Desyatnikov, A. S. and Maimistov, A. I. 1998. Interaction of two spatially separated light beams in a nonlinear Kerr medium. *JETP*, **86**(6), 1101–16.

[31] Schjodt-Eriksen, J., Schmidt, M. R., Rasmussen, J. J., Christiansen, P. L., Gaididei, Y. B., and Berge, L. 1998. Two-beam interaction in saturable media. *Phys. Lett. A*, **246**(5), 423–8.

[32] Garcia-Ripoll, J. J., Perez-Garcia, V. M., Ostrovskaya, E. A., and Kivshar, Yu. S. 2000. Dipole-mode vector solitons. *Phys. Rev. Lett.*, **85**(1), 82–5.

[33] Carmon, T., Uzdin, R., Pigier, C., Musslimani, Z. H., Segev, M., and Nepomnyashchy, A. 2001. Rotating propeller solitons. *Phys. Rev. Lett.*, **87**(14), 143901–4.

[34] Skryabin, D. V., McSloy, J. M., and Firth, W. J. 2002. Stability of spiralling solitary waves in Hamiltonian systems. *Phys. Rev. E*, **66**(5), 055602–4.

[35] Stepken, A., Belic, M. R., Kaiser, F., Krolikowski, W., and Luther-Davies, B. 1999. Three dimensional trajectories of interacting incoherent photorefractive solitons. *Phys. Rev. Lett.*, **82**(3), 540–3.

[36] Belic, M. R., Stepken, A., and Kaiser, F. 1999. Spiraling behavior of photorefractive screening solitons. *Phys. Rev. Lett.*, **82**(3), 544–7.

[37] Krolikowski, W., Luther-Davies, B., Denz, C. *et al.* 1999. Interaction of two-dimensional spatial incoherent solitons in photorefractive medium. *Appl. Phys. B*, **68**(5), 975–82.

[38] Denz, C., Krolikowski, W., Petter, J. *et al.* 1999. Dynamics of formation and interaction of photorefractive screening solitons. *Phys. Rev. E*, **60**(5), 6222–65.

[39] Desyatnikov, A. S., and Kivshar, Yu. S. 2002. Rotating optical soliton clusters. *Phys. Rev. Lett.*, **88**(5), 053901.

[40] Desyatnikov, A., Denz, C., and Kivshar, Y. 2004. Nonlinear optical beams carrying phase dislocations. *J. Opt. A-Pure Appl. Opt.*, **6**(5), S209–12.

[41] Kartashov, Y. V., Molina-Terriza, G., and Torner, L. 2002. Multicolor soliton clusters. *J. Opt. Soc. Am. B*, **19**(11), 2682–91.

[42] Kartashov, Y. V., Crasovan, L. C., Mihalache, D., and Torner, L. 2002. Robust propagation of two-color soliton clusters supported by competing nonlinearities. *Phys. Rev. Lett.*, **89**(27), 273902–4.

[43] Mihalache, D., Mazilu, D., Crasovan, L. C., Malomed, B. A., Lederer, F., and Torner, L. 2003. Robust soliton clusters in media with competing cubic and quintic nonlinearities. *Phys. Rev. E*, **68**(4), 046612–8.

[44] Crasovan, L. C., Kartashov, Y. V., Mihalache, D., Torner, L., Kivshar, Yu. S., and Perez-Garcia, V. M. 2003. Soliton "molecules": robust clusters of spatiotemporal optical solitons. *Phys. Rev. E*, **67**(4), 046610–5.

[45] Mihalache, D., Mazilu, D., Crasovan, L. C., Malomed, A., Lederer, F., and Torner, L. 2004. Soliton clusters in three-dimensional media with competing cubic and quintic nonlinearities. *J. Opt. B: Quantum Semicl. Opt.*, **6**(5), S333–40.

[46] Berge, L. 1998. Wave collapse in physics: principles and applications to light and plasma waves. *Phys. Rep.*, **303**(5–6), 260–370.

[47] Barthelemy, A., Froehly, C., Shalaby, M., Donnat, P., Paye, J., and Migus, A. 1993. In Martin, J. J. *et al.* (ed), *Ultrafast Phenomena VIII*. Berlin: Springer, pp. 299–305.

[48] Barthelemy, A., Froehly, C., and Shalaby, M. 1994. *Proc. SPIE Int. Soc. Opt. Eng.*, **2041**, 104.

[49] Soljacic, M., Sears, S., and Segev, M. 1998. Self-trapping of "necklace" beams in self-focusing Kerr media. *Phys. Rev. Lett.*, **81**(22), 4851–4.

[50] Soljacic, M. and Segev, M. 2000. Self-trapping of "necklace-ring" beams in self-focusing Kerr media. *Phys. Rev. E*, **62**(2), 2810–20.

[51] Soljacic, M. and Segev, M. 2001. Integer and fractional angular momentum borne on self-trapped necklace-ring beams. *Phys. Rev. Lett.*, **86**(3), 420–3.

[52] Desyatnikov, A. S. and Kivshar, Yu. S. 2002. Spatial optical solitons and soliton clusters carrying an angular momentum. *J. Opt. B: Quantum Semicl. Opt.*, **4**(2), S58–65.

[53] Abramochkin, E. G. and Volostnikov, V. G. 2004. Spiral light beams. *Phys. Usp.*, **74**, 1177–203.
[54] Daria, V. R., Palima, D. Z., and Glückstad, J. 2011. Optical twists in phase and amplitude. *Opt. Express*, **19**, 476–81.
[55] Desyatnikov, A. S., Sukhorukov, A. A., and Kivshar, Yu. S. 2005. Azimuthons: spatially modulated vortex solitons. *Phys. Rev. Lett.*, **95**(20), 203904.
[56] Carr, L. D., Clark, C. W., and Reinhardt, W. P. 2000. Stationary solutions of the one-dimensional nonlinear Schrödinger equation. II. Case of attractive nonlinearity. *Phys. Rev. A*, **62**(6), 063611.
[57] Berry, M. V. 2009. Optical currents. *J. O. A: Pure Appl. Opt.* **11**(9), 094001.
[58] Bekshaev, A. and Soskin, M. 2006. Rotational transformations and transverse energy flow in paraxial light beams: linear azimuthons. *Opt. Lett.*, **31**(14), 2199–201.
[59] Maucher, F., Buccoliero, D., Skupin, S., Grech, M., Desyatnikov, A., and Krolikowski, W. 2009. Tracking azimuthons in nonlocal nonlinear media. *Opt. Quantum Electr*, **41**, 337–48. doi: 10.1007/s11082-009-9351-9.
[60] Vuong, L. T., Grow, T. D., Ishaaya, A. *et al.* 2006. Collapse of optical vortices. *Phys. Rev. Lett.*, **96**(13), 133901.
[61] Krolikowski, W., Ostrovskaya, E. A., Weilnau, C. *et al.* 2000. Observation of dipole-mode vector solitons. *Phys. Rev. Lett.*, **85**(7), 1424–17.
[62] Zhang, P., Huang, S., Hu, Y., Hernandez, D., and Chen, Z. 2010. Generation and nonlinear self-trapping of optical propelling beams. *Opt. Lett.*, **35**(18), 3129–31.
[63] Minovich, A., Neshev, D. N., Desyatnikov, A. S., Krolikowski, W., and Kivshar, Yu. S. 2009. Observation of optical azimuthons. *Opt. Express*, **17**(26), 23610–16.
[64] Rochester, S. M., Hsiung, D. S., Budker, D., Chiao, R. Y., Kimball, D. F., and Yashchuk, V. V. 2001. Self-rotation of resonant elliptically polarized light in collision-free rubidium vapor. *Phys. Rev. A*, **63**(4), 043814.
[65] Briedis, D., Petersen, D., Edmundson, D., Krolikowski, W., and Bang, O. 2005. Ring vortex solitons in nonlocal nonlinear media. *Opt. Express*, **13**, 435–43.
[66] Yakimenko, A. I., Zaliznyak, Yu. A., and Kivshar, Yu. 2005. Stable vortex solitons in nonlocal self-focusing nonlinear media. *Phys. Rev. E*, **71**(6), 065603.
[67] Rotschild, C., Cohen, O., Manela, O., Segev, M., and Carmon, T. 2005. Solitons in nonlinear media with an infinite range of nonlocality: first observation of coherent elliptic solitons and of vortex-ring solitons. *Phys. Rev. Lett.*, **95**(21), 213904.
[68] Rotschild, C., Segev, M., Xu, Z., Kartashov, Y. V., Torner, L., and Cohen, O. 2006. Two-dimensional multipole solitons in nonlocal nonlinear media. *Opt. Lett.*, **31**(22), 3312–14.
[69] Conti, C., Peccianti, M., and Assanto, G. 2003. Route to nonlocality and observation of accessible solitons. *PRL*, **91**(7).
[70] Conti, C., Peccianti, M., and Assanto, G. 2004. Observation of optical spatial solitons in a highly nonlocal medium. *Phys. Rev. Lett.*, **92**(11), 113902.
[71] Krolikowski, W., Bang, O., Nikolov, N. I. *et al.* 2004. Modulational instability, solitons and beam propagation in spatially nonlocal nonlinear media. *J. Opt. B*, **6**(5), S288–94.
[72] Snyder, A. W. and Mitchell, D. J. 1997. Accessible solitons. *Science*, **276**(5318), 1538–41.
[73] Buccoliero, D., Desyatnikov, A. S., Krolikowski, W., and Kivshar, Yu. S. 2007. Laguerre and Hermite soliton clusters in nonlocal nonlinear media. *Phys. Rev. Lett.*, **98**(5), 053901.

[74] Lopez-Aguayo, S., Desyatnikov, A. S., Kivshar, Yu. S., Skupin, S., Krolikowski, W., and Bang, O. 2006. Stable rotating dipole solitons in nonlocal optical media. *Opt. Lett.*, **31**(8), 1100–2.

[75] Skupin, S., Bang, O., Edmundson, D., and Krolikowski, W. 2006. Stability of two-dimensional spatial solitons in nonlocal nonlinear media. *Phys. Rev. E*, **73**(6), 066603.

[76] Skupin, S., Saffman, M., and Królikowski, W. 2007. Nonlocal stabilization of nonlinear beams in a self-focusing atomic vapor. *Phys. Rev. Lett.*, **98**(26), 263902.

[77] Ye, F., Kartashov, Y. V., and Torner, L. 2008. Stabilization of dipole solitons in nonlocal nonlinear media. *Phys. Rev. A*, **77**(4), 043821.

[78] Lopez-Aguayo, S., Desyatnikov, A. S., and Kivshar, Yu. S. 2006a. Azimuthons in nonlocal nonlinear media. *Opt. Express*, **14**(17), 7903–78.

[79] Skupin, S., Grech, M., and Królikowski, W. 2008. Rotating soliton solutions in nonlocal nonlinear media. *Opt. Express*, **16**(12), 9118–31.

[80] Zhang, Y., Skupin, S., Maucher, F., Pour, A. G., Lu, K., and Królikowski, W. 2010. Azimuthons in weakly nonlinear waveguides of different symmetries. *Opt. Express*, **18**(26), 27846–57.

[81] Buccoliero, D., Desyatnikov, A. S., Krolikowski, W., and Kivshar, Yu. S. 2008. Spiraling multivortex solitons in nonlocal nonlinear media. *Opt. Lett.*, **33**(2), 198–200.

[82] Abramochkin, E. G. and Volostnikov, V. G. 2004. Generalized Gaussian beams. *J. Opt. A: Pure Appl. Opt.*, **6**(5), S157.

[83] Rozanov, N. 2004. On the translational and rotational motion of nonlinear optical structures as a whole. *Opt. Spectroscopy*, **96**, 405–8. doi: 10.1134/1.1690033.

[84] Buccoliero, D. and Desyatnikov, A. S. 2009. Quasi-periodic transformations of nonlocal spatial solitons. *Opt. Express*, **17**(12), 9608–13.

[85] Kaminer, I., Rotschild, C., Manela, O., and Segev, M. 2007. Periodic solitons in nonlocal nonlinear media. *Opt. Lett.*, **32**(21), 3209–11.

[86] Yakimenko, A. I., Lashkin, V. M., and Prikhodko, O. O. 2006. Dynamics of two-dimensional coherent structures in nonlocal nonlinear media. *Phys. Rev. E*, **73**(6), 066605.

[87] Kartashov, Y. V., Torner, L., Vysloukh, V. A., and Mihalache, D. 2006. Multipole vector solitons in nonlocal nonlinear media. *Opt. Lett.*, **31**(10), 1483–15.

[88] Buccoliero, D., Desyatnikov, A. S., Krolikowski, W., and Kivshar, Yu. S. 2009. Boundary effects on the dynamics of higher-order optical spatial solitons in nonlocal thermal media. *J. Opt. A*, **11**(9), 094014.

[89] Flach, S. and Gorbach, A. V. 2008. Discrete breathers – advances in theory and applications. *Phys. Rep.*, **467**(1–3), 1–116.

[90] Buccoliero, D. 2009. *PhD thesis*. The Australian National University.

[91] Desyatnikov, A. S., Buccoliero, D., Dennis, M. R., and Kivshar, Yu. S. 2010. Suppression of collapse for spiraling elliptic solitons. *Phys. Rev. Lett.*, **104**(5), 053902.

[92] Chiao, R. Y., Garmire, E., and Townes, C. H. 1964. Self-trapping of optical beams. *Phys. Rev. Lett.*, **13**, 479–80.

[93] Moll, K. D., Gaeta, A. L., and Fibich, G. 2003. Self-similar optical wave collapse: observation of the townes profile. *Phys. Rev. Lett.*, **90**(20), 203902.

[94] Boyd, R. W., Lukishova, S. G., and Shen, Y. R. 2009. *Self-focusing, Past and Present: Fundamentals and Prospects*. Berlin: Springer.

[95] Fibich, G. and Ilan, B. 2003. Self-focusing of circularly polarized beams. *Phys. Rev. E*, **67**(3), 036622.

[96] Courtial, J., Dholakia, K., Allen, L., and Padgett, M. J. 1997. Gaussian beams with very high orbital angular momentum. *Opt. Commun.*, **144**, 210–13.

[97] Crosignani, B. and Porto, P. Di. 1993. Nonlinear propagation in Kerr media of beams with unequal transverse widths. *Opt. Lett.*, **18**(17), 1394–16.

[98] Goncharenko, A. M., Logvin, Yu. A., Samson, A. M., Shapovalov, P. S., and Turovets, S. I. 1991. Ermakov Hamiltonian systems in nonlinear optics of elliptic Gaussian beams. *Phys. Lett. A*, **160**(2), 138–42.

[99] Goncharenko, A. M, Logvin, Yu. A, Samson, A. M, and Shapovalov, P. S. 1991. Rotating elliptical gaussian beams in nonlinear media. *Opt. Commun.*, **81**(3–4), 225–30.

[100] Abdullaev, J., Desyatnikov, A. S., and Ostrovskaya, E. A. 2011. Suppression of collapse for matter waves with orbital angular momentum. *J. Opt.: Special issue on orbital angular momentum*, **13**(6), 064023.

5

Ray optics, wave optics and quantum mechanics

G. NIENHUIS

5.1 Introduction

A monochromatic light beam with frequency ω propagating in the z direction in free space with wave number $k = \omega/c$ can be described by a slowly varying amplitude $u(\rho, z)$, that is a solution of the paraxial wave equation [1]

$$\frac{\partial}{\partial z}u(\rho, z) = -\frac{1}{2ik}\frac{\partial^2}{\partial \rho^2}u(\rho, z). \tag{5.1}$$

Here the two-dimensional vector $\rho = (x, y)$ denotes the transverse coordinates, and z is the coordinate along the propagation direction. The paraxial wave equation follows from the Helmholtz wave equation $\nabla^2\mathbf{E} = -k^2\mathbf{E}$ for the monochromatic electric field

$$\mathbf{E}(\mathbf{r}, t) = \text{Re } E_0\epsilon u(\rho, z)\exp(ikz - i\omega t), \tag{5.2}$$

in the approximation that the variation of the envelope function $u(\rho, z)$ in the transverse direction is small over a wavelength. The uniform polarization is specified by the unitary vector ϵ in the ρ-plane. For convenience we assume that $u(\rho, z)$ is normalized in the sense that $\int d_2\rho|u(\rho, z)|^2 = 1$, so that $\epsilon_0 c|E_0|^2/2$ is the beam intensity (power per unit area). It is noteworthy that the form of the paraxial wave equation (5.1) is identical to that of the Schrödinger equation for a free particle in two-dimensional ρ-space, where the propagation coordinate z plays the role of time [2].

Paraxial beams of light can have an intensity pattern that is invariant under free propagation. Well-known examples are the Hermite–Gaussian modes. They resemble the factorized eigenstates of the two-dimensional quantum harmonic oscillator. We denote as $\psi_n(\xi)$ the real eigenfunction of the dimensionless harmonic-oscillator Hamiltonian

$$\hat{H}_\xi = \frac{1}{2}\left(\xi^2 - \frac{\partial^2}{\partial \xi^2}\right), \tag{5.3}$$

The Angular Momentum of Light, ed. D. L. Andrews and M. Babiker. Published by Cambridge University Press.

with eigenvalue $n + 1/2$. Then the Hermite–Gaussian modes can be expressed as [3, 4]

$$u_{nm}^{\mathrm{HG}}(\rho, z) = \frac{1}{\gamma}\psi_n(\xi)\psi_m(\eta)\exp\left(-\mathrm{i}\chi(n+m+1)\right)\exp\left(\frac{\mathrm{i}k}{2R}\rho^2\right). \tag{5.4}$$

The dimensionless transverse coordinates ξ and η are defined as $\xi = x/\gamma$, $\eta = y/\gamma$. The three z-dependent parameters γ, R and χ are defined by the two relations

$$\frac{1}{\gamma^2} - \frac{\mathrm{i}k}{R} = \frac{k}{b+\mathrm{i}z}, \quad \tan\chi = z/b, \tag{5.5}$$

where b has the significance of the diffraction length or Rayleigh range of the beam. Notice that these parameters vary with z over a distance of the order of the diffraction length. Equation (5.4) shows that $\gamma(z)$ is the width of the beam. The smallest width $\gamma_0 \equiv \gamma(0) = \sqrt{b/k}$ is the beam waist, which occurs at the focus of the beam at $z = 0$. For each value of z along the beam, the planes of constant phase are defined by the requirement that $z + \rho^2/2R$ is constant, so that $R(z)$ is the curvature radius of the wave fronts. At focus this radius is infinite, and the wavefronts are flat. When z is much larger than the diffraction length, we find that $R \approx z$. The phase $\chi(z)$ is the Gouy phase, which changes by an amount π over a few diffraction lengths across focus. The Hermite–Gaussian modes (5.3) are exact solutions of the paraxial wave equation (5.1). A well-known characteristic is that their transverse mode pattern is invariant as a function of z. As a function of z, its scale is measured by the width parameter $\gamma(z)$.

Less well known is that the analogy between paraxial beams of light and wave functions of the two-dimensional harmonic oscillator holds generally, and not just for stationary states and shape-invariant beams. The Schrödinger equation for the harmonic-oscillator wave functions is

$$\frac{\partial}{\partial\chi}\Psi(\sigma, \chi) = -\mathrm{i}\hat{H}\Psi(\sigma, \chi), \tag{5.6}$$

where χ can be viewed as dimensionless time, and we combined the dimensionless coordinate into the two-dimensional vector $\sigma = (\xi, \eta)$. The Hamiltonian is $\hat{H} = \hat{H}_\xi + \hat{H}_\eta$. The energy eigenvalue of the factorized state is $n + m + 1$. An arbitrary solution $u(\rho, z)$ of the paraxial wave equation can be written as a linear combination of the Hermite–Gaussian modes. After taking this linear combination of Eq. (5.4), one identifies in the right-hand side a general solution of the Schrödinger equation (5.6). This produces the correspondence relation [5]

$$u(\rho, z) = \frac{1}{\gamma}\Psi(\sigma, \chi)\exp\left(\frac{\mathrm{i}k}{2R}\rho^2\right) \tag{5.7}$$

between an arbitrary solution $u(\rho, z)$ of the paraxial wave equation and an arbitrary solution $\Psi(\sigma, \chi)$ of the Schrödinger equation for the harmonic oscillator. Half a cycle of the wave function $\Psi(\sigma, \chi)$ of the oscillator, in the interval $-\pi/2 < \chi < \pi/2$, corresponds to the infinite range of z, and yields the complete transverse mode pattern u. The factor $1/\gamma$ in (5.7) ensures that the mode function u is normalized in ρ, when the wave function Ψ is normalized in σ.

These analogies are just special cases of general properties of Hamiltonian systems with a Hamiltonian that is quadratic in generalized coordinates and momenta. Classical realizations of such systems are governed by the Hamilton equations. For quadratic Hamiltonians, these take the form of linear differential equations for coordinates and momenta, so that the classical dynamics can be described in terms of matrices. We shall demonstrate that basis sets of quantum-mechanical solutions are generated by bosonic ladder operators, with the simple commutator algebra that is best-known in the case of the quantum harmonic oscillator. A paraxial beam of light, or, equivalently, a free quantum particle, is another example. In the paraxial wave equation, the quantum nature is replaced by the wave nature, while the role of coordinates and momenta is taken over by the transverse position coordinates and the direction angles. These variables define a ray of light, and they vary in a linear fashion during propagation through an optical system, as described by ray matrices [3].

5.2 Evolution with a quadratic Hamiltonian

In this section, we consider dynamical systems with a quadratic Hamiltonian. This includes a number of standard quantum systems, as well as systems in classical wave optics. We demonstrate that basis sets of generalized Gaussian solutions are generated by bosonic ladder operators. These operators in turn are specified by trajectories in classical phase space.

5.2.1 Quantum and classical evolution

We point out that for systems with a quadratic Hamiltonian the classical and the quantum evolution are described by the same linear equations. We consider a dynamical system with D degrees of freedom. In the quantum-mechanical case, the position operators are denoted in a dimensionless form as operators $\hat{\sigma}_i$, with $i = 1, 2, \ldots, D$. For ease of notation, we combine these position operators into the D-dimensional column vector operator $\hat{\sigma}$. The corresponding momentum operators $\hat{\pi}_i$ take the form $-i\partial/\partial\sigma_i$ in the coordinate representation, and they compose the D-dimensional column vector operator $\hat{\pi}$ of momentum. It is convenient to combine these two vectors into the $2D$-dimensional vector operator $\hat{\mathfrak{v}}$, so that

$$\hat{\mathfrak{v}} = \begin{pmatrix} \hat{\sigma} \\ \hat{\pi} \end{pmatrix}, \tag{5.8}$$

which is the combination of the two D-dimensional vector operators for position and momentum. The corresponding classical vector ν with the $2D$ components $\{\sigma_i\}$ and $\{\pi_i\}$ is a point in classical phase space of the system. A Hamiltonian that is has quadratic form in both coordinates and momenta can be written as

$$\hat{H} = \frac{1}{2}\hat{\mathfrak{v}}\mathbb{K}\hat{\mathfrak{v}} \equiv \frac{1}{2}\left(\hat{\sigma}\mathsf{K}_{\sigma\sigma}\hat{\sigma} + \hat{\sigma}\mathsf{K}_{\sigma\pi}\hat{\pi} + \hat{\pi}\mathsf{K}_{\pi\sigma}\hat{\sigma} + \hat{\pi}\mathsf{K}_{\pi\pi}\hat{\pi}\right), \tag{5.9}$$

where $\mathbb{K}$ is a $2D$-dimensional matrix that can be separated into four D-dimensional submatrices. Since the Hamiltonian is Hermitian, the matrix $\mathbb{K}$ can be assumed to be real and symmetric. This implies that the matrices $\mathsf{K}_{\sigma\sigma}$ and $\mathsf{K}_{\pi\pi}$ are symmetric, while the matrices $\mathsf{K}_{\sigma\pi}$ and $\mathsf{K}_{\pi\sigma}$ are each other's transpose. We denote dimensionless time as χ. We allow the Hamiltonian to vary with time, so that the matrix $\mathbb{K}$ may depend on χ. The Schrödinger equation for the wave function $\Psi(\sigma, \chi)$ in coordinate representation is

$$\frac{\partial}{\partial \chi}\Psi(\sigma, \chi) = -\mathrm{i}\hat{H}(\chi)\Psi(\sigma, \chi). \tag{5.10}$$

The evolution operator that transforms an arbitrary initial wave function $\Psi(\sigma, 0)$ to the wave function at a later time χ is denoted by $\hat{U}(\chi)$, so that

$$\Psi(\sigma, \chi) = \hat{U}(\chi)\Psi(\sigma, 0). \tag{5.11}$$

A characteristic feature of quadratic Hamiltonians is that the classical evolution equations for position and momentum are linear. This is obvious from the classical Hamilton equations $\partial\sigma_i/\partial\chi = \partial H/\partial\pi_i$, $\partial\pi_i/\partial\chi = -\partial H/\partial\sigma_i$, which for the present quadratic Hamiltonian can be summarized as

$$\frac{\partial}{\partial \chi}\nu = \begin{pmatrix} \mathsf{K}_{\pi\sigma} & \mathsf{K}_{\pi\pi} \\ -\mathsf{K}_{\sigma\sigma} & -\mathsf{K}_{\sigma\pi} \end{pmatrix}\nu = \mathbb{G}\mathbb{K}\nu, \tag{5.12}$$

where $\mathbb{G}$ is the real antisymmetric matrix in $2D$ dimensions

$$\mathbb{G} = \begin{pmatrix} \mathsf{0} & \mathsf{1} \\ -\mathsf{1} & \mathsf{0} \end{pmatrix}, \tag{5.13}$$

with $\mathsf{1}$ and $\mathsf{0}$ the $D \times D$ unit and zero matrices. The Hamilton equations (5.12) are identical in form to the equations for the quantum operators in the Heisenberg picture, in accordance with the Ehrenfest theorem. Their explicit form is

$$\mathrm{i}[\hat{H}(\chi), \hat{\nu}] = \mathbb{G}\mathbb{K}(\chi)\hat{\nu}. \tag{5.14}$$

As a result, the evolution of an initial point $\nu(0)$ in classical phase space to the phase point $\nu(\chi)$ at a later time is also described by the same linear transformation as the Heisenberg evolution of the operator $\hat{\nu}$ from time 0 to time χ. This transformation is also described by a $2D$-dimensional matrix, so that $\nu(\chi) = \mathbb{N}(\chi)\nu(0)$, and likewise

$$\hat{U}^{\dagger}(\chi)\hat{\nu}\hat{U}(\chi) = \mathbb{N}(\chi)\hat{\nu}, \tag{5.15}$$

with $\mathrm{d}\mathbb{N}/\mathrm{d}\chi = \mathbb{G}\mathbb{K}\mathbb{N}$. The matrix $\mathbb{N}$ is real at all times. Moreover, the operators $\hat{\sigma}$ and $\hat{\pi}$ obey the canonical commutation rules $[\hat{\pi}_i, \hat{\sigma}_j] = -\mathrm{i}\delta_{ij}$, which must be preserved under the unitary transformation (5.15). This means that the matrix $\mathbb{N}(\chi)$ obeys the general condition

$$\mathbb{N}^{\mathrm{T}}(\chi)\mathbb{G}\mathbb{N}(\chi) = \mathbb{G}, \tag{5.16}$$

where $\mathbb{N}^{\mathrm{T}}$ is the transpose of the (real) matrix $\mathbb{N}$. It is easily checked that the product of two matrices $\mathbb{N}$ with this property (5.16) again has this property. Therefore, the property defines a group of real matrices in $2D$ dimensions, which is termed the symplectic group $Sp(2D, \mathbb{R})$. In the special case that the Hamiltonian does not vary with χ, the classical evolution matrix has the exponential form

$$\mathbb{N}(\chi) = \exp(\mathbb{G}\mathbb{K}\chi). \tag{5.17}$$

5.2.2 Ladder operators generating basis sets of Gaussian solutions

Bosonic ladder operators originally arose in the algebraic description of the stationary states of the quantum harmonic oscillator. This is a special case of a quadratic Hamiltonian. These operators are linear combinations of a position and a momentum operator, so that their commutation properties follow from the canonical commutation rules. We wish to generalize these operators for an arbitrary quadratic Hamiltonian, and also allow them to generate basis sets of non-stationary states. Thus, we look for lowering operators $\hat{b}_p(\chi)$ that obey the standard bosonic commutation rules

$$[\hat{b}_{p'}(\chi), \hat{b}_p(\chi)] = 0, \;\; [\hat{b}_{p'}(\chi), \hat{b}_p^\dagger(\chi)] = \delta_{p'p}, \tag{5.18}$$

and that define basis sets of solutions of the Schrödinger equation for the system with a quadratic Hamiltonian. Just as for the quantum harmonic oscillator, these commutation rules guarantee the existence of eigenfunctions of the generalized number operators $\hat{b}_p^\dagger(\chi)\hat{b}_p(\chi)$, with the natural numbers $n_p = 0, 1, \ldots$ as eigenvalues. As we shall show, D lowering operators are necessary and sufficient to determine the basis of solutions completely. Moreover, each of these lowering operators is specified by a trajectory $\nu_p(\chi)$ in classical phase space.

We require that for any solution $\Psi(\sigma, \chi)$ of the Schrödinger equation (5.10), the wave function $\hat{b}_p(\chi)\Psi(\sigma, \chi)$ is again a solution. Equation (5.11) then demonstrates that the χ-dependence of $\hat{b}_p$ is determined by the relation

$$\hat{b}_p(\chi) = \hat{U}(\chi)\hat{b}_p(0)\hat{U}^\dagger(\chi). \tag{5.19}$$

With this equation, it is sufficient to define the lowering operators for the initial time $\chi = 0$. In view of the linear dependence on position and momentum, it is always possible to denote the operators as

$$\hat{b}_p(0) = \frac{\mathrm{i}}{\sqrt{2}}\left(\sigma_p(0)\hat{\pi} - \pi_p(0)\hat{\sigma}\right) = \frac{\mathrm{i}}{\sqrt{2}}\,\nu_p(0)\mathbb{G}\hat{\nu}, \tag{5.20}$$

where σ_p and π_p are subvectors of ν_p, as expressed by

$$\nu_p(\chi) = \begin{pmatrix} \sigma_p(\chi) \\ \pi_p(\chi) \end{pmatrix}, \tag{5.21}$$

for all values of χ. The χ-dependence of $\hat{b}_p$ as defined by Eq. (5.19) now leads to the expression

$$\hat{b}_p(\chi) = \frac{\mathrm{i}}{\sqrt{2}}\, \nu_p(\chi)\mathbb{G}\hat{\mathfrak{v}} \;, \quad \text{with} \quad \nu_p(\chi) = \mathbb{N}(\chi)\nu_p(0), \tag{5.22}$$

where we also used the χ-dependence (5.15) of the operator $\hat{\mathfrak{v}}$, and the symplectic relation (5.16). This demonstrates that the position and momentum components of the trajectory $\nu_p(\chi)$ in phase space obey the Hamilton equations of motion.

In order that the commutation rules for the ladder operators (5.22) have the bosonic form (5.18), the trajectories $\nu_p(\chi)$ must obey the conditions

$$\nu_p(\chi)\mathbb{G}\nu_{p'}(\chi) = 0, \;\; \nu_p^*(\chi)\mathbb{G}\nu_{p'}(\chi) = 2i\delta_{p'p}, \tag{5.23}$$

for all values of χ. The first relation (5.23) can be viewed as a symplectic version of orthogonality, while the second relation determines the normalization. Obviously, the trajectories $\nu_p(\chi)$ must be complex. For a system with D degrees of freedom, D vectors ν_p obeying these conditions together with their complex conjugates form a complete set of vectors in classical phase space.

For later use it is convenient to summarize these conditions (5.23) in matrix form. We introduce two $D \times D$ matrices $\mathsf{S}(\chi)$ and $\mathsf{P}(\chi)$, with columns that are equal to the coordinate vectors σ_p and the momentum vectors π_p respectively, so that

$$\mathsf{S}(\chi) = \begin{pmatrix}\sigma_1(\chi) & \sigma_2(\chi) & \ldots & \sigma_D(\chi)\end{pmatrix}, \; \mathsf{P}(\chi) = \begin{pmatrix}\pi_1(\chi) & \pi_2(\chi) & \ldots & \pi_D(\chi)\end{pmatrix}, \tag{5.24}$$

where $\sigma_1, \sigma_2, \ldots$, and $\pi_1, \pi_2, \ldots$, should be read as D-dimensional column vectors. These two matrices $\mathsf{S}(\chi)$ and $\mathsf{P}(\chi)$ describe the D vectors ν_p in phase space. The relations (5.23) can be rewritten in terms of the matrix relations

$$\mathsf{S}^{\mathsf{T}}\mathsf{P} - \mathsf{P}^{\mathsf{T}}\mathsf{S} = 0, \; \mathsf{S}^{\dagger}\mathsf{P} - \mathsf{P}^{\dagger}\mathsf{S} = 2i\mathsf{1}, \tag{5.25}$$

for all χ. The Hamilton equations of motion (5.12) for the vectors ν_p give rise to corresponding equations for the matrices $\mathsf{S}(\chi)$ and $\mathsf{P}(\chi)$, which read

$$\frac{\mathrm{d}}{\mathrm{d}\chi}\mathsf{S} = \mathsf{K}_{\pi\sigma}\mathsf{S} + \mathsf{K}_{\pi\pi}\mathsf{P}, \; \frac{\mathrm{d}}{\mathrm{d}\chi}\mathsf{P} = -\mathsf{K}_{\sigma\sigma}\mathsf{S} - \mathsf{K}_{\sigma\pi}\mathsf{P}. \tag{5.26}$$

The formal solution of these differential equations equations is

$$\mathsf{S}(\chi) = \mathsf{N}_{\pi\sigma}(\chi)\mathsf{S}(0) + \mathsf{N}_{\pi\pi}(\chi)\mathsf{P}(0), \; \mathsf{P}(\chi) = \mathsf{N}_{\sigma\sigma}(\chi)\mathsf{S}(0) + \mathsf{N}_{\sigma\pi}(\chi)\mathsf{P}(0), \tag{5.27}$$

in terms of the submatrices of $\mathbb{N}(\chi)$.

5.2.3 Fundamental solution

A basis set of solutions of the Schrödinger equation corresponding to a set of ladder operators is generated by repeated action of the raising operators $\hat{b}_p^{\dagger}(\chi)$ on a fundamental solution $\Psi_{\{0\}}(\sigma, \chi)$. The symbol $\{0\}$ denotes the special case of the set of D natural numbers $\{n_p\}$ that they are all equal to zero. A lowering operator $\hat{b}_p(\chi)$ must give zero when acting on this fundamental solution. This means that for any value of χ, $\Psi_{\{0\}}$ must be proportional to a general Gaussian of the form $\exp(-\sigma\mathsf{A}\sigma/2)$, with A a symmetric $D \times D$ matrix. The condition that all lowering operators $\hat{b}_p$ give zero when acting on this Gaussian can be summarized in the identity $\mathsf{AS} + \mathrm{i}\mathsf{P} = 0$, which implies that

$$\mathsf{A} = -\mathrm{i}\mathsf{PS}^{-1}. \tag{5.28}$$

This expression indeed gives a symmetric matrix, since according to the first equality (5.25) the matrix $\mathsf{S}^{\mathrm{T}}\mathsf{AS} = -\mathrm{i}\mathsf{S}^{\mathrm{T}}\mathsf{P}$ is symmetric. The normalization factor of the state $\Psi_{\{0\}}$ depends only on the real part A_{R} of A, which is given by the expression

$$2\mathsf{A}_{\mathrm{R}} = -\mathrm{i}\mathsf{PS}^{-1} + \mathrm{i}(\mathsf{S}^{\dagger})^{-1}\mathsf{P}^{\dagger}. \tag{5.29}$$

Combined with the second condition (5.25), this demonstrates that $\mathsf{S}^{\dagger}\mathsf{A}_{\mathrm{R}}\mathsf{S} = \mathsf{1}$, which is identical to the equality

$$\mathsf{A}_{\mathrm{R}}^{-1} = \mathsf{SS}^{\dagger}. \tag{5.30}$$

This ensures that the matrix A_{R} is positive definite. Normalization then defines the fundamental solution $\Psi_{\{0\}}$, apart from a phase factor. It requires an independent check to verify that the normalized expression

$$\Psi_{\{0\}}(\sigma, \chi) = \frac{1}{\sqrt{\pi^{D/2} \det \mathsf{S}}} \exp\left(-\sigma\mathsf{A}\sigma/2\right) \tag{5.31}$$

indeed has the correct χ-dependent phase, in that it solves the Schrödinger equation (5.10). The left-hand side of (5.10) can be evaluated by using

$$\frac{\mathrm{d}}{\mathrm{d}\chi}\Psi_{\{0\}} = -\frac{1}{2}\left(\frac{1}{\det \mathsf{S}}\frac{\mathrm{d}}{\mathrm{d}\chi}\det \mathsf{S} + \sigma\frac{\mathrm{d}\mathsf{A}}{\mathrm{d}\chi}\sigma\right)\Psi_{\{0\}}. \tag{5.32}$$

The derivative of the determinant can be rewritten by using the linearized expression $\mathsf{S}(\chi + \mathrm{d}\chi) \approx [\mathsf{1} + \mathrm{d}\chi(\mathrm{d}\mathsf{S}/\mathrm{d}\chi)\mathsf{S}^{-1}(\chi)](\mathsf{S}(\chi))$. Taking the determinant and using (5.26) and (5.28) gives the identity

$$\frac{1}{\det \mathsf{S}}\frac{\mathrm{d}}{\mathrm{d}\chi}\det \mathsf{S} = \mathrm{Tr}\left(\frac{\mathrm{d}\mathsf{S}}{\mathrm{d}\chi}\mathsf{S}^{-1}\right) = \mathrm{Tr}\left(\mathsf{K}_{\pi\sigma} + \mathrm{i}\mathsf{K}_{\pi\pi}\mathsf{A}\right). \tag{5.33}$$

The last term in (5.32) can be evaluated using (5.28) and (5.26) again, which gives the obviously symmetric result

$$\frac{\mathrm{d}\mathsf{A}}{\mathrm{d}\chi} = \mathrm{i}\left(\mathsf{K}_{\sigma\sigma} + \mathrm{i}\mathsf{AK}_{\pi\sigma} + \mathrm{i}\mathsf{K}_{\sigma\pi}\mathsf{A} - \mathsf{AK}_{\pi\pi}\mathsf{A}\right). \tag{5.34}$$

Substitution of these results into Eq. (5.32) gives an explicit expression for the time derivative of the normalized wave function $\Psi_{\{0\}}(\sigma, \chi)$ as expressed by Eq. (5.31). It is then easy to verify that this wave function does indeed obey the Schrödinger equation (5.10), with the quadratic Hamiltonian (5.9).

5.2.4 Basis sets of the solutions of the Schrödinger equation

Since the χ-dependent ladder operators (5.22) obey the bosonic commutation rules, they generate an orthonormal basis set of solutions $\Psi_{\{n_p\}}(\sigma, \chi)$ of the Schrödinger equation (5.10), with $n_1, n_2, \ldots, n_D = 0, 1, 2, \ldots$. Just as for a multidimensional harmonic oscillator, these χ-dependent states arise from the fundamental solution by repeated application of the raising operators $\hat{b}_p^\dagger(\chi)$, with the standard result

$$\Psi_{\{n_p\}}(\sigma, \chi) = \frac{1}{\sqrt{n_1! n_2! \cdots n_D!}} \left(\hat{b}_1^\dagger(\chi)\right)^{n_1} \left(\hat{b}_2^\dagger(\chi)\right)^{n_2} \cdots \left(\hat{b}_D^\dagger(\chi)\right)^{n_D} \Psi_{\{0\}}(\sigma, \chi). \quad (5.35)$$

For each value of χ, these states are normalized eigenstates of the operators $\hat{b}_p^\dagger(\chi)\hat{b}_p(\chi)$ with eigenvalues n_p. The total number of raising operators $N = \sum_p n_p$ is the total order of a solution. The χ-dependence of these states arises from the χ-dependence of the phase points ν_p, which determine the raising operators $\hat{b}_p^\dagger(\chi)$, as well as the matrices $\mathsf{S}(\chi)$ and $\mathsf{P}(\chi)$.

5.2.5 Stationary states and eigenenergies

The states (5.35) are stationary only in special cases. First, the Hamiltonian must be constant, so that the classical evolution matrix $\mathbb{N}(\chi)$ has the exponential form (5.17). In that case, a basis of normalizable stationary states exists, provided that the matrix $\mathbb{N}(\chi)$ (or, equivalently, the product matrix $\mathbb{GK}$) has $2D$ complex eigenvectors in phase space. This can be demonstrated by using the symplectic relation (5.16). Since $\mathbb{N}$ is real, its eigenvectors and eigenvalues come in pairs which are each other's complex conjugate, so that we may write $\mathbb{N}\nu_p = \lambda_p \nu_p$, $\mathbb{N}\nu_p^* = \lambda_p^* \nu_p^*$. When we take matrix elements of the relation (5.16) between the eigenvectors, we obtain the result

$$\lambda_p^* \lambda_{p'} \nu_p^* \mathbb{G} \nu_p' = \nu_p^* \mathbb{G} \nu_p', \quad \lambda_p \lambda_{p'} \nu_p \mathbb{G} \nu_p' = \nu_p \mathbb{G} \nu_p'. \quad (5.36)$$

So when $\lambda_p^* \lambda_{p'}$ or $\lambda_p \lambda_{p'}$ is not equal to 1, the corresponding matrix element must vanish. When the matrix element $\nu_p^* \mathbb{G} \nu_p$ is non-zero, the eigenvalue λ_p is unitary, and we may write $\lambda_p = \exp(\mathrm{i}\gamma_p \chi)$. Moreover, this matrix element is imaginary, because of the antisymmetry of $\mathbb{G}$. Without loss of generality we can assume that the elements $\nu_p^* \mathbb{G} \nu_p$ are positive imaginary, when needed after interchanging ν_p and ν_p^*, and by proper normalization we can ensure that they are equal to 2i. In conclusion, when the matrix $\mathbb{N}$ has $2D$ complex eigenvectors ν_p, we can assume that these obey the conditions (5.22). Their χ-dependence is $\nu_p(\chi) = \nu_p(0) \exp(\mathrm{i}\gamma_p \chi)$, with $\mathrm{i}\gamma_p$ the imaginary eigenvalue of $\mathbb{GK}$.

When we choose the ladder operators as specified by these eigenvectors ν_p, the raising operators $\hat{b}_p^\dagger$ are proportional to $\exp(-\mathrm{i}\chi\gamma_p)$. Moreover, the matrix A is independent of

χ, while the χ-dependence of det S is $\exp(i\chi \sum_p \gamma_p)$. Therefore, both the fundamental solution (5.31), and the higher-order solutions (5.35) vary with χ as a single-frequency oscillation. This proves that in this special case, the basis set (5.35) is a basis of normalized stationary states $\Psi_{\{n_p\}}$ with energy $\sum_p \gamma_p(n_p + 1/2)$, and with the fundamental solution as ground state. As we shall see, however, even in cases where normalized stationary states do not exist, we can still obtain basis sets of solutions of the type (5.35).

5.2.6 Basis transformations

From one basis set, others can be obtained by linear transformations. A trivial realization of the identities (5.25) is given by the choice $\mathsf{S}(0) = \mathbb{1}$, and $\mathsf{P}(0) = i\mathbb{1}$, at the initial time $\chi = 0$. This gives a fundamental wave function at the initial time with the simple Gaussian form $\Psi_{\{0\}}(\sigma, 0) = \exp(-\sigma^2/2)/\pi^{D/4}$, and the lowering operators are just the components of $(\sigma + \partial/\partial\sigma)/\sqrt{2}$. The higher-order solutions at this initial time $\chi = 0$ are then products of energy eigenfunctions $\psi_{n_i}(\sigma_i)$ of the quantum harmonic oscillator, which are real wave functions. However, these states are not necessarily stationary. The basis set at later times is generated by the matrices $\mathsf{S}(\chi)$ and $\mathsf{P}(\chi)$, and their χ-dependence is determined by the Hamilton equations (5.26).

A simple transformation between basis sets arises when the matrices $\mathsf{S}(\chi)$ and $\mathsf{P}(\chi)$ are both multiplied from the right by the same unitary matrix. We define transformed matrices S' and P' by the requirement that $\mathsf{S}'(\chi) = \mathsf{S}(\chi)\mathsf{U}^\dagger$ and $\mathsf{P}'(\chi) = \mathsf{P}(\chi)\mathsf{U}^\dagger$. Here U is a unitary matrix in D dimensions, with matrix elements $U_{p'p}$. This transformed pair of matrices still solves the Hamilton equations (5.26), and it also obeys the conditions (5.25), so that they indeed generate a basis set of solutions. Since the matrix $\mathsf{A} = -i\mathsf{P}\mathsf{S}^{-1}$ is invariant under this transformation, these two basis sets have the same fundamental solution $\Psi_{\{0\}}$, apart from a constant overall phase factor $(\det \mathsf{U})^{1/2}$. Using the unitarity, one finds that the relation between the two sets of vectors in phase space is related by the transformation $\nu'_p = \sum_{p'} U^*_{pp'} \nu_{p'}$, and the same expansion holds for the lowering operators, according to Eq. (5.22). The two corresponding sets of D raising operators are related by the transformation

$$(\hat{b}'_p)^\dagger = \sum_{p'} U_{pp'} \hat{b}^\dagger_{p'}. \tag{5.37}$$

Hence the basis transformation arises from a simple mixing of the raising operators, acting on the same fundamental state. It is obvious that the solutions of a specific order N in one basis set are a linear combination of the solutions of the same order in the other basis set.

More interesting transformations are those which also change the fundamental solution. When the first set is generated by the D vectors $\nu_p(0)$ in the $2D$-dimensional phase space, we introduce a transformed set $\nu'_p(0) = \mathbb{N}_0 \nu_p(0)$ at one point in time, which we take as the initial time $\chi = 0$. Here $\mathbb{N}_0$ is an arbitrary element of the symplectic group $Sp(2D, \mathbb{R})$. Since $\mathbb{N}_0$ obeys the symplectic relation (5.16), the transformed vectors ν'_p still obey the relations (5.23), and therefore they also generate a basis set of solutions. The χ-dependence of the vectors is then determined by $\nu'_p(\chi) = \mathbb{N}(\chi)\nu'_p(0) = \mathbb{N}(\chi)\mathbb{N}_0\nu_p(0)$.

The general procedure for obtaining basis sets of solutions of the Schrödinger equation for a quantum system with a quadratic Hamiltonian should now be clear. First we select a set of D complex conjugate pairs of vectors $\nu_p(0)$ and $\nu_p^*(0)$ in classical phase space, which obey the conditions (5.23). This is equivalent to specifying the matrices $\mathsf{S}(0)$ and $\mathsf{P}(0)$ at the initial time, which obey the relations (5.25). Their variation with χ is then simply determined by the classical evolution matrix $\mathbb{N}(\chi)$. The vectors $\nu_p(\chi)$ define a set of D pairs of lowering and raising operators, in accordance with Eq. (5.22). The fundamental solution in the basis is directly expressed by (5.31) in terms of the matrices $\mathsf{S}(\chi)$ and $\mathsf{P}(\chi)$, and the higher-order solutions follow from these by application of the raising operators. The calculations mainly consist of linear algebra in D or $2D$ dimensions.

5.3 Quantum harmonic oscillator in two dimensions

In view of the analogy between freely propagating paraxial modes and the two-dimensional quantum harmonic oscillator, we summarize relevant properties of the oscillator eigenstates and of the general time-dependent solution. This is the simplest realization of a dynamical system with a quadratic Hamiltonian and two degrees of freedom.

The quantum harmonic oscillator is a prototype of an exactly solvable quantum system in terms of ladder operators. We consider the two-dimensional isotropic case, with the Hamiltonian expressed as

$$\hat{H} = \frac{1}{2}(\hat{\sigma}^2 + \hat{\pi}^2). \tag{5.38}$$

The components of the two-dimensional coordinate vector σ are now called $\sigma_1 = \xi$ and $\sigma_2 = \eta$, so that $\hat{\sigma} = (\hat{\xi}, \hat{\eta})$ is the corresponding vector operator. The operators $\hat{\pi}_\xi = -\mathrm{i}\partial/\partial\xi$ and $\hat{\pi}_\eta = -\mathrm{i}\partial/\partial\eta$, which are the components of the two-dimensional momentum operator $\hat{\pi}$ in coordinate representation. The Hamiltonian separates into two one-dimensional Hamiltonians as $\hat{H} = \hat{H}_\xi + \hat{H}_\eta$. In the language of Section 5.2, the number of degrees of freedom is $D = 2$, and the matrix $\mathbb{K}$ is equal to the unit matrix in $2D = 4$ dimensions, so that the 2×2 submatrices are $\mathsf{K}_{\sigma\sigma} = \mathsf{K}_{\pi\pi} = \mathsf{1}$ and $\mathsf{K}_{\sigma\pi} = \mathsf{K}_{\pi\sigma} = \mathsf{0}$.

As can be found in any textbook on elementary quantum mechanics, an attractive feature of the harmonic oscillator is that the eigenfunctions and eigenvalues follow directly from the commutation rules for ladder operators. In the general formalism of Section 5.2, these operators are obtained in the special case that the vectors ν_p defining them are eigenvectors of $\mathbb{N}$. They generate a basis set of stationary states. As explained above, this matrix $\mathbb{N}(\chi)$ defines the linear evolution both of a classical phase point ν and of the quantum operator $\hat{\nu}$. In the present case where $\mathbb{K}$ is the four-dimensional unit matrix, the transformation matrix is given by Eq. (5.17), which yields

$$\mathbb{N}(\chi) = \begin{pmatrix} \mathsf{1}\cos\chi & \mathsf{1}\sin\chi \\ -\mathsf{1}\sin\chi & \mathsf{1}\cos\chi \end{pmatrix}, \tag{5.39}$$

with $\mathsf{1}$ the two-dimensional unit matrix. We see that only a single frequency occurs in the time behavior of $\mathbb{N}$. As denoted in Eq. (5.24), the two vectors ν_p at time $\chi = 0$ are specified by the matrices S and P at $\chi = 0$. Explicitly, the χ-dependence of these matrices is given by

$$\mathsf{S}(\chi) = \mathsf{S}(0)\cos\chi + \mathsf{P}(0)\sin\chi\,,\ \mathsf{P}(\chi) = -\mathsf{S}(0)\sin\chi + \mathsf{P}(0)\cos\chi\,. \tag{5.40}$$

5.3.1 Hermite–Gaussian eigenstates

First we choose $\mathsf{S}(0) = \mathsf{S}^{\mathrm{HG}}(0) = \mathsf{1}$, $\mathsf{P}(0) = \mathsf{P}^{\mathrm{HG}}(0) = \mathrm{i}\mathsf{1}$. They trivially obey the conditions (5.25). The index HG refers to the Hermite–Gaussian eigenstates that they produce. Their χ-dependence takes the simple form $\mathsf{S}(\chi) = \mathsf{1}\exp(\mathrm{i}\chi)$, $\mathsf{P}(\chi) = \mathrm{i}\mathsf{1}\exp(i\chi)$. As indicated in the general case in Section 5.2.3, the wave function of the ground state is determined by these two matrices. We find from Eq. (5.28) that $\mathsf{A} = \mathsf{1}$ is independent of χ, while $\det\mathsf{S} = \exp(2\mathrm{i}\chi)$, so that

$$\Psi_{00}(\sigma,\chi) = \frac{1}{\sqrt{\pi}}\exp(-\sigma^2/2)\exp(-\mathrm{i}\chi). \tag{5.41}$$

The two phase-space vectors ν_p refer to the two spatial dimensions ξ and η, which are now indicated as ν_ξ and ν_η. They are both eigenvectors of $\mathbb{N}(\chi)$ with the same eigenvalue $\exp(\mathrm{i}\chi)$. As discussed in Section 5.2.5, these vectors produce a basis of stationary states. Substitution in Eq. (5.22) shows that the two lowering operators are

$$\hat{b}_\xi(\chi) = \frac{1}{\sqrt{2}}\left(\hat{\xi} + \mathrm{i}\hat{\pi}_\xi\right)e^{\mathrm{i}\chi},\ \hat{b}^\dagger_\eta(\chi) = \frac{1}{\sqrt{2}}\left(\hat{\eta} + \mathrm{i}\hat{\pi}_\eta\right)\mathrm{e}^{\mathrm{i}\chi}. \tag{5.42}$$

These expressions are identical to the lowering operators for the oscillator in the ξ and η directions. The excited states are obtained by repeated application of the raising operators, as indicated in Eq. (5.35), with the result

$$\begin{aligned}\Psi^{\mathrm{HG}}_{nm}(\sigma,\chi) &= \frac{1}{\sqrt{n!m!}}\left(\hat{b}^\dagger_\xi(\chi)\right)^n\left(\hat{b}^\dagger_\eta(\chi)\right)^n\Psi_{00}(\sigma,\chi)\\ &= \psi_n(\xi)\psi_m(\eta)\exp\left(-\mathrm{i}\chi(n+m+1)\right),\end{aligned} \tag{5.43}$$

Where the functions ψ_n are energy eigenfunctions of the one-dimensional harmonic oscillator

$$\psi_n(\xi) = \frac{1}{(2^n n!\sqrt{\pi})^{1/2}}\mathrm{e}^{-\xi^2/2}H_n(\xi). \tag{5.44}$$

The solutions (5.43) are stationary states, with the energy eigenvalue $n + m + 1$. They have the form of Hermite–Gaussian functions, which are products of functions of the Cartesian coordinates ξ and η. The strength pattern of the lowest Hermite–Gaussian states

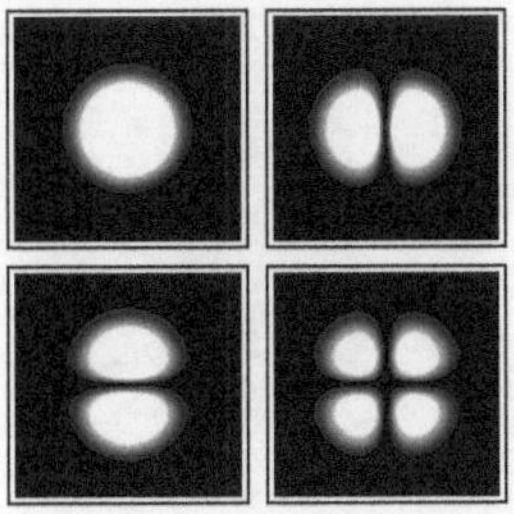

Figure 5.1 Profiles of the Hermite–Gaussian eigenstates, with the quantum numbers $(0, 0)$, $(1, 0)$, $(0, 1)$, $(1, 1)$. The same profiles also apply to the Hermite–Gaussian modes described in Section 5.5.3.

is displayed in Fig. 5.1. These patterns are identical to the intensity pattern of Hermite–Gaussian light beams.

5.3.2 Laguerre–Gaussian eigenstates

Since the eigenstates $\psi_n(\xi)\psi_m(\eta)$ are degenerate, other choices of a basis set of stationary are possible. As described at the end of Section 5.2.4, a natural way to generate alternative basis sets is to select pairs of raising operators that are unitary linear combinations of $\hat{b}_\xi^\dagger$ and $\hat{b}_\eta^\dagger$. As a first example we introduce the circular raising operators

$$\hat{b}_\pm^\dagger = \frac{1}{\sqrt{2}}\left(\hat{b}_\xi^\dagger \pm \mathrm{i}\hat{b}_\eta^\dagger\right). \tag{5.45}$$

This transformation is identical in form to the relation between circular and linear polarization vectors. As a function of χ, both raising operators are proportional to $\exp(-\mathrm{i}\chi)$. The higher-order states are

$$\Psi_{n_+n_-}^{\mathrm{LG}}(\sigma,\chi) = \frac{1}{\sqrt{n_+!n_-!}}\left(\hat{b}_+^\dagger(\chi)\right)^{n_+}\left(\hat{b}_-^\dagger(\chi)\right)^{n_-}\Psi_{00}(\sigma,\chi), \tag{5.46}$$

where the ground state Ψ_{00} is the same as for the Hermite–Gaussian states. The energy eigenvalue of the solution $\Psi_{n_+n_-}^{\mathrm{LG}}(\sigma,\chi)$ is $n_+ + n_- + 1$. For a fixed value of the excitation number $N = n_+ + n_-$, these states are linear combinations of the factorized states $\Psi_{nm}^{\mathrm{HG}}(\sigma)$ of the same order $n + m = N$. These Laguerre–Gaussian states factorize into polar coordinates rather than into Cartesian coordinates. The dependence on the azimuthal angle ϕ is proportional to $\exp(\mathrm{i}(n_+ - n_-)\phi$, while the functional dependence on the radial coordinate contains a Laguerre polynomial. These states are more familiar in paraxial wave optics than in quantum mechanics [3]. Their transverse strength patterns are illustrated in Fig. 5.2. They have a vortex in the center. The transformation between the Hermite–Gaussian and Laguerre–Gaussian states is directly found by substitution of the expressions (5.45) for the circular raising operators in the linear ones.

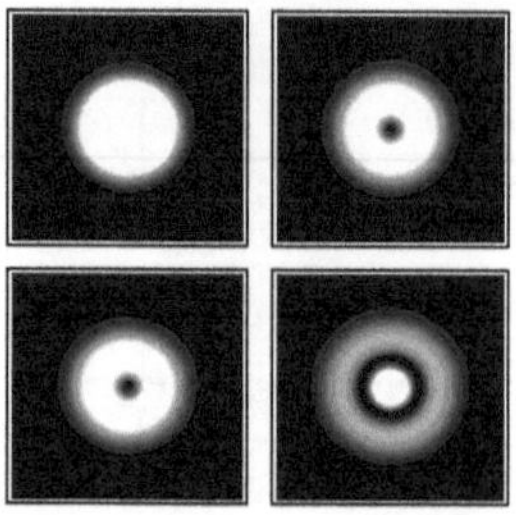

Figure 5.2 Profiles of the Laguerre–Gaussian eigenstates, with the quantum numbers $(0, 0)$, $(1, 0)$, $(0, 1)$, $(1, 1)$. The same profiles also apply to the Laguerre–Gaussian modes described in Section 5.5.3.

5.3.3 General Hermite–Laguerre states

In optics, it is customary to specify a pair of orthogonal polarization vectors as a unitary transformation of the two opposite circular polarization vectors. As any two-state system, this is mathematically identical to a spin $1/2$. An arbitrary spin state can be represented by the direction of the expectation value of the spin vector, which defines a point on a sphere. Each point on this Bloch sphere represents a linear combination of the states with spin up and spin down. In the same way, an arbitrary polarization vector can always be expressed as a linear combination of left- and right-hand circular polarization, which is represented by a point on the Poincaré sphere.

The same representation on a sphere applies to a pair of raising operators as a unitary transformation of $\hat{b}_+^\dagger$ and $\hat{b}_-^\dagger$, in the form

$$\hat{b}_1^\dagger(\chi|\theta,\phi) = \hat{b}_+^\dagger(\chi)\cos\frac{\theta}{2} + \hat{b}_-^\dagger(\chi)\mathrm{e}^{\mathrm{i}\phi}\sin\frac{\theta}{2},$$
$$\hat{b}_2^\dagger(\chi|\theta,\phi) = -\hat{b}_+^\dagger(\chi)\mathrm{e}^{-\mathrm{i}\phi}\sin\frac{\theta}{2} + \hat{b}_-^\dagger(\chi)\cos\frac{\theta}{2}. \tag{5.47}$$

In the present case of a harmonic oscillator, each point on the sphere specifies a pair of raising operators, and thereby a basis set of stationary states. We call this the Hermite–Laguerre sphere. In spin language, the transformation (5.47) corresponds to the two-dimensional unitary transformation $\mathsf{U} = \exp(-\mathrm{i}\phi S_3)\exp(-\mathrm{i}\theta S_2)\exp(\mathrm{i}\phi S_3)$ in $SU(2)$, which rotates the states up and down by a spin $1/2$ into the spin states along the axis specified by the polar angle θ and the azimuthal angle ϕ. Acting on the Hilbert space of wave functions of the harmonic oscillator, these spin matrices S_i correspond to the operators

$$\hat{S}_1 = (\hat{b}_-^\dagger\hat{b}_+ + \hat{b}_+^\dagger\hat{b}_-)/2,\ \hat{S}_2 = \mathrm{i}(\hat{b}_-^\dagger\hat{b}_+ - \hat{b}_+^\dagger\hat{b}_-)/2,\ \hat{S}_3 = (\hat{b}_+^\dagger\hat{b}_+ - \hat{b}_-^\dagger\hat{b}_-)/2, \tag{5.48}$$

which obey the $SU(2)$ commutation rules, just as the three components of angular momentum. These operators (5.48) generate the transformation (5.47), and they have the form of the Schwinger representation of a pair of boson modes [6].

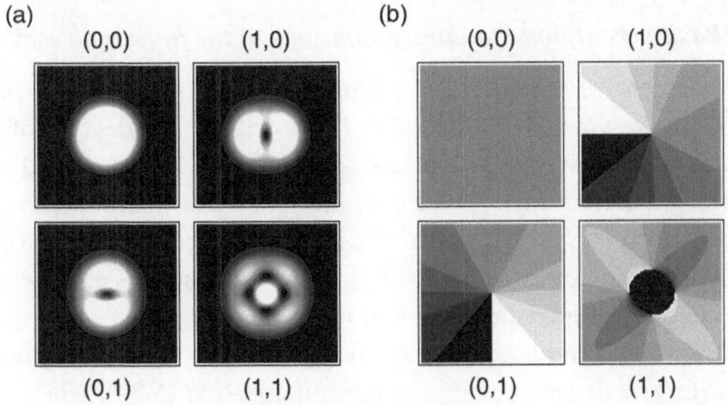

Figure 5.3 Intensity profiles (a) and phase profiles (b) of Hermite–Laguerre states, represented by a point on the sphere with $\theta = \pi/6$, $\phi = 0$. The quantum numbers (n_1, n_2) are $(0, 0)$, $(1, 0)$, $(0, 1)$ and $(1, 1)$. The same profiles also apply to the Hermite–Laguerre modes described in Section 5.5.3.

Acting on the ground state $\Psi_{00}(\sigma, \chi)$, for each value of ϕ and θ these raising operators (5.47) generate an orthogonal basis of states

$$\Psi^{\mathrm{HL}}_{n_1 n_2}(\sigma, \chi) = \frac{1}{\sqrt{n_1! n_2!}} \left(\hat{b}_1^\dagger(\chi|\theta, \phi)\right)^{n_1} \left(\hat{b}_2^\dagger(\chi|\theta, \phi)\right)^{n_2} \Psi_{00}(\sigma, \chi). \tag{5.49}$$

These states are eigenstates of $\hat{b}_1^\dagger \hat{b}_1$ with eigenvalue n_1, and of $\hat{b}_2^\dagger \hat{b}_2$ with eigenvalue n_2, so that they are stationary states with energy eigenvalue is $n_1 + n_2 + 1$. Each point on the Hermite–Laguerre sphere represents a basis set of stationary states. The north and the south poles of the sphere (where $\theta = 0$ or π) correspond to the basis of Laguerre–Gaussian states (5.46), while points on the equator ($\theta = \pi/2$) reproduce the Hermite–Gaussian basis set Ψ_{nm} rotated over an angle $\phi/2$. For each point on the sphere, labeled by the angles θ and ϕ, there are $N + 1$ states for which $n_1 + n_2 = N$. These $N + 1$ states span the subspace of Hermite–Gaussian states Ψ_{nm} with $n + m = N$, which is also the subspace spanned by the Laguerre–Gaussian states (5.46).

Equation (5.47) represents the unitary transformations of the raising operators that connect two basis sets of states with the same fundamental state, and that mix the higher-order solutions of the same order N. This set of transformations was described at the end of Section 5.2.4. In the present case, one basis of stationary states $\Psi^{\mathrm{LG}}_{n_+ n_-}$ is transformed into another basis of stationary states $\Psi^{\mathrm{HL}}_{n_1 n_2}$. The matrices S and P are specified by $\mathsf{S}(0) = \mathsf{U}^\dagger$, $\mathsf{P}(0) = \mathrm{i}\mathsf{U}^\dagger$. As a function of χ, they are both proportional to $\exp(\mathrm{i}\chi)$. An example of Hermite–Laguerre states is illustrated in Fig. 5.3. Both the intensity pattern and the phase pattern are much richer than for the Hermite–Gaussian and Laguerre–Gaussian states, and more vortices appear.

5.3.4 Basis sets of non-stationary solutions of the harmonic oscillator

Now we apply the general scheme of Section 5.2.2 to the present case of the $2D$ harmonic oscillator. So we take the Hamiltonian (5.9), with $D = 2$ degrees of freedom, and the matrix $\mathbb{K}$ equal to the unit matrix. The basis set is defined by choosing the two matrices $\mathsf{S}(0)$ and $\mathsf{P}(0)$ for the initial time $\chi = 0$, with the properties (5.25). This is equivalent to two four-dimensional vectors $\nu_1(0)$ and $\nu_2(0)$ in phase space, with the properties (5.23). The variation of the matrices S and P with χ is now given by Eq. (5.40). The general expression (5.31) gives the fundamental solution $\Psi_{00}^{\mathrm{NS}}(\sigma, \chi)$, which is a generalized Gaussian function, determined by the matrix A. In general, this is not the ground state. The variation of A with χ is determined by the general equation (5.34), which now takes the simple form

$$\frac{\mathrm{d}\mathsf{A}}{\mathrm{d}\chi} = \mathrm{i}(1 - \mathsf{A}^2). \tag{5.50}$$

By using Eq. (5.40), the solution is easily found in the form

$$\mathsf{A}(\chi) = (\mathsf{A}(0)\cos\chi + \mathrm{i}\sin\chi)\,(\cos\chi + \mathrm{i}\mathsf{A}(0)\sin\chi)^{-1}\,. \tag{5.51}$$

The matrices $\mathsf{S}(\chi)$ and $\mathsf{P}(\chi)$ define the two vectors $\nu_1(\chi)$ and $\nu_2(\chi)$ in phase space. As in the general case (5.22), the ladder operators take the form

$$\hat{b}_p(\chi) = \frac{\mathrm{i}}{\sqrt{2}}\,\nu_p(\chi)\mathbb{G}\hat{\mathfrak{v}} = \frac{\mathrm{i}}{\sqrt{2}}\,(\sigma_p(\chi)\hat{\pi} - \pi_p(\chi)\hat{\sigma})\,, \tag{5.52}$$

with $p = 1, 2$. The formalism of Section 5.2 now gives an orthonormal basis set of solutions $\Psi_{n_1 n_2}(\sigma, \chi)$ of the harmonic oscillator, in the form

$$\Psi_{n_1 n_2}^{\mathrm{NS}}(\sigma, \chi) = \frac{1}{\sqrt{n_1! n_2!}}\left(\hat{b}_1^\dagger(\chi)\right)^{n_1}\left(\hat{b}_2^\dagger(\chi)\right)^{n_2}\Psi_{00}^{\mathrm{NS}}(\sigma, \chi), \tag{5.53}$$

with $n_1, n_2 = 0, 1, 2, \ldots$. For each value of χ, these states are eigenstates of the operator $\hat{b}_1^\dagger(\chi)\hat{b}_1(\chi)$ with eigenvalue n_1, and of the operator $\hat{b}_2^\dagger(\chi)\hat{b}_2(\chi)$ with eigenvalue n_2. In this general case the two vectors $\nu_1(\chi)$ and $\nu_2(\chi)$ are not linear combinations of $\nu_\xi(\chi)$ and $\nu_\eta(\chi)$ alone. Likewise, the raising operators cannot be expressed as linear combinations of $\hat{b}_\xi^\dagger$ and $\hat{b}_\eta^\dagger$ alone, but also the lowering operators $\hat{b}_\xi$ and $\hat{b}_\eta$ are needed. As a result, the states (5.53) are not stationary in general.

5.3.5 Angular momentum

In an arbitrary state of the harmonic oscillator, the expectation value of the dimensionless angular momentum $\hat{L} = \hat{\sigma} \times \hat{\pi}$ (in units of $\hbar$) takes the general form

$$\langle L \rangle = \int \mathrm{d}_2\sigma\,\Psi^*(\sigma)(\hat{\sigma} \times \hat{\pi})\,\Psi(\sigma). \tag{5.54}$$

The cross product is taken between the two two-dimensional transverse vector operators $\hat{\sigma}$ and $\hat{\pi}$. In order to express the ray operator $\hat{\mathfrak{v}}$ (which is equivalent to the combination of $\hat{\sigma}$ and $\hat{\pi}$) in terms of the ladder operators, we have to invert the expressions (5.22), with the result

$$\hat{\nu} = \frac{1}{\sqrt{2}} \sum_{p=1,2} \left(\nu_p \hat{b}_p^\dagger + \nu_p^* \hat{b}_p \right). \tag{5.55}$$

This expansion is directly checked by using the properties (5.23). The vector operators for position $\hat{\sigma}$ and momentum $\hat{\pi}$ can be extracted from this expression. In the non-stationary basis state $\Psi_{n_1 n_2}^{\mathrm{NS}}$, which is specified in Eq. (5.53), the expectation value of the angular momentum is

$$\langle L \rangle_{n_1 n_2}^{\mathrm{NS}} = \sum_{p=1,2} \left(n_p + \frac{1}{2} \right) \mathrm{Re}(\sigma_p^* \times \pi_p). \tag{5.56}$$

This expression does not vary with χ, since the angular momentum commutes with the Hamiltonian. For the special case of the Hermite–Laguerre basis set (5.49), this expression simplifies as

$$\langle L \rangle_{n_1 n_2}^{\mathrm{HL}} = (n_1 - n_2) \cos\theta. \tag{5.57}$$

The angular momentum vanishes for Hermite–Gaussian states, when $\theta = \pi/2$. Only in the special case that $\theta = 0$ or π is the state $\Psi_{n_1 n_2}$ an eigenstate of the operator $\hat{L}$, with eigenvalue ± 1.

5.3.6 Oscillation between Fourier transforms

A characteristic property of the harmonic oscillator is that it has the same dependence on the coordinates $\hat{\xi}$ and $\hat{\eta}$ and on the momenta $\hat{\pi}_\xi$ and $\hat{\pi}_\eta$. A natural consequence is that the eigenfunctions in coordinate and momentum representation must have the same form. Since these two representations are related by Fourier transforms, the harmonic-oscillator eigenfunctions are also eigenfunctions of the Fourier transformation. The eigenfunctions in momentum representation are

$$\phi_n(\pi_\xi) = \frac{1}{\sqrt{2\pi}} \int_{\infty}^{\infty} \mathrm{d}\xi \, \exp\left(-\mathrm{i}\xi\pi_\xi\right) \psi_n(\xi). \tag{5.58}$$

For the ground state one directly finds that $\phi_0(\pi_\xi) = \psi_0(\pi_\xi)$. Expressions for the higher-order modes in momentum representation follow from Eq. (5.44), when we apply the raising operator in the same representation, which has the form

$$\hat{b}_\xi^\dagger = \frac{-\mathrm{i}}{\sqrt{2}} \left(\pi_\xi - \frac{\partial}{\partial \pi_\xi} \right). \tag{5.59}$$

For the higher-order eigenfunctions in momentum representation this gives

$$\phi_n(\pi_\xi) = \frac{1}{\sqrt{n!}} \left(\hat{b}_\xi^\dagger\right)^n \phi_0(\pi_\xi) = (-\mathrm{i})^n \psi_n(\pi_\xi), \tag{5.60}$$

which confirms that the harmonic-oscillator wavefunctions are eigenfunctions of the Fourier transformation. Of course, the same relations in terms of η and π_η hold for the other dimension.

We expand an arbitrary solution of the Schrödinger equation in the basis of the Hermite–Gaussian solutions (5.43)

$$\Psi(\sigma, \chi) = \sum_{n,m} c_{nm} \Psi_{nm}^{\mathrm{HG}}(\sigma, \chi), \tag{5.61}$$

where $\Psi_{nm}(\sigma) \equiv \psi_n(\xi)\psi_m(\eta)$ is a factorized eigenstate of the total Hamiltonian $\hat{H}$. After a quarter period, when $\chi = \pi/2$, the phase factor in (5.43) is $(-\mathrm{i})^{n+m+1}$. When we use the identity (5.60), we find that

$$\Psi(\sigma, \pi/2) = -\mathrm{i} \sum_{n,m} c_{nm} \phi_n(\xi) \phi_m(\eta) = -\mathrm{i}\Phi(\sigma, 0), \tag{5.62}$$

where Φ is the double Fourier transform of Ψ with respect to both coordinates. We conclude that an arbitrary initial wave function of the harmonic oscillator evolves into its Fourier transform after a quarter period. This is obviously true for any number of dimensions. Using this argument twice, one finds that after half a period, the initial state has evolved into

$$\Psi(\sigma, \pi) = -\Psi(-\sigma, 0), \tag{5.63}$$

since the Fourier transform of the Fourier transform of any function $\Psi(\sigma)$ is equal to $\Psi(-\sigma)$.

5.4 Paraxial wave optics as a Hamiltonian system

5.4.1 Propagation and ray operators

We now apply the general formalism of Section 5.2 to the propagation of paraxial beams of light through an optical system. The system consists of regions of free propagation, separated by optical elements such as thin lenses or mirrors. In the regions of free propagation, the beam profile is described by the paraxial wave equation (5.1), which has the structure of the Schrödinger equation with a quadratic Hamiltonian. Therefore, the general formalism of Section 5.2 applies, with $D = 2$ degrees of freedom. The classical version of the Hamiltonian describes in this case the propagation of a ray of light, which is characterized in each transverse plane by the two-dimensional transverse position $\rho = (x, y)$, and two direction angles $\theta = (\vartheta_x, \vartheta_y)$. Since the direction angle is just the ratio of the transverse and the longitudinal momentum, the corresponding wave-optical operator takes the form $\hat{\theta} = (-\mathrm{i}/k)\partial/\partial\rho$ [7]. The canonical commutation rules of the ray operators translate into the identities $[\hat{x}, \hat{\vartheta}_x] = \mathrm{i}/k$, etc.

Since the paraxial wave equation (5.1) has the operator form $\partial u/\partial z = -ik\hat{\theta}^2 u/2$, the operator for free propagation over a distance L is

$$\hat{U}_{\mathrm{prop}}(L) = \exp(-ikL\hat{\theta}^2/2). \tag{5.64}$$

A spherical lossless lens is characterized by its focal length f. Its effect on the beam profile u is to impose a phase factor $\exp(-ik\rho^2/(2f))$. We want to allow the lenses to be astigmatic, which means that they have different focal lengths along two orthogonal axes. For instance, the effect of a thin astigmatic lens that is aligned along the x and y axes corresponds to a phase factor $\exp\left[-ik(x^2/f_x + y^2/f_y)/2\right]$. For an arbitrary orientation of the astigmatic lens its effect on a mode profile $u(\rho, z)$ takes the form $u(\rho, z_+) = \hat{U}_{\mathrm{lens}}(\mathsf{F})$ $u(\rho, z_-)$, with [7]

$$\hat{U}_{\mathrm{lens}}(\mathsf{F}) = \exp\left(\frac{-ik}{2}\hat{\rho}\mathsf{F}^{-1}\hat{\rho}\right). \tag{5.65}$$

Here we denote z_- as the input plane, and z_+ as the output plane of the lens, while F is a real and symmetric 2×2 matrix, with the two focal lengths as eigenvalues, and the corresponding mutually orthogonal real eigenvectors directed along the symmetry axes of the mirror in the transverse plane. Whereas the exponent in the propagation operator (5.64) is quadratic in the direction angles, the exponent in the lens operator (5.65) is quadratic in the coordinates. The propagation operator $\hat{U}(z)$ from a reference plane at $z = 0$ to a plane $z > 0$ through regions of free propagation that are separated by lenses is then the product of operators $\hat{U}_{\mathrm{prop}}(L_i)$ for free propagation alternating with operators $\hat{U}_{\mathrm{lens}}(\mathsf{F}_j)$ for the lenses, in the correct order from right to left, so that

$$\hat{U}(z) = \cdots \hat{U}_{\mathrm{prop}}(L_2)\hat{U}_{\mathrm{lens}}(\mathsf{F}_1)\hat{U}_{\mathrm{prop}}(L_1). \tag{5.66}$$

This operator transforms the transverse mode profile as $u(\rho, z) = \hat{U}(z)u(\rho, 0)$. These operators may be viewed as evolution operators of a system with a quadratic Hamiltonian, and the formalism of Section 5.2 applies.

The analogue of the operator $\hat{\mathfrak{v}}$ introduced in (5.8) in the special case of paraxial propagation is the four-dimensional ray operator [7]

$$\hat{\mu} = \begin{pmatrix} \hat{\rho} \\ \hat{\theta} \end{pmatrix}. \tag{5.67}$$

In full analogy to the Heisenberg–Ehrenfest identity (5.15), the transformation of a ray is described by a matrix $\mathbb{M}(z)$, defined by the transformation

$$\hat{U}^\dagger(z)\hat{\mu}\hat{U}(z) = \mathbb{M}(z)\hat{\mu}. \tag{5.68}$$

In the case of free propagation, this four-dimensional ray matrix is given by

$$\mathbb{M}_{\mathrm{prop}}(L) = \begin{pmatrix} \mathsf{1} & L\mathsf{1} \\ \mathsf{0} & \mathsf{1} \end{pmatrix}. \tag{5.69}$$

This corresponds to the obvious transformation of the ray parameters $\rho(L) = \rho(0) + L\theta(0)$, $\theta(z) = \theta(0)$ in free space. These parameters are combined into the four-dimensional ray vector μ, analogous to Eq. (5.67). In the same manner the lens operator (5.65) gives rise to the ray transformation matrix

$$\mathbb{M}_{\text{lens}}(\mathsf{F}) = \begin{pmatrix} 1 & 0 \\ -\mathsf{F}^{-1} & 1 \end{pmatrix}, \tag{5.70}$$

which expresses the transformation of the ray parameters across a lens $\rho(z_+) = \rho(z_-)$, $\theta(z_+) = \theta(z_-) - \mathsf{F}^{-1}\rho(z_-)$, or $\mu(z_+) = \mathbb{M}_{\text{lens}}\mu(z_-)$. One directly checks that the propagation operator (5.66) corresponds to the ray matrix

$$\mathbb{M}(z) = \cdots \mathbb{M}_{\text{prop}}(L_2)\mathbb{M}_{\text{lens}}(\mathsf{F}_1)\mathbb{M}_{\text{prop}}(L_1). \tag{5.71}$$

In the simplified case of a single transverse dimension, the matrices $\mathbb{M}$ would be two-dimensional, and they are commonly called ABCD-matrices [3, 8, 9]. For this case, the connection between quadratic propagation operators acting on mode space and the ray matrices has been analyzed before in terms of group theory [10]. With two transverse dimensions, and in the presence of spherical lenses only, the description is basically the same. We include the case of astigmatic lenses. When the symmetry axes of the astigmatic lenses are parallel, the beam profile factorizes in components corresponding to the two transverse dimensions. This is called simple astigmatism. The situation is considerably more complex when the axes of the astigmatic lenses (or mirrors) are not aligned. In this case of general astigmatism, the beam profile no longer factorizes, and the analytical description is cumbersome [11]. In this case, the operator description is particularly advantageous [7, 12]. Since the transformation (5.68) conserves the canonical commutation rules, all ray matrices necessarily obey the symplectic relation

$$\mathbb{M}^{\mathrm{T}}(z)\mathbb{G}\mathbb{M}(z) = \mathbb{G}, \tag{5.72}$$

which is just a special realization of Eq. (5.16).

5.4.2 Ladder operators and basis sets of modes

From the general formalism of quadratic Hamiltonians in Section 5.2 we know that basis sets of paraxial modes are generated by ladder operators. In the present case of paraxial beams they are specified by two rays μ_p with $p = 1, 2$, which are composed of a position part ρ_p and a direction part θ_p, so that

$$\mu_p(z) = \begin{pmatrix} \rho_p(z) \\ \theta_p(z) \end{pmatrix}, \tag{5.73}$$

in analogy to Eq. (5.21). Naturally, the two-dimensional vectors ρ_p have the dimension of length, while the vectors θ_p denote angles, and are dimensionless. Their variation with z is determined by the ray matrix, so that

$$\mu_p(z) = \mathbb{M}(z)\mu_p(0). \tag{5.74}$$

The two lowering operators can be expressed in analogy to Eq. (5.22) as

$$\hat{a}_p(z) = \frac{\mathrm{i}k}{\sqrt{2}} \left(\rho_p(z)\hat{\theta} - \theta_p(z)\hat{\rho} \right) = \frac{\mathrm{i}k}{\sqrt{2}} \mu_p(z)\mathbb{G}\hat{\mu}, \tag{5.75}$$

where the wave number k ensures that these operators are dimensionless. The bosonic commutation rules for these ladder operators are equivalent to the conditions

$$\mu_p(z)\mathbb{G}\mu_{p'}(z) = 0, \; \mu_p^*(z)\mathbb{G}\mu_{p'}(z) = \frac{2\mathrm{i}}{k}\delta_{p'p}. \tag{5.76}$$

These relations are a symplectic version of orthogonality, just as the general relations Eq. (5.23). The expansion of an arbitrary ray vector μ in terms of these eigenrays has the form,

$$\mu = \frac{k}{2\mathrm{i}} \sum_p \left[\mu_p \left(\mu_p^* \mathbb{G}\mu \right) - \mu_p^* \left(\mu_p \mathbb{G}\mu \right) \right] \tag{5.77}$$

as is easily checked by using the general properties (5.76). This confirms the completeness of the basis of four rays μ_p and μ_p^*.

Just as the matrices S and P are composed out of the column vectors σ_p and π_p respectively, according to Eq. (5.24), we now define two-dimensional matrices R and T

$$\mathsf{R}(z) = \left(\rho_1(z) \quad \rho_2(z) \right), \; \mathsf{T}(z) = \left(\theta_1(z) \quad \theta_2(z) \right). \tag{5.78}$$

The relations (5.76) can be rewritten in terms of the matrix relations

$$\mathsf{R}^{\mathsf{T}}\mathsf{T} - \mathsf{T}^{\mathsf{T}}\mathsf{R} = 0, \; \mathsf{R}^{\dagger}\mathsf{T} - \mathsf{T}^{\dagger}\mathsf{R} = \frac{2\mathrm{i}}{k}\mathsf{1}, \tag{5.79}$$

for all z. In analogy to Eqs. (5.26) and (5.27), and in accordance with (5.74), the z-dependence of these matrices is governed by the ray matrix $\mathbb{M}(z)$.

Once the matrices R and T are chosen at the transverse reference plane at $z = 0$, they are known for all z. Moreover, they specify the two lowering operators (5.75), and thereby the basis of modes as a function of z. When R and T are known, the fundamental mode of the basis is defined by the general expression (5.31), which now takes the form

$$u_{00}(\rho, z) = \frac{1}{\sqrt{\pi \det \mathsf{R}}} \exp\left(-k\rho\mathsf{B}\rho/2\right) \tag{5.80}$$

with $\mathsf{B} = -\mathrm{i}\mathsf{T}\mathsf{R}^{-1}$. An astigmatic lens that is passed by an isotropic input beam will impose its astigmatism on the output beam. As always, the ladder operators generate the higher-order modes, according to

$$u_{n_1 n_2}(\rho, z) = \frac{1}{\sqrt{n_1! n_2!}} \left(\hat{a}_1^\dagger(z) \right)^{n_1} \left(\hat{a}_2^\dagger(z) \right)^{n_2} u_{00}(\rho, z). \tag{5.81}$$

The variation of the two vectors $\mu_p(z)$ with z, which is governed by the ray matrix $\mathbb{M}(z)$, defines the ladder operators, and thereby the modes of all orders by Eq. (5.81). The formalism guarantees that these expressions are solutions of the paraxial wave equation in the regions in between lenses, and that they obey the correct lens transformation (5.65) at all lenses.

Paraxial modes can carry orbital angular momentum in the propagation direction, owing to the helical nature of the transverse momentum density [13, 14]. It is convenient to express this orbital angular momentum L per photon, so that the expression only depends on the profile, and not on the intensity [15]. The operator for orbital angular momentum is now $\hat{L} = \hbar k \hat{\rho} \times \hat{\theta}$, and for a normalized beam profile $u(\rho)$, the angular momentum per photon is

$$\langle L \rangle = \hbar k \int \mathrm{d}\rho u^*(\rho)(\hat{\rho} \times \hat{\theta})\, u(\rho). \tag{5.82}$$

Expressions for the ray operators $\hat{\rho}$ and $\hat{\theta}$ follow from the expansion in terms of the ladder operators

$$\hat{\mu} = \frac{1}{\sqrt{2}} \sum_{p=1,2} \left(\mu_p \hat{a}_p^\dagger + \mu_p^* \hat{a}_p \right), \tag{5.83}$$

which is the inversion of the expressions (5.75). For the mode $u_{n_1 n_2}$ from the basis (5.81), the result is found in analogy to Eq. (5.56)

$$\langle L \rangle_{n_1 n_2} = \hbar k \sum_p \left(n_p + \frac{1}{2} \right) \mathrm{Re} \left(\rho_p^* \times \theta_p \right). \tag{5.84}$$

This expression is valid for each transverse plane z. In a region of free propagation, it does not vary with z. However, when the beam crosses an astigmatic element, $\langle L \rangle_{n_1 n_2}$ can change its value. In this situation, the beam exerts a torque on the lens [7, 16].

5.5 Basis sets of paraxial beams and harmonic oscillators

5.5.1 Equivalence of free modes and oscillator states

A freely propagating light beam as described by the paraxial wave equation (5.1) does not have normalizable stationary states. However, the formalism of Section 5.2 still applies, which allows us to obtain discrete basis sets of solutions. A basis set is obtained when we choose the matrices $\mathsf{R}(0)$ and $\mathsf{T}(0)$ in a reference plane $z = 0$. Only the free-propagation matrix (5.69) applies, so that

$$\mathbb{M}(z) = \begin{pmatrix} 1 & z1 \\ 0 & 1 \end{pmatrix}. \tag{5.85}$$

Therefore the z-dependence of the matrices R and T is simply

$$\mathsf{R}(z) = \mathsf{R}(0) + z\mathsf{T}(0), \;\; \mathsf{T}(z) = \mathsf{T}(0), \tag{5.86}$$

which just reflects the straight-line behavior of a ray. A basis of modes $u_{n_1n_2}(\rho, z)$ is determined once we have chosen the two matrices $\mathsf{R}(0)$ and $\mathsf{T}(0)$ for $z = 0$, with the properties (5.79). This determines the two vectors $\mu_1(z)$ and $\mu_2(z)$, and thereby the basis (5.81).

As we argued in Section 5.1, an arbitrary solution $\Psi(\sigma, \chi)$ of the Schrödinger equation of the harmonic oscillator gives rise to a corresponding solution $u(\rho, z)$ of the paraxial wave equation. This correspondence, which is expressed in Eq. (5.7), was first derived by inspection of the similarity between the Hermite–Gaussian states of the oscillator, and the Hermite–Gaussian paraxial modes [5]. Of course, it can also be confirmed by an analytical calculation [17]. This general relation can also be applied to basis sets of solutions $\Psi_{n_1n_2}(\sigma, \chi)$ of the harmonic oscillator and of freely propagating modes $u_{n_1n_2}(\rho, z)$. The modes are generated by ladder operators $\hat{a}_p(z)$ of the form (5.75), which are defined as a function of z, in terms of the position and angle operators $\hat{\rho}$ and $\hat{\theta}$. The harmonic-oscillator wave functions are generated by ladder operators $\hat{b}_p(\chi)$ of the form (5.52). These are defined in terms of the dimensionless operators $\hat{\sigma}$ and $\hat{\pi}$, and are functions of χ. From the general correspondence (5.7) it is obvious that the ladder operators for these two basis sets must obey the relation

$$\hat{a}_p(z) = \exp\left(\frac{ik}{2R}\hat{\rho}^2\right)\hat{b}_p(\chi)\exp\left(-\frac{ik}{2R}\hat{\rho}^2\right). \tag{5.87}$$

When this operator acts on the expression (5.7), the exponential curvature term is first removed, so that the operator $\hat{b}_p$ acts on the wave function Ψ, after which the curvature term is restored. As specified in Section 5.1, the dimensionless position vector σ is related to the transverse position ρ by $\sigma = \rho/\gamma$. The parameters R, γ and χ are functions of z, as defined in Eq. (5.5). This gives the operator relations $\hat{\rho} = \gamma(z)\hat{\sigma}$, and $\hat{\theta} = \hat{\pi}/(k\gamma(z))$. The relation between the vector operators $\hat{\mu}$ and $\hat{\nu}$ can therefore be expressed in the matrix form $\hat{\mu} = \mathbb{D}(z)\hat{\nu}$, in terms of the diagonal 4×4 matrix

$$\mathbb{D}(z) = \begin{pmatrix} \mathsf{1}\gamma(z) & \mathsf{0} \\ \mathsf{0} & \mathsf{1}\frac{1}{k\gamma(z)} \end{pmatrix}. \tag{5.88}$$

At $z = 0$, $\chi = 0$, the width is $\gamma(0) = \gamma_0 = \sqrt{b/k}$, while the radius of curvature $R(0)$ is infinite, so that the exponential curvature terms vanish in Eqs. (5.87) and (5.7). This means that the two corresponding basis sets are specified by the phase-space vectors $\mu_p(0)$ and $\nu_p(0)$, which are related by $\mu_p(0) = \mathbb{D}(0)\nu_p(0)$. This is equivalent to the matrix relations

$$\mathsf{R}(0) = \mathsf{S}(0)\gamma_0, \ \mathsf{T}(0) = \mathsf{P}(0)/(k\gamma_0). \tag{5.89}$$

The conditions (5.25) for the matrices S and P imply the corresponding relations (5.79) for the matrices R and T. It is important to notice that $\mu_p(z)$ is not equal to $\mathbb{D}(z)\nu_p(\chi)$ for non-zero values of χ (and z). The variation of these matrices with χ (or z) is expressed by Eqs. (5.40) and (5.86), and equivalently the variation of μ_p with z is governed by the matrix $\mathbb{M}(z)$, and the variation of ν_p with χ is governed by the matrix $\mathbb{N}(\chi)$.

This general correspondence is determined by the Rayleigh range b, which can be freely chosen. A different choice of b leads to different modes $u(\rho, z)$ for a single harmonic-oscillator solution. These different modes just differ in their degree of focussing. Conversely, a single mode leads to different harmonic oscillator solutions for different values of b. These two solutions may be viewed as squeezed versions of each other. In addition, a shift in "time" χ of a harmonic-oscillator solution selects a different half of the cycle that is mapped into a mode. The resulting modes can be quite different in nature. In particular, the wave function $\Psi(\sigma, \chi_0)$ at any time χ_0 can be taken as the focal plane, or, for that matter, as the far field.

5.5.2 Classical oscillations and rays of light

We have derived a correspondence between basis sets of harmonic-oscillator wave functions and basis sets of paraxial modes. They are generated by pairs of ladder operators $\hat{a}_p(z)$ for the modes and ladder operators $\hat{b}_p(\chi)$ for the wave functions. These ladder operators are characterized by pairs of complex vectors $\mu_p(z)$ in ray space and complex vectors $\nu_p(\chi)$ in phase space. The variation of the vectors ν_p with χ has the oscillatory behavior of the matrix $\mathbb{N}$ described by Eq. (5.39), which reflects the behavior of the classical harmonic oscillator. On the other hand, the variation of the vectors $\mu_p(z)$ with z is linear. This is reflected by the matrix $\mathbb{M}$, Eq. (5.85), which describes the straight-line behavior of a ray of light. Therefore, the correspondence that was found on the level of quantum mechanics and wave optics must also hold on the level of the classical oscillator and ray optics. In view of the rather different behavior of a classical oscillator and a straight ray, this is a little surprising, and it is worthwhile to understand how this happens.

We start from the relation (5.87) between ladder operators, with $\hat{a}_p$ and $\hat{b}_p$ defined by (5.75) and (5.52). The effect of the curvature term in Eq. (5.87) can be accounted for by using the relation

$$\exp\left(\frac{ik}{2R}\hat{\rho}^2\right)\hat{\mathfrak{v}}\exp\left(-\frac{ik}{2R}\hat{\rho}^2\right) = \mathbb{C}^{-1}(\chi)\hat{\mathfrak{v}}, \tag{5.90}$$

where we define

$$\mathbb{C}(\chi) = \begin{pmatrix} 1 & 0 \\ 1\tan\chi & 1 \end{pmatrix}. \tag{5.91}$$

In deriving the relation (5.90), we used the substitution $k\gamma^2(z)/R(z) = z/b = \tan\chi$, which directly follows from the definitions (5.5). The relation (5.87) between ladder operators implies a relation between the vectors $\mu_p(z) = \mathbb{M}(z)\mu_p(0)$ and $\nu_p(\chi) = \mathbb{N}(\chi)\nu_p(0)$. Combined with the initial relation $\mu_p(0) = \mathbb{D}(0)\nu_p(0)$, this yields the matrix relation

$$k\mathbb{D}(0)\mathbb{M}^{\mathrm{T}}(z)\mathbb{G}\mathbb{D}(z) = \mathbb{N}^{\mathrm{T}}(\chi)\mathbb{G}\mathbb{C}^{-1}(\chi). \tag{5.92}$$

Since $\mathbb{C}$ results from the transformation of the operator $\hat{\mathfrak{v}}$ by a quadratic form, it is an element of the group $Sp(4,\mathbb{R})$, and the symplectic relation (5.16) gives the identity

$\mathbb{G}\mathbb{C}^{-1} = \mathbb{C}^{\mathrm{T}}\mathbb{G}$. Moreover, the matrix $\mathbb{D}$ obeys the identity $k\mathbb{G}\mathbb{D}(z) = \mathbb{D}^{-1}(z)\mathbb{G}$. After these substitutions and taking the transpose, one notices that the identity (5.92) is equivalent to the relation

$$\mathbb{D}^{-1}(z)\mathbb{M}(z)\mathbb{D}(0) = \mathbb{C}(\chi)\mathbb{N}(\chi) \tag{5.93}$$

between the straight-line behavior of the ray matrix $\mathbb{M}(z)$, and the oscillatory behavior of the matrix $\mathbb{N}(\chi)$. Indeed, one can easily check explicitly that this relation holds for all values of $z = b \tan \chi$. It may be viewed as the ray analogue of the correspondence relation (5.7), and it links a trajectory in phase space of the classical harmonic oscillator to a ray of light in ray space. On the left-hand side, the matrix $\mathbb{D}(0)$ translates a point in the oscillator phase space into the ray in the plane $z = 0$. After propagation over a distance z, the inverse matrix $\mathbb{D}^{-1}(z)$ maps the ray back into the phase space. The matrix $\mathbb{C}$ represents the effect of the exponential curvature term. On a ray (or a paraxial beam) this term has the same effect as a lens with focal length R.

5.5.3 Shape-invariant modes

A mode $u(\rho, z)$ that corresponds to a stationary state $\Psi(\sigma, \chi)$ of the harmonic oscillator has an intensity profile that only changes its scale during propagation, while its shape remains invariant. A z-dependent curvature is added, but otherwise the only change is that the size of the mode pattern scales with the z-dependent width $\gamma(z)$. The intensity pattern is the same as for the oscillator states, and it keeps its form under propagation. The Hermite–Gaussian modes u_{nm}^{HG} defined in Eq. (5.4) form a well-known basis of shape-invariant modes. These modes correspond to the Hermite–Gaussian stationary states (5.4) of the harmonic oscillator. The modes are generated by the ladder operators

$$\begin{aligned}\hat{a}_x(z) &= \exp\left(\frac{ik}{2R}\hat{\rho}^2\right)\hat{b}_\xi(\chi)\exp\left(-\frac{ik}{2R}\hat{\rho}^2\right), \quad \hat{a}_y(z)\\ &= \exp\left(\frac{ik}{2R}\hat{\rho}^2\right)\hat{b}_\eta(\chi)\exp\left(-\frac{ik}{2R}\hat{\rho}^2\right),\end{aligned} \tag{5.94}$$

which arise from the ladder operators (5.42) by the transformation (5.87). These ladder operators can also be expressed as

$$\hat{a}_x(z) = \sqrt{\frac{k}{2b}}(\hat{x} + \mathrm{i}(b + \mathrm{i}z)\hat{\theta}_x), \quad \hat{a}_y(z) = \sqrt{\frac{k}{2b}}(\hat{y} + \mathrm{i}(b + \mathrm{i}z)\hat{\theta}_y), \tag{5.95}$$

which have the form (5.75). It is easy to see that these operators correspond to the matrices $\mathsf{R}^{\mathrm{HG}}(z) = \mathbb{1}(b + \mathrm{i}z)/\sqrt{kb}$, $\mathsf{T}^{\mathrm{HG}}(z) = \mathrm{i}\mathbb{1}/\sqrt{kb}$. The fundamental mode (5.80) is then found as

$$u_{00}(\rho, z) = \sqrt{\frac{kb}{\pi}}\frac{1}{b + \mathrm{i}z}\exp\left(-\frac{k\rho^2}{2(b + \mathrm{i}z)}\right) = \frac{1}{\gamma}\Psi_{00}(\sigma, \chi)\exp\left(\frac{ik}{2R}\rho^2\right), \tag{5.96}$$

in terms of the harmonic-oscillator ground state (5.41). The first expression in (5.96) may be viewed as the radiation emitted by a point source at a complex location. Repeated action with the raising operators $\hat{a}_x^\dagger$ and $\hat{a}_y^\dagger$ produces the Hermite–Gaussian modes (5.4). The intensity pattern of the Hermite–Gaussian modes is the same as the strength pattern of the corresponding stationary states, and is illustrated in Fig. 5.1.

Any linear combination of Hermite–Gaussian modes with the same value of the total mode order $N = n + m$ is a shape-invariant mode, and corresponds to a stationary state of the harmonic oscillator. Therefore, a pair of raising operators $\hat{a}_p^\dagger$ that is a unitary linear transformation of $\hat{a}_x^\dagger$ and $\hat{a}_y^\dagger$ produces a basis set of shape-invariant modes. These are the analogue of the basis sets (5.49) of Hermite–Laguerre stationary states of the harmonic oscillator. The intensity and phase pattern of these modes is the same as for the harmonic oscillator, and is illustrated in Fig. 5.3. To the circular Laguerre–Gaussian states $\Psi_{n_+n_-}$ defined in (5.46) correspond the Laguerre–Gaussian beams $u_{n_+n_-}$, which are now labeled by the two mode numbers n_+ and n_-. The Laguerre–Gaussian eigenstates of the oscillator are also eigenstates of the operator $\hat{L}$ of angular momentum, with eigenvalue $n_+ - n_-$. The customary labeling of the Laguerre–Gaussian modes is by the azimuthal mode index $l = n_+ - n_-$ and the radial mode index $p = \min(n_+, n_-)$. When the mode profile is written in polar coordinates, the azimuthal dependence is proportional to $\exp(\mathrm{i}l\phi)$, while the radial mode index p determines the number of nodes in the radial direction. The Laguerre–Gaussian modes carry an orbital angular momentum that equals $\hbar l$ per photon. Moreover, the density of angular momentum is proportional to the local intensity. Their intensity pattern modes are illustrated in Fig. 5.2.

The transformation between these basis sets of Hermite–Gaussian, Laguerre–Gaussian and generalized Hermite–Laguerre beams is identical to the transformation between the corresponding stationary solutions of the harmonic oscillator. It follows directly from the relations (5.45) and (5.47) between the ladder operators. This algebraic representation of the transformation is much more transparent than the analytical representations of the transformations [18, 19]. The basis sets of shape-invariant modes are generated by matrices of the form $\mathsf{R}(z) = (b + \mathrm{i}z)\mathsf{U}^\dagger/\sqrt{kb}$, $\mathsf{T}(z) = \mathrm{i}\mathsf{U}^\dagger/\sqrt{kb}$, with U a unitary 2×2-matrix.

Other basis sets of shape-invariant modes have been derived that are intermediate between the Hermite–Gaussian and Laguerre–Gaussian sets. One example is the set of Ince–Gaussian modes, which have an inherent elliptical symmetry [20]. Also these modes are linear combinations of Hermite–Gaussian modes of the same total order $N = n + m$. This implies that they correspond in the sense of Eq. (5.7) to stationary states of the harmonic oscillator. The general conclusion that only linear combinations of Hermite–Gaussian beams of the same total order are shape-invariant has also been reached by purely analytical considerations [21].

5.5.4 Astigmatic modes

Modes corresponding to non-stationary solutions $\Psi^{\mathrm{NS}}_{n_1n_2}(\sigma, \chi)$ of the harmonic oscillator are characterized by matrices R and T that are not proportional to one another. Then the modes are not shape-invariant under propagation. The fundamental mode (5.80) is characterized by the matrix $\mathsf{B} = -\mathrm{i}\mathsf{T}\mathsf{R}^{-1}$. According to (5.86), this matrix has a z-dependence $\mathsf{B}(0)\,(1 + \mathrm{i}z\mathsf{B}(0))^{-1}$. The intensity profile and the wave fronts have an elliptical, not a circular shape. This characterizes an astigmatic Gaussian mode. The astigmatism is simple when the axes of the two ellipses are aligned. In the more general case that these two ellipses are not aligned, the intensity ellipse changes its orientation and its shape during propagation. This is characteristic of general astigmatism [11, 12]. This reorientation of the shape of the profile represents an orbital angular momentum. The amount of angular momentum per photon in a mode of the basis set is given by Eq. (5.84). This amount does not change during propagation [16].

5.5.5 Fourier relations of general paraxial beams

As shown in Eq. (5.62), a characteristic of the harmonic oscillator is that its wave functions evolve into their Fourier transform after a quarter of a cycle. Because of the correspondence between harmonic-oscillator wave functions and paraxial beams, an analogous Fourier relation holds between the transverse profile of a paraxial beam between two points with a difference $\pi/2$ in Gouy phase. We shall discuss two examples.

The first example is the far field of an arbitrary paraxial beam. We consider the beam profile at one transverse, which we take as $z = 0$. We start with the general correspondence relation (5.7), and consider the asymptotic behavior of the beam profile $u(\rho, z)$ in the far-field limit $z \to \infty$. In this limit, the curvature radius and the width behave as $R \to z$, and $\gamma \to z/\sqrt{kb}$, while the Gouy phase χ approaches $\pi/2$. From Eqs. (5.7) and (5.62) we then find the asymptotic behavior for $z \to \infty$

$$u(\rho, z) \to -\mathrm{i}\frac{\sqrt{kb}}{z} \exp\left(\frac{\mathrm{i}k}{2z}\rho^2\right) \Phi(\sigma, 0), \tag{5.97}$$

with $\sigma = \rho\sqrt{kb}/z$. The Fourier transform $\Phi(\sigma, 0)$ of the wave function at $\chi = 0$ is related to the Fourier transform v of the beam profile $u(\rho, 0)$ at $z = 0$ by the expression

$$\Phi(\sigma, 0) = \frac{1}{\gamma_0} v\left(\frac{\sigma}{\gamma_0}, 0\right), \tag{5.98}$$

with $\gamma_0 = \sqrt{b/k}$ the value of γ at $z = 0$. Substitution in (5.97) gives the explicit asymptotic behavior

$$u(\rho, z) \to -\frac{\mathrm{i}k}{z} \exp\left(\frac{\mathrm{i}k}{2z}\rho^2\right) v\left(\frac{k\rho}{z}, 0\right), \tag{5.99}$$

for $z \to \infty$. This shows how the far-field limit of a paraxial beam is related to the Fourier transform of the transverse profile at $z = 0$. The far-field intensity pattern has the shape of the modulus of the Fourier transform of the mode in the near field. The asymptotic width γ_{as} is related to the width γ_0 at $z = 0$ by the relation $\gamma_{\mathrm{as}} \to z/(k\gamma_0)$. Moreover, a curvature term with a radius of the order of z is added. The result does not depend on the value

b of the diffraction length. This is to be expected, since this parameter could be chosen arbitrarily.

As a second example, we apply the correspondence (5.7) for the two values $\chi = \pm\pi/4$. The two transverse planes of the reference basis for the paraxial beam correspond to the values $z = \pm b$, so that their separation is $L = 2b$. The two opposite radii of curvature are $R = \pm 2b = \pm L$, and the two identical widths $\gamma = \sqrt{2b/k} = \sqrt{L/k}$. Since the phase difference is $\pi/2$, the two oscillator wave functions are connected by a Fourier transform as in Eq. (5.62), so that $\Psi(\xi, \eta, \pi/4) = -\mathrm{i}\Phi(\xi, \eta, -\pi/4)$. When the mode profiles are denoted as $u(\rho, \pm b) = u_\pm(\rho)$, this connection implies a relation between $u_+ \exp(-\mathrm{i}k\rho^2/2L)$ and the Fourier transform of $u_- \exp(\mathrm{i}k\rho^2/2L)$, in the form

$$u_+(\rho)\exp(-\mathrm{i}k\rho^2/2L) = -\frac{\mathrm{i}k}{2\pi L}\int \mathrm{d}_2\rho' \exp\left(-\mathrm{i}k\rho\rho'/L\right) u_-(\rho')\exp(\mathrm{i}k\rho'^2/2L). \tag{5.100}$$

In fact, since we are free in our choice of the reference mode basis in which to expand the mode u, the value of b and the position of $z = 0$ are arbitrary. This makes it understandable that the relation (5.100) between the mode profiles $u_\pm$ only depends on the separation L between the two transverse planes, and that it is independent of the chosen value of b and of the location of the reference plane at $z = 0$. We can rewrite this relation in the more familiar form

$$u_+(\rho) = -\frac{\mathrm{i}k}{2\pi L}\int \mathrm{d}_2\rho' \exp[\mathrm{i}k(\rho - \rho')^2/2L]u_-(\rho'). \tag{5.101}$$

This describes the effect of free propagation of an arbitrary input profile u_- over a distance L, with the profile u_+ as output result. The propagator is just the well-known Huygens–Fresnel integration kernel for paraxial propagation over a distance L [3, 22].

5.6 Astigmatic optical resonators

5.6.1 *Equivalent lens guide*

In the case of a two-mirror optical resonator of length L, the light is bouncing back and forth between the mirrors. In that case the resonator can be unfolded into a periodic lens guide with period $2L$, so that each period is equivalent to a round-trip in the resonator. An astigmatic mirror, characterized by the focal matrix F, is represented by a lens with the same focal matrix. The lens-guide picture of a resonator is shown in Fig. 5.4.

We allow the mirrors to be astigmatic, so that they have different focal lengths in two orthogonal planes through the optical axis. The output plane of mirror 1 is taken as the reference plane, so that a round trip consists successively of propagation over a distance L, reflection by mirror 2, propagation back over a distance L, reflection by mirror 1. The mode structure and the stability properties of astigmatic resonators have been analyzed by Habraken and Nienhuis [7]. The round-trip operator for an input mode profile u is the product of the operators for free propagation and for the mirrors in the correct order, so that

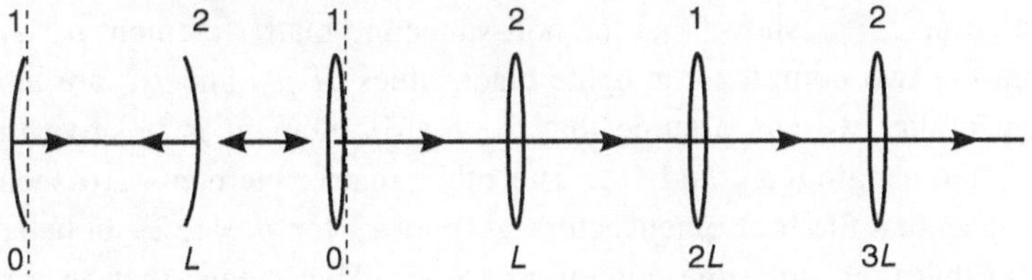

Figure 5.4 A resonator can be unfolded into an equivalent lens guide by replacing the mirrors by lenses with the same focal lengths. One round trip inside the resonator corresponds to one period of length $2L$ of the lens guide and the dashed line indicates the reference plane at $z=0$.

$$\hat{U}_{\rm rt} = \hat{U}_{\rm mirror}(\mathsf{F}_1)\hat{U}_{\rm prop}(L)\hat{U}_{\rm mirror}(\mathsf{F}_2)\hat{U}_{\rm prop}(L), \tag{5.102}$$

where the operator $\hat{U}_{\rm mirror}(\mathsf{F})$ is identical to the lens operator (5.65). Likewise the ray matrix over a round trip is

$$\mathbb{M}_{\rm rt} = \mathbb{M}_{\rm mirror}(\mathsf{F}_2)\mathbb{M}_{\rm prop}(L)\mathbb{M}_{\rm mirror}(\mathsf{F}_2)\mathbb{M}_{\rm prop}(L). \tag{5.103}$$

In the lens guide, the reference plane is represented by the coordinate $z=0$. The operator $\hat{U}(z)$ represents propagation in the lens guide from the reference plane to the transverse plane z, and the corresponding ray matrix is $\mathbb{M}(z)$. For each value of z, the operator $\hat{U}$ and the ray matrix $\mathbb{M}$ are related by the identity (5.68). In particular, this is the case for a round trip, described for a mode by $\hat{U}(2L)=\hat{U}_{\rm rt}$, and for a ray by $\mathbb{M}(2L)=\mathbb{M}_{\rm rt}$.

5.6.2 Stability condition of an optical resonator

As outlined in Section 5.4, the formalism of quadratic Hamiltonians can be applied to obtain basis sets of modes propagating through the lens guide. A basis set is specified by two rays $\mu_1(0)$ and $\mu_2(0)$ obeying the conditions (5.76). They define the ladder operators that generate the basis set. We now apply this technique to obtain expressions for the complete set of resonant modes in the resonator. This method has been applied before to modes with general astigmatism [7]. A basis of lens-guide modes translates into a basis of modes in the resonator when the light bouncing back and forth between the mirrors reproduces itself after a round trip, apart from an overall phase factor. In the lens-guide picture, a resonant mode must be periodic with period $2L$, apart from a phase factor. This is analogous to the properties of a Bloch wave of electrons in a periodic crystal. This requires that the ladder operators $\hat{a}_p(z)$, and therefore the complex ray vectors $\mu_p(z)$ have the same periodicity. In other words, in order that the ray vectors $\mu_p(0)$ in the reference plane define a resonant mode, they must be eigenvectors of $\mathbb{M}_{\rm rt}$, and the eigenvalues must be unitary in order to preserve the bosonic commutation rules of the ladder operators. When the eigenvalues are complex, so must be the eigenvectors.

Since the round-trip ray matrix $\mathbb{M}_{\rm rt}$ is real, the eigenvectors and eigenvalues come in two pairs that are each other's complex conjugate, and we can use an argumentation as in Section 5.2.5. Taking the matrix elements of $\mathbb{G}$ between eigenvectors, while using the symplectic relation (5.72), shows that the non-vanishing matrix element $\mu_p^* \mathbb{G} \mu_p$ is imaginary, and that the two complex conjugate eigenvalues of μ_p and μ_p^* are unitary. Without loss of generality we can assume that the matrix element is positive imaginary, if needed after interchanging μ_p and μ_p^*. The other matrix elements are seen to vanish. This demonstrates that the four eigenvectors μ_p and μ_p^* for $p=1, 2$ can be arranged and normalized, so that they obey the conditions (5.76). This means that an arbitrary two-mirror optical resonator has a basis set of resonant modes, provided that the round-trip ray matrix $\mathbb{M}_{\rm rt}$ has four complex eigenvalues. In physical terms, the existence of these complex unitary eigenvalues is the condition for the resonator to be stable. The basis of lens-guide modes is then defined by the ray vectors $\mu_p(z) = \mathbb{M}(z)\mu_p(0)$, which ensures that $\mu_p(2L) = \mu_p(0) \exp(\mathrm{i}\chi_p)$. For the ladder operators this means that

$$\hat{a}_p(2L) = \hat{a}_p(0) \exp(\mathrm{i}\chi_p), \ \ \hat{a}_p^\dagger(2L) = \hat{a}_p^\dagger(0) \exp(-\mathrm{i}\chi_p). \tag{5.104}$$

The two phases χ_p may be viewed as round-trip Gouy phases.

5.6.3 Structure and frequencies of resonator modes

In Section 5.4.2 a basis set of orthonormal modes in a paraxial optical system is defined in terms of the complex ray vectors $\mu_p(z)$. Periodic ray vectors produce periodic mode functions. The fundamental mode is given by Eq. (5.80), while higher-order modes are expressed in Eq. (5.81). As a result, the complete algebraic expressions for these modes are fully determined by the properties of the round-trip matrix $\mathbb{M}_{\rm rt}$, in particular their eigen-rays and eigenvalues. The effect of a round trip on the matrices R and T is that they are multiplied from the right by the diagonal matrix with matrix elements $\exp(i\chi_p)$. After a round trip, the matrix B does not acquire an additional phase, while the determinant $\det \mathsf{R}$ is multiplied by $\exp(\mathrm{i}(\chi_1+\chi_2))$. The resulting phase change of the fundamental mode after a round trip is then

$$u_{00}(\rho, 2L) = u_{00}(\rho, 0) \exp\left(-\mathrm{i}(\chi_1 + \chi_2)/2\right). \tag{5.105}$$

Because of the properties (5.104) of the raising operators, this gives for the higher-order modes (5.81) the round-trip property

$$u_{n_1 n_2}(\rho, 2L) = u_{n_1 n_2}(\rho, 0) \exp\left[-\mathrm{i}(n_1 + 1/2)\chi_1 - \mathrm{i}(n_2 + 1/2)\chi_2\right]. \tag{5.106}$$

The structure of the modes is illustrated in Fig. 5.5, in the case of two identical astigmatic mirrors.

For a resonant mode, the field (5.2), which is proportional to $u(\rho, z)\exp(ikz)$, must be periodic over a round trip, so the total phase difference $2kL - (n_1+1/2)\chi_1 - (n_2+1/2)\chi_2$ must be an integer multiple of 2π. The possible frequency values of a mode with transverse mode numbers n_1 and n_2 are

$$\omega = \frac{c}{2L}\left[2\pi q + \left(n_1 + \frac{1}{2}\right)\chi_1 + \left(n_2 + \frac{1}{2}\right)\chi_2\right], \tag{5.107}$$

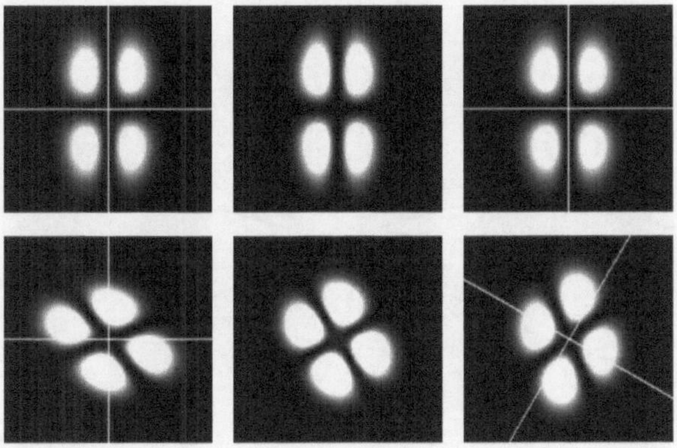

Figure 5.5 Intensity distribution of the $(n_1, n_2) = (1, 1)$ mode of an optical cavity between two identical astigmatic mirrors for two different relative orientations of the mirrors. The upper pictures show the modes in the case of parallel alignment and the lower pictures show the modes in the case of non-parallel alignment. The plots on the left and the right give the intensity patterns on the left and the right mirror, and the white crosses indicate the mirror orientation. The middle plots give the intensity pattern halfway the cavity. The two mirrors have radii of curvature $2L$ and $20L$, where L is the mirror separation.

with the integer q the longitudinal mode index. The spectrum depends only on the two generalized Gouy phases χ_1 and χ_2, and not on the eigenrays. This spectrum has the same structure for simple or general astigmatism. The behavior of the Gouy phases as a function of the relative orientation angle of the mirrors is sketched in Fig. 5.6, both in the case of identical mirrors, and for slightly different mirrors. For identical mirrors, the Gouy phases are identical for symmetry reasons when the mirrors are aligned perpendicular to each other. The crossing is avoided when the mirrors are slightly different.

Since the mode structure in the resonator returns to itself after a round trip, the corresponding mode transformation has a cyclic nature. This allows the phases χ_p that determine the spectrum of mode frequencies to be interpreted as geometric phases, which arise after a cyclic transformation in mode space [23].

5.6.4 Rays as displaced modes

In this section, we consider resonators where a ray of light can form a closed path in the resonator, so that it returns to itself after a limited number of round trips. Then real rays μ must exist that are eigenvectors of an integer power $\mathbb{M}_{\mathrm{rt}}^N$ of the round-trip matrix with eigenvalue 1. It was noticed a long time ago that resonators with closed-ray trajectories must have degenerate eigenfrequencies [24]. These degeneracies have been studied from the point of view of periodic orbits and their maps [25, 26]. These geometric modes have been observed in a laser with a non-astigmatic resonator that is pumped off-axis in single-frequency operation [27]. The off-axis geometric modes have been described by an expansion in terms of Hermite–Gaussian modes with a fixed total order [28], in analogy

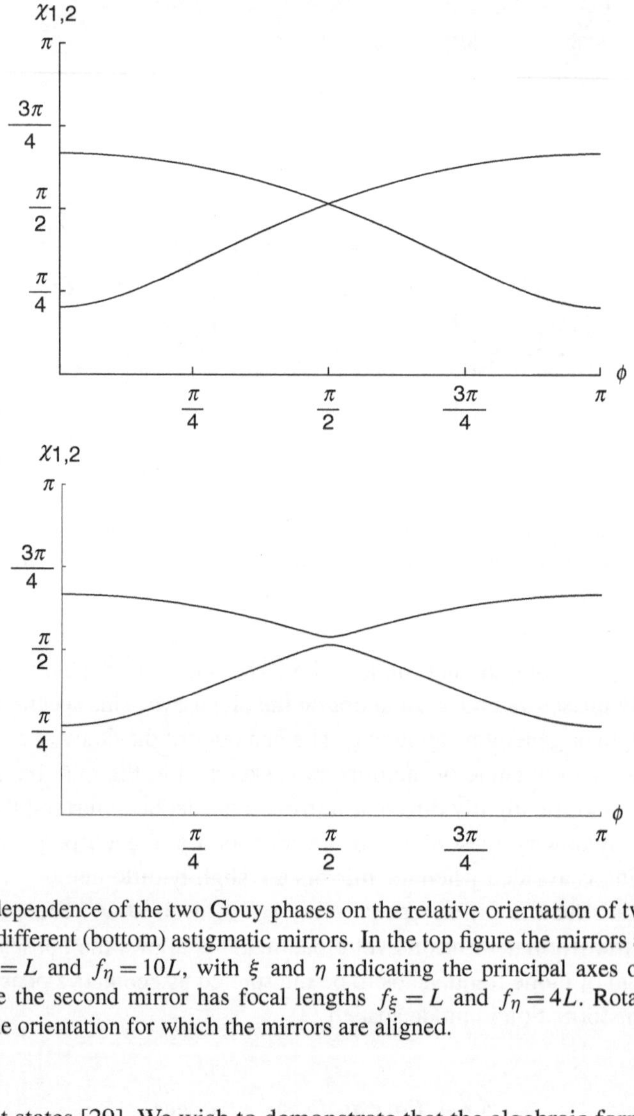

Figure 5.6 The dependence of the two Gouy phases on the relative orientation of two identical (top) and two slightly different (bottom) astigmatic mirrors. In the top figure the mirrors are identical with focal lengths $f_\xi = L$ and $f_\eta = 10L$, with ξ and η indicating the principal axes of the mirrors. In the bottom figure the second mirror has focal lengths $f_\xi = L$ and $f_\eta = 4L$. Rotation angle $\phi = 0$ corresponds to the orientation for which the mirrors are aligned.

to spin-coherent states [29]. We wish to demonstrate that the algebraic formalism outlined here allows a simple description of the consequences of frequency degeneracy of optical resonators, including the effects of astigmatism. A natural representation of an off-axis ray is a displaced fundamental mode, where the displacement operator is transforms the ground state of an oscillator (or the vacuum state of a quantized mode) into a Glauber coherent state [30]. In this way, we present a description of closed-ray trajectories in terms of the general astigmatic eigenmodes (5.81).

In ray optics, a ray propagating through the resonator, or through the equivalent lens guide, is defined by a real ray vector μ in the reference plane $z = 0$. The four complex ray eigenvectors μ_p and μ_p^* of the round-trip ray matrix $\mathbb{M}_{\mathrm{rt}}$ form a complete set, and the

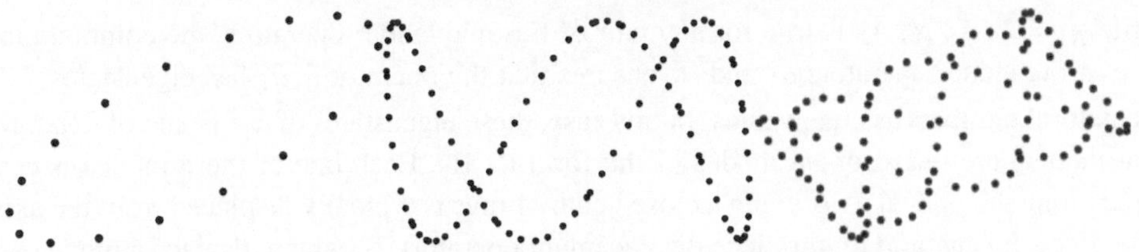

Figure 5.7 Hit points at a mirror of a ray in a resonator with degeneracy, in the case of a non-astigmatic resonator (left), a resonator with simple astigmatism (middle) and a resonator with general astigmatism (right).

expansion of an arbitrary ray vector μ in terms of these eigenrays has the general form of Eq. (5.77). In analogy to the expansion (5.83), we write for an arbitrary real ray vector in the reference plane

$$\mu(0) = \frac{1}{\sqrt{2}} \sum_p (\mu_p \alpha_p^* + \mu_p^* \alpha_p), \tag{5.108}$$

in terms of two complex parameters $\alpha_p = \mathrm{i}k\mu_p \mathbb{G}\mu/\sqrt{2}$. This is in accordance with the expansion (5.77). After a single round trip, this initial ray (5.108) is transformed into

$$\mu(2L) = \frac{1}{\sqrt{2}} \sum_p (\mu_p \alpha_p^* \exp(\mathrm{i}\chi_p) + \mu_p^* \alpha_p \exp(-\mathrm{i}\chi_p)). \tag{5.109}$$

Therefore, any ray in the reference plane forms a closed path after N round trips when the Gouy phases χ_1 and χ_2 are both a rational multiple of 2π with the same integer denominator N, so that $\chi_1 = 2\pi K_1/N$, $\chi_2 = 2\pi K_2/N$, with K_1, K_2 and N integers. Then $\mathbb{M}_{\mathrm{rt}}^N = \mathbb{M}(2NL)$ is the unit matrix. For simplicity we assume that the path of the ray closes for the first time after N round trips. This implies that the integers K_1, K_2 and N have no common divisor. In that case, the frequency spectrum (5.107) has degeneracies: for any mode with mode numbers n_1, n_2 and q, other modes with the same frequency are obtained after the replacement $n_1 \to n_1 + N$, and $q \to q - K_1$, or the replacement $n_2 \to n_2 + N$, $q \to q - K_2$. Figure 5.7 illustrates the closed path of a ray in a resonator with degeneracy for a resonator without astigmatism, with simple astigmatism and with general astigmatism [7]. The figure shows the hit points of the closed-ray path on one of the mirrors.

We wish to give a wave-optical description of closed paths of rays. For ease of notation, we restrict ourselves to closed paths that arise from a degeneracy condition of only one of the Gouy phases. An initial ray μ in the reference plane of the form

$$\mu = \frac{1}{\sqrt{2}} \left(\mu_1 \alpha_1^* + \mu_1^* \alpha_1\right), \tag{5.110}$$

is an eigenray of $\mathbb{M}_{\mathrm{rt}}^N = \mathbb{M}(2NL)$, with N integer, provided that $\chi_1 = 2\pi\mathrm{i}K/N$, with integer K and N. This eigenvalue must correspond to one of the Gouy phases, so that we can write $\chi_1 = 2\pi K/N$. As is true for any pair of bosonic ladder operators, the commutation rule of the ladder operators $\hat{a}_1$ and $\hat{a}_1^\dagger$ ensures that the operator $\hat{a}_1^\dagger \hat{a}_1$ has eigenstates with the natural numbers as eigenvalues. In this case, these eigenstates of $\hat{a}_1^\dagger \hat{a}_1$ (and of $\hat{a}_2^\dagger \hat{a}_2$, for that matter) are the resonant modes of the form (5.81). Each one of these modes is centered along the optical axis, while a closed path of rays is typically displaced from the axis. Therefore, it is natural to introduce displacement operators $\hat{D}$, which displace a mode over a given distance in phase space. This concept of displacement operators is well known in quantum optics, where they transform the vacuum state of a quantized mode in a coherent state [31]. These coherent states are eigenstates of the lowering operator. In the present case of ladder operators $\hat{a}_1$ and $\hat{a}_1^\dagger$ acting on mode space, we parameterize this displacement by the real ray μ within the eigenspace of $\mathbb{M}_{\mathrm{rt}}^N$. We denote the displacement operator as

$$\hat{D}(\mu) = \exp(-ik\mu\mathbb{G}\hat{\mu}) = \exp(\alpha_1 \hat{a}_1^\dagger - \alpha_1^* \hat{a}_1). \tag{5.111}$$

The equivalence of these two expressions (5.111) follows from the expansions (5.110) and (5.83), in combination with the properties (5.76).

The defining property of the displacement operator is expressed by the compact relation

$$\hat{D}^\dagger(\mu)\hat{\mu}\hat{D}(\mu) = \hat{\mu} + \mu, \tag{5.112}$$

for any ray μ in the eigenspace, or, in other words, for any ray μ of the form (5.110). This shows that when a mode u is transformed into $\hat{D}(\mu)u$ by the displacement operator, the expectation value of the ray operator $\hat{\mu}$ is shifted by the amount μ. Therefore, the operator $\hat{D}$ transforms a mode centered on the axis into a mode with a non-vanishing average value of position and/or direction angle.

5.6.5 Geometric mode

The mode propagating through the lens guide is defined by its profile $u(\rho, 0)$ in the reference plane $z = 0$. Consider now the case that this initial profile is the displaced fundamental mode, so that

$$u(\rho, 0) = \hat{D}(\mu)u_{00}(\rho, 0). \tag{5.113}$$

Since the displacement operator only contains the ladder operators $\hat{a}_1$ and $\hat{a}_1^\dagger$, the displaced mode is a linear combination of the modes $u_{n_1 0}$ out of the class (5.81), with $n_2 = 0$.

One round-trip downstream, the mode is transformed by the round-trip propagation operator, so that $u(\rho, 2L) = \hat{U}_{\mathrm{rt}} u(\rho, 0)$. The transformation property of the displacement operator under propagation can be expressed by the suggestive relation

$$\hat{U}_{\mathrm{rt}}\hat{D}(\mu)\hat{U}_{\mathrm{rt}}^\dagger = \hat{D}(\mathbb{M}_{\mathrm{rt}}\mu). \tag{5.114}$$

It can be checked by using the transformation property (5.68), and the symplectic property (5.72). Since $\mu = \mu(0)$ as expressed by (5.110) specifies the ray at $z = 0$, and μ_1 is the eigenray of the round-trip operator with eigenvalue $\exp(i\chi_1)$, the ray after one round trip is

$$\mu(2L) = \mathbb{M}_{rt}\mu(0) = \frac{1}{\sqrt{2}}(\mu_1\alpha_1^* e^{i\chi_1} + \mu_1^*\alpha_1 e^{-i\chi_1}). \tag{5.115}$$

The effect of the round-trip ray matrix on a ray μ is apparently that the complex parameter α_1 is multiplied by the factor $\exp(-i\chi_1)$. The result for the mode profile after one round trip is

$$u(\rho, 2L) = \hat{D}(\mathbb{M}_{rt}\mu)\hat{U}_{rt}u_{00}(\rho, 0) = \exp(-i\chi_0)\,\hat{D}(\mu(2L))u_{00}(\rho, 0), \tag{5.116}$$

where we write $\chi_0 = (\chi_1 + \chi_2)/2$ as the average of the two Gouy phases for short. In words, after each round trip, the mode picks up an additional phase factor $\exp(-i\chi_0)$, and the complex parameter α_1 that defines the ray acquires an additional factor $\exp(-i\chi_1)$. Recall that the Gouy phase χ_1 has the form $\chi_1 = 2\pi K/N$, while χ_2 is unconditioned.

After N phase changes χ_1, the parameter α_1 that specifies the ray has acquired the phase factor $\exp(-iN\chi_1) = 1$, so that $\mu(2NL) = \mu(0)$. Therefore, the displacement operator is periodic with period $2NL$, and the path of the ray is closed. The resonance frequencies of the mode now follow from the resonance requirement that $u_{00}(\rho, z)\exp(ikz)$ is periodic with the super-period $2NL$, which corresponds to N round trips. This gives the condition that $2NkL - N\chi_0$ must be equal to $2q\pi$, with integer q. The frequency condition is then

$$\omega = \frac{c}{2L}\left[\frac{2\pi q}{N} + \chi_0\right]. \tag{5.117}$$

The resonance frequencies for this lens-guide mode form an equidistant mesh separated by $c\pi/(NL)$, which is a factor $1/N$ smaller than the free spectral range of the resonator with length L.

The geometric mode in the resonator is obtained by refolding the lens guide into the resonator that is located between $z = 0$ and $z = L$. Each period $2L$ from the super-period $2NL$ gives a different contribution. The field in the resonator consists of N rays traveling back and forth between the mirrors. Each of these N rays is determined by its location and direction in the reference plane. The mode profile in this plane is proportional to the sum over these N rays, with the result

$$u(\rho, 0) + \exp(2ikL)u(\rho, 2L) + \cdots + \exp(2ikL(N-1))u(\rho, 2ikL(N-1)). \tag{5.118}$$

In view of Eq. (5.116), this summation can be written as an operator sum of the form [32]

$$\hat{D}(\mu(0)) + \exp(2\pi iq/N)\hat{D}(\mu(2L)) + \cdots + \exp(2\pi iq(N-1)/N)\hat{D}[\mu(2(N-1)L)], \tag{5.119}$$

acting on the mode $u_{00}(\rho, 0)$. One checks that each of the frequencies (5.117) corresponds to a value out of the set (5.107), for a limited set of mode indices n_1, that are separated by a multiple of N. In fact, the summation in (5.119) ensures that for each value of q the proper subset of modes $u_{n_1 0}$ survives.

When both Gouy phases have the rational form $\chi_1 = 2\pi K_1/N$, $\chi_2 = 2\pi K_2/N$ with the common denominator N, a similar line of reasoning holds. In this case, an arbitrary initial ray μ as expressed by (5.108) in the reference plane at $z = 0$ returns to itself after N round trips. The corresponding wave-optical description of the geometric mode still has the form (5.118), where now the displacement operator has the unrestricted form

$$\hat{D}(\mu) = \exp\left[\sum_p (\alpha_p \hat{a}_p^\dagger - \alpha_p^* \hat{a}_p)\right] = \exp(-ik\mu\mathbb{G}\hat{\mu}). \tag{5.120}$$

This treatment generalizes earlier work, where the geometric modes of a stable optical resonator were described in terms of displaced modes [32]. In the present work we have included the effect of astigmatism. A natural wave description of the geometric mode is that the shape around its center is just the fundamental mode $u_{00}(\rho, z)$ for the astigmatic resonator. The location of the center in ray space of the mode follows a ray $\mu(z)$, described by the ray matrix $\mathbb{M}(z)$ for the resonator. The displacement is simply described by the displacement operator $\hat{D}(\mu(z))$, which is a standard tool in quantum optics. This description in terms of displaced fundamental modes seems more natural than a description in terms of spin-coherent states [28].

5.7 Summary and conclusions

A central aspect of this contribution is the use of bosonic ladder operators for the description of optical modes. These operators are well known in the theory of harmonic oscillators in elementary quantum mechanics, and more generally in quantum field theory of bosonic particles, including photons. Once these operators have been identified, the commutator algebra of these operators is sufficient to define basis sets of states of the system. These states are analogs of the particle number states in field theory (or the photon number states in quantum optics). It is noteworthy, however, that ladder operators can be used to define basis sets of non-stationary states. We point out that ladder operators arise naturally for any system that is described by a quadratic Hamiltonian. The evolution of the dynamical variables of such a system is linear.

The common underlying structure of systems with a quadratic Hamiltonian is revealed in Section 5.2. As a simple relevant application the quantum harmonic oscillator in two dimensions is discussed in Section 5.3. This case is particularly useful for paraxial wave optics, because of an exact correspondence between solutions of the Schrödinger equation for the harmonic oscillator and solutions of the paraxial wave equation. The ladder-operator description is applied in Section 5.4 for an arbitrary optical system, and in Section 5.6 for astigmatic resonators. The correspondence between harmonic oscillators and paraxial modes is applied in Section 5.5.

Moreover, it is a well-known fact that the paraxial wave equation is mathematically equivalent to the Schrödinger equation for a free particle in two spatial dimensions. For instance, a Gaussian paraxial mode before focus is analogous to a particle wave function that first contracts to a smaller extension in space, before diffracting to its expanding asymptotic long-time behavior. This initial contraction arises from an inward momentum distribution, which corresponds to the negative curvature before focus. For a free atom, these states have been termed contractive states [33]. The technique presented in this paper leads to explicit expressions for basis sets of time-dependent wave functions for a free quantum particle in three dimensions.

The description that we presented here applies both to quantum-mechanical systems and to wave optics. In the former case, the ladder operators that generate the basis sets of solutions are determined by classical trajectories in complex phase space. In the latter case, the ladder operators are determined by rays of light with complex values of the position and the direction angles.

References

[1] M. Lax, W. H. Louisell and W. B. McKnight, From Maxwell to paraxial wave optics, *Phys. Rev. A* **11**, 1365 (1975).

[2] D. J. Griffiths, *Introduction to Quantum Mechanics* (Upper Saddle River, NJ: Pearson Prentice Hall, 2005).

[3] A. E. Siegman, *Lasers* (Mill Valley, CA: University Science Books, 1986).

[4] G. Nienhuis and L. Allen, Paraxial wave optics and harmonic oscillators, *Phys. Rev. A* **48**, 656 (1993).

[5] G. Nienhuis and J. Visser, Angular momentum and vortices in paraxial beams, *J. Opt. A: Pure Appl. Opt.* **6**, 248 (2004).

[6] J. Schwinger, On angular momentum, in: L. C. Biedenharn and H. van Dam (eds.), *Quantum Theory of Angular Momentum* (New York: Academic Press, 1965), p. 229.

[7] S. J. M. Habraken and G. Nienhuis, Modes of a twisted optical cavity, *Phys. Rev. A* **75**, 033819 (2007).

[8] H. Kogelnik and T. Li, Laser beams and resonators, *Appl. Opt.* **5**, 1550 (1966).

[9] H. Kogelnik, On the propagation of Gaussian beams of light through lens-like media including those with a loss or gain variation, *Appl. Opt.* **4**, 1562 (1965).

[10] R. Simon and N. Mukunda, Iwasawa decomposition in first-order optics: universal treatment of shape-invariant propagation for coherent and partially coherent beams, *J. Opt. Soc. Am. A* **15**, 2146 (1998).

[11] J. A. Arnaud and H. Kogelnik, Gaussian light beams with general astigmatism, *Appl. Opt* **8**, 1687 (1969).

[12] J. Visser and G. Nienhuis, Orbital angular momentum of general astigmatic modes, *Phys. Rev. A* **70**, 013809 (2004).

[13] L. Allen, M. W. Beijersbergen, R. J. C. Spreeuw and J. P. Woerdman, Orbital angular momentum of light and the transformation of Laguerre–Gaussian laser modes, *Phys. Rev. A* **45**, 8185 (1992).

[14] J. Courtial, K. Dholakia, L. Allen and M. J. Padgett, Gaussian beams with very high orbital angular momentum, *Opt. Commun.* **144**, 210 (1997).

[15] G. Nienhuis, Angular momentum and vortices in optics, in: D. L. Andrews (ed.), *Structured Light and its Applications* (Upper Saddle River, NJ: Academic Press, 2008), p. 19.
[16] S. J. van Enk and G. Nienhuis, Eigenfunction description of laser beams and orbital angular momentum of light, *Opt. Commun.* **94**, 147 (1992).
[17] O. Steuernagel, Equivalence between focused paraxial beams and the quantum harmonic oscillator, *Am. J. Phys.* **73**, 625 (2005).
[18] C. Tamm and C. O. Weiss, Bistability and optical switching of spatial patterns in a laser, *J. Opt. Soc. Am. B* **7**, 1034, (1990).
[19] E. G. Abramochkin and V. G. Volostnikov, Generalized Gaussian beams, *J. Opt. A: Pure Appl. Opt.* **6**, 157 (2004).
[20] M. A. Bandres and J. C. Gutiérrez-Vega, Ince-Gaussian beams, *Opt. Lett.* **29**, 144 (2004).
[21] R. Borghi, M. Santarsiero and R. Simon, Shape invariance and a universal form for the Gouy phase, *J. Opt. Soc. Am. A* **21**, 572 (2004).
[22] M. Born and E. Wolf, Principles of Optics (New York: Pergamon, 1980).
[23] S. J. M. Habraken and G. Nienhuis, Geometric phases in astigmatic optical modes of arbitrary order, *J. Math. Phys.* **51**, 082702 (2010).
[24] I. A. Ramsay and J. J. Degnan, A ray analysis of optical resonators formd by two spherical mirrors, *Appl. Opt.* **9**, 385 (1970).
[25] C. C. Palma, Complex dynamics of a beam in a Gaussian cavity, *Opt. Commun.* **129**, 120 (1996).
[26] L. M. Sánchez and A. A. Hnilo, Optical cavities as iterative maps in the complex plane, *Opt. Commun.* **166**, 229 (1999).
[27] J. Dingjan, M. P. van Exter and J. P. Woerdman, Geometric modes in a single-frequency Nd:YVO_4 laser, *Opt. Commun.* **188**, 345 (2001).
[28] Y. F. Chen, Y. P. Lan and K. F. Huang, Observation of quantum-classical correspondence from high-order transverse patterns, *Phys. Rev. A* **68**, 043803.
[29] F. T. Arecchi, E. Courtens, R. Gilmore and H. Thomas, Atomic coherent states in quantum optics, *Phys. Rev. A* **6**, 2146 (1972).
[30] R. J. Glauber, Coherent and incoherent states of the radiation field, *Phys. Rev.* **131**, 2766 (1963).
[31] L. Mandel and E. Wolf, *Optical Coherence and Quantum Optics* (Cambridge: Cambridge University Press, 1995).
[32] J. Visser, N. J. Zelders and G. Nienhuis, Wave description of geometric modes of a resonator, *J. Opt. Soc. Am. A* **8**, 1559 (2005).
[33] P. Storey, T. Sleator, M. Collett and D. Walls, Contractive states of a free atom, *Phys. Rev. A* **49**, 2322 (1994).

6
Quantum formulation of angle and orbital angular momentum

J. B. GÖTTE AND S. M. BARNETT

The quantum mechanical description of the azimuthal rotation angle and the orbital angular momentum around the polar axis differs greatly from the description of position and momentum. This has led to a controversy over the existence of a self-adjoint angle operator in the literature. It is possible to circumvent the problems of defining a self-adjoint operator for a periodic variable by using trigonometric functions of operators as a basis for the quantum mechanical description. This approach, however, does not allow us to study the properties of the angle operator itself. By using a state space of an arbitrarily large yet finite number of dimensions, it is possible to introduce angle and orbital angular momentum as a conjugate pair of variables both represented by Hermitian matrices. After physical and measurable quantities have been calculated in this state space the number of dimensions is allowed to tend to infinity in a limiting procedure.

The first part of this chapter reviews the difficulties associated with the quantum mechanical formulation of angle and orbital angular momentum and presents a way to overcome these problems using a finite-dimensional state space. In the second part this formulation is generalised to include the phenomenon of non-integer orbital angular momentum. This part is based on a chapter of the PhD thesis of Götte [1], where further details may be found.

6.1 Introduction

The correct description of a periodic variable such as a rotation angle or the optical phase has been a long standing problem in quantum mechanics. At the root of the problem is the question of whether the variable itself is restricted in a 2π radian range or whether it evolves continuously without bound. In the latter case the periodicity of the system would be contained in derived physical quantities, which would be periodic in the variable. The difficulties in the quantum mechanical description, however, are directly connected to the periodicity of the variable, and similar problems arise for the conjugate pair of photon number and optical phase. Naturally both settings are related to the correct description of

The Angular Momentum of Light, ed. D. L. Andrews and M. Babiker. Published by Cambridge University Press.

the phase of a harmonic oscillator. This is why most of the literature we are referring to in this chapter covers the question of the phase of the harmonic oscillator or equivalently the phase of a single mode optical field.

If the phase of the harmonic oscillator χ and the rotation angle φ are treated as continuously evolving variables, the classical Poisson brackets with the Hamiltonian H, or the orbital angular momentum along the rotation axis L_z, have a standard form for a conjugate pair of variables [2, 3]. But a problem arises if these Poisson brackets are translated into quantum mechanical commutators according to the correspondence principle. Louisell [4] noted that the commutator $[\hat{\chi}, \hat{N}] = \mathrm{i}$ shows inconsistencies. This commutator corresponds to the Poisson bracket $\{\chi, H\}$ because the Hamiltonian can be written in terms of the number operator $\hat{N}$ [5]. In the number state basis $\{|n\rangle\}_{n\in\mathbb{N}_0}$, where n can take on every positive integer value and zero, the matrix elements of the commutator are given by

$$\langle n|[\hat{\chi}, \hat{N}]|n'\rangle = \mathrm{i}\delta_{nn'} \,. \tag{6.1}$$

This requires that all diagonal matrix element of the phase operator are undefined, as the number states are the eigenstates of the number operator with $\hat{N}|n\rangle = n|n\rangle$ and the matrix elements of the phase operator $\hat{\chi}$ are therefore given by

$$(n' - n)\langle n|\hat{\chi}|n'\rangle = \delta_{nn'} \,. \tag{6.2}$$

This would make the off-diagonal elements zero, which makes $\hat{\chi}$ and $\hat{N}$ commuting variables. A similar inconsistency can be derived for orbital angular momentum and angle [3]. The commutator $[\hat{\varphi}, \hat{L}_z] = \mathrm{i}\hbar$ corresponding to the classical Poisson bracket $\{\varphi, L_z\} = 1$ leads to undetermined diagonal matrix elements of the angle operator:

$$(m' - m)\langle m|\hat{\varphi}|m'\rangle = \mathrm{i}\hbar\delta_{mm'} \,, \tag{6.3}$$

where $|m\rangle, m \in \mathbb{Z}$ are eigenstates of $\hat{L}_z$ with eigenvalues $\hbar m$. It is important to note that despite all similarities in the problem of finding the correct quantum mechanical description for either the phase or the rotation angle, there is a difference in the eigenvalue spectrum of the respective conjugate variable. Whereas the angular momentum eigenstates $|m\rangle$ can take on all integer values, positive, negative and zero, the number states $|n\rangle$ are only defined for positive values of n and zero.

A different commutator is obtained if one assumes that the periodic variable is strictly bound within a 2π radian interval from θ_0 to $\theta_0 + 2\pi$. The angle or phase operator acts then not as simple multiplication, but is modified by a series of step functions [6]:

$$Y(\varphi) = \varphi - 2\pi \sum_{p=1}^{\infty} \Theta(\varphi - \theta_0 - 2p\pi) + 2\pi \sum_{p=1}^{\infty} \Theta(-\varphi - \theta_0 - 2p\pi), \tag{6.4}$$

where $\Theta(x)$ is the unit step function with $\Theta(x) = 0$ for $x < 0$ and $\Theta(x) = 1$ for $x > 0$. The Judge–Lewis commutator $[\hat{Y}(\varphi), \hat{L}_z]$ exhibits a series of Dirac-δ functions:

$$[\hat{Y}(\varphi), \hat{L}_z] = \mathrm{i}\hbar[1 - 2\pi\delta_{2\pi}(\varphi - \theta_0)], \tag{6.5}$$

where $\delta_{2\pi}$ denotes the 2π periodic δ-function [7]. Similar results can be obtained for the periodic phase variable [8]. Interestingly, the correspondence between Poisson bracket and commutator is restored, if the classical phase or angle variable is treated similarly [9]; instead of a smoothly evolving variable, the angle or the phase show discontinuities whenever the end of the 2π radian is reached. We shall discuss this further in the following section.

While this approach yields correct results for the matrix elements of the angle or phase operator in the basis states of the conjugate variable [10], it fails to give well-defined matrix elements for the angular momentum operator $\hat{L}_z$ or number operator $\hat{N}$ in the respective basis states of the angle or phase operator [3, 9]. For the matrix elements of the number operator $\hat{N}$ in the phase state basis this is a clear consequence of the missing negative numbers in the spectrum of the number operator [8]. But even in the presence of negative quantum numbers, as in the case of the angular momentum operator $\hat{L}_z$, a careful analysis shows that the matrix elements in the angle state basis are undefined [3]. This is because phase and angle states, as eigenstates of a Hermitian phase or angle operator, cannot be represented in an infinite-dimensional Hilbert space. This shortcoming of the Hilbert space to accommodate angle or phase states has been seen as an indication that a Hermitian angle or phase operator does not exist [11]. However, eigenstates of the linear momentum and position operator also are not elements of the Hilbert space $L^2(-\infty, \infty)$, on which their wavefunctions in position or momentum representation are defined.

The search for a quantum mechanical description of the angle or phase operator has attracted some controversy (see [8] and the extensive bibliography therein). A very different approach exists which does not consider a Hermitian phase operator, but trigonometric operators [12]:

$$\widehat{\exp}(i\chi) = \sum_{n=0}^{\infty} |n\rangle\langle n+1|, \tag{6.6a}$$

$$\widehat{\exp}(-i\chi) = \sum_{n=0}^{\infty} |n+1\rangle\langle n|. \tag{6.6b}$$

This approach has been developed for the phase variable. For a unique determination in the form given above Susskind and Glogower [12] had to add the postulate that the projector onto the vacuum state $|\psi\rangle\langle 0|$ is absent in the expressions in Eq. (6.6). These operators are not unitary and do not commute. It is therefore not possible to deduce the existence of a unique phase operator in this approach, as the operators in Eq. (6.6) cannot be functions of a common phase operator. The approach by Susskind and Glogower can also be applied to angle and orbital angular momentum [11]. In this case the operators are unitary, because of the inclusion of negative quantum numbers for the angular momentum. However, the angle operator deduced from the exponential operator acts on the infinite-dimensional space and the pair of angle and angular momentum operator in this approach suffers from the same difficulties with the indefiniteness of matrix elements as mentioned above.

The problem with a phase or angle operator is that their eigenstates cannot be accommodated in an infinite-dimensional space. It is, however, possible to define a Hermitian phase and angle operator on a finite-dimensional space. To this end, Pegg and Barnett introduced a state space Ψ of $2L + 1$ dimensions, where L is arbitrary but finite [2, 3, 9, 13]. Within this state space the eigenstates of the operators form an orthonormal basis and all matrix elements are well defined. The operators are Hermitian and owing to the finite dimension also self-adjoint. After physical results have been calculated, such as expectation values or matrix elements, L is allowed to tend to infinity, restoring an infinite, but countable set of orbital angular momentum (OAM) or number states and a respective dense set of angle or phase states. The limiting procedure of starting in a space with a finite number of dimensions, which is allowed to tend to infinity only after measurable results have been calculated, is therefore crucial to the approach by Barnett and Pegg [9]. The mathematical reasoning behind this approach is that the convergence behaviour for operators is different from that of matrix elements or expectation values [14]. The former are defined by their actions on elements of a Hilbert space whereas the latter are elements of the space of complex numbers. The question of convergence of the phase operator has been addressed by Vaccaro [15] who introduced an infinite Hilbert space on which a unitary and Hermitian phase operator can be defined. The phase operator introduced by Barnett and Pegg [9] converges strongly on this particular Hilbert space.

In the following we present the quantum mechanical description of rotation angles in the approach by Barnett and Pegg [3] in more detail. This is a very natural formulation which allows us to investigate properties of the angle and OAM operators directly.

6.2 Quantum theory of rotation angles

The essential simplicity in the finite-dimensional approach lies in the fact that all the desired properties of a pair of conjugate operators occur naturally within this method. Angle and OAM are complementary variables: they act as mutual generators and they also give the correct commutator analogous to Eq. (6.5) [10]. Obviously an elegant formulation alone is not a proof for the validity of a theory. A consistent formulation though is a necessary requirement and in this section we show that the finite-dimensional approach by Barnett and Pegg [3] describes angle and OAM consistently.

In this section we look first at the definition of angle and OAM states in finite-dimensional state space. With these states we can then proceed to calculate the commutator for angle and OAM. Angle states cannot be prepared in an experiment, and we therefore review the concept of physical states at the end of this section.

6.2.1 Angle and orbital angular momentum states

The finite-dimensional space state Ψ is spanned by $2L + 1$ eigenstates $|m\rangle$ of the OAM operator $\hat{L}_z$. The action of $\hat{L}_z$ on one of the eigenstates is defined by $\hat{L}_z|m\rangle = m\hbar|m\rangle$. The positive integer L is arbitrary and the quantum number m takes on the values $-L$,

$-L+1, \ldots, L$. Angle states can be unambiguously decomposed in the basis of the OAM states by writing:

$$|\varphi\rangle = \frac{1}{\sqrt{2L+1}} \sum_{m=-L}^{L} \exp(-im\varphi)|m\rangle. \tag{6.7}$$

In this form angle states have a periodic structure; the state $|\varphi\rangle$ is the same as $|\varphi + 2\pi\rangle$. This is only natural, as rotation angles differing by 2π radians are indistinguishable from each other. All distinct angle states are therefore contained in an interval ranging from an arbitrary starting angle, which we denote θ_0, up to, but not including, $\theta_0 + 2\pi$, that is $\varphi \in [\theta_0, \theta_0 + 2\pi)$. This interval includes a continuum of angle states that are not all mutually orthogonal:

$$\langle\varphi|\varphi'\rangle = \frac{1}{2L+1} \frac{\sin[(2L+1)(\varphi' - \varphi)/2]}{\sin[(\varphi' - \varphi)/2]}. \tag{6.8}$$

The states $|\varphi\rangle$ and $|\varphi'\rangle$ are orthogonal when the difference $\varphi' - \varphi$ is a non-zero, integer multiple of $2\pi/(2L+1)$. It is, therefore, possible to form an orthonormal basis of $2L+1$ angle states $|\theta_n\rangle$ by selecting values $|\theta_n\rangle$ as

$$\theta_n = \theta_0 + \frac{2\pi n}{2L+1}, \quad n = 0, 1, \ldots, 2L. \tag{6.9}$$

With this choice for the allowed angles all states in the set $\{|\theta_n\rangle\}_{n=0,\ldots,2L}$ are mutually orthonormal to each other:

$$\langle\theta_n|\theta_{n'}\rangle = \delta_{nn'}. \tag{6.10}$$

It is very important to note that the choice of θ_0 is arbitrary and determines a particular basis set. Angle states from two different basis sets, with two different values for θ_0, will not be orthogonal in general. The choice of θ_0 determines a particular angle operator which will not, in general, commute with other angle operators for a different starting angle θ_0. The angle operator $\hat{\varphi}_{\theta_0}$ thus has to be labelled with θ_0 indicating that the eigenvalues are given by Eq. (6.9):

$$\hat{\varphi}_{\theta_0} = \sum_{n=0}^{2L} \theta_n |\theta_n\rangle\langle\theta_n| = \theta_0 + \sum_{n=0}^{2L} \frac{2\pi n}{2L+1} |\theta_n\rangle\langle\theta_n|. \tag{6.11}$$

The angle and OAM are both defined on a space of $2L+1$ dimensions. In this formulation both basis sets can be used to represent any element of the state space Ψ equally well. In particular the angle basis can be represented in the basis states of the OAM operator and vice versa. It is perhaps surprising that in the finite-dimensional approach the angle and OAM bases span the same state space, while the two state spaces for OAM and angle are different in character when working directly in infinite dimensions. There, the OAM is characterised as a discrete and unbounded variable, while the angle is bounded and continuous. However, in mathematics a number of projections are known which map an unbounded domain on to a compact range. For example, the whole complex plane can be projected onto a Riemann sphere [16]. With increasing L the interval for $m \in [-L, L]$ is expanded, while the interval for the angle $\theta_n \in [\theta_0, \theta_0 + 2\pi)$ remains the same. The spacing of two consecutive angles, however, decreases with increasing L. In the limit of $L \to \infty$ the OAM eigenvalues take on all numbers $m \in \mathbb{Z}$, while the angle eigenvalues form a dense set on the circle $[\theta_0, \theta_0 + 2\pi]$. The starting angle θ_0 retains its importance in the limit of $L \to \infty$. As mentioned above, θ_0 determines a particular angle operator and the uncertainty and variance of this operator will be different from other angle operators for different values of θ_0.

6.2.2 Commutator for angle and orbital angular momentum

The expectation value of the commutator determines the uncertainty relation, which will be studied in more detail in the next section. We therefore derive the commutator for the angle and OAM operator in both representations. In order to express the angle operator in the OAM basis we have to calculate the matrix elements in the expansion

$$\hat{\varphi}_{\theta_0} = \sum_{m,m'=-L}^{L} |m'\rangle\langle m'|\hat{\varphi}_{\theta_0}|m\rangle\langle m|. \tag{6.12}$$

On using the completeness of the angle states we can calculate the matrix elements $\langle m'|\hat{\varphi}_{\theta_0}|m\rangle$ as a sum over exponentials:

$$\langle m'|\hat{\varphi}_{\theta_0}|m\rangle = \sum_{n=0}^{2L} \theta_n \langle m'|\theta_n\rangle\langle\theta_n|m\rangle = \frac{1}{2L+1}\sum_{n=0}^{2L} \theta_n \exp[\mathrm{i}(m-m')\theta_n]. \tag{6.13}$$

The diagonal matrix elements involve a summation of the positive integer n from zero to $2L$. This sum is given by $(2L + 1)L$, which allows us to evaluate the diagonal matrix elements to

$$\langle m|\hat{\varphi}_{\theta_0}|m\rangle = \theta_0 + \frac{2\pi L}{2L+1}. \tag{6.14}$$

For the off-diagonal elements we can use the geometrical progression to evaluate the sum

$$\langle m'|\hat{\varphi}_{\theta_0}|m\rangle = \frac{2\pi}{2L+1}\frac{\exp[\mathrm{i}(m-m')\theta_0]}{\exp[\mathrm{i}(m-m')2\pi/(2L+1)]-1}. \tag{6.15}$$

With the help of these matrix elements we can express the angle operator in the OAM basis as

$$\hat{\varphi}_{\theta_0} = \theta_0 + \frac{2\pi}{2L+1}\left[L + \sum_{\substack{m,m'=-L \\ m\neq m'}}^{L} \frac{\exp[\mathrm{i}(m-m')\theta_0]|m'\rangle\langle m|}{\exp[\mathrm{i}(m-m')2\pi/(2L+1)]-1}\right]. \tag{6.16}$$

Using the expression of the angle operator in the OAM basis in Eq. (6.16) we can write the commutator in the same basis as

$$[\hat{\varphi}_{\theta_0}, \hat{L}_z] = \frac{2\pi\hbar}{2L+1} \sum_{\substack{m,m'=-L \\ m\neq m'}}^{L} \frac{(m-m')\exp[\mathrm{i}(m-m')\theta_0]|m'\rangle\langle m|}{\exp[\mathrm{i}(m-m')2\pi/(2L+1)]-1}. \tag{6.17}$$

To express the commutator in the angle state basis we can use the representation of $|m\rangle$ in terms of angle states:

$$|m\rangle = \sum_{n=0}^{2L} |\theta_n\rangle\langle\theta_n|m\rangle = \frac{1}{\sqrt{2L+1}} \sum_{n=0}^{2L} \exp(\mathrm{i}m\theta_n)|\theta_n\rangle, \tag{6.18}$$

and write the OAM operator in the basis of angle states:

$$\hat{L}_z = -\frac{i\hbar}{2} \sum_{\substack{n,n'=0 \\ n\neq n'}}^{2L} \frac{(-1)^{n-n'}|\theta_{n'}\rangle\langle\theta_n|}{\sin[(n-n')\pi/(2L+1)]}. \tag{6.19}$$

The commutator can then be written in the same basis as:

$$[\hat{\varphi}_{\theta_0}, \hat{L}_z] = \mathrm{i}\frac{\hbar\pi}{2L+1} \sum_{\substack{n,n'=0 \\ n\neq n'}}^{2L} \frac{(n-n')(-1)^{n-n'}|\theta_{n'}\rangle\langle\theta_n|}{\sin[(n-n')\pi/(2L+1)]}. \tag{6.20}$$

This commutator has well-defined matrix elements and does not suffer from the difficulties mentioned in the introduction to this chapter.

6.2.3 Physical states

Angle states are impossible to realise experimentally. This is because in the limit as L tends to infinity, we would require a superposition of states with unbounded OAM. For all practical purposes it is more interesting to consider physical states $|p\rangle$ for which all the moments of the OAM operator $\hat{L}_z$ are finite. Within the finite state space Ψ this condition seems to be fulfilled trivially. As long as the sum of the coefficients $\langle m|p\rangle$ in the decomposition of $|p\rangle$ into the angular momentum basis

$$|p\rangle = \sum_{m=-L}^{L} \langle m|p\rangle|m\rangle \tag{6.21}$$

is finite, any finite moment $\langle p|\hat{L}_z^n|p\rangle$, $n \in \mathbb{N}$ will be finite. A moment of an operator, however, is a physical quantity for which we can allow L to tend to infinity. If we require all moments to be finite even in the limit of $L \to \infty$, the coefficients have to fall off sufficiently quickly [2]. Such states may be approximated to any degree by an expansion in the form of Eq. (6.21) where the coefficients $\langle m|p\rangle$ are zero for $|m| > M$. The bound M is as large as required for the desired accuracy in the approximation but less than L. If we restrict the domain of $\hat{\varphi}_{\theta_0}$ to physical states, we can employ a physical angle operator in the limit of large L by expanding the exponential in the denominator:

$$(\hat{\varphi}_{\theta_0})_p = \theta_0 + \pi - \mathrm{i} \sum_{\substack{m,m \\ m\neq m'}} \frac{\exp[\mathrm{i}(m-m')\theta_0]}{m-m'} |m'\rangle\langle m|. \tag{6.22}$$

We should stress that this is a deviation from the rule that the limit of $L \to \infty$ is only to be taken after physical results have been calculated. This is why the approximate form in Eq. (6.22) can only replace $\hat{\varphi}_{\theta_0}$ when operating on physical states, and not on the angle states themselves, which cannot be so approximated. Using the physical form of the angle operator we can also give a specialised expression for the commutator when acting on physical states:

$$[\hat{\varphi}_{\theta_0}, \hat{L}_z]_p = -\mathrm{i}\hbar \sum_{\substack{m,m' \\ m\neq m'}} \exp[\mathrm{i}(m-m')\theta_0]|m'\rangle\langle m|. \tag{6.23}$$

It is possible to calculate matrix elements for this physical commutator without another limiting procedure. Again, this is a deviation from the normal procedure, where the limit is taken only after expectation values and matrix elements have been calculated. For the matrix elements in the OAM basis the relevant limit has been applied already by forming the physical angle operator. The physical commutator has matrix elements given by

$$\langle m'|[\hat{\varphi}_{\theta_0}, \hat{L}_z]_p|m\rangle = -\mathrm{i}\hbar(1-\delta_{mm'})\exp[\mathrm{i}(m-m')\theta_0]. \tag{6.24}$$

Orbital angular momentum eigenstates are physical states and this is why we can use the physical form of the commutator to calculate the matrix elements in Eq. (6.24). However, the coefficients in the decomposition of an OAM eigenstate into the angle basis do not fall off with increasing $|m|$. Therefore the commutator in Eq. (6.20) needs to contain values of $n - n'$ up to $2L + 1$. It is possible to substitute the definition of angle states in terms of OAM states (see Eq. (6.7)),

$$|\theta_n\rangle = \frac{1}{\sqrt{2L+1}} \sum_{m=-L}^{L} \exp(-\mathrm{i}m\theta_n)|m\rangle, \tag{6.25}$$

into the expression for the physical commutator in the OAM basis:

$$[\hat{\varphi}_{\theta_0}, \hat{L}_z]_p = \mathrm{i}\hbar[1 - (2L+1)|\theta_0\rangle\langle\theta_0|]. \tag{6.26}$$

For any physical state $|p\rangle$ the expectation value of the commutator will be

$$\langle p|[\hat{\varphi}_{\theta_0}, \hat{L}_z]_p|p\rangle = \mathrm{i}\hbar[1 - (2L+1)|\langle p|\theta_0\rangle|^2]. \tag{6.27}$$

If we introduce an angle probability measure $P(\theta_n)$ within the finite state space Ψ, the probability that the angle is found within $\delta\theta$ from θ_n is given by $P(\theta_n)\delta\theta$. The distance between two consecutive angles $\delta\theta$ is given by $\delta\theta = 2\pi/(2L+1)$. With this probability measure the expectation value of the commutator can be written as

$$\langle p|[\hat{\varphi}_{\theta_0}, \hat{L}_z]_p|p\rangle = \mathrm{i}\hbar[1 - 2\pi P(\theta_0)]. \tag{6.28}$$

This is reminiscent of the expectation value of the Judge–Lewis commutator, Eq. (6.5), which we now see as an approximation valid for physical states. There is one crucial difference, however, the operator $\hat{Y}(\phi)$ is assumed to correspond to an unbounded angle variable but the more rigorous $\hat{\phi}_{\theta_0}$ has eigenvalues restricted to lie within a 2π interval.

In the limit of $L \to \infty$ the spacing between two consecutive angles tends to zero and the probability measure turns into a probability density of the continuous variable φ, which is normalised according to

$$\int_{\theta_0}^{\theta_0+2\pi} P(\varphi)\mathrm{d}\varphi = 1. \tag{6.29}$$

In the following sections we will be using the continuous probability density when we deal with physical states. The expectation value of the commutator also dictates the lower bound for the uncertainty product. For physical states we can write

$$(\Delta L_z)_p(\Delta\varphi)_p \geq \frac{\hbar}{2}|1 - 2\pi P(\theta_0)|, \tag{6.30}$$

but it is important to keep in mind that this is an approximate form, which is derived from the exact form in Eq. (6.20). We note that this uncertainty relation was first proposed in [17] but not the associated angle operator. The angle uncertainty $(\Delta\varphi)_p$ is given by the square root of the angle variance for a physical state $(\Delta\varphi)_p^2$:

$$(\Delta\varphi)_p^2 = \int_{\theta_0}^{\theta_0+2\pi} \varphi^2 P(\varphi)\mathrm{d}\varphi - \left(\int_{\theta_0}^{\theta_0+2\pi} \varphi P(\varphi)\mathrm{d}\varphi\right)^2. \tag{6.31}$$

For the OAM uncertainty the formulation in terms of a continuous angle representation requires more care. A direct representation of $\hat{L}_z$ as derivative with respect to φ can lead to problems even for physical states [3]. A physical state was defined above as a state for which all moments of the OAM operator $\hat{L}_z$ are finite. This includes the second moment $\langle\hat{L}_z^2\rangle_p$, which forms part of the OAM variance. The coefficients $\langle m|p\rangle$ in the decomposition in Eq. (6.21) thus have to fall off faster than m^{-2} for large m. As pointed out by Pegg *et al.* [18], states with this fall-off characteristic show a discontinuity in the gradient of the angle representation $p(\varphi) = \langle\varphi|p\rangle$. Acting with $\hat{L}_z^2$ on such a state can lead to singularities which render the states not normalisable. By using the self-adjointness of the OAM operator we can represent $\langle\hat{L}_z^2\rangle_p$ by

$$\langle\hat{L}_z^2\rangle_p = \int_{\theta_0}^{\theta_0+2\pi} \left|-\mathrm{i}\hbar\frac{\mathrm{d}p(\varphi)}{\mathrm{d}\varphi}\right|^2 \mathrm{d}\varphi. \tag{6.32}$$

The OAM uncertainty is thus given by the square root of

$$(\Delta L_z)_p^2 = \hbar^2 \int_{\theta_0}^{\theta_0+2\pi} \left|\frac{\mathrm{d}p}{\mathrm{d}\varphi}\right|^2 \mathrm{d}\varphi + \hbar^2 \left(\int_{\theta_0}^{\theta_0+2\pi} \frac{\mathrm{d}p}{\mathrm{d}\varphi}\mathrm{d}\varphi\right)^2. \tag{6.33}$$

In this class of physical states fall the intelligent states, which satisfy the equality in the uncertainty relation. Another class of physical states are the minimum uncertainty product states, which give a constrained or global minimum in the uncertainty product.

6.3 Intelligent and minimum uncertainty states

An uncertainty relation places an intrinsic restriction on the possibility of simultaneously defining the values of a pair of incompatible observables. For the (linear) position x and momentum p_x this takes the familiar form

$$\Delta x \Delta p_x \geq \frac{\hbar}{2}. \tag{6.34}$$

For the angle and angular momentum we have seen that with the restriction to physical states the uncertainty relation is

$$\Delta L_z \Delta\varphi \geq \frac{\hbar}{2}|1 - 2\pi P(\theta_0)|. \tag{6.35}$$

As we cannot fix both variables precisely, it is natural to ask how well we can specify them. There are two obvious ways to address this question and these lead to the intelligent states and the minimum uncertainty product states [19]. The intelligent states are those for which the equality holds in the uncertainty relation. The minimum uncertainty product states, however, are those for which the uncertainty in one observable is minimised for a given uncertainty in the other. For the $x - p_x$ uncertainty relation these two sets of states are the same and are simple Gaussian functions of x or p_x. For the angular variables, however, the state-dependent lower bound associated with $P(\theta_0)$ means that they are distinct states. We address each type in turn.

6.3.1 Intelligent states

Consider the uncertainty relations for a pair of observables A and B [20]

$$\Delta A \Delta B \geq \frac{1}{2}|\langle[\hat{A}, \hat{B}]\rangle. \tag{6.36}$$

An intelligent state, if it exists, will satisfy the *equality* in this uncertainty relation. It is straightforward to show that such a state $|g\rangle$ will satisfy an eigenvalue equation of the form [21]

$$[\hat{A} - \langle\hat{A}\rangle - \mathrm{i}\lambda(\hat{B} - \langle\hat{B}\rangle)]|g\rangle = 0, \tag{6.37}$$

where λ is a real constant. Hence, for our angle and angular momentum, we seek intelligent states, g, satisfying the equation

$$\left[\hat{L}_z - \langle\hat{L}_z\rangle\right]|g\rangle = \mathrm{i}\hbar\lambda\left[\hat{\varphi}_\theta - \langle\hat{\varphi}_\theta\rangle\right]|g\rangle, \qquad (6.38)$$

where $\hbar$ is introduced to make λ dimensionless.

Projecting Eq. (6.38) onto $\langle\varphi|$ yields a differential equation if we identify in the angle representation the OAM operator $\hat{L}_z$ as a derivative with respect to φ and $\hat{\varphi}_{\theta_0}$ as the multiplicative operator $Y(\varphi) = \varphi + 2\pi n$:

$$\left[\mathrm{i}\hbar\frac{\partial}{\partial\varphi} + \langle\hat{L}_z\rangle + \mathrm{i}\hbar\lambda(Y(\varphi) - \langle\hat{\varphi}_\theta\rangle)\right]g(\varphi) = 0. \qquad (6.39)$$

Only for a continuous wavefunction is the representation of $\hat{L}_z$ as a derivative with respect to φ well defined. Owing to the periodic character this continuous behaviour is also required for the end points of the interval $[\theta_0, \theta_0 + 2\pi]$. It is more convenient to work in units where $\hbar = 1$, but to the same effect we can introduce a dimensionless OAM operator $\hat{m} = \hat{L}_z/\hbar$. In terms of this operator the differential equation may be written without $\hbar$ as

$$\left[\mathrm{i}\frac{\partial}{\partial\varphi} + \langle\hat{m}\rangle + \mathrm{i}\lambda(Y(\varphi) - \langle\hat{\varphi}_\theta\rangle)\right]g(\varphi) = 0. \qquad (6.40)$$

Solutions to this equation are the intelligent states for OAM and angle. The sign of λ determines the curvature of the wavefunction at $\theta_0 + \pi$, the centre of the allowed range. If we restrict this analysis to cases where the mean values are equal to zero, we find that the second derivative of $g(\varphi)$ is given by

$$\frac{\partial^2 g}{\partial\varphi^2} = \lambda^2 Y^2(\varphi)g(\varphi) - \lambda g(\varphi). \qquad (6.41)$$

For a positive λ this leads to a peak in the centre of the allowed range and turning points at $\varphi = \pm 1/\sqrt{\lambda}$. For negative λ the wavefunction peaks at the edges and has no turning points.

A family of solutions to the differential equation (6.40) is given by truncated exponentials, which are characterised by the parameter λ [22, 23]:

$$g(\varphi) = N_g \exp(\mathrm{i}\langle\hat{m}\rangle\varphi)\exp\left(-\frac{\lambda}{2}(Y(\varphi) - \langle\hat{\varphi}\rangle)^2\right), \qquad (6.42)$$

where N_g is the normalisation constant for the intelligent states. The normalisation integral is calculated on the interval $[\theta_0, \theta_0 + 2\pi]$ and $Y(\varphi)$ can thus be replaced with φ. The resulting normalisation constant can be given in terms of the error function erf [24]:

$$\frac{1}{N_g^2} = \int_{\theta_0}^{\theta_0+2\pi} \mathrm{d}\varphi\, \exp\left(-\lambda(\varphi - \langle\hat{\varphi}\rangle)^2\right) = \frac{\mathrm{erf}(\pi\sqrt{\lambda})}{\sqrt{\lambda/\pi}}. \qquad (6.43)$$

Discontinuities in the wavefunction lead to infinite uncertainties for the OAM, we therefore require the wavefunction $g(\varphi)$ to be continuous at the ends of the 2π radian interval such that $g(\theta_0)=g(\theta_0+2\pi)$. This ensures that the mean angle lies in the middle of its definition range so that $\langle\hat{\varphi}\rangle=\theta_0+\pi$ and that the OAM mean is an integer. The mean $\langle\hat{m}\rangle$ contributes only to the phase of the wavefunction. We can therefore restrict the analysis of intelligent states to the case of zero OAM mean. This is because the unitary OAM shift operator $\exp(\mathrm{i}\hat{\varphi}_{\theta_0}k)$ shifts the angular spectrum uniformly by an integer number $k\in\mathbb{Z}$, but this does not change ΔL_z [3]. Also, owing to the unitary character of $\exp(\mathrm{i}\hat{\varphi}_{\theta_0}k)$, the angle probability density $P(\varphi)=|g(\varphi)_{\theta_0}|^2$ is left invariant and so is $\Delta\varphi$. Therefore, $\exp(\mathrm{i}\hat{\varphi}_{\theta_0}k)|g\rangle$ is also an intelligent state with a shifted OAM mean. Until now we have not specified the starting point of the 2π radian interval θ_0 and the wavefunction g is given in complete general terms. In the following, however, we will specifically set $\theta_0=-\pi$, such that the intelligent states have $\langle\hat{\varphi}\rangle=0$. Jointly with choosing $\langle\hat{m}\rangle=0$ this simplifies the notation in subsequent calculations and graphs without a loss of generality.

The angle variance is given by $(\Delta\varphi)^2=\langle\hat{\varphi}^2\rangle-\langle\hat{\varphi}\rangle^2$. As our solutions are symmetric about $\varphi=0$, the mean value of $\hat{\varphi}$ is zero and the angle variance is simply the expectation value of $\hat{\varphi}^2$:

$$(\Delta\varphi)^2=\int_{-\pi}^{\pi}\varphi^2P(\varphi)\mathrm{d}\varphi=\frac{1}{N_g^2}\int_{-\pi}^{\pi}\varphi^2\exp(-\lambda\varphi^2)\mathrm{d}\varphi$$
$$=\left(\frac{1}{2\lambda}-\frac{\sqrt{\pi}\exp(-\pi^2\lambda)}{\sqrt{\lambda}\mathrm{erf}(\pi\sqrt{\lambda})}\right). \tag{6.44}$$

The square of a self-adjoint operator is itself self-adjoint and hence the variance of $\hat{\varphi}$ is real.

Using the continuous wavefunction in Eq. (6.42) and the representation of $\hat{L}_z$ as a derivative requires some care. For intelligent states the first derivative of $g(\varphi)$ with respect to φ is well behaved, while the second derivative is not. This is the reason why we can use $-\mathrm{i}\hbar(\mathrm{d}/\mathrm{d}\varphi)$ as a valid representation of $\hat{L}_z$ in the eigenvalue equation for the intelligent states.

To calculate the OAM uncertainty ΔL_z from its variance $(\Delta L_z)^2=\langle\hat{L}_z^2\rangle-\langle\hat{L}_z\rangle^2$ we have to operate with $\hat{L}_z^2$ on an intelligent state. Identifying $\hat{L}_z^2$ with $-\hbar^2(\mathrm{d}^2/\mathrm{d}\varphi^2)$ does not lead to a well-behaved result because of the discontinuity of the first derivative of the wavefunction at the edges of the 2π radian range [18]. It is possible, however, to express $\langle\hat{L}_z^2\rangle$ in terms of the first derivative with respect to φ as follows:

$$\langle\hat{L}_z^2\rangle=\sum_{m=-L}^{L}|g_m|^2\langle m|\hat{L}_z^2|m\rangle=\sum_{m=-L}^{L}|g_m|^2\langle\hat{L}_zm|\hat{L}_zm\rangle=\sum_{m=-L}^{L}|g_m|^2\|\hat{L}_z|m\rangle\|^2, \tag{6.45}$$

where $\|\hat{L}_z|m\rangle\|^2$ denotes the norm of the the state vector $\hat{L}_z|m\rangle$. As $|m\rangle$ is an eigenstate of $\hat{L}_z$ and obeys the orthogonality condition $\langle m|m'\rangle = \delta_{mm'}$, we can exchange the order in which we take the norm and evaluate the sum over m. The equality in Eq. (6.45) may thus be continued by

$$\langle \hat{L}_z^2 \rangle = \sum_{m=-L}^{L} |g_m|^2 \|\hat{L}_z|m\rangle\|^2 = \left| \sum_{m=-L}^{L} \hat{L}_z g_m |m\rangle \right|^2 = \|\hat{L}_z|g\rangle\|^2. \tag{6.46}$$

If we now revert to the infinite-dimensional space we can represent $\hat{L}_z$ by $-\mathrm{i}\hbar(\mathrm{d}/\mathrm{d}\varphi)$ in the angle representation. Taking the norm for the continuous wavefunction corresponds to the integral of the modulus square of the wavefunction over the full 2π radian interval:

$$\langle \hat{L}_z^2 \rangle = \int_{-\pi}^{\pi} \left| \frac{\hbar}{\mathrm{i}} \frac{\mathrm{d}}{\mathrm{d}\varphi} g(\varphi) \right|^2 \mathrm{d}\varphi. \tag{6.47}$$

On substituting the wavefunction in Eq. (6.42) into this expression we find that the dimensionless OAM uncertainty Δm can be expressed in terms of the angle uncertainty $\Delta\varphi$:

$$(\Delta m)^2 = \frac{1}{N_g^2} \int_{-\pi}^{\pi} |\lambda|^2 \varphi^2 \exp(|\lambda|\varphi^2)\mathrm{d}\varphi = |\lambda|^2 (\Delta\varphi)^2, \tag{6.48}$$

so that the angular momentum uncertainty is proportional to the angle uncertainty as it should be.

The uncertainty product is

$$\Delta m \Delta\varphi = |\lambda|(\Delta\varphi)^2 = \left| \lambda \left(\frac{1}{2\lambda} - \frac{\sqrt{\pi}\exp(-\pi^2\lambda)}{\sqrt{\lambda}\mathrm{erf}(\pi\sqrt{\lambda})} \right) \right|. \tag{6.49}$$

For our choice of θ_0 the left-hand side of the uncertainty relation for $\Delta m \Delta\varphi$ is given by $|1 - 2\pi P(\pi)|/2$ as required.

An experiment was performed to test the uncertainty relation for angle and angular momentum using the orbital angular momentum of light [23]. By passing a Gaussian mode through a suitably defined angular aperture it was possible to make a beam of light with a truncated Gaussian profile in angle, of the form given in Eq. (6.42). This distribution changes on propagation, of course, but because the OAM is conserved, the spectrum of the OAM is unchanged. This means that it is possible to measure the probability for each angular momentum in the beam and, in particular, to obtain the width of this distribution, or equivalently, the angular uncertainty.

The results of the experiment by Franke-Arnold *et al.* [23] are presented in Figs. 6.1 and 6.2. In Fig. 6.1 we show the angular profiles for three widths ranging from a small angular width to a large one. In each case these are accompanied by the measured angular

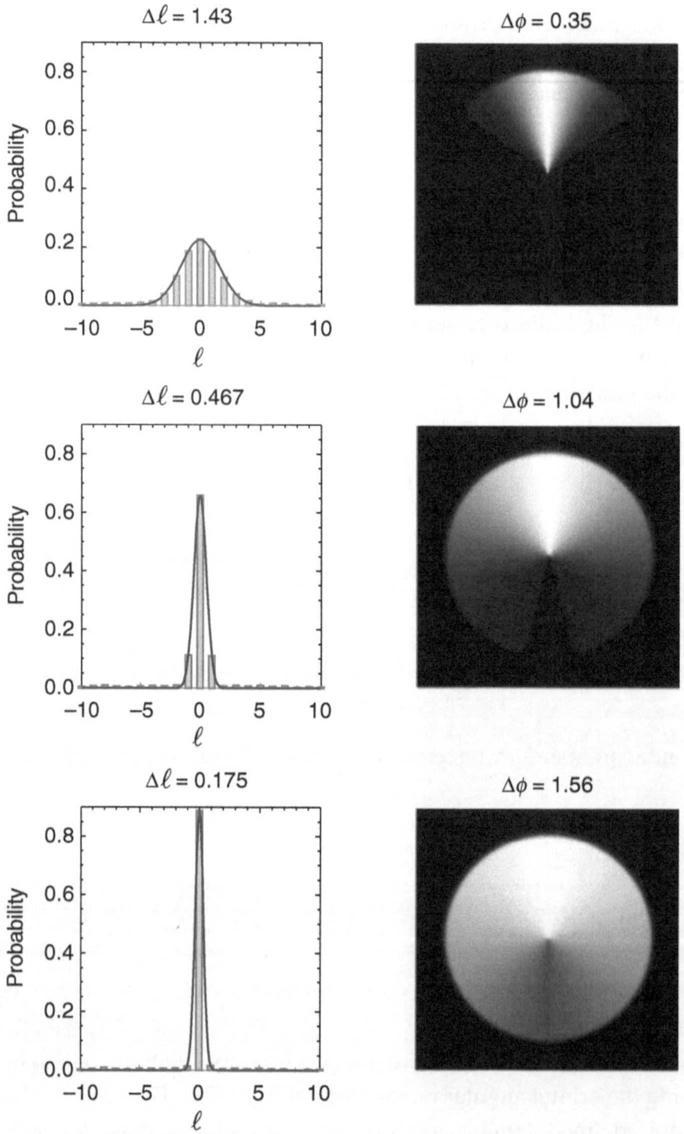

Figure 6.1 Examples of experimental observation. The observed distribution of the orbital angular momentum (left panel) after transmission through the angular apertures (right panel). The solid curve shows the best fit to the theoretically predicted distribution. Reproduced with permission from [23].

distribution. The solid curves present, in each case, the best fit to the theoretically predicted distribution. The measured product of the uncertainties in the angle and angular momentum is given in Fig. 6.2 as a function of the imposed angular uncertainty. The curve corresponds to the equality in the uncertainty relation, which is the anticipated value for an intelligent state.

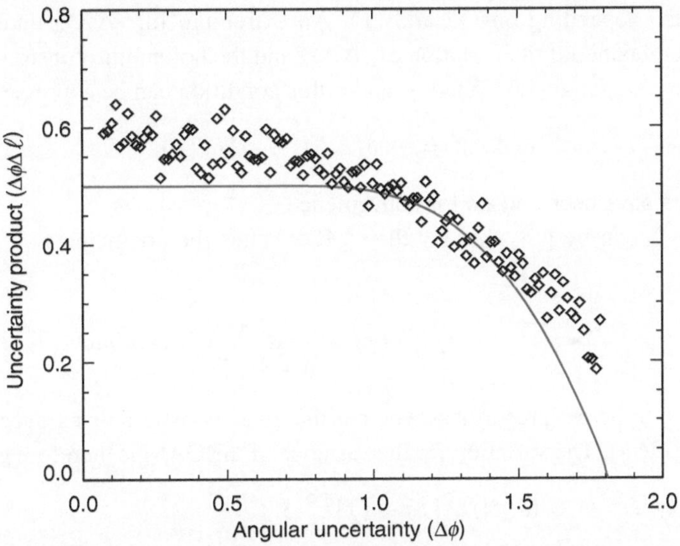

Figure 6.2 Experimental measurement and theoretical prediction of the angular uncertainty relation. The observed (diamonds) and predicted (line) relationship between the product of the uncertainties in angular position and angular momentum in units of $\hbar$ plotted against the uncertainty in angular position. For small angles, the uncertainty product is constant and equal to 0.5. In the absence of any restricting aperture, there is no uncertainty in angular momentum and the uncertainty product decreases to zero. Reproduced with permission from [23].

6.3.2 Constrained minimum uncertainty product states

The constrained minimum uncertainty product (CMUP) states minimise the uncertainty product for the additional constraint of a given uncertainty in the angle or in the OAM. The form of the CMUP states is the same, whether the additional constraint is a given uncertainty in the angle or in the OAM. This is because the defining equation is derived using a variation of the uncertainty product $\Delta L_z \Delta\varphi$ and the given uncertainty in either the angle or the OAM is taken into account by Lagrange multipliers. Whether an uncertainty in the angle is given or an uncertainty in the OAM, we have the same linear combination of variances with undetermined multipliers in both cases.

We can employ a variational method to determine minimum uncertainty product states as first introduced by Jackiw [25]. Seeking states which give a minimum in the uncertainty product for a given uncertainty in the angle or for a given uncertainty in the OAM is equivalent to finding a minimum for the uncertainty that is not given as a constraint. In the following we will use $(\Delta A)^2$ for the variance which is to be minimised and $(\Delta B)^2$ for the variance given as a constrained in the variation. We thus require that a CMUP state $|f\rangle$ minimises the free variance $(\Delta A)^2$, but with the constraint of keeping the given variance $(\Delta B)^2$ constant and obeying the normalisation condition $\langle f|f\rangle = 1$.

The variables in the variation are $\langle f|$ and $|f\rangle$ and they are treated as independent in the variation of $(\Delta A)^2$. The additional requirements can be expressed formally as holonomic constraints, also depending on $\langle f|$ and $|f\rangle$. An extremum for $(\Delta A)^2$ under these constraints will be obtained if the variation of $(\Delta A)^2$ and the holonomic constraints is stationary. Introducing Lagrange multipliers μ and ν this condition can be expressed as

$$\delta(\Delta A)^2 = -\nu\delta(\Delta B)^2 + \mu\delta\langle f|f\rangle, \tag{6.50}$$

where the signs have been chosen for convenience.

It has been shown by [18] that for the CMUP states the coefficients f_m in the OAM decomposition

$$|f\rangle = \sum_m f_m|m\rangle \quad f(\varphi) = \frac{1}{\sqrt{2\pi}}\sum_m f_m \exp(im\varphi) \tag{6.51}$$

can be assumed to be real, regardless whether the given constraint is an uncertainty in the angle or in the OAM. The variation for the variance of the OAM is therefore given by [26]:

$$\begin{aligned}\delta\langle f|\hat{L}_z^2|f\rangle &= (\delta\langle f|)\hat{L}_z^2|f\rangle + \langle f|\hat{L}_z^2(\delta|f\rangle) \\ &= \sum_m \delta f_m\big(\langle m|\hat{L}_z^2|f\rangle + \langle f|\hat{L}_z^2|m\rangle\big) = 2(\delta\langle f|)\hat{L}_z^2|f\rangle.\end{aligned} \tag{6.52}$$

The same reasoning can be applied to the variation $\delta\langle f|f\rangle$ and, by using the wavefunction in Eq. (6.51), to the variation of the angle variance $\delta\langle f|\hat{\varphi}^2|f\rangle$. The linear combination of variations in Eq. (6.52) may thus be turned into

$$2(\delta\langle f|)(\hat{L}_z^2 + \nu\hat{\varphi}^2 - \mu\hat{I})|f\rangle = 0. \tag{6.53}$$

For $|f\rangle$ to minimise the variance in the angle or OAM under the constraints of normalisation and keeping the other variance constant, this equation has to vanish for all variations $(\delta\langle f|)$. This gives a condition on the linear combination of operators applied to $|f\rangle$ and we find as eigenvalue equation for the CMUP states

$$\hat{L}_z^2|f\rangle = (-\nu\hat{\varphi}^2 + \mu\hat{I})|f\rangle. \tag{6.54}$$

The identification of the OAM operator $\hat{L}_z$ as derivative with respect to φ sets additional requirements on the wavefunction representing CMUP states [18, 27]. If these are treated carefully then we are left with a simple differential equation for $f(\varphi) = \langle\phi|f\rangle$.

$$\frac{\mathrm{d}^2}{\mathrm{d}\varphi^2}f(\varphi) = (\nu\varphi^2 - \mu)f(\varphi), \tag{6.55}$$

which is valid for $-\pi < \varphi < \pi$. From Eq. (6.55) follows that the curvature at $\varphi = 0$ is determined by μ. Owing to the undetermined multipliers, however, the curvature is not constant over the whole 2π radian interval. To satisfy the continuous boundary condition $\mathrm{d}f/\mathrm{d}\varphi = 0$ at $\varphi = \pm\pi$ the wavefunction has to have a turning point with $\mathrm{d}^2 f/\mathrm{d}\varphi^2 = 0$ and thus the Lagrange multipliers are either both positive or both negative. The case of both multipliers being positive leads to the small-uncertainty CMUP states with a peak in the middle of 2π radian interval.

The differential equation for the CMUP states has an exact solution in the form [18]

$$f(\varphi) \propto \exp\left(-\frac{\sqrt{\nu}}{2}\varphi^2\right) M\left(-\frac{1}{4}\frac{\mu}{\sqrt{\nu}} + \frac{1}{4}, \frac{1}{2}, \sqrt{\nu}\varphi^2\right), \quad \mu, \nu > 0, \tag{6.56}$$

where M is the the confluent hypergeometric Kummer function [24, Section 13.1]. Like the intelligent states this wavefunction has a single peak at $\phi = 0$, it is symmetrical about $\varphi = 0$ and falls off monotonically to minima at $\varphi = \pm\pi$. Unlike the intelligent state, however, its first derivative is everywhere continuous.

6.3.3 Large angular uncertainties

An orbital angular momentum eigenstate with uncertainty $\Delta m = 0$ is both an intelligent state and the global minimum uncertainty product state. Its angle wavefunction is flat with an angle uncertainty of

$$(\Delta\varphi)^2 = 2\int_0^{\pi} \varphi^2 \frac{1}{2\pi} \mathrm{d}\varphi = \frac{\pi^2}{3}. \tag{6.57}$$

This is not, however, the largest possible angle uncertainty. It is easy to see from the estimate

$$(\Delta\varphi)^2 = \int_{-\pi}^{\pi} \varphi^2 P(\varphi)\, \mathrm{d}\varphi \leq 2\pi^2 \int_0^{\pi} P(\varphi)\, \mathrm{d}\varphi = \pi^2, \tag{6.58}$$

that the angle uncertainty is bounded from above by π and we call states with an angular uncertainty $\pi/\sqrt{3} \leq \Delta\varphi \leq \pi$ large angular uncertainty states. It may seem an unfamiliar concept that states can exist which have a larger angle uncertainty than angular momentum eigenstates with a flat wavefunction. It is, however, the compactness of the interval from $-\pi$ to π, or more generally from θ_0 to $\theta_0 + 2\pi$, on which the angle is defined which makes it possible that wavefunctions which increase towards the edges at $\varphi = \pm\pi$ are perfectly valid and normalisable representations of quantum states.

Both intelligent states [27] and CMUP states [28] exist as large angular uncertainty states. The wavefunction of large angular uncertainty intelligent states is given by Eq. (6.42) with a negative λ. The upper bound in the angle uncertainty is reached in the limit of $\lambda \to \infty$. Figure 6.3 shows that the wavefunctions for intelligent states with small uncertainties have a cusp at $\varphi = 0$ whereas for large angular uncertainties the wavefunctions increase towards the edges. Constrained minimum uncertainty product states in Eq. (6.56) have a large angular uncertainty of $\Delta\varphi > \pi/\sqrt{3}$ when both Lagrange multipliers μ and ν are negative. A plot of wavefunctions for CMUP states constrained by several different angular uncertainties $\Delta\varphi$ is shown in Fig. 6.4.

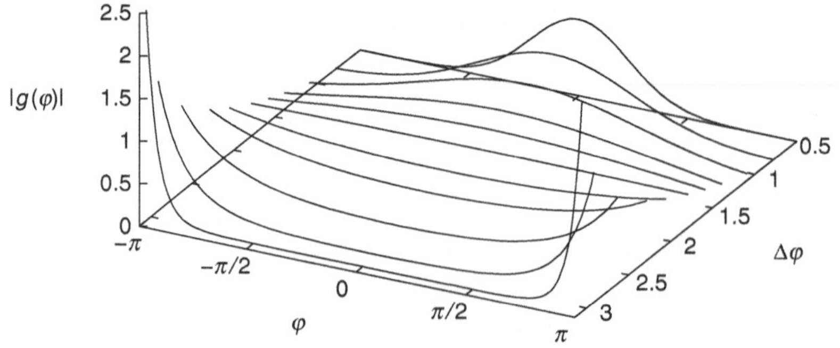

Figure 6.3 Plot of the modulus of the wavefunction for intelligent states in the angle representation. The wavefunction (6.42) is plotted for different values of angle uncertainty $\Delta\varphi$ and shows the transition from the small uncertainty to the large uncertainty regime. Reproduced with permission from [27].

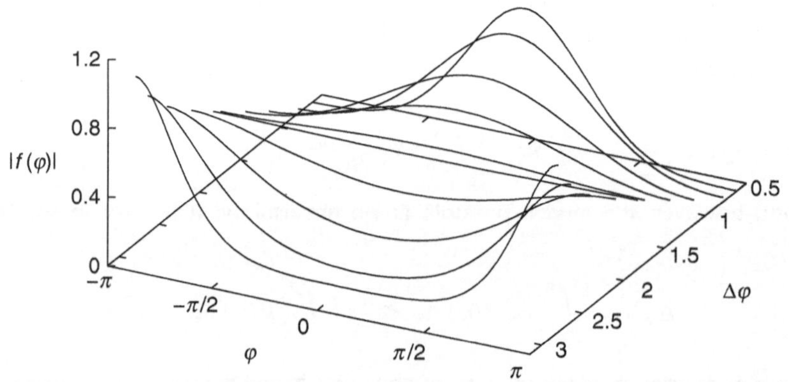

Figure 6.4 Plot of the modulus of the wavefunction for constrained minimum uncertainty product states in the angle representation. The plot shows the transition from the small uncertainty regime to the large uncertainty regime. This distinction refers to the angle uncertainty $\Delta\varphi$, and the dividing point is the flat wavefunction for $\Delta\varphi = \pi/\sqrt{3}$. Reproduced with permission from [28].

6.4 Fractional orbital angular momentum

The eigenstates or eigenmodes of OAM are all of the form $\exp(im\varphi)$, where m is an integer. For these states the associated angular momentum is $m\hbar$ and each mode of this form carries an orbital angular momentum of $m\hbar$ per photon [29]. The requirement that m be an integer is a consequence of the necessity for the wavefunction to be single-valued and the natural desire that it should be continuous. If it were not continuous, of course, the differentiation with respect to φ would not be well defined. It is reasonable to ask, however, how close we can get to a state with fractional OAM, or equivalently what happens if we impose such a phase structure onto a well-behaved mode. It should be emphasised that these states are not in any sense eigenstates of an OAM operator.

6.4.1 Construction of fractional orbital angular momentum states

Until now we have only considered OAM states $|m\rangle$ for which m is an integer. An optical representation of such states are the Laguerre–Gaussian modes [29]. As we have seen, a Laguerre–Gaussian mode with integer OAM index m can be generated from a fundamental Gaussian beam by inducing an azimuthal phase ramp of step height $2\pi m$. It is perfectly possible, however, to induce an azimuthal phase variation with a step height which is not an integer, but rather a fractional multiple of 2π [30]. The light emerging from such a fractional spiral phase plate no longer has the rotationally symmetric intensity structure of the Laguerre–Gaussian modes [31, 32], but shows a line of reduced intensity along the direction of the phase discontinuity. In this section we examine the properties of fractional OAM sates within the rigorous quantum formulation of angle and orbital angular momentum.

A quantum state with fractional OAM is denoted by $|M\rangle$, where $M = m+\mu$ with m as the integer part and $\mu \in [0, 1)$ as the fractional part. Choosing 'fractional' over 'non-integer' follows the terminology in earlier work, but μ can take on every real number between zero and one and not just rational numbers. This distinction, however, is not relevant for all practical purposes. The angle decomposition of $|M\rangle$ is given by:

$$\begin{aligned} |M\rangle &= (2L+1)^{1/2} \sum_{n=0}^{2L} \exp(\mathrm{i}M\theta_n)|\theta_n\rangle \\ &= (2L+1)^{1/2} \sum_{n=0}^{2L} \exp(\mathrm{i}m\theta_n)\exp(\mathrm{i}\mu\theta_n)|\theta_n\rangle. \end{aligned} \tag{6.59}$$

In general, $\exp(\mathrm{i}\mu\theta_n)$ will be a multivalued function, but a common way to render a multivalued function single-valued is the introduction of a branch cut [33], which restricts the range of the function. In our case we would like to choose the position of the branch cut α, that is the position of the discontinuity in the function $\exp(\mathrm{i}\mu\theta_n)$, independently of θ_0, the starting angle of the 2π radian interval. For that we introduce an integer-valued function $f_\alpha(\theta_n)$, which takes on the values 1 and 0, depending whether θ_n is smaller or greater than α:

$$f_\alpha : \{\theta_n\}_{n=0,\ldots,2L} \to \{0, 1\}, \quad \theta_n \mapsto \begin{cases} 1 & \theta_0 < \theta_n < \theta_0 + \alpha, \\ 0 & \theta_0 + \alpha < \theta_n < \theta_0 + 2\pi, \end{cases} \tag{6.60}$$

where $\alpha \in [0, 2\pi)$. The effect can be seen in Fig. 6.5. We should note that according to the mathematical definition we do not create a branch of the multivalued function, as we

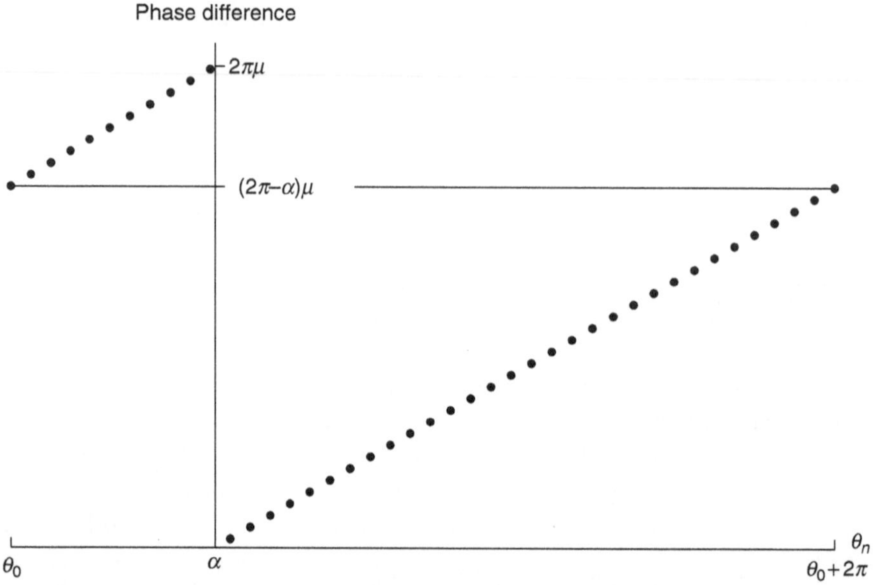

Figure 6.5 Graph of the phase discontinuity in fractional orbital angular momentum states. The position of the discontinuity α is chosen independently from the starting angle of the 2π radian interval, on which the angle is defined. In the state space Ψ the angle is a discrete variable with $2L+1$ angle eigenstates $|\theta_n\rangle$, $n=0,\ldots,2L$. Reproduced with permission from [34].

explicitly include the discontinuity. With this construction the fractional state $|M\rangle$ can be written as

$$|M(\alpha)\rangle = \frac{\exp(-\mathrm{i}\mu\alpha)}{\sqrt{2L+1}}\sum_{n=0}^{2L}\exp(\mathrm{i}M\theta_n)\exp[\mathrm{i}2\pi\mu f_\alpha(\theta_n)]|\theta_n\rangle. \tag{6.61}$$

The factor $\exp(-\mathrm{i}\mu\alpha)$ ensures that the fractional OAM state remains in the same 'branch' independent of the value of α. The fractional OAM state $|M(\alpha)\rangle$ depends now on the orientation of the discontinuity α.

6.4.2 Overlap of fractional orbital angular momentum states

The overlap of two states with fractional values of OAM and two different orientations is given by $\langle M'(\alpha')|M(\alpha)\rangle$. It is possible to introduce a unitary operator $\hat{U}_m(\beta)$, $\beta\in[0,2\pi)$ which formally rotates the position of the discontinuity by writing

$$\hat{U}_m(\beta)|M'(\alpha)\rangle = \exp[\mathrm{i}(m-m')\beta]|M'(\alpha\oplus\beta)\rangle, \tag{6.62}$$

where $\alpha\oplus\beta$ denotes an addition of α and β modulo 2π. This is simply a formal way of signifying that only the differences in angle matter [34].

On applying the operator $\hat{U}_{m'}(-\alpha')$ to the state $|M(\alpha)\rangle$ and $\hat{U}^{\dagger}_{m'}(-\alpha')$ to $\langle M'(\alpha')|$ the orientation of the states in the overlap is altered such that one orientation is zero and the other is equal to the difference of the original orientations. This yields

$$\begin{aligned}\langle M'(\alpha')|M(\alpha)\rangle &= \langle M'(\alpha')|\hat{U}^{\dagger}_{m'}(-\alpha')\hat{U}_{m'}(-\alpha')|M(\alpha)\rangle \\ &= \exp[(\mathrm{i}(m-m')\alpha']\langle M'(0)|M(\beta)\rangle, \end{aligned} \tag{6.63}$$

where we have set $\beta = \alpha \oplus (-\alpha')$. The moduli of the rotated and unrotated overlaps are the same so we proceed in calculating $\langle M'(0)|M(\beta)\rangle$ only. With the definition of the fractional OAM state in Eq. (6.61) the overlap is given by

$$\langle M'(0)|M(\beta)\rangle = \frac{\exp(-\mathrm{i}\mu\beta)}{2L+1}\sum_{n=0}^{2L}\exp[\mathrm{i}(M-M')\theta_n]\exp[\mathrm{i}\mu 2\pi f_\beta(\theta_n)], \tag{6.64}$$

where we have used the fact that $\{|\theta_n\rangle\}_{n=0,\ldots,2L}$ is an orthonormal set (see Eq. (6.10)).

On substituting the expression for the angle θ_n in Eq. (6.9) it is possible to evaluate the sums using the geometric progression [34]. In the limit $L \to \infty$ we find

$$\begin{aligned}\langle M'(0)|M(\beta)\rangle = \exp(-\mathrm{i}\mu\beta)&\frac{\mathrm{i}\exp[\mathrm{i}(M-M')\theta_0]}{(M-M')2\pi} \\ &\times \left\{\exp(\mathrm{i}\mu 2\pi)\left[1-\exp[\mathrm{i}(M-M')\beta]\right]\right. \\ &\left.+\left[\exp[\mathrm{i}(M-M')\beta]-\exp[\mathrm{i}(M-M')2\pi]\right]\right\}. \end{aligned} \tag{6.65}$$

From this expression it is possible to calculate the modulus square of the overlap $\langle M'(0)|M(\beta)\rangle$:

$$\begin{aligned}|\langle M'(0)|M(\beta)\rangle|^2 = &\frac{1}{(M-M')^2\pi^2}\Big[\sin^2(M\pi)+\sin^2(M'\pi) \\ &-2\cos[(M-M')(\pi-\beta)]\sin(M\pi)\sin(M'\pi)\Big]. \end{aligned} \tag{6.66}$$

The overlap probability is plotted for different values of β in Fig. 6.6. For $\beta = 0$ the plot shows that there are parallel lines of equal overlap probability for $M - M' = K \in \mathbb{R}$. The overlap probability along these lines is given by $\sin^2(\pi K)/(\pi^2 K^2)$ and does not depend on the particular values of μ and μ'. This is in contrast to the case of $\beta = \pi$, where the overlap probability reaches zero for $M = M'$ with $\mu, \mu' = 1/2$.

This can be seem more clearly by considering the particular case of $M = M'$. On expanding the exponentials in Eq. (6.65) for $M - M' \to 0$, the overlap amplitude is given by

$$\langle M(0)|M(\beta)\rangle = \frac{\beta}{2\pi}(\exp(\mathrm{i}\mu 2\pi)-1)+1, \tag{6.67}$$

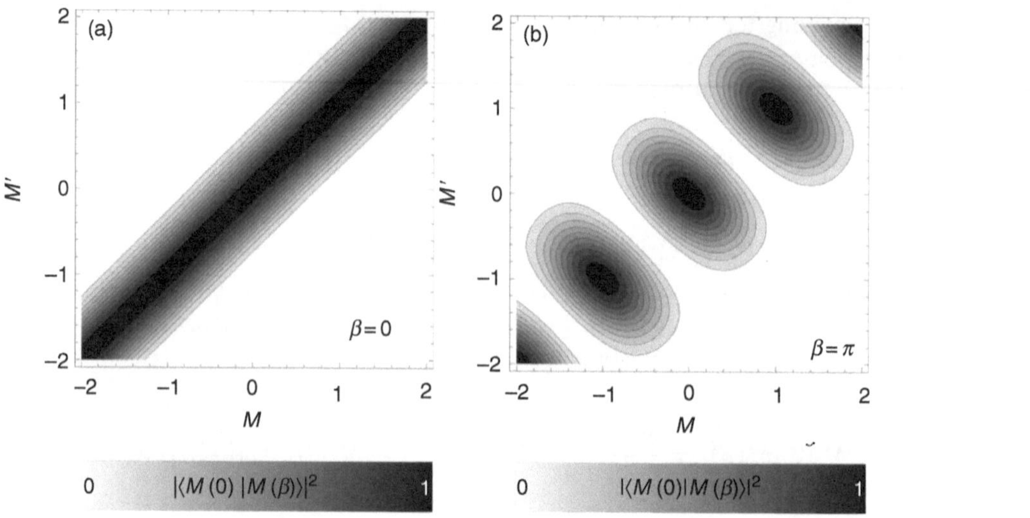

Figure 6.6 Plot of the overlap probability for two general OAM states $|M\rangle$ and $|M'\rangle$ for two different values of the difference in the orientation β. (a) $\beta = 0$: for $M - M' = K \in \mathbb{Z}$ the overlap probability is independent of the particular values M and M'. (b) $\beta = \pi$: for $M = M'$ and $\mu = \mu' = 1/2$, the overlap probability reaches zero.

which gives the probability density [35]

$$|\langle M(0)|M(\beta)\rangle|^2 = \left(1 - \frac{\beta}{\pi}\right)^2 \sin^2(\mu\pi) + \cos^2(\mu\pi). \tag{6.68}$$

The overlap probability only depends on the fractional part μ and the relative orientation β and not the integer part m. In Fig. 6.7 the overlap probability is plotted as a function of β for different values of μ. Only for $\mu = 1/2$ does the overlap probability reach zero for $\beta = \pi$.

6.4.3 Orbital angular momentum distribution of fractional states

The expression for the general overlap of two fractional angular momentum states can be specialised to yield the coefficients of fractional states decomposed into an integer basis. By choosing $M' = m' \in \mathbb{Z}$ we can evaluate the probability amplitudes $c_{m'}[M(\beta)] = \langle m'(0)|M(\beta)\rangle$ from Eq. (6.65):

$$c_{m'}[M(\beta)] = \langle m'|M(\beta)\rangle = \exp(-\mathrm{i}\mu\beta)\frac{\mathrm{i}\exp[\mathrm{i}(M - m')\theta_0]}{2\pi(M - m')} \times \left[\exp[\mathrm{i}(M - m')\beta]\,(1 - \exp(\mathrm{i}\mu 2\pi))\right]. \tag{6.69}$$

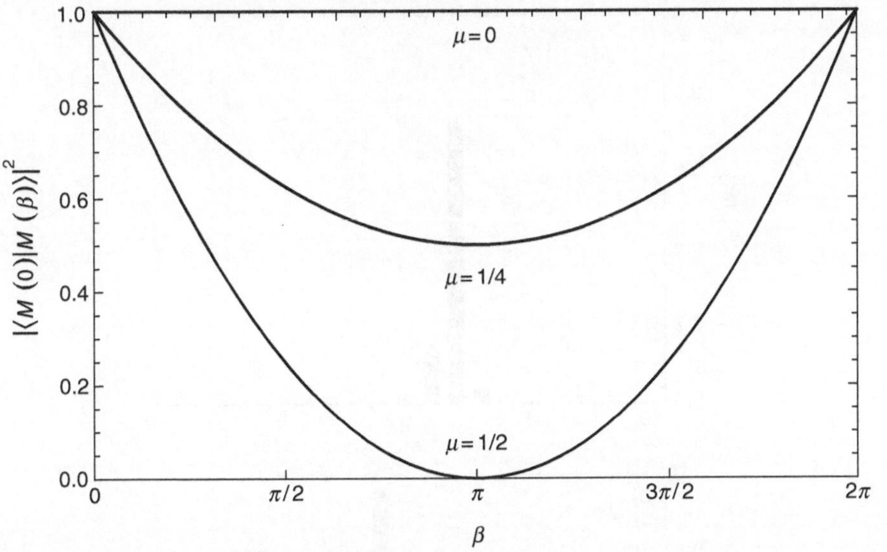

Figure 6.7 Plot of the overlap probability for two fractional orbital angular momentum states with $M = M'$ as a function of the relative orientation β. The overlap probability reaches zero for $\mu = 1/2$ and $\beta = \pi$.

Only the complex argument of the probability amplitudes $c_{m'}[M(\beta)]$ depends on the relative orientation; the probabilities $P_{m'}(M)$, given by the modulus square of the probability amplitudes, are independent of β:

$$P_{m'}(M) = |c_{m'}[M(\beta)]|^2 = \frac{1 - \cos(2\mu\pi)}{(M - m')^2 2\pi^2} = \frac{\sin^2(\mu\pi)}{(M - m')^2\pi^2}. \tag{6.70}$$

For integer values of M the OAM distribution is thus singular, consisting of a single non-vanishing probability at $M = m'$. For fractional values of M, however, the probabilities are peaked around the nearest integer to M, as can be seen in Fig. 6.8.

The OAM mean of a fractional state $|M\rangle$ is [32]

$$\begin{aligned}\bar{M} &= \sum_{m'=-\infty}^{\infty} m' P_{m'}(M) = \frac{1 - \cos(2M\pi)}{2\pi^2} \sum_{m'=-\infty}^{\infty} \frac{m'}{(M - m')^2} \\ &= M - \frac{\sin(2M\pi)}{2\pi}.\end{aligned} \tag{6.71}$$

A plot of the OAM mean as a function of M is given in Fig. 6.9. The OAM variance, however, tends to infinity:

$$(\Delta M)^2 = \sum_{m'=-\infty}^{\infty} m'^2 P_{m'}(M) - \left(\sum_{m'=-\infty}^{\infty} m' P_{m'}(M) \right)^2 \to \infty. \tag{6.72}$$

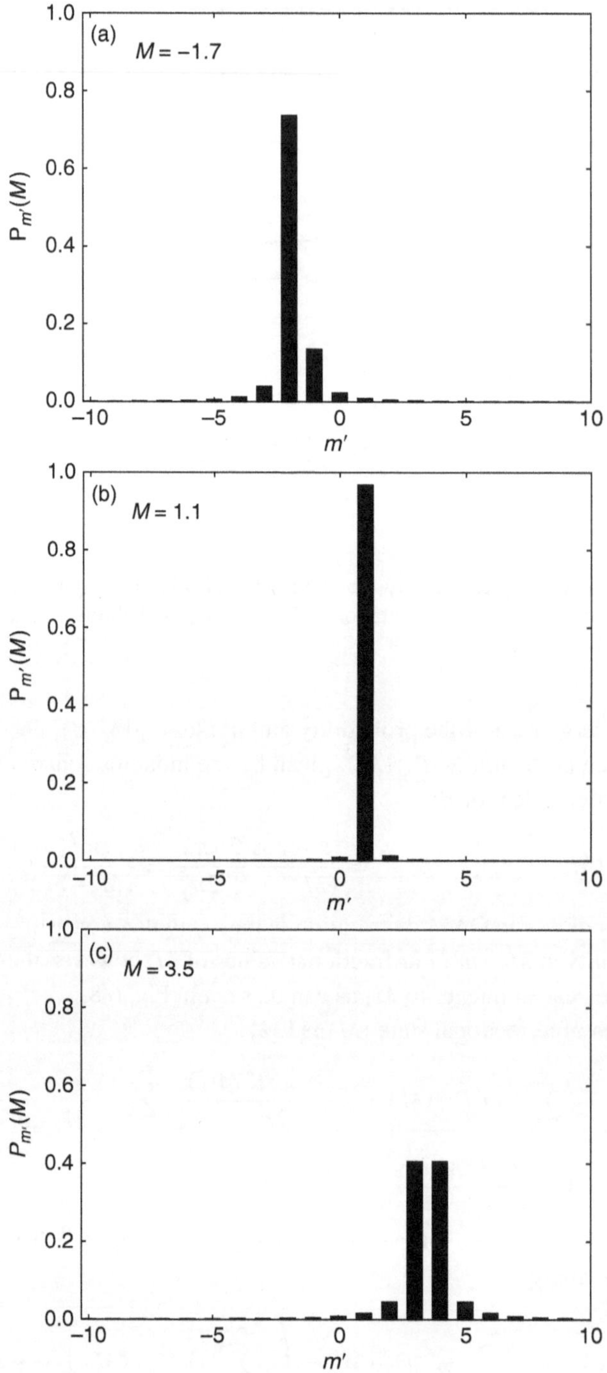

Figure 6.8 Plot of the angular momentum distribution for fractional states. The distribution has a peak at the nearest integer to M. For $\mu = 0.5$ this results in two peaks of equal height at the two neighbouring integers. Reproduced with permission from [34].

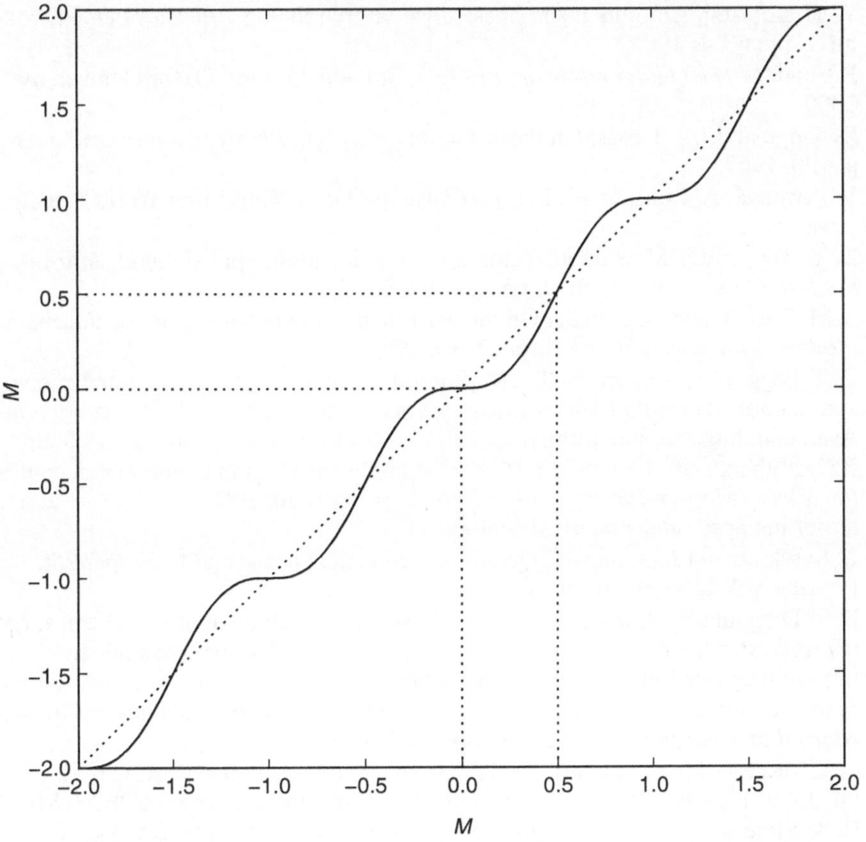

Figure 6.9 Plot of the angular momentum mean value for fractional angular momentum states. Only at integer and half integer values of M is the angular momentum mean $\bar{M}$ equal to M. In general, the relation between $\bar{M}$ and M is given in Eq. (6.71). Reproduced with permission from [34].

While the second term is the square of the mean value and therefore finite, the summation in the first term is divergent. This is not surprising as the discontinuity in the phase leads also to a discontinuity in the angle wavefunction. For such a discontinuous wavefunction the mean square OAM will be divergent as demonstrated by [18].

References

[1] J. B. Götte, Integral and fractional orbital angular momentum of light, Ph.D. dissertation, University of Strathclyde, Glasgow, 2006.

[2] D. T. Pegg and S. M. Barnett, Phase properties of the quantized singe-mode eletromagnetic field, *Physical Review A*, vol. 39, no. 4, pp. 1665–75, 1989. [Online]. Available: http://link.aps.org/abstract/PRA/v39/p1665.

[3] S. M. Barnett and D. T. Pegg, Quantum theory of rotation angles, *Physical Review A*, vol. 41, no. 7, pp. 3427–35, 1990. [Online]. Available: http://link.aps.org/abstract/PRA/v41/p3427.

[4] W. H. Louisell, Amplitude and phase uncertainty relations, *Physics Letters*, vol. 7, no. 1, pp. 60–1, 1963.

[5] R.Loudon, *The Quantum Theory of Light*, 3rd edn. Oxford: Oxford University Press, 2000.

[6] D. Judge and J. T. Lewis, On the commutator $[l_z, \varphi]$, *Physics Letters*, vol. 5, no. 3, p. 190, 1963.

[7] V. Peřinová, A. Lukš, and J. Peřina, *Phase in Optics*. Singapore: World Scientific, 1998.

[8] D. T. Pegg and S. M. Barnett, Tutorial review: quantum optical phase, *Journal of Modern Optics*, vol. 44, no. 2, pp. 225–64, 1997.

[9] S. M. Barnett and D. T. Pegg, On the hermitian optical phase operator, *Journal of Modern Optics*, vol. 36, no. 1, pp. 7–19, 1989.

[10] D. T. Pegg, J. A. Vaccaro, and S. M. Barnett, Quantum-optical phase and canonical conjugation, *Journal of Modern Optics*, vol. 37, no. 11, pp. 1703–10, 1990. [Online]. Available: http://taylorandfrancis.metapress.com/link.asp?id=dj2ug1jw55hdpx34.

[11] P. Carruthers and M. M. Nieto, Phase and angle variables in quantum mechanics, *Reviews of Modern Physics*, vol. 40, no. 2, pp. 411–40, 1968. [Online]. Available: http://link.aps.org/abstract/RMP/v40/p411.

[12] L. Susskind and J. Glogower, Quantum mechanical phase and time operator, *Physics*, vol. 1, pp. 49–61, 1964.

[13] D. T. Pegg and S. M. Barnett, Unitary phase operator in quantum mechanics, *EPL (Europhysics Letters)*, vol. 6, no. 6, pp. 483–7, 1988. [Online]. Available: http://iopscience.iop.org/0295-5075/6/6/002.

[14] S. M. Barnett and D. T. Pegg, Limiting procedure for the optical phase operator, *Journal of Modern Optics*, vol. 39, pp. 2121–9, 1992.

[15] J. A. Vaccaro, Phase operators on hilbert space, *Physical Review A*, vol. 51, no. 4, pp. 3309–17, 1995. [Online]. Available: http://link.aps.org/abstract/PRA/v51/p3309.

[16] R. A. Silverman, *Complex Analysis with Applications*. Mineola, NY: Dover Publications, Inc., 1984, reprint. Originally published by Prentice-Hall, Inc., Englewood Cliffs, NJ, 1973.

[17] K. Kraus, Complementary observables and uncertainty relations, *Physical Review D*, vol. 35, no. 10, pp. 3070–5, 1987. [Online]. Available: http://dx.doi.org/10.1103/PhysRevD.35.3070.

[18] D. T. Pegg, S. M. Barnett, R. Zambrini, S. Franke-Arnold, and M. Padgett, Minimum uncertainty states of angular momentum and angular position, *New Journal of Physics*, vol. 7, p. 62, 2005. [Online]. Available: http://www.iop.org/EJ/abstract/1367-2630/7/1/062/.

[19] C. Aragone, G. Guerri, S. Salamó, and J. L. Tani, Intelligent spin states, *Journal of Physics A*, vol. 7, no. 15, pp. L149–51, 1974.

[20] H. P. Robertson, The uncertainty principle, *Physial Review*, vol. 34, pp. 163–4, 1929.

[21] E. Merzbacher, *Quantum Mechanics*, 3rd edn. New York: John Wiley & Sons, Inc., 1998.

[22] A. Galindo and P. Pascual, *Quantum Mechanics*, vol. 1. Berlin: Springer Verlag, 1990.

[23] S. Franke-Arnold, S. M. Barnett, E. Yao, J. Leach, J. Courtial, and M. Padgett, Uncertainty principle for angular position and angular momentum, *New Journal of Physics*, vol. 6, p. 103, 2004. [Online]. Available: www.iop.org/EJ/abstract/1367-2630/6/1/103.

[24] M. Abramowitz and I. S. Stegun, *Handbook of Mathematical Functions*. Mineola, NY: Dover Publications, Inc., 1974, reprint. Originally published: National Bureau of Standards, corrected edition, 1964.

[25] R. Jackiw, Minimum uncertainty product, number-phase uncertainty product and coherent states, *Journal of Mathematical Physics*, vol. 9, no. 3, pp. 339–46, 1968.

[26] G. S. Summy and D. T. Pegg, Phase optimized quantum states of light, *Optics Communications*, vol. 77, no. 1, pp. 75–9, 1990.

[27] J. B. Götte, R. Zambrini, S. Franke-Arnold, and S. M. Barnett, Large-uncertainty intelligent states for angular momentum and angle, *Journal of Optics B*, vol. 7, pp. S563–71, 2005.

[28] J. B. Götte, P. M. Radmore, R. Zambrini, and S. M. Barnett, Angular minimum uncertainty states with large uncertainties, *Journal of Physics B*, vol. 39, pp. 2791–801, 2006. [Online]. Available: www.iop.org/EJ/abstract/0953-4075/39/12/013/.

[29] L. Allen, M. W. Beijersbergen, R. J. C. Spreeuw, and J. P. Woerdman, Orbital angular momentum of light and the transformation of laguerre-gaussian modes, *Physical Review A*, vol. 45, no. 11, pp. 8185–90, June 1992, reprinted in [36, Paper 2.1]. [Online]. Available: http://link.aps.org/abstract/PRA/v45/p8185.

[30] S. S. R. Oemrawsingh, E. R. Eliel, J. P. Woerdman, E. J. K. Verstegen, J. G. Kloosterboer, and G. W. t' Hooft, Half-integral spiral phase plates for optical wavelengths, *Journal of Optics A*, vol. 6, pp. S288–90, 2004. [Online]. Available: www.iop.org/EJ/abstract/1464-4258/6/5/029.

[31] M. V. Berry, Optical vortices evolving from helicoidal integer and fractional phase steps, *Journal of Optics A*, vol. 6, no. 2, pp. 259–69, 2004. [Online]. Available: www.iop.org/EJ/abstract/1464-4258/6/2/018.

[32] J. Leach, E. Yao, and M. J. Padgett, Observation of the vortex structure of a non-integer vortex beam, *New Journal of Physics*, vol. 6, p. 71, 2004. [Online]. Available: www.iop.org/EJ/abstract/1367-2630/6/1/071/.

[33] J. B. Conway, *Functions of One Complex Variable*, 2nd edn. Springer Verlag, 1978.

[34] J. B. Götte, S. Franke-Arnold, R. Zambrini, and S. M. Barnett, Quantum formulation of fractional orbital angular momentum, *Journal of Modern Optics*, vol. 54, no. 12, pp. 1723–38, August 2007.

[35] S. S. R. Oemrawsingh, A. Aiello, E. R. Eliel, G. Nienhaus, and J. P. Woerdman, How to observe high-dimensional two-photon entanglement with only two detectors, *Physical Review Letters*, vol. 92, p. 217901, 2004. [Online]. Available: http://link.aps.org/abstract/PRL/v92/e217901.

[36] L. Allen, S. M. Barnett, and M. J. Padgett, *Optical Angular Momentum*. Bristol: Institute of Physics Publishing, 2003.

7

Dynamical rotational frequency shift

I. BIALYNICKI-BIRULA AND Z. BIALYNICKA-BIRULA

7.1 Introduction

The term rotational frequency shift (RFS) has been used in different contexts and given different meanings [1]–[8]. Other terms have also been used (e.g. azimuthal Doppler shift, angular Doppler shift) to describe various related phenomena. In this article we stick to the meaning of the rotational frequency shift given by us in [9]. In order to make a clear distinction between our RFS and other related shifts we use the term dynamical RFS (DRFS). We will study the spectral properties of radiation emitted by rotating quantum sources.

Radiation emitted by sources in motion looks different when observed in the laboratory frame. Frequency shift can only be determined for monochromatic waves. In general, a monochromatic field will lose this property when observed from a moving frame. Therefore, to observe a frequency shift we have to restrict ourselves to some special forms of radiation, which fully preserve their monochromaticity when the frame of reference is changed.

When the source moves with constant velocity, one observes the well-known Doppler shift. In this case a special role is played by plane waves characterized by their wave vectors. We may invoke the relativistic transformation properties of a wave vector to derive the change of frequency. As a result, all inertial observers see such waves as monochromatic plane waves with a shifted frequency and a transformed direction of propagation. This effect can also be deduced from the transformation laws of the photon energy-momentum four-vector $p_\mu = \hbar k_\mu$. For uniform motion this kinematical Doppler shift is the only effect. According to the principle of relativity there is no preferred inertial frame of reference. Therefore, the properties of the moving source are not affected by its uniform motion.

When the source rotates with respect to an inertial frame, the symmetry between the two frames is lost. The source will "know" that it is moving because it "feels" inertial forces. These forces affect the dynamics of the radiation process. In particular, the energy levels of a rotating source are modified. This is the first contribution to the dynamical rotational

The Angular Momentum of Light, ed. D. L. Andrews and M. Babiker. Published by Cambridge University Press.

frequency shift. The second contribution is due to the transformation properties of the electromagnetic field. The radiation emitted by a rotating source looks different when observed in the laboratory frame because when the frame is changed, the electric and magnetic field vectors must be transformed accordingly. In the case of rotation, the role of the photon momentum is played by angular momentum. Monochromatic waves with a given projection of the angular momentum on the axis of rotation will remain monochromatic for all observers rotating around this axis. Their frequency will be shifted by $M\Omega$, where M is a natural number that determines the projection of the angular momentum on rotation axis and Ω is the frequency of rotation. This kinematical shift of frequency is the counterpart of the Doppler shift and it has been predicted and observed in the [1]–[8]. Our aim here is to study an interplay between this purely kinematical shift and the shift caused by the rotation of the source. The combined effect is called the dynamical rotation frequency shift (DRFS).

The dynamical rotational frequency shift in its pure form so far has not been seen. However, its close counterpart was detected in an experiment [10] on thioformaldehide molecules that were forced to rotate by an applied circularly polarized microwave.

The study of the DRFS involves the energy levels of the emitting system. Therefore, it requires a quantum mechanical setting and we begin with a suitable form of description which will be applied first to the Doppler shift.

7.2 Doppler shift

In order to describe in the next section the DRFS shift in a fully quantum-mechanical setting, we shall first introduce the analogous formulation for the standard Doppler shift, even though in this simple case this description may look overly complicated. Let us imagine an atom moving with constant velocity $\boldsymbol{v}$. The radiation process in the laboratory frame is described by the Schrödinger equation,

$$\mathrm{i}\hbar\partial_t|\psi(t)\rangle = H(t)|\psi(t)\rangle, \tag{7.1}$$

with the following Hamiltonian (for simplicity we shall consider one electron as the source of radiation):

$$H(t) = H_{\mathrm{A}}(t) + H_{\mathrm{F}} + H_{\mathrm{I}}, \tag{7.2}$$

$$H_{\mathrm{A}}(t) = \frac{\boldsymbol{p}^2}{2m} + V(\boldsymbol{r}(t)), \tag{7.3}$$

$$H_{\mathrm{F}} = \tfrac{1}{2}\int \mathrm{d}^3r : [\epsilon_0\boldsymbol{E}^2(\boldsymbol{r}) + \boldsymbol{B}^2(\boldsymbol{r})/\mu_0] := \hbar\sum_i \omega_i a_i^\dagger a_i, \tag{7.4}$$

$$H_{\mathrm{I}} = -\frac{e\boldsymbol{p}\cdot\boldsymbol{A}(\boldsymbol{r})}{m} + \frac{e^2\boldsymbol{A}^2(\boldsymbol{r})}{2m}, \tag{7.5}$$

where $\boldsymbol{E}(\boldsymbol{r})$, $\boldsymbol{B}(\boldsymbol{r})$, and $\boldsymbol{A}(\boldsymbol{r})$ are the operators of the electric field, magnetic field, and the vector potential. The sum over the modes of the electromagnetic field may also include integration. The time dependence of the potential V is due to a motion of its center, $\boldsymbol{r}(t) = \boldsymbol{r} - \boldsymbol{R}(t)$. For a uniform motion $\boldsymbol{R}(t) = \boldsymbol{v}t$. For future reference, we have split the Hamiltonian into the atomic part, the field part, and the interaction part. The coupling of the electron to the radiation field will be treated as a perturbation.

We will now eliminate the time dependence of H by a transformation to the moving frame. To this end, we transform the state vectors according to the following unitary transformation $U_{\mathrm{V}}(t)$:

$$U_{\mathrm{V}}(t) = \mathrm{e}^{-\mathrm{i}\boldsymbol{v}\cdot\boldsymbol{P}_{\mathrm{T}}t/\hbar}, \tag{7.6}$$

$$|\psi(t)\rangle = U_{\mathrm{V}}(t)|\tilde{\psi}(t)\rangle, \tag{7.7}$$

where $\boldsymbol{P}_{\mathrm{T}}$ is the operator of the total momentum of the system. This transformation removes the time dependence of the Hamiltonian since

$$U_{\mathrm{V}}^{\dagger}(t)\boldsymbol{r}(t)U_{\mathrm{V}}(t) = U_{\mathrm{V}}^{\dagger}(t)(\boldsymbol{r} - \boldsymbol{v}t)U_{\mathrm{V}}(t) = \boldsymbol{r}, \tag{7.8}$$

but it also produces an additional term in the Schrödinger equation,

$$\mathrm{i}\hbar\partial_t|\tilde{\psi}(t)\rangle = [H(t)|_{t=0} - \boldsymbol{v}\cdot\boldsymbol{P}_{\mathrm{T}}]|\tilde{\psi}(t)\rangle. \tag{7.9}$$

The additional term in the Hamiltonian comes from the time dependence of the transformation.

The atomic part $\boldsymbol{p}$ of the total momentum modifies the atomic Hamiltonian,

$$\tilde{H}_{\mathrm{A}} = H_{\mathrm{A}}(t)|_{t=0} - \boldsymbol{v}\cdot\boldsymbol{p}. \tag{7.10}$$

It is important to note that this modification does not lead to a change of the energy level separation. It can be eliminated (apart from an additive constant $mv^2/2$) by an optional unitary transformation

$$|\tilde{\psi}(t)\rangle = \mathrm{e}^{im\boldsymbol{v}\cdot\boldsymbol{r}/\hbar}|\phi(t)\rangle, \tag{7.11}$$

which restores the standard form, $\boldsymbol{p}^2/2m + V(\boldsymbol{r})$, of the atomic Hamiltonian. Thus, the separations of the atomic energy levels in a uniformly moving frame are not modified – an expression of the relativity principle.

The interaction Hamiltonian is not modified by the transformation $U_{\mathrm{V}}(t)$, since it commutes with the total momentum. However, it acquires an additional term $e\boldsymbol{v}\cdot\boldsymbol{A}(\boldsymbol{r})$ after the transformation (7.11). This only changes slightly the photon emission probability but not the photon spectrum.

The field part $\boldsymbol{P}$ of the total momentum modifies the field Hamiltonian,

$$\tilde{H}_{\mathrm{F}} = H_{\mathrm{F}} - \boldsymbol{v}\cdot\boldsymbol{P} = \hbar\sum_{\lambda}\int \mathrm{d}^3k\,(\omega(\boldsymbol{k}) - \boldsymbol{v}\cdot\boldsymbol{k})a_{\lambda}^{\dagger}(\boldsymbol{k})a_{\lambda}(\boldsymbol{k}). \tag{7.12}$$

We have chosen here plane waves as the appropriate modes of the electromagnetic field. The modification of the field Hamiltonian is responsible for the Doppler shift. This follows from the energy conservation in the moving frame where the Hamiltonian is time independent. The eigenvalue of the photon Hamiltonian must match the energy difference of the atomic system $\omega - \boldsymbol{v} \cdot \boldsymbol{k} = \Delta E/\hbar$. Therefore, the frequency of the photon ω in the laboratory frame undergoes the Doppler shift,

$$\omega = \Delta E/\hbar + \boldsymbol{v} \cdot \boldsymbol{k}. \tag{7.13}$$

The quadratic Doppler shift does not appear here, since we are using the nonrelativistic theory.

7.3 Dynamical rotational frequency shift

We shall now repeat the procedure that led to the standard Doppler effect, replacing translation by rotation in the time-dependent potential. In order to remove the time dependence of the Hamiltonian, we apply the unitary transformation $U_\Omega(t)$,

$$U_\Omega(t) = \mathrm{e}^{-\mathrm{i}\boldsymbol{\Omega}\cdot\boldsymbol{J}t/\hbar}, \tag{7.14}$$

where $\boldsymbol{\Omega}$ is the angular velocity vector and $\boldsymbol{J}$ is the total angular momentum. This transformation is the counterpart of the transformation $U_{\mathrm{V}}(t)$. It removes the time dependence of the potential since $\boldsymbol{\Omega} \cdot \boldsymbol{J}$ is the generator of rotations around the $\boldsymbol{\Omega}$ axis:

$$U_\Omega^\dagger(t)\boldsymbol{r}(t)U_\Omega(t) = \boldsymbol{r}. \tag{7.15}$$

The additional term in the Schrödinger equation,

$$\mathrm{i}\hbar\partial_t|\tilde{\psi}(t)\rangle = [H(t)|_{t=0} - \boldsymbol{\Omega} \cdot \boldsymbol{J}]|\tilde{\psi}(t)\rangle, \tag{7.16}$$

is the sum of the atomic part $\boldsymbol{\Omega} \cdot \boldsymbol{L}$ and the field part $\boldsymbol{\Omega} \cdot \boldsymbol{M}$.

The Hamiltonian in the rotating frame cannot be identified with the energy operator. However, its eigenvalues–the quasi-energies–are time independent and they may be used to follow the energy transfers between the atom and the field. Any decrease in the quasi-energy of the atom results in a corresponding increase in the photon quasi-energy.

In order to write down the field Hamiltonian in an explicit form we choose the decomposition of the electromagnetic field into different field modes than in the previous section. We choose the axis of rotation along the z axis and instead of the plane-wave modes we employ the cylindrical modes labeled by the frequency ω, the projection M of the total angular momentum on the z direction, the z component of the wave vector k_z, and the helicity χ. With this choice, the field Hamiltonian after the unitary transformation is

$$\tilde{H}_{\mathrm{F}} = \hbar \sum_{M,\chi} \int_0^\infty \mathrm{d}\omega(\omega - \Omega M) \int_{-\infty}^\infty \mathrm{d}k_z\, a^\dagger(\omega, M, k_z, \chi)a(\omega, M, k_z, \chi). \tag{7.17}$$

As in the case of the Doppler shift, we may say that in the rotating frame each photon acquires additional energy. In the present case the energy difference is equal to $\hbar\Omega M$. Again there is no change in the interaction Hamiltonian since it is invariant under rotations, H_{I} commutes with the total angular momentum.

Despite formal similarities, there is a big difference between the translational and the rotational motion that affect the atomic Hamiltonian. The energy levels of this Hamiltonian are now modified and not just by a trivial constant term as was the case for the Doppler shift. We shall analyze these modification in two steps.

First, let us assume that the potential energy V has cylindrical symmetry. In this case, the Ω-dependent part of the atomic Hamiltonian $\tilde{H}_{\mathrm{A}}$,

$$\tilde{H}_{\mathrm{A}} = H_{\mathrm{A}}(t)|_{t=0} - \boldsymbol{\Omega} \cdot \boldsymbol{L} = H_{\mathrm{A}}(t)|_{t=0} - \Omega(xp_y - yp_x), \tag{7.18}$$

commutes with the remaining part but cannot be removed by a unitary transformation. The quasi-energy spectrum is modified and resembles the atomic spectrum in a constant magnetic field. Each degenerate level characterized by the maximal value $m_{\max}$ of the z component of angular momentum is split into $2m_{\max} + 1$ equally spaced Zeeman-like sublevels separated by $\hbar\Omega$. This level splitting, however, will never be seen for systems with cylindrical symmetry. The explanation of this fact within our model involves a cancelation of two effects. The quasi-energy spectrum of the atomic electron is $E_q - \hbar\Omega m_z$, where q stands for a set of quantum numbers characterizing different eigenstates of the Hamiltonian $H_{\mathrm{A}}(t)|_{t=0}$. Owing to the quasi-energy conservation, the photon emitted in an atomic transition between two states with quantum numbers (q, m_z) and (q', m_z') has its quasi-energy $\hbar(\omega - \Omega M)$ equal to the difference $E_q - E_{q'} - \hbar\Omega(m_z - m_z')$. Thus, the photon frequency is given by the formula

$$\omega = (E_q - E_{q'})/\hbar - \Omega(m_z - m_z') + \Omega M. \tag{7.19}$$

For systems with cylindrical symmetry the last two terms cancel due to the conservation of angular momentum and we are left with the standard Bohr formula. This confirms a general rule that a rotation of a cylindrically symmetric *quantum object* around its symmetry axis is unobservable.

Next, we consider the potential $V(\boldsymbol{r})$ that is not invariant under rotations. In this case, the angular momentum operator does not commute with the atomic Hamiltonian. Therefore, the change of the spectrum of $\tilde{H}_{\mathrm{A}}$ induced by rotation does not have such a simple structure, as in the rotationally symmetric case. In order to find the spectrum of quasi-energies we must diagonalize the complete Hamiltonian and not separately its two commuting parts. The spectrum of quasi-energies, in general, is described by some nonlinear function $\mathcal{E}(q, m_z, \Omega)$ of Ω. Therefore, the dependence on Ω will not be canceled by the frequency shift ΩM of the photon, as was the case for cylindrically symmetric systems. The resulting frequency of the observed photon,

$$\omega(\Omega) = \mathcal{E}(q, m_z, \Omega) - \mathcal{E}(q', m_z', \Omega))/\hbar + \Omega M, \tag{7.20}$$

differs from the frequency of a photon emitted by an atom at rest. The difference $\omega(\Omega) - \omega(0)$ is the dynamical frequency shift of [9]. In the next section we illustrate these properties with the help of specific examples.

7.4 Atom on a turntable

The title of this section is not to be taken literally (turntables rotate too slowly) but it conveys the right meaning. We shall study the changes of the atomic spectrum due to the orbital motion of its center. In this case, an atomic electron is moving in the potential $V(\boldsymbol{r} - \boldsymbol{R}(t))$, where $\boldsymbol{R}(t)$ traces a circle with constant velocity.

We shall start by considering first an exactly soluble case: the harmonic oscillator. The Hamiltonian (7.3) in this case is

$$\frac{\boldsymbol{p}^2}{2m} + \frac{m\omega_0^2}{2}(\boldsymbol{r} - \boldsymbol{R}(t))^2, \tag{7.21}$$

where ω_0 characterizes the spring tension of the oscillator. Due to the invariance of the distance under rotations,

$$|\boldsymbol{r} - \boldsymbol{R}(-t)| = |\boldsymbol{r}(t) - \boldsymbol{R}|, \tag{7.22}$$

we can transfer time dependence to the position of the electron. The time-independent atomic Hamiltonian (7.18) obtained after the unitary transformation (7.15) is

$$\tilde{H}_{\mathrm{A}} = \frac{\boldsymbol{p}^2}{2m} + \frac{m\omega_0^2}{2}(\boldsymbol{r} - \boldsymbol{R})^2 - \boldsymbol{\Omega}\cdot\boldsymbol{L}. \tag{7.23}$$

It is convenient to move the origin of the coordinate system from the rotation axis to the center of the potential. This is accomplished by the unitary transformation U_{S} that simply shifts the origin of the coordinate system by $\boldsymbol{R}$,

$$U_{\mathrm{S}} = \exp[-\mathrm{i}\boldsymbol{R}\cdot\boldsymbol{P}_{\mathrm{T}}/\hbar]. \tag{7.24}$$

After this transformation the three parts of the Hamiltonian of the system take on the form

$$U_{\mathrm{S}}^\dagger \tilde{H}_{\mathrm{A}} U_{\mathrm{S}} = \frac{\boldsymbol{p}^2}{2m} + \frac{m\omega_0^2}{2}\boldsymbol{r}^2 - \boldsymbol{\Omega}\cdot\boldsymbol{L} - \boldsymbol{v}_{\mathrm{c}}\cdot\boldsymbol{p}, \tag{7.25}$$

$$U_{\mathrm{S}}^\dagger \tilde{H}_{\mathrm{I}} U_{\mathrm{S}} = -\frac{e\boldsymbol{p}\cdot\boldsymbol{A}(\boldsymbol{r})}{m} + \frac{e^2\boldsymbol{A}^2(\boldsymbol{r})}{2m}, \tag{7.26}$$

$$U_{\mathrm{S}}^\dagger \tilde{H}_{\mathrm{F}} U_{\mathrm{S}} = H_{\mathrm{F}} - \boldsymbol{\Omega}\cdot\boldsymbol{M} - \boldsymbol{v}_{\mathrm{c}}\cdot\boldsymbol{P}, \tag{7.27}$$

where $\boldsymbol{v}_{\mathrm{c}}$ is the orbital velocity of the center of the potential,

$$\boldsymbol{v}_{\mathrm{c}} = \boldsymbol{\Omega}\times\boldsymbol{R}. \tag{7.28}$$

Additional terms proportional to $\boldsymbol{v}_{\mathrm{c}}$ in the atomic and the field Hamiltonians arise because the total momentum $\boldsymbol{P}_{\mathrm{T}}$ does not commute with angular momentum: from the commutation relations between momentum and angular momentum we obtain for the atomic part

$$U_{\mathrm{S}}^\dagger \boldsymbol{\Omega}\cdot\boldsymbol{L} U_{\mathrm{S}} = \boldsymbol{\Omega}\cdot\boldsymbol{L} - \mathrm{i}\hbar[\boldsymbol{R}\cdot\boldsymbol{p}, \boldsymbol{\Omega}\cdot\boldsymbol{L}] = \boldsymbol{v}_{\mathrm{c}}\cdot\boldsymbol{p}. \tag{7.29}$$

An analogous relation holds for the field part. The atomic and field Hamiltonians (7.25) and (7.27) reduce to the Hamiltonians (7.10) and (7.12) in the limit, when $\Omega \to 0$ and $R \to \infty$, while the orbital velocity $\boldsymbol{v}_{\mathrm{c}}$ is kept fixed. It must be so, because in this limit, as in the case of the Doppler shift, the motion of the atom takes place along a straight line with constant velocity. In fact, Doppler broadenings in the spectra of rotating stars rely on this limiting case.

The atomic Hamiltonian can be simplified by an additional unitary transformation U that removes the last term in (7.25) and reduces the problem to the standard case of a harmonic oscillator:

$$U = \exp[\mathrm{i}(\boldsymbol{a} \cdot \boldsymbol{r} - \boldsymbol{b} \cdot \boldsymbol{P}_{\mathrm{T}})/\hbar]. \tag{7.30}$$

When the vectors $\boldsymbol{a}$ and $\boldsymbol{b}$ are chosen as

$$\boldsymbol{a} = \frac{m\omega_0^2}{\omega_0^2 - \Omega^2}\boldsymbol{\Omega} \times \boldsymbol{R}, \quad \boldsymbol{b} = \frac{\Omega^2}{\omega_0^2 - \Omega^2}\boldsymbol{R}, \tag{7.31}$$

the transformed Hamiltonian $\tilde{H}_{\mathrm{A}}$ takes on the standard, rotationally symmetric form,

$$\hat{H}_{\mathrm{A}} = U^\dagger \tilde{H}_{\mathrm{A}} U = \frac{\boldsymbol{p}^2}{2m} + \frac{m\omega_0^2}{2}\boldsymbol{r}^2 - \boldsymbol{\Omega} \cdot \boldsymbol{L} - \frac{m\omega_0^2\Omega^2 R^2}{2(\omega_0^2 - \Omega^2)}. \tag{7.32}$$

The field Hamiltonian is not modified by the transformation (7.30) but the interaction Hamiltonian changes as follows:

$$\hat{H}_{\mathrm{F}} = U^\dagger \tilde{H}_{\mathrm{F}} U = \tilde{H}_{\mathrm{F}}, \tag{7.33}$$

$$\hat{H}_{\mathrm{I}} = U^\dagger \tilde{H}_{\mathrm{I}} U = -\frac{e\boldsymbol{p} \cdot \boldsymbol{A}(\boldsymbol{r})}{m} + \frac{e^2\boldsymbol{A}^2(\boldsymbol{r})}{2m} - \frac{\omega_0^2}{\omega_0^2 - \Omega^2}e\boldsymbol{v}_{\mathrm{c}} \cdot \boldsymbol{A}(\boldsymbol{r}). \tag{7.34}$$

After the unitary transformations the atomic and field Hamiltonians are the same as for a harmonic oscillator placed on the rotation axis. If it were not for the last term in the interaction Hamiltonian, this system would have exact rotational symmetry and no effect of rotation would be seen. This additional term does not change the spectrum but merely changes the photon transition probabilities. The absence of any changes in the spectrum is exceptional and it is specific to the harmonic potential.

We shall consider now a more realistic case of an atom on a turntable by replacing the harmonic potential by the Coulomb potential. This will lead to significant changes in the spectrum. Following the procedure applied in the previous case, after the first few steps, we obtain the following Hamiltonian:

$$U_{\mathrm{S}}^\dagger \tilde{H}_{\mathrm{A}} U_{\mathrm{S}} = \frac{\boldsymbol{p}^2}{2m} - \frac{\gamma}{|\boldsymbol{r}|} - \boldsymbol{\Omega} \cdot \boldsymbol{L} - \boldsymbol{v}_{\mathrm{c}} \cdot \boldsymbol{p}, \tag{7.35}$$

where $\gamma = Ze^2/4\pi\epsilon_0$. The interaction and the field Hamiltonians are given by the same formulas (7.26) and (7.27). The quasi-energy spectrum of this Hamiltonian cannot be determined exactly. However, the unitary transformation

$$U = \exp[im\boldsymbol{r} \cdot \boldsymbol{v}_c/\hbar], \tag{7.36}$$

produces the familiar expression

$$U^\dagger \tilde{H}_A U = \frac{\boldsymbol{p}^2}{2m} - \frac{\gamma}{|\boldsymbol{r}|} - e\boldsymbol{\mathfrak{E}} \cdot \boldsymbol{r} - \frac{e}{2m}\boldsymbol{\mathfrak{B}} \cdot \boldsymbol{L} - \frac{mR^2\Omega^2}{2}, \tag{7.37}$$

where we have introduced fictitious electric and magnetic fields that mimic the action of inertial forces

$$e\boldsymbol{\mathfrak{E}} = m\Omega^2\boldsymbol{R}, \quad e\boldsymbol{\mathfrak{B}} = 2m\boldsymbol{\Omega}. \tag{7.38}$$

In this way, we can reduce our task to the thoroughly studied problem of an electron moving in a Coulomb potential in the presence of additional crossed fields: the electric field $\boldsymbol{\mathfrak{E}}$ and the magnetic field $\boldsymbol{\mathfrak{B}}$. The quadratic in $\boldsymbol{\mathfrak{B}}$ (diamagnetic) term is absent here, which means that we may use the analogy between the rotation and the magnetic field only for relatively weak fields. The additive constant (last term) shifts the whole spectrum but does not change the level separation.

The spectrum of the Hamiltonian (7.37) in the weak-field regime has been found already in old quantum theory [11]. In quantum mechanics, the solution of this problem is an example of perturbation theory with degenerate unperturbed spectrum. The energy levels of the Hamiltonian (7.37) in the lowest order in $\boldsymbol{\mathfrak{E}}$ and $\boldsymbol{\mathfrak{B}}$ are (cf., for example, [12])

$$\mathcal{E}(n, m_z, \Omega) = -m\left(\frac{Ze^2}{4\pi\epsilon_0\hbar}\right)^2\frac{1}{2n^2} - \hbar m_z\sqrt{(e\boldsymbol{\mathfrak{B}}/2m)^2 + \left(\frac{12\pi\epsilon_0\hbar}{2Zem}\right)^2 n^2\boldsymbol{\mathfrak{E}}^2}. \tag{7.39}$$

Thus, the inertial forces in a rotating frame lead to modifications of the energy levels similar to the dynamical Stark shift caused by an electromagnetic wave. The quantum number $m_z = -m_{\max}, \ldots, m_{\max}$ plays the same role as in the rotationally symmetric case.

Upon inserting the expressions (7.38) into the formula (7.39), we obtain the following atomic spectrum modified by rotation:

$$\mathcal{E}(n, m_z, \Omega) = -\frac{mv_a^2}{2n^2} - \hbar\Omega m_z\sqrt{1 + \left(\frac{3nR\Omega}{2v_a}\right)^2}, \tag{7.40}$$

where v_a is the characteristic velocity of the electron in an atom,

$$v_a = \frac{Ze^2}{4\pi\epsilon_0\hbar}. \tag{7.41}$$

For $Z = 1$ this velocity is equal to the speed of light multiplied by the fine-structure constant, $v_a = \alpha c \approx 2.2 \cdot 10^6$ m/s.

The DRFS of a photon emitted by an orbiting atom in a transition $(n, m_z) \to (n', m'_z)$, observed in the laboratory frame, according to the formula (7.20), is

$$\omega(\Omega) - \omega(0) = -\Omega m_z \sqrt{1 + \left(\frac{3nR\Omega}{2v_a}\right)^2} + \Omega m'_z \sqrt{1 + \left(\frac{3n'R\Omega}{2v_a}\right)^2} + \Omega M$$
$$\approx \frac{9\pi^2(n'^2 m'_z - n^2 m_z)}{2} \Omega \left(\frac{v_c}{v_a}\right)^2. \tag{7.42}$$

The smallness of the velocity v_c of the potential center as compared to v_a justifies the approximation of the square root by the first two terms of the Taylor expansion. The linear dependence of the DRFS on the quantum numbers m_z that label the projections of the angular momentum on the axis of rotation means that the observed radiation can be red-shifted or blue-shifted. These quantum numbers are the analogues of the projections of the photon momentum on the direction of motion in the standard Doppler shift.

In the case of an orbiting atom one may also consider the linear Doppler shift due to the orbital velocity v_c. However, a direct comparison of the DRFS and the Doppler shift is not possible for two reasons. First, the orbital velocity is continuously changing its direction. In order to measure the standard Doppler shift of an orbiting atom, we must know its position on the orbit at the moment of the photon emission. This information is not available (except in the limiting case of $\Omega \to 0$, as described above) since the description of states according to the their quasi-energy is complementary (owing to the time-energy uncertainty relation) to the description that involves time. Second, these two effects rely on complementary descriptions of the radiation process. Photon eigenstates of M_z contain the whole range of wave vectors and, vice versa, the eigenstates of momentum contain the whole range of angular momenta.

Even though we cannot compare directly the linear Doppler shift with the DRFS, we can do it for the quadratic Doppler shift. The quadratic Doppler shift in the direction perpendicular to the orbital plane and far from this plane is not sensitive to the position of the atom on its orbit. Therefore, the observation of the quadratic (transverse) Doppler shift on the axis of rotation $\omega(v/c)^2/2$ can be directly compared with the DRFS. Their ratio depends only on the ratio of two frequencies and not on the orbital velocity. Within the approximation employed in the formula (7.42) it can be estimated as

$$9\pi^2(n'^2 m'_z - n^2 m_z)\frac{1}{\alpha^2}\frac{\Omega}{\omega}, \tag{7.43}$$

where α is fine structure constant. This ratio may vary over a very wide range, from 10^{-10} for experiments using the Mössbauer effect [13, 14] to as much as 10^3 for rotating molecules emitting infrared photons.

7.5 Observation of the dynamical rotational frequency shift

The DRFS would be very hard to observe when the atom is placed on a macroscopic mechanical device. The fastest rotating devices, an ultracentrifuge, a gas turbine or a turbocharger, may reach at most a few thousand revolutions per second. For R of the order of 0.1 m this translates into the linear velocity of the order of 1000 m/s. Thus, the value of $(v/v_a)^2$ that appears in the formula (7.42) is of the order of 10^{-6}.

Can we make the measurement of the DRFS feasible, by resorting to microscopic rotating devices? Nature has provided us with such objects in the form of various molecules. Molecules rotate very quickly. Unfortunately, we also have to replace the macroscopic value of R with the molecular bond length and this gives *the same effect* as a macroscopic "turntable." Indeed, with the values $\Omega \approx 10^{13}$/s and $R \approx 10^{-10}$ m we again obtain $(v/v_a)^2 \approx 10^{-6}$. A large prefactor in (7.42) increases the shift only by a factor of about 100 to the value of about 10^{-4} of the rotational frequency Ω. Thus, in both cases the DRFS is very small and to detect it would require great ingenuity and determination. What has then been measured in the experiment described in [10]? To answer this question we have to extend our model to include an external electromagnetic field.

Molecules in gaseous phase at a given rotational temperature rotate with a random distribution of directions and a thermal distribution of absolute values of $\boldsymbol{\Omega}$. In order to organize this motion so that the signal would become detectable, the authors of the experiment [10] used a circularly polarized microwave. The rotating electric field of the wave couples to the large electric dipoles of the thioformaldehyde molecules and makes them rotate in unison. To understand what frequency shift has been measured in this experiment, we shall take into account the role of a rotating electric field in our turntable model. Such a field rotating with frequency Ω is represented by a dipole-coupling term, leading to the following form of the potential:

$$V(\boldsymbol{r}(t)) = -\frac{\gamma}{|\boldsymbol{r} - \boldsymbol{R}(t)|} - e\boldsymbol{r} \cdot \boldsymbol{E}(t). \tag{7.44}$$

In the nonrelativistic regime, the influence of the magnetic field of the wave can be neglected. After the transformation (7.15) the rotating electric field becomes frozen. In the case when our atom is embedded in a molecule, the electric field will be aligned with the molecular dipole. Thus, the time-independent Hamiltonian (7.37) in the presence of a rotating electric field has the form:

$$H_{ff} = \frac{\boldsymbol{p}^2}{2m} - \frac{\gamma}{|\boldsymbol{r}|} - e(\mathfrak{E} + \boldsymbol{E}) \cdot \boldsymbol{r} - \frac{e}{m}\mathfrak{B} \cdot \boldsymbol{L} - \frac{mR^2\Omega^2}{2} - e\boldsymbol{E} \cdot \boldsymbol{R}. \tag{7.45}$$

In general, the vectors $\mathfrak{E}$ and $\boldsymbol{E}$ may have different directions. In the special case of a linear molecule they are collinear. Depending on whether they point in the same or in opposite directions, the effect will be enhanced or reduced. The two terms might even cancel each other. The spectrum of quasi-energies in the presence of a circularly polarized plane wave is still described by the formula (7.39) but one has to add to the pseudo-electric field $\mathfrak{E}$ the true electric field $\boldsymbol{E}$. In this way we obtain for the quasi-energy spectrum

$$\mathcal{E}(n, m_z, \Omega) = -\frac{m v_{a^2}}{2n^2} + \hbar\Omega m_z \sqrt{1 + \left(\frac{3n(m\Omega^2\boldsymbol{R} + e\boldsymbol{E})}{2mv_a\Omega}\right)^2}. \tag{7.46}$$

Under the square-root sign in this expression there appears the sum of two forces: the centrifugal force $m\Omega^2\boldsymbol{R}$ and the electric force $e\boldsymbol{E}$. Their ratio for the values of the parameters of the experiment described in [10] ($E = 30\,\text{kV/m}$, $\Omega/2\pi = 80\,\text{MHz}$) is

$$\frac{m\Omega^2 R}{eE} = 2.24 \cdot 10^{-10}\frac{(\Omega/2\pi\,[\text{Hz}])^2 R[\text{m}]}{E[\text{V/m}]} \approx 2.4 \cdot 10^{-9}, \tag{7.47}$$

where the value of R was chosen as equal to the atomic unit (Bohr radius) $5 \cdot 10^{-11}$ m. Thus, in this experiment the DRFS is nine orders of magnitude smaller than the dynamical Stark shift caused by the rotating electric field. It seems that in any experiment where a rotating electric field is used to control rotation, the influence of the electric field will completely overshadow the DRFS.

Of course, the turntable model does not represent correctly rotating molecules of thioformaldehyde in the experiment but it enables one to classify various terms in the formula for the DRFS. There are three contributions under the square-root sign in (7.46) that differ in their dependence on the frequency of rotation Ω: quadratic, quartic, and independent. It is the quartic term that describes the effect of quasi-energy level modifications in our model caused by the centrifugal force. Such a term (or any other expression) was absent in equations (7) and (8) of [10]. Obviously, this omission was fully justified, as seen from our order of magnitude estimate.

7.6 Conclusions

We identified two sources of the dynamical rotational frequency shifts. The first part ΩM comes from the transformation properties the electromagnetic radiation under the change from the rotating frame to the inertial frame. The second part is due to the inertial forces acting on the emitter (or the absorber). For cylindrically symmetric emitters these two effects cancel each other. For asymmetric emitters the DRFS is in most cases very small but might be observable in laboratory experiments.

Acknowledgement

We acknowledge support from the Polish Ministry of Science and Higher Education under a grant for the years 2010–2012.

References

[1] Garetz, B. A. and Arnold, S. 1979. Variable frequency shift of circularly polarized laser radiation via a rotating half-way retardation plate. *Opt. Comm.*, **31**, 1.
[2] Garetz, B. A. 1981. Angular Doppler shift. *J. Opt. Soc. Am.*, **71**, 609.
[3] Allen, L., Babiker, M., and Power, W. L. 1994. Azimuthal Doppler shift in light beams with orbital angular momentum. *Opt. Comm.*, **112**, 141.
[4] Nienhuis, G. 1996. Doppler effect induced by rotating lenses. *Opt. Comm.*, **132**, 8.

[5] Courtial, J., Robertson, D. A., Dholakia, K., Allen, L., and Padgett, M. J. 1998. Rotational frequency shift of a light beam. *Phys. Rev. Lett.*, **81**, 4828.

[6] Padgett M. J. and Courtial, J. 1999. Poincaré-sphere equivalent for light beams containing orbital angular momentum. *Opt. Lett.*, **24**, 430.

[7] Bekshaev A. Ya, Soskin, M. S., and Vasnetsov, M. V. 2005 Angular momentum of a rotating light beam. *Opt. Comm.*, **249**, 367.

[8] Nienhuis, G. 2006. Polychromatic and rotating beams of light. *J. Phys. B: Atom. Mol. Opt. Phys.*, **39**, S529.

[9] Bialynicki-Birula, I. and Bialynicka-Birula, Z. 1997. Rotational frequency shift. *Phys. Rev. Lett.*, **78**, 2539.

[10] Michalski, M., Hüttner W., and Schimming, H. 2005 Experimental demonstration of the rotational frequency shift in a molecular system. *Phys. Rev. Lett.* **95** 203005.

[11] Epstein, P. S. 1923. Simultaneous action of an electric and a magnetic field on a hydrogen-like atom. *Phys. Rev.*, **22**, 202.

[12] Penent, F., Delande, D., and Gay, J. C. 1988 Rydberg states of rubidium in crossed electric and magnetic fields. *Phys. Rev. A*, **37**, 4707.

[13] Kündig, W. 1963. Measurement of the transverse Doppler effect in accelerated systems. *Phys. Rev.*, **129**, 2371.

[14] Kholometskii, A. L., Yarman, T., Missevitch, O. V., and Rogozev, B. I. 2009. A Mössbauer experiment in a rotating system on the second-order Doppler shift: confirmation of the corrected result by Kündig, *Phys. Scr.*, **79**, 065007.

8

Spin-orbit interactions of light in isotropic media

K. Y. BLIOKH, A. AIELLO AND M. A. ALONSO

8.1 Introduction

8.1.1 Spin-orbit interaction in quantum physics

The term *spin-orbit interaction* (SOI) is known from quantum physics, where it describes the coupling between the *spin* and *orbital angular momentum* (AM) of electrons or other quantum particles [1]. The SOI is usually interpreted as an electromagnetic interaction of the moving magnetic moment of the electron with an external electric field. However, in central fields the SOI Hamiltonian becomes proportional to the product of the spin AM (SAM) and orbital AM (OAM). Furthermore, the geometric *Berry-phase* description of the SOI uncovered that it is basically related to the intrinsic AM properties of the particle and is largely independent of the particular character of interaction with the external field [2–4]. The SOI can take various forms in different systems, but the unifying feature in all cases is the coupling between the *spin* and *momentum* of the particle [5].

There are two basic manifestations of the SOI of electrons. First, the SOI brings about the *fine splitting of the energy levels* in the finite orbital motions, e.g., in potential wells or magnetic fields [1, 6, 7]. This can be regarded as the Berry-phase contribution to the *quantization of orbits* [8–11]. Second, due to the SOI upon free motion in an external field the electron undergoes a transverse spin-dependent deflection known as the *spin Hall effect* [4, 5, 12–14]. This effect is a dynamical (transport) manifestation of the Berry phase closely related to the Coriolis effect [15, 16], and *conservation of the total AM* of the particle [12].

8.1.2 Angular momenta and spin-orbit interactions of light

Since classical light and quantum photons also possess both SAM and OAM [17–21], one can expect that a coupling between spin and orbital AM degrees of freedom of light may occur at certain conditions. This is the *SOI of light*, which represents a variety of fine

The Angular Momentum of Light, ed. D. L. Andrews and M. Babiker. Published by Cambridge University Press.

(typically, subwavelength) phenomena originating from the spin-momentum coupling in optical fields.

A convenient classification can be introduced based on the different types of AM of paraxial light. It is well-known that the SAM is the intrinsic (i.e., independent of the origin) AM associated with the degree of *circular polarization* of light. At the same time, the two types of the OAM can be distinguished [22–26]: the *intrinsic OAM* (IOAM) typically associated with the *optical vortex* structure of the field and circulating Poynting flow; and the origin-dependent *extrinsic OAM* (EOAM) produced by the *trajectory of motion* of the centre of gravity (centroid) of a light beam or wave packet. Thus, the three types of AM of light – SAM, IOAM, and EOAM – can be exemplified, respectively, by circular polarization, optical vortices, and the trajectory of a localized light field as a whole, see Fig. 8.1.

According to this division, the following pair AM interactions may take place: (i) the SAM–IOAM ('polarization-vortex') interaction; (ii) the SAM–EOAM ('polarization-trajectory') interaction; and (iii) the IOAM–EOAM ('vortex-trajectory') interaction[1] (see Table 8.1). Specifically:

(i) The SAM–IOAM interaction between the two intrinsic AM occurs typically in cylindrically or spherically symmetric systems – e.g., upon focusing or scattering of light [27–31] and upon propagation in optical fibers [32–36]. This coupling results in the generation of spin-induced vortices in the field, which is known as *spin-to-orbit AM conversion* of light. Besides, the SAM–IOAM interaction is accompanied by a fine polarization *splitting of longitudinal propagation constants and transverse caustics*, akin to the fine energy-level splitting of confined electrons [24, 32, 36, 37].

(ii) The SAM–EOAM interaction is quite similar to the SOI of free electrons and produces the *spin Hall effect* of light. This is a wavelength-order polarization-dependent *transverse shift* of the optical beam, which occurs, e.g., upon propagation or reflection/refraction in inhomogeneous media [38–44] and in asymmetric focusing or scattering [31, 45–47].

(iii) Finally, the IOAM–EOAM interaction produces a similar *orbital Hall effect* of light, i.e., a vortex-dependent transverse shift of higher-order beams in analogous systems [48–54].

As we show below, the *geometric Berry phases* and *AM dynamics and conservation* essentially underlie *all* the optical SOI phenomena in locally isotropic media. This establishes a close link between the SOI of light and quantum electrons, and also reveals the fundamentally dual *geometrodynamical* nature of the SOI. It should be noted that strong optical SOI also occurs in media with space-variant anisotropies [55–65]. The dynamical phase difference between the ordinary and extraordinary modes and the Pancharatnam–Berry geometric phase underlies this extrinsic (produced by the medium) effect, which is not considered here (for reviews, see [66, 67]).

[1] In actual fact, case (iii) represents the *orbit-orbit interaction*, but we also attribute it to the SOI group of phenomena owing to their close similarities.

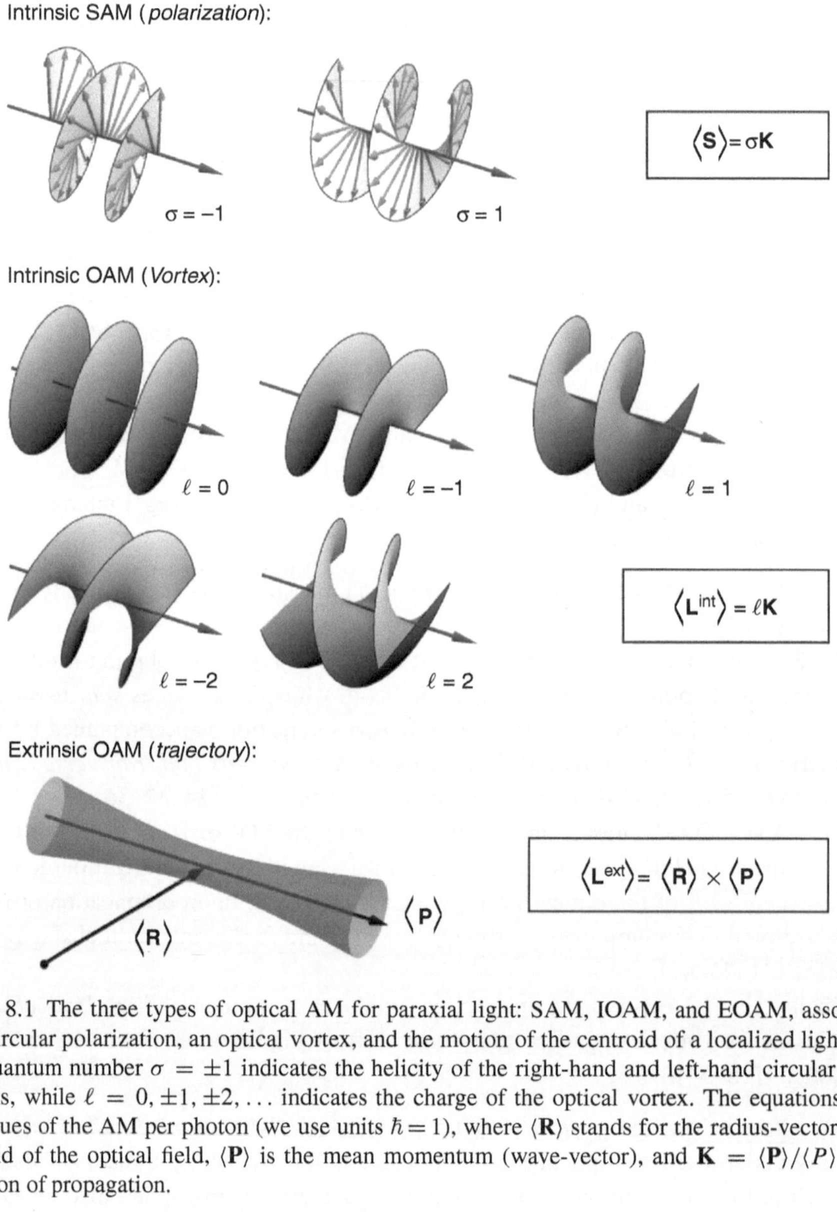

Figure 8.1 The three types of optical AM for paraxial light: SAM, IOAM, and EOAM, associated with circular polarization, an optical vortex, and the motion of the centroid of a localized light field. The quantum number $\sigma = \pm 1$ indicates the helicity of the right-hand and left-hand circular polarizations, while $\ell = 0, \pm 1, \pm 2, \ldots$ indicates the charge of the optical vortex. The equations show the values of the AM per photon (we use units $\hbar = 1$), where $\langle \mathbf{R} \rangle$ stands for the radius-vector of the centroid of the optical field, $\langle \mathbf{P} \rangle$ is the mean momentum (wave-vector), and $\mathbf{K} = \langle \mathbf{P} \rangle / \langle P \rangle$ is the direction of propagation.

8.1.3 History and current motivation

The concept of the SOI of light was first introduced by Boris Zel'dovich and co-authors in the 1990s in a series of seminal papers studying fine polarization-dependent effects upon the evolution of light in multimode fibers, inhomogeneous media, and lenses [32, 33, 38, 45, 46]. Before this, an example of the spin-Hall effect of light arising at plane

Table 8.1 *Classification of the AM interactions and their main manifestations. The final column indicates the presence of the axial symmetry about the propagation direction.*

<table>
<tr><th>Interaction</th><th>Manifestation</th><th>Systems (paraxiality of the field)</th><th>Symmetry</th></tr>
<tr><td>SAM–IOAM (polarization-vortex)</td><td>Spin-to-orbit AM conversion

Fine splitting of levels</td><td>Focusing and scattering (non-paraxial)

Propagation in optical fibers (paraxial)

Propagation and diffraction in anisotropic media (paraxial)</td><td>+</td></tr>
<tr><td>SAM–EOAM (polarization-position)</td><td>Spin Hall effects (polarization-dependent transverse shift)</td><td rowspan="2">Asymmetric focusing and scattering (non-paraxial)

Reflection/refraction at interfaces (locally paraxial)

Propagation in gradient-index media (locally paraxial)</td><td rowspan="2">–</td></tr>
<tr><td>IOAM–EOAM (vortex-position)</td><td>Orbital Hall effect (vortex-dependent transverse shift)</td></tr>
</table>

dielectric interfaces was known as the transverse *Imbert–Fedorov shift* [68–73]. In those days, however, the theoretical description of the fine subwavelength effects was somewhat controversial and the accuracy and conclusiveness of experiments were not sufficient to verify the theories. The SOI of light caught a second wind in 2004–8, when the unifying Berry-phase and AM description of SOI phenomena were developed and the experimental abilities allowed their precise measurements [27–31, 39–44, 47–54].

Nowadays, the SOI of light represents a rapidly growing area of research, which is of both *fundamental* and *practical* interest. On the one hand, the optical SOI possesses a fundamental physical importance, because it reveals universal intrinsic geometrodynamical properties of the evolution of relativistic spin-1 particles (photons) [4, 74–76]. In this manner, classical optics offers a unique opportunity to directly observe fine relativistic effects which *cannot* be detected in *high-energy* and *condensed matter* physics [43]. On the other hand, being of the order of a fraction of the wavelength (i.e., typically neglected in classical optics), the fine SOI phenomena have become very important for modern optical applications. Miniaturization of optical devices and the fast development of *nano-optics*, *photonics*, *plasmonics*, and *singular optics* (all essentially dealing with *subwavelength* scales) make accounting for the SOI indispensable. Indeed, it turns out that most if not all basic optical processes–propagation, reflection, focusing, scattering, diffraction–are strongly influenced by the SOI at subwavelength scales and should be carefully revisited.

In the present review, we aim to give a unifying description of the main SOI phenomena in optics of isotropic media and reveal their physical origins and interrelations. Our approach is based on the geometric Berry-phase interpretations and the dynamical AM aspects. This provides a thorough understanding and allows an efficient exchange of ideas between classical optics and quantum physics. To clarify the underlying physical mechanisms and cover a diversity of the basic optical processes, we restrict ourselves to the simplest configurations for each type of phenomenon.

8.2 Geometric phases, angular momenta, and energy flows

8.2.1 Transversality, rotations, and Berry phase

Maxwell equations

We consider propagating monochromatic light in free space, which is characterized by the constant frequency ω, wave number $k = \omega$ (we use units $\hbar = c = 1$), and complex electric and magnetic fields $\mathbf{E}(\mathbf{r})$ and $\mathbf{H}(\mathbf{r})$, while the real fields are $\mathcal{E}(\mathbf{r}, t) = \mathrm{Re}[\mathbf{E}(\mathbf{r})\mathrm{e}^{-\mathrm{i}\omega t}]$ and $\mathcal{H}(\mathbf{r}, t) = \mathrm{Re}[\mathbf{H}(\mathbf{r})\mathrm{e}^{-\mathrm{i}\omega t}]$. The complex fields satisfy the Maxwell equations

$$\mathrm{i}\omega\mathbf{H} = \nabla \times \mathbf{E}, \quad -\mathrm{i}\omega\mathbf{E} = \nabla \times \mathbf{H}, \quad \nabla \cdot \mathbf{E} = \nabla \cdot \mathbf{H} = 0, \tag{8.1}$$

and can equivalently be characterized by their Fourier spectra $\tilde{\mathbf{E}}(\mathbf{k})$ and $\tilde{\mathbf{H}}(\mathbf{k})$:

$$\mathbf{E}(\mathbf{r}) \propto \int \tilde{\mathbf{E}}(\mathbf{k})\mathrm{e}^{\mathrm{i}\mathbf{k}\cdot\mathbf{r}}[\mathrm{d}\mathbf{k}], \quad \mathbf{H}(\mathbf{r}) \propto \int \tilde{\mathbf{H}}(\mathbf{k})\mathrm{e}^{\mathrm{i}\mathbf{k}\cdot\mathbf{r}}[\mathrm{d}\mathbf{k}]. \tag{8.2}$$

Here we neglect the evanescent waves and imply the proper integration measure in the $\mathbf{k}$-space, which depends on the particular problem being considered. The plane-wave Fourier amplitudes $\tilde{\mathbf{E}}(\mathbf{k})$ and $\tilde{\mathbf{H}}(\mathbf{k})$ determine the *momentum representation* of the fields. We will also use the unit polarization vectors of the fields: e.g., $\mathbf{e} = \mathbf{E}/E$ and $\tilde{\mathbf{e}} = \tilde{\mathbf{E}}/\tilde{E}$.

In most cases below we explore the plane-wave properties of the field, for which the electric and magnetic contributions are equivalent, so it is sufficient to consider the electric field only. The Maxwell equations (8.1) for the electric field can be written as

$$\hat{\mathbf{p}} \times (\hat{\mathbf{p}} \times \mathbf{E}) + \omega^2\mathbf{E} = 0, \tag{8.3}$$

where $\hat{\mathbf{p}} = -\mathrm{i}\nabla$ is the operator of momentum. In the momentum representation $\hat{\mathbf{p}} = \mathbf{k}$, and Eq. (8.3) takes the form

$$\mathbf{k} \times (\mathbf{k} \times \tilde{\mathbf{E}}) + \omega^2\tilde{\mathbf{E}} = 0. \tag{8.3a}$$

From Eq. (8.3a) it follows that

$$\mathbf{k} \cdot \tilde{\mathbf{E}} = 0. \tag{8.4}$$

This is the *transversality condition*, which compels the electric field of a plane wave to be orthogonal to its wave vector. This constraint couples *polarization* and *momentum* of light, which is a typical feature of the SOI. The condition (8.4) has a convenient geometrical representation: the wave electric field must be tangent to the surface of the *unit sphere of directions*, $\mathrm{S}^2 = \{\boldsymbol{\kappa}\}$, $\boldsymbol{\kappa} = \mathbf{k}/k$, in momentum $\mathbf{k}$-space, (see Fig. 8.2).

Coordinate frames

Alongside the 'absolute' field vectors $\mathbf{E}$ and $\tilde{\mathbf{E}}$, which are independent of the choice of the stationary coordinate frame, we introduce 'vectors of components', $|\mathbf{E})_A$ and $|\tilde{\mathbf{E}})_A$, i.e., field components in the given coordinate frame A. For instance, in the global Cartesian frame (x, y, z), the *laboratory frame*, we have

$$\mathbf{E} = E_x \mathbf{u}_x + E_y \mathbf{u}_y + E_z \mathbf{u}_z, \quad |\mathbf{E})_{\mathrm{L}} = (E_x, E_y, E_z)^{\mathrm{T}}, \tag{8.5}$$

where $\mathbf{u}_a$, $a = x, y, z$, stand for the basic vectors of the corresponding axes. We will also use the transverse vector components considered with respect to the z-axis of the frame in use and denote by the '$\perp$' subscript: $|\mathbf{E})_{\perp\mathrm{L}} = (E_x, E_y)^{\mathrm{T}}$, $\mathbf{r}_\perp = (x, y)$, etc. Components of the same vector in different bases are related by a *unitary transformation* which links the basic vectors of the frames: $(\mathbf{u}'_x, \mathbf{u}'_y, \mathbf{u}'_z) = (\mathbf{u}_x, \mathbf{u}_y, \mathbf{u}_z)\hat{U}$ and $(E'_x, E'_y, E'_z)^{\mathrm{T}} = \hat{U}^\dagger (E_x, E_y, E_z)^{\mathrm{T}}$. For instance, in the *'laboratory circular basis'* $(\mathbf{u}_0^+, \mathbf{u}_0^-, \mathbf{u}_z)$ of the circular polarizations in the (x, y) plane, $\mathbf{u}_0^\pm = (\mathbf{u}_x \pm i\mathbf{u}_y)/\sqrt{2}$, one has $E_0^\pm = (E_x \mp iE_y)/\sqrt{2}$, i.e.,

$$|\mathbf{E})_{\mathrm{C}} = \hat{V}^\dagger |\mathbf{E})_{\mathrm{L}}, \quad \hat{V} = \frac{1}{\sqrt{2}} \begin{pmatrix} 1 & 1 & 0 \\ \mathrm{i} & -\mathrm{i} & 0 \\ 0 & 0 & \sqrt{2} \end{pmatrix}. \tag{8.6}$$

Since the transversality condition (8.4) attaches the electric field $\tilde{\mathbf{E}}$ to the surface of the κ-sphere, it is natural to describe the polarization evolution in the *spherical coordinates* (θ, ϕ, k) in momentum space (Fig. 8.2). Their basic vectors are

$$\mathbf{u}_\theta(\mathbf{k}) = \mathbf{u}_\phi \times \boldsymbol{\kappa}, \quad \mathbf{u}_\phi(\mathbf{k}) = \frac{\mathbf{u}_z \times \boldsymbol{\kappa}}{|\mathbf{u}_z \times \boldsymbol{\kappa}|}, \quad \mathbf{u}_k(\mathbf{k}) = \boldsymbol{\kappa}. \tag{8.7}$$

Transformation between the laboratory and spherical frames is a product of two rotations by the angles θ and ϕ, $\hat{U}_{\mathrm{S}} = \hat{R}_z(-\phi)\hat{R}_y(-\theta)$:

$$|\tilde{\mathbf{E}})_{\mathrm{S}} = \hat{U}_{\mathrm{S}}^\dagger(\mathbf{k})|\tilde{\mathbf{E}})_{\mathrm{L}}, \quad \hat{U}_{\mathrm{S}} = \begin{pmatrix} \cos\theta\cos\phi & -\sin\phi & \sin\theta\cos\phi \\ \cos\theta\sin\phi & \cos\phi & \sin\theta\sin\phi \\ -\sin\theta & 0 & \cos\theta \end{pmatrix}, \tag{8.8}$$

where we have introduced the operators of SO(3) rotations: $\hat{R}_a(\alpha) = \exp(\mathrm{i}\alpha\hat{S}_a)$, $a = x, y, z$, where $\hat{S}_a$ denote the generators of rotations, i.e., spin-1 matrices. In spherical coordinates, the transversality condition (8.4) becomes particularly simple: $\tilde{E}_k = 0$, so that the field becomes two-component $|\tilde{\mathbf{E}})_{\mathrm{S}} = (\tilde{E}_\theta, \tilde{E}_\phi, 0)^{\mathrm{T}}$.

Spherical coordinates provide a description of a plane wave of the field spectrum of the local basis attached to its $\mathbf{k}$-vector. The circular polarizations defined with respect to the spherical vectors form the *helicity basis*

$$\mathbf{u}^\sigma = \frac{\mathbf{u}_\theta + \mathrm{i}\sigma\mathbf{u}_\phi}{\sqrt{2}} \mathrm{e}^{\mathrm{i}m\sigma\phi}, \tag{8.9}$$

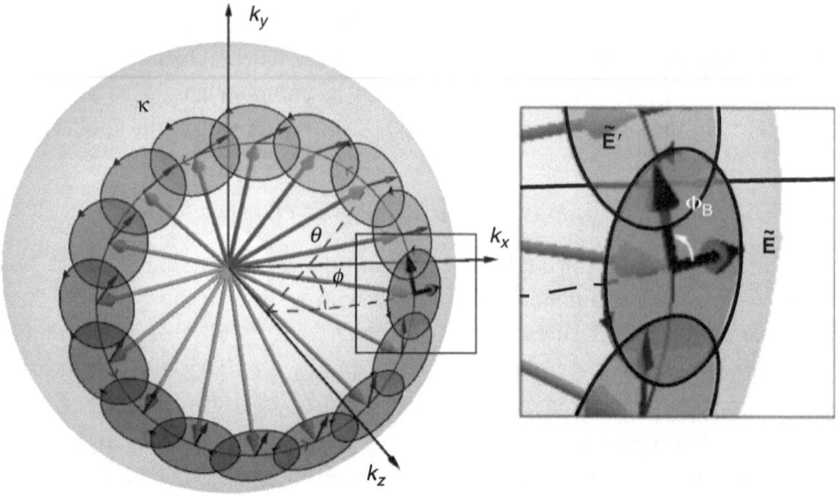

Figure 8.2 The unit sphere $S^2 = \{\boldsymbol{\kappa}\}$ in the momentum space represents the directions of propagation of plane waves specified by spherical angles (θ, ϕ). The electric field $\tilde{\mathbf{E}}$ is tangent to this sphere. Parallel transport of the field along the contour C brings about rotation $\tilde{\mathbf{E}} \to \tilde{\mathbf{E}}'$ by the angle numerically equal to the Berry phase $\Phi_B(C)$, Eqs. (8.17) and (8.19). Correspondingly, the circular-polarization components in a global coordinate frame acquire the phase factors $\exp(-i\sigma\Phi_B)$, Eq. (8.17).

where $\sigma = \pm 1$ is the helicity and the $e^{im\sigma\phi}$ is an arbitrary gauge factor [24, 77]. In particular, the choice $m = 1$ subtracts the azimuthal rotation of the spherical basic vectors removing the pole singularity at $\theta = 0$. Then, in the paraxial limit $\theta \to 0$ (propagation along the z-axis), the basic vectors $\mathbf{u}^\sigma$, Eq. (8.9), tend to the laboratory circular-polarization basis: $\mathbf{u}^\sigma \to \mathbf{u}_0^\sigma$. Transformation from the laboratory circular frame to the helicity basis (8.9) is given by $\hat{U}_H(\mathbf{k}) = \hat{V}^\dagger \hat{R}_z(-\phi)\hat{R}_y(-\theta)\hat{R}_z(m\phi)\hat{V}$, which at $m = 1$ yields

$$|\tilde{\mathbf{E}})_H = \hat{U}_H^\dagger(\mathbf{k})|\tilde{\mathbf{E}})_C, \quad \hat{U}_H(\mathbf{k}) = \begin{pmatrix} a & -be^{-2i\phi} & \sqrt{2ab}e^{-i\phi} \\ -be^{2i\phi} & a & \sqrt{2ab}e^{i\phi} \\ -\sqrt{2ab}e^{i\phi} & -\sqrt{2ab}e^{-i\phi} & a-b \end{pmatrix}, \tag{8.10}$$

where $a = \cos^2(\theta/2)$ and $b = \sin^2(\theta/2)$. The off-diagonal elements of the matrix (8.10) contain the *geometric phase* factors $e^{\pm i\phi}$ and $e^{\pm 2i\phi}$, which appear due to the noncommutativity of rotations. Indeed, if the rotations $\hat{R}_z(-\phi)$ and $\hat{R}_y(-\theta)$ commuted, the transformation would be $\hat{U}_H = \hat{V}^\dagger \hat{R}_y(-\theta)\hat{V}$, which is independent of ϕ. The pure helicity states, right- and left-circularly polarized with respect to their wave vectors, will be denoted as $|\tilde{\mathbf{E}}^\sigma)_H \propto |e^\sigma)$

$$|e^+) = (1, 0, 0)^T, \quad |e^-) = (0, 1, 0)^T. \tag{8.11}$$

They are eigenvectors of the diagonal *helicity operator* $\hat{\sigma}$:

$$\hat{\sigma} = \mathrm{diag}(1, -1, 0), \quad \hat{\sigma}|e^{\sigma}) = \sigma|e^{\sigma}), \quad \sigma = \pm 1. \tag{8.12}$$

The third eigenvector $|e_k) = (0, 0, 1)^{\mathrm{T}}$ with eigenvalue $\sigma = 0$ corresponds to the prohibited longitudinal polarization. Note that for the pure helicity states the magnetic field differs from the electric one by a constant phase factor: $\tilde{\mathbf{H}}^{\sigma}(\mathbf{k}) = -\mathrm{i}\sigma\tilde{\mathbf{E}}^{\sigma}(\mathbf{k})$. This leads to $\mathbf{H}^{\sigma}(\mathbf{r}) = -\mathrm{i}\sigma\mathbf{E}^{\sigma}(\mathbf{r})$, and, hence, the electric and magnetic contributions to the observable quantities are always equal, both in momentum and real spaces.

Transition to the helicity basis *diagonalizes* the equation (8.3a). Indeed, the double vector product $\tilde{\mathbf{E}}' = -\mathbf{k} \times (\mathbf{k} \times \tilde{\mathbf{E}}) = k^2[\tilde{\mathbf{E}} - \boldsymbol{\kappa}(\boldsymbol{\kappa} \cdot \tilde{\mathbf{E}})]$ represents a *spherical projection* of $\tilde{\mathbf{E}}$ onto the $\boldsymbol{\kappa}$-sphere and can be written, in the laboratory circular basis, as the action of a nondiagonal operator:

$$|\tilde{\mathbf{E}}')_{\mathrm{C}} = k^2\hat{\Pi}(\mathbf{k})|\tilde{\mathbf{E}})_{\mathrm{C}}, \quad \hat{\Pi}(\mathbf{k}) = \hat{U}_{\mathrm{H}}(\mathbf{k})\mathsf{P}_z\hat{U}_{\mathrm{H}}^{\dagger}(\mathbf{k}), \tag{8.13}$$

where $\mathsf{P}_z = \mathrm{diag}(1, 1, 0)$ is the projector onto the orthogonal plane. Naturally, the operator $\hat{\Pi}(\mathbf{k})$ becomes a simple projection in the spherical or helicity bases (8.7)–(8.10): $|\tilde{\mathbf{E}})_{\mathrm{C}} = \hat{U}_{\mathrm{H}}(\mathbf{k})|\tilde{\mathbf{E}})_{\mathrm{H}}$,

$$(\hat{\Pi})_{\mathrm{H}} = \hat{U}_{\mathrm{H}}^{\dagger}(\mathbf{k})\hat{\Pi}(\mathbf{k})\hat{U}_{\mathrm{H}}(\mathbf{k}) = \mathsf{P}_z. \tag{8.14}$$

This is another manifestation of the fact that the polarization transversality is naturally adopted in the momentum representation and basis attached to the $\boldsymbol{\kappa}$-sphere.

Geometric phase

Any polarization state is defined up to a common phase. In the case of circular polarization, the phase can be induced by a rotation of the coordinate frame about the $\mathbf{k}$-vector. For instance, if a plane wave propagates along the z-axis, rotation of the coordinates (x, y) by the angle $-\Phi$, $|\mathbf{E}')_{\mathrm{L}} = \hat{R}_z(-\Phi)|\mathbf{E})_{\mathrm{L}}$, induces the following change in the circular field components:

$$\mathbf{u}_0^{\sigma'} = \mathrm{e}^{\mathrm{i}\sigma\Phi}\mathbf{u}_0^{\sigma}, \quad |\mathbf{E}')_{\mathrm{C}} = \exp(-\mathrm{i}\hat{\sigma}\Phi)|\mathbf{E})_{\mathrm{C}}, \quad \exp(-\mathrm{i}\hat{\sigma}\Phi) = \begin{pmatrix} \mathrm{e}^{-\mathrm{i}\Phi} & 0 & 0 \\ 0 & \mathrm{e}^{\mathrm{i}\Phi} & 0 \\ 0 & 0 & 1 \end{pmatrix}, \tag{8.15}$$

where the helicity operator appears from the spin matrix $\hat{S}_z$ in the circular-polarization basis, $(\hat{S}_z)_{\mathrm{C}} = \hat{V}^{\dagger}\hat{S}_z\hat{V} = \hat{\sigma}$. Equation (8.15) demonstrates the simplest two-dimensional case of the coupling between the helicity and rotation yielding the *geometric phase*, $\sigma\Phi$. This is a particular case of the *AM-rotation coupling* [15, 78–80].

In the general three-dimensional case, polarizations of different waves are defined in the *different* planes orthogonal to the wave vectors $\mathbf{k}$. In other words, the electric field belongs to the manifold of vectors tangent to the curved surface of the κ-sphere (a tangent fiber bundle with the S^2 base), Fig. 8.2. Hence, to compare the phases and polarizations of plane waves with different **k**-vectors, one has to transport their electric fields over the surface of the sphere in order to bring them to the same plane. The geometric *parallel transport* of vectors tangent to the sphere is defined in such a way that the vector does not experience *local* rotation about the normal κ-vector. Therefore, in a coordinate frame with its axes parallel-transported over the sphere, the circularly polarized waves do not gain additional phases (8.15). However, the parallel-transport frame cannot be defined *globally* on the sphere, and any globally defined coordinate system on the sphere inevitably experiences local rotations of its axes about the κ-vector. The differential angle of rotation or the phase of the helicity basic vectors $\mathbf{u}^\sigma$ reads

$$-\sigma \mathrm{d}\Phi = \mathrm{i}\mathbf{u}^{\sigma *} \cdot \mathrm{d}\mathbf{u}^\sigma = \mathrm{i}\sum_j \left[\mathbf{u}^{\sigma *}(\mathbf{k}) \cdot (\partial_{k_j})\mathbf{u}^\sigma(\mathbf{k})\right]\mathrm{d}k_j. \tag{8.16}$$

For the transport of the field along a contour C between $\mathbf{k}$ and $\mathbf{k}'$ in momentum space (owing to the transversality, it can always be projected onto the κ-sphere), integration of Eq. (8.16) results in the accumulation of the geometric phase (8.15):

$$|\tilde{\mathbf{E}}')_{\mathrm{H}} = \exp(-\mathrm{i}\hat{\sigma}\,\Phi_{\mathrm{B}})|\tilde{\mathbf{E}})_{\mathrm{H}}, \quad \Phi_{\mathrm{B}}(C) = \int_C \mathbf{A}_{\mathrm{B}}(\mathbf{k}) \cdot \mathrm{d}\mathbf{k}. \tag{8.17}$$

Here Φ_{B} is the *Berry phase* [90] described by the momentum-space contour integral of the *Berry connection* (vector-potential) $\mathbf{A}_{\mathrm{B}}(\mathbf{k})$, which is determined from Eqs. (8.7), (8.9), and (8.16):

$$\sigma \mathbf{A}_{\mathrm{B}} = -\mathrm{i}\mathbf{u}^{\sigma *} \cdot (\partial_{\mathbf{k}})\mathbf{u}^\sigma = \sigma \frac{m - \cos\theta}{k \sin\theta}\mathbf{u}_\phi. \tag{8.18}$$

In most cases, we will use the gauge $m = 1$, which provides a non-singular paraxial limit $\theta \to 0$. The Berry connection can be represented in matrix form as a transverse part of the pure gauge field $\hat{\mathbf{A}} = -\mathrm{i}\hat{U}_{\mathrm{H}}^\dagger \partial_{\mathbf{k}} \hat{U}_{\mathrm{H}}$ induced by the momentum-dependent transformation (8.10):

$$\hat{\sigma}\mathbf{A}_{\mathrm{B}} = -\mathrm{i}\mathsf{P}(\hat{U}_{\mathrm{H}}^\dagger \partial_{\mathbf{k}} \hat{U}_{\mathrm{H}}). \tag{8.18a}$$

Here $\mathsf{P}(\cdots) = \hat{\mathsf{P}}_z(\cdots)\hat{\mathsf{P}}_z$ provides the projection onto the subspace orthogonal to $\mathbf{k}$ (the upper left 2×2 sector of matrices in the spherical or helicity basis).

Thus, the parallel transport of the electric field constrained by the transversality condition (8.4), brings about an additional Berry phase (8.17) in the field components in a global coordinate frame. This phenomenon have been extensively studied [74, 81–89] and reviewed [90–97] in optics, and we only briefly recall its main properties. The Berry phase depends on the path of evolution on the unit κ-sphere and yields the opposite phase shifts for opposite circular polarizations. For a general elliptical polarization, this is equivalent to the rotation of the polarization ellipse by the angle $\Phi_{\rm B}$, (Fig. 8.2). The Berry phase (8.17) manifests coupling between the spin (helicity) σ and momentum $\mathbf{k}$, signifying the *SOI of light.*

Importantly, the Berry phase does not vanish for closed contours of evolution C; in this case Eq. (8.17) takes the form

$$\Phi_{\rm B}(C) = \oint_C \mathbf{A}_{\rm B} \cdot {\rm d}\mathbf{k} = 2\pi(m-1) + \int_\Sigma \mathbf{F}_{\rm B} \cdot {\rm d}\boldsymbol{\Sigma} = 2\pi(m-1) + \Sigma, \tag{8.19}$$

where Σ is the segment of the $\boldsymbol{\kappa}$-sphere bounded by the contour C ($C = \partial\Sigma$), ${\rm d}\boldsymbol{\Sigma}$ is the oriented element of its area, and $\mathbf{F}_{\rm B}(\mathbf{k})$ is the gauge-invariant *Berry curvature* given by

$$\mathbf{F}_{\rm B} = \partial_{\mathbf{k}} \times \mathbf{A}_{\rm B} = \frac{\mathbf{k}}{k^3}. \tag{8.20}$$

On the one hand, the Berry curvature corresponds to the unit Gaussian curvature of the $\boldsymbol{\kappa}$-sphere: $\mathbf{F}_{\rm B}(\boldsymbol{\kappa}) = \boldsymbol{\kappa}$. On the other hand, it has the form of a '*magnetic monopole*' located at the origin of momentum space. This topological monopole underlies the evolution of massless spinning particles in high-energy, condensed-matter, and optical systems [4, 12, 39, 40, 74, 75, 98]. For the circular contour with a fixed polar angle shown in Fig. 8.2: $C = \{\theta = \text{const}, \phi \in (0, 2\pi)\}$, the Berry phase (8.19) takes explicit form ($m = 1$)

$$\Phi_{\rm B}(C) = 2\pi(1 - \cos\theta). \tag{8.19a}$$

Alongside the geometrical parallel-transport explanation, the Berry phase allows dynamical interpretations. Upon the evolution of the $\mathbf{k}$-vector, the electric field $\tilde{\mathbf{E}}(\mathbf{k})$ possesses a sort of *inertia* and remains locally non-rotating about $\mathbf{k}$. Observation in a globally defined spherical frame produces a *Coriolis effect* caused by local rotations of the basis. This is expressed in the simple 'dynamical' formulae for the Berry phase [47, 88, 97, 99]:

$$\sigma\Phi_{\rm B} = -\sigma \int \boldsymbol{\kappa} \cdot \boldsymbol{\Omega}_\zeta {\rm d}\zeta, \tag{8.21}$$

where ζ is a parameter underlying the evolution of waves in the momentum space (this can be time, a coordinate in real space, etc.) and $\boldsymbol{\Omega}_\zeta$ is the angular velocity of rotation of the coordinate frame defined with respect to this parameter ζ. On the one hand, $-\Phi_{\rm B}$ is the total angle of rotation of the frame about the $\mathbf{k}$-vector. On the other hand, Eq. (8.21) represents the Berry phase as a result of the coupling between the SAM $\sigma\boldsymbol{\kappa}$ and the rotation $\boldsymbol{\Omega}_\zeta$. In this spirit, it is completely analogous to the rotation of the Foucault pendulum [100, 101], the Coriolis effect in rotating systems [15, 78–80], and the rotational Doppler effect for waves carrying intrinsic AM [20, 102–106].

It is important to note that the Berry phase associated with the variations of the $\mathbf{k}$-vector (in contrast to the Pancharatnam–Berry phase) becomes significant in globally *non-paraxial* optical systems involving 3D distributions of the wave vectors. One can distinguish the two typical situations: (i) successive non-planar variations of the direction of propagation of a polarized wave (in multiple reflections, refractions, curved optical fibers, etc.) [81–83, 85, 89] (Section 8.4.3 below); and (ii) simultaneous interference involving multiple polarized waves with different **k**-vectors (e.g., tightly focused or scattered non-paraxial fields) [24, 27, 31, 47, 107, 108] (Section 8.3 below). As we will see, the Berry phase of the type (ii) underlies all the AM interactions summarized in Table 8.1.

8.2.2 Coordinate, momentum, and angular momentum

General formalism

While the Berry phase describes the SOI phenomena on the level of the geometric interference of *plane waves*, the angular momenta represent the integral dynamical characteristics of the *localized wave packets (photons)* or *beams*. The latter are naturally described within a quantum-like operator formalism.

The *operators* for the coordinate, momentum, and energy of a photon are written in momentum representation as

$$\hat{\mathbf{r}} = \mathrm{i}\partial_{\mathbf{k}}, \quad \hat{\mathbf{p}} = \mathbf{k}, \quad \hat{w} = \omega, \tag{8.22}$$

whereas the AM operator is known to be [109]

$$\hat{\mathbf{J}} = \hat{\mathbf{r}} \times \hat{\mathbf{p}} + \hat{\mathbf{S}} \equiv \hat{\mathbf{L}} + \hat{\mathbf{S}}. \tag{8.23}$$

Here $\hat{\mathbf{L}} = -\mathrm{i}(\mathbf{k} \times \partial_{\mathbf{k}})$, whereas $\hat{\mathbf{S}}$ is the spin-1 operator given by the SO(3) rotation generators $(\hat{S}_a)_{ij} = -\mathrm{i}\varepsilon_{aij}$ (ε_{aij} is the Levi–Civita symbol) that act on the field components in the laboratory frame, $|\tilde{\mathbf{E}})_{\mathrm{L}}$. The canonical OAM and SAM operators, $\hat{\mathbf{L}}$ and $\hat{\mathbf{S}}$, satisfy SO(3) algebra and act in the momentum and polarization spaces, respectively. The centre of gravity, momentum, energy, OAM, and SAM of a one-photon localized state of light can be written as *expectation values* of the operators (8.22) and (8.23):

$$\langle \mathbf{R} \rangle = N^{-1} \langle \tilde{\mathbf{E}} | \mathrm{i}\partial_{\mathbf{k}} | \tilde{\mathbf{E}} \rangle_{\mathrm{L}}, \quad \langle \mathbf{P} \rangle = N^{-1} \langle \tilde{\mathbf{E}} | \mathbf{k} | \tilde{\mathbf{E}} \rangle_{\mathrm{L}}, \quad \langle W \rangle = N^{-1} \langle \tilde{\mathbf{E}} | \hat{w} | \tilde{\mathbf{E}} \rangle_{\mathrm{L}}, \tag{8.24}$$

$$\langle \mathbf{L} \rangle = N^{-1} \langle \tilde{\mathbf{E}} | \hat{\mathbf{L}} | \tilde{\mathbf{E}} \rangle_{\mathrm{L}}, \quad \langle \mathbf{S} \rangle = N^{-1} \langle \tilde{\mathbf{E}} | \hat{\mathbf{S}} | \tilde{\mathbf{E}} \rangle_{\mathrm{L}}. \tag{8.25}$$

where $N = \langle \tilde{\mathbf{E}} | \tilde{\mathbf{E}} \rangle$ is the norm and the state vector $|\tilde{\mathbf{E}}\rangle$ is formally proportional to the polarization vector of components $|\tilde{\mathbf{E}})$ but it is also considered as a vector in the Hilbert space. Correspondingly, the 'bra-ket' inner product implies the scalar product of complex polarizations together with the integration over the momentum (or coordinate) space: $\langle \tilde{\mathbf{E}} | \tilde{\mathbf{E}}' \rangle = \int (\tilde{\mathbf{E}} | \tilde{\mathbf{E}}') [d\mathbf{k}]$.[2]

[2] There is some difference in application of this formalism to different problems: e.g., non-monochromatic 3D-localized wave packets (photons) and 2D-localized monochromatic wave beams. In the latter case, where the z-coordinate plays the role of the independent variable, all the quantities are considered as functions of $\mathbf{k}_\perp = (k_x, k_y)$, and the z-dependent state vector is $|\tilde{\mathbf{E}}\rangle = |\tilde{\mathbf{E}}\rangle e^{\mathrm{i}k_z z}$, $k_z = \sqrt{k^2 - k_\perp^2}$ [24, 110, 111]. Then, the expectation values (8.24) and (8.25) mean the linear densities of the corresponding quantities per unit z-length, and the centre of gravity yields $\langle \mathbf{R} \rangle = [\langle X \rangle, \langle Y \rangle, z]$. For the sake of simplicity, we avoid discussing these issues here, consider the beam properties in the $z = 0$ plane, and specify the integration measure in each particular problem.

Equations (8.24) and (8.25) describe the main observable characteristics of the localized light fields. However, the operators (8.22) and (8.23) face some fundamental difficulties. In particular, the operator $\hat{\mathbf{J}}$ looks rather natural as a sum of the orbital and spin parts. This division is indeed meaningful for paraxial fields, where the eigenmodes of $\hat{L}_z = -\mathrm{i}\partial_\phi$ and $\hat{S}_z$ represent paraxial circularly polarized vortex beams with the corresponding quantum numbers $\ell = 0, \pm 1, \pm 2, \ldots$ (topological charge of vortex $\mathrm{e}^{\mathrm{i}\ell\phi}$) and $\sigma = \pm 1$ (helicity), see Fig. 8.1 [18–20]. The situation becomes more complicated in the general, non-paraxial case, where 'The separation of the total AM into orbital and spin parts has restricted physical meaning. ... States with definite values of OAM and SAM do not satisfy the condition of transversality in the general case.' [109] and 'In the general non-paraxial case there is no simple separation into ℓ-dependent orbital and σ-dependent spin component of AM' [112]. The essence of the problem is that the operators $\hat{\mathbf{L}}$ and $\hat{\mathbf{S}}$, as well as the coordinate operator $\hat{\mathbf{r}}$, when acting on a transverse field $|\tilde{\mathbf{E}}\rangle_L$, result in a field with a non-zero longitudinal component which violates the transversality condition (8.4). In other words, these operators do not keep the manifold of vectors tangent to the $\boldsymbol{\kappa}$-sphere invariant [113–115].[3]

To make the operators $\hat{\mathbf{r}}$, $\hat{\mathbf{L}}$, and $\hat{\mathbf{S}}$ consistent with the transversality constraint, they can be *projected* onto the corresponding subspace of transverse waves. This results in the modified coordinate and AM operators [24]:

$$\hat{\mathbf{r}}' = \hat{\mathbf{r}} + \frac{\mathbf{k} \times \hat{\mathbf{S}}}{k^2}, \tag{8.26}$$

$$\hat{\mathbf{L}}' = \hat{\mathbf{r}}' \times \mathbf{p} = \hat{\mathbf{L}} - \boldsymbol{\kappa} \times (\boldsymbol{\kappa} \times \hat{\mathbf{S}}), \quad \hat{\mathbf{S}}' = \boldsymbol{\kappa}(\boldsymbol{\kappa} \cdot \hat{\mathbf{S}}). \tag{8.27}$$

The total AM $\hat{\mathbf{J}} = \hat{\mathbf{L}}' + \hat{\mathbf{S}}'$, remains unchanged because it is fully consistent with the transversality condition. One can see that all the modified operators (8.26) and (8.27) act in both momentum and polarization spaces which cannot be globally factorized due to the transversality constraint (8.4) – these spaces form a non-trivial vector fiber bundle shown in Fig. 8.2. Equations (8.27) suggest that part of the SAM is transferred to the OAM, and coordinate (8.26) also acquires a SAM-dependent correction. This hints at the basic SOI effects: *spin-to-orbit AM conversion* and *spin Hall effect*. The spin-dependent coordinate operator (8.26) is the Pryce coordinate known in the context of photon localization and Berry phase [4, 74, 75, 77, 116, 117]. It brings about the space non-commutativity with the monopole term in $\mathbf{k}$-space:

$$[\hat{r}'_i, \hat{r}'_j] = -\mathrm{i}\varepsilon_{ijl}(\boldsymbol{\kappa} \cdot \hat{\mathbf{S}})\frac{k_l}{k^3}, \tag{8.28}$$

where $(\boldsymbol{\kappa} \cdot \hat{\mathbf{S}})$ is the helicity operator in the laboratory basis. At the same time, the modified OAM and SAM operators (8.27) correspond to those considered in the second-quantization

[3] Note, however, that this fundamental problem does not affect the expectation values (8.24) and (8.25). The scalar product of the non-transverse field $\{\hat{\mathbf{L}}, \hat{\mathbf{S}}, \hat{\mathbf{r}}\}|\tilde{\mathbf{E}}\rangle_L$ with the transverse field $\langle\tilde{\mathbf{E}}|$ nulls the longitudinal component.

formalism by van Enk and Nienhuis [113, 114]. They also obey unusual commutation relations:

$$[\hat{S}'_i, \hat{S}'_j] = 0, \quad [\hat{L}'_i, \hat{L}'_j] = \mathrm{i}\varepsilon_{ijl}(\hat{L}'_l - \hat{S}'_l)_k, \quad [\hat{L}'_i, \hat{S}'_j] = \mathrm{i}\varepsilon_{ijl}\hat{S}'_l. \tag{8.29}$$

All operators (8.26) and (8.27) transform as vectors under rotations, $[\hat{J}_i, \hat{O}'_j] = \mathrm{i}\varepsilon_{ijl}\hat{O}'_l$, $\hat{\mathbf{O}}' = \hat{\mathbf{r}}', \hat{\mathbf{L}}'$, and $\hat{\mathbf{S}}'$, and yield the same expectation values (8.24) and (8.25) as for the canonical operators (8.22) and (8.23).

Importantly, the above operators are diagonalized in the helicity basis (8.9) and (8.10). Making the transformations to the laboratory circular and then to the helicity basis, $\hat{\boldsymbol{\mathcal{O}}} \equiv (\hat{\mathbf{O}}')_{\mathrm{H}} = \hat{U}_{\mathrm{H}}^{\dagger}\hat{V}^{\dagger}\hat{\mathbf{O}}'\hat{V}\hat{U}_{\mathrm{H}}$, the helicity becomes diagonal, $\hat{U}_{\mathrm{H}}^{\dagger}\hat{V}^{\dagger}(\boldsymbol{\kappa}\cdot\hat{\mathbf{S}})\hat{V}\hat{U}_{\mathrm{H}} = \hat{\sigma}$, Eq. (8.12), and we derive [24]

$$\hat{\boldsymbol{r}} = \mathrm{i}\partial_{\mathbf{k}} - \hat{\sigma}\mathbf{A}_B, \tag{8.30}$$

$$\hat{\boldsymbol{\mathcal{L}}} = -\mathrm{i}\mathbf{k}\times\partial_{\mathbf{k}} - \hat{\sigma}\mathbf{A}_B\times\mathbf{k}, \quad \hat{\boldsymbol{\mathcal{S}}} = \boldsymbol{\kappa}\hat{\sigma}. \tag{8.31}$$

Here the Berry connection (8.18a) is obtained from

$$\hat{\sigma}\mathbf{A}_{\mathrm{B}} = -\mathrm{i}\hat{U}_{\mathrm{H}}^{\dagger}\partial_{\mathbf{k}}\hat{U}_{\mathrm{H}} - \hat{U}_{\mathrm{H}}^{\dagger}\frac{\mathbf{k}\times\hat{\mathbf{S}}}{k^2}\hat{U}_{\mathrm{H}}, \tag{8.32}$$

which shows that the term $(\mathbf{k}\times\hat{\mathbf{S}})/k^2$ in Eqs. (8.26) and (8.27) is indeed equivalent to the local projection onto the two-dimensional transversality subspace. Equations (8.30) and (8.31) represent the main result of this subsection: they describe the SOI of light in the operator form through the helicity-dependent Berry-connection terms. These SOI terms arise from transition to the helicity basis and the projection onto the transversality subspace. Obviously, the energy and momentum operators $\hat{w}$ and $\hat{\mathbf{p}}$, Eq. (8.22), are not affected by the above transformations. The space noncommutativity (8.28) in terms of coordinates (8.30) becomes

$$[\hat{r}_i, \hat{r}_j] = -\mathrm{i}\varepsilon_{ijl}\hat{\sigma}F_{\mathrm{B}l}, \tag{8.28a}$$

where $\mathbf{F}_{\mathrm{B}}(\mathbf{k})$ is the Berry curvature (8.20). Using Eqs. (8.7) and Eq. (8.18) with $m = 1$, the transverse coordinates (8.30) and the z-components of the AM operators (8.31) take the form

$$\hat{x} = \mathrm{i}\partial_{k_x} + \hat{\sigma}\frac{1-\cos\theta}{k\sin\theta}\sin\phi, \quad \hat{y} = \mathrm{i}\partial_{k_y} - \hat{\sigma}\frac{1-\cos\theta}{k\sin\theta}\cos\phi, \tag{8.30a}$$

$$\hat{\mathcal{L}}_z = -\mathrm{i}\partial_{\phi} + \hat{\sigma}(1-\cos\theta), \quad \hat{\mathcal{S}}_z = \hat{\sigma}\cos\theta. \tag{8.31a}$$

These equations describe the spin-Hall effect and spin-to-orbit AM conversion in the explicit operator form and show that the $\hat{\sigma}$-dependent SOI effects vanish in the paraxial limit $\theta \to 0$.

The expectation values (8.24) and (8.25) can now be re-written using the operators (8.30) and (8.31) together with the helicity representation of the field:

$$\langle \mathbf{R} \rangle = N^{-1}[\langle \tilde{\mathbf{E}} | i\partial_{\mathbf{k}} | \tilde{\mathbf{E}} \rangle_{\mathrm{H}} - \langle \tilde{\mathbf{E}} | \hat{\sigma} \mathbf{A}_{\mathrm{B}} | \tilde{\mathbf{E}} \rangle_{\mathrm{H}}], \quad \langle \mathbf{P} \rangle = N^{-1} \langle \tilde{\mathbf{E}} | \mathbf{k} | \tilde{\mathbf{E}} \rangle_{\mathrm{H}}, \quad \langle W \rangle = N^{-1} \langle \tilde{\mathbf{E}} | \hat{w} | \tilde{\mathbf{E}} \rangle_{\mathrm{H}}, \tag{8.33}$$

$$\langle \mathbf{L} \rangle = N^{-1}[\langle \tilde{\mathbf{E}} | \hat{\mathbf{L}} | \tilde{\mathbf{E}} \rangle_{\mathrm{H}} - \langle \tilde{\mathbf{E}} | \hat{\sigma} \mathbf{A}_{\mathrm{B}} \times \mathbf{k} | \tilde{\mathbf{E}} \rangle_{\mathrm{H}}], \quad \langle \mathbf{S} \rangle = N^{-1} \langle \tilde{\mathbf{E}} | \hat{\sigma} \boldsymbol{\kappa} | \tilde{\mathbf{E}} \rangle_{\mathrm{H}}. \tag{8.34}$$

In the laboratory-frame equations (8.24) and (8.25), the complex SOI properties originating from the transversality constraint (8.4) were hidden in the state vector $|\tilde{\mathbf{E}}\rangle_{\mathrm{L}}$. The use of the modified operators in the helicity representation makes these features explicit in the operators, whereas the state vector can often be simplified. (This is similar to the transition between the Schrödinger and Heisenberg pictures of time-dependent quantum mechanics, but here it is applied to the momentum-dependent properties.) In particular, if the polarization is uniform in the helicity basis, the electric field $|\tilde{\mathbf{E}}(\mathbf{k})\rangle_{\mathrm{H}} = |\mathbf{e})_{\mathrm{H}} |\tilde{E}(\mathbf{k})\rangle$ represents a product of the constant polarization vector $|\mathbf{e})_{\mathrm{H}}$ and the scalar amplitude distribution $|\tilde{E}(\mathbf{k})\rangle = \tilde{E}(\mathbf{k})$. In this case, the polarization- and Hilbert-space calculations are effectively separated in the problem, and Eqs. (8.33) and (8.34) are simplified as

$$|\tilde{\mathbf{E}}(\mathbf{k})\rangle \to |\tilde{E}(\mathbf{k})\rangle, \quad \hat{\sigma} \to \bar{\sigma} = (\mathbf{e}|\hat{\sigma}|\mathbf{e})_H = \text{const}, \tag{8.35}$$

where $\bar{\sigma}$ stands for the mean helicity.

Finally, the OAM of light can be divided into *extrinsic* (origin-dependent) and *intrinsic* (origin-independent) contributions [22–24, 26, 118]. Assuming that the extrinsic contribution is given by the centre of gravity and the mean momentum of the field (Fig. 8.1), we have

$$\langle \mathbf{L}^{\mathrm{ext}} \rangle = \langle \mathbf{R} \rangle \times \langle \mathbf{P} \rangle, \quad \langle \mathbf{L}^{\mathrm{int}} \rangle = \langle \mathbf{L} \rangle - \langle \mathbf{L}^{\mathrm{ext}} \rangle. \tag{8.36}$$

Paraxial case

The SOI Berry-connection terms in Eqs. (8.30), (8.31), (8.33), and (8.34) vanish in the paraxial approximation $\theta \ll 1$. The helicity basis (8.10) nearly coincides with the laboratory basis of circular polarizations (8.6). (The difference is only in the longitudinal z-components of the field of the order of θ, which make contributions of order θ^2 to the expectation values.) Therefore, in most cases one can neglect the z-components of the fields and consider the transverse (x, y)-components denoted by $|\tilde{\mathbf{E}}\rangle_\perp$ and $|\mathbf{e})_\perp$. As an example, let us consider paraxial circularly polarized Laguerre–Gaussian beams with the vortex charge ℓ and radial quantum number $\mathrm{p} = 0$. The electric field in the laboratory basis of circular polarizations can be written as

$$|\tilde{\mathbf{E}}_\ell^\sigma(\mathbf{k})\rangle_{\perp C} = |e^\sigma)_\perp |\tilde{E}_\ell(\mathbf{k})\rangle, \quad |\tilde{E}_\ell(\mathbf{k})\rangle = \theta^{|\ell|} \mathrm{e}^{i\ell\phi} G(\theta), \tag{8.37}$$

where $G(\theta) \propto \exp[-(kw_0)^2\theta^2/4]$ is the Gaussian envelope function with $w_0 \gg k^{-1}$ being the beam waist. It is easy to see that paraxial beams (8.37) are eigenmodes of $\hat{L}_z = -i\partial_\phi$ and $(\hat{S}_z)_C = \hat{\sigma}$ with the discrete eigenvalues $\sigma = \pm 1$ and $\ell = 0, \pm 1, \pm 2, \ldots$. The expectation values (8.24) and (8.25) or (8.33) and (8.34) yield for Eq. (8.37), up to $O(\theta^2)$,

$$\langle \mathbf{R} \rangle = z\mathbf{K}, \quad \langle \mathbf{P} \rangle = k\mathbf{K}, \quad \langle W \rangle = \omega, \quad \langle \mathbf{L} \rangle = \ell\mathbf{K}, \quad \langle \mathbf{S} \rangle = \sigma\mathbf{K}, \tag{8.38}$$

where $\mathbf{K} = \mathbf{u}_z$ is the direction of propagation of the beam. These are known results for the well-defined IOAM and SAM of the paraxial vortex beams [18–20], Fig. 8.1. For the beam with an arbitrary homogeneous polarization state $|\mathbf{e})_\perp = (e^+, e^-)^{\mathrm{T}}$, the helicity σ in Eqs. (8.38) is replaced by the mean helicity $\bar{\sigma} = (\mathbf{e}|\hat{\sigma}|\mathbf{e})_{\perp C} = |e^+|^2 - |e^-|^2$.

Note that transverse part of the helicity operator represents the Pauli matrix, $\hat{\sigma}_\perp = \mathrm{diag}(1, -1) = \hat{\sigma}_3$, with eigenvectors of circular polarizations, $|e^+)_\perp = (1, 0)^{\mathrm{T}}$ and $|e^-)_\perp = (0, 1)^{\mathrm{T}}$. An arbitrary homogeneous polarization state is characterized by the complex unit *Jones vector* $|\mathbf{e})_\perp = (e^+, e^-)^{\mathrm{T}}$ obeying the SU(2) symmetry of a two-level system. In this manner, the Pauli matrix $\hat{\sigma}_3$ describes the degree of the circular polarization responsible for the SAM, but the complete characterization of the polarization requires the use of all the Pauli matrices $\hat{\vec{\sigma}} = (\hat{\sigma}_1, \hat{\sigma}_2, \hat{\sigma}_3)$, where $\hat{\sigma}_1$ and $\hat{\sigma}_2$ are unrelated to the SAM [22]. Thus, a uniform polarization state is completely characterized by the expectation values

$$\vec{\mathfrak{S}} = (\tilde{\mathbf{e}}|\hat{\vec{\sigma}}|\tilde{\mathbf{e}})_{\perp C}, \tag{8.39}$$

which form the normalized *Stokes vector* [119–121]. It can be considered as a *pseudo-spin* in the two-level system, which represents the SU(2) polarization state on the SO(3) Poincaré sphere – an analogue of the Bloch sphere. The pure helicity states $|e^\sigma)_\perp$ correspond to the poles of the sphere. To emphasize that the Poincaré sphere is an abstract space different from the real polarization space, we indicate the Stokes vectors by arrows. The third component of the Stokes vectors is the mean helicity; for space-variant polarizations it also determines the local *SAM density* of light, s_z:

$$\mathfrak{S}_3 = \bar{\sigma} = s_z. \tag{8.40}$$

8.2.3 Poynting energy flows

So far, we considered the properties of the electromagnetic field in the Fourier, i.e., momentum representation. This is because the transversality condition (8.4) essentially represents a local constraint on the spectrum of the field. And, indeed, most of the SOI phenomena are intimately related to the evolution in the momentum space. However, there are also many problems involving local spatial properties of light. They naturally require *coordinate representation* of the field and the corresponding description of the main dynamical quantities. In that case one deals with the spatial energy density and *Poynting energy flow* [122].

Using the complex electric and magnetic fields $\mathbf{E}(\mathbf{r})$ and $\mathbf{H}(\mathbf{r})$, which satisfy the Maxwell equations (8.1), the time-averaged energy density and Poynting energy flow are given by [122]

$$w \propto \frac{1}{2}(|\mathbf{E}|^2 + |\mathbf{H}|^2), \tag{8.41}$$

$$\boldsymbol{\pi} \propto \mathrm{Re}(\mathbf{E}^* \times \mathbf{H}) = \frac{1}{2\omega}\mathrm{Im}[\mathbf{E}^* \times (\nabla \times \mathbf{E}) + \mathbf{H}^* \times (\nabla \times \mathbf{H})]. \tag{8.42}$$

Here the Poynting flow represents the *momentum density* of the field. Note that, in contrast to the momentum helicity representation, where the electric- and magnetic-field components are equal up to a constant phase, in the real space the local electric and magnetic contributions are different in the generic case. Therefore, it is important to keep explicitly the 'electric-magnetic' symmetry between the two fields [115, 123]. It was recently suggested that using Maxwell equations (8.1), the Poynting flow can be divided into *spin* and *orbital* parts [25, 123–125], $\boldsymbol{\pi} = \boldsymbol{\pi}^{\mathrm{O}} + \boldsymbol{\pi}^{\mathrm{S}}$:

$$\boldsymbol{\pi}^{\mathrm{O}} = \frac{1}{2\omega}\mathrm{Im}[\mathbf{E}^* \cdot (\nabla)\mathbf{E} + \mathbf{H}^* \cdot (\nabla)\mathbf{H}], \tag{8.43}$$

$$\boldsymbol{\pi}^{\mathrm{S}} = \frac{1}{4\omega}\mathrm{Im}[\nabla \times (\mathbf{E}^* \times \mathbf{E}) + \nabla \times (\mathbf{H}^* \times \mathbf{H})], \tag{8.44}$$

where $\mathbf{E}^* \cdot (\nabla)\mathbf{E} \equiv \sum_i E_i^* \nabla E_i$. Using the linear momentum densities (8.42)–(8.44), the total AM, SAM, and OAM densities are defined as

$$\mathbf{j} = \mathbf{r} \times \boldsymbol{\pi}, \quad \mathbf{l} = \mathbf{r} \times \boldsymbol{\pi}^{\mathrm{O}}, \quad \mathbf{s} = \mathbf{r} \times \boldsymbol{\pi}^{\mathrm{S}}. \tag{8.45}$$

The one-photon integral values of the energy, momentum, OAM, and SAM of a localized state of light are determined by the integrals of Eqs. (8.41)–(8.45) over the proper real-space measure $[\mathrm{d}\mathbf{r}]$:

$$\langle W \rangle = N^{-1} \int w[\mathrm{d}\mathbf{r}], \quad \langle \mathbf{P} \rangle = N^{-1} \int \boldsymbol{\pi}[\mathrm{d}\mathbf{r}], \quad \langle \mathbf{L} \rangle = N^{-1} \int \mathbf{r} \times \boldsymbol{\pi}^{\mathrm{O}}[\mathrm{d}\mathbf{r}],$$
$$\langle \mathbf{S} \rangle = N^{-1} \int \mathbf{r} \times \boldsymbol{\pi}^{\mathrm{S}}[\mathrm{d}\mathbf{r}], \tag{8.46}$$

where $N = \int w/\omega[\mathrm{d}\mathbf{r}]$ is the real-space norm. Performing the Fourier transform (8.2) and transformation of the fields to the helicity basis (8.10), one can show that quantities (8.46) exactly coincide with those given by the momentum-representation expectation values (8.24) and (8.25) or (8.33) and (8.34) [24, 25]. In a similar manner, the centre of gravity of the field is given by

$$\langle \mathbf{R} \rangle = N^{-1} \int \mathbf{r} \frac{w}{\omega} \mathrm{d}^2\mathbf{r}_{\perp}, \tag{8.47}$$

which is also equivalent to the expectation value of the coordinates in Eqs. (8.24) and (8.33).

Depending on the problem, both the momentum and real-space approaches summarized in Sections 8.2.2 and 8.2.3 can equally be applied for calculations of the dynamical characteristic of the wave field. Note that in the coordinate representation the Riemann–Silberstein formalism provides a convenient approach which enables separation of the helicity states and account for the electric-magnetic symmetry [126, 127].

8.3 Spin-orbit interactions in non-paraxial fields

8.3.1 Free-space solutions

The Berry-phase and AM concepts introduced in Section 8.2 reveal themselves in a rich variety of situations. To begin, we consider exact solutions of Maxwell equations in free space. Bessel beams [128–132] provide the simplest example of non-paraxial vortex beams. They possess the angular spectrum consisted of plane waves uniformly distributed along a circle C on the unit κ-sphere with a fixed polar angle: $C = \{\theta = \theta_0, \phi \in (0, 2\pi)\}$, exactly as shown in Fig. 8.2. The phases of the waves grow linearly with the azimuthal angle forming the vortex $e^{i\ell\phi}$. We consider *pure helicity states* with uniform circular polarizations in the helicity basis, Eqs. (8.10)–(8.12), so that the momentum representation of such vector Bessel beams is

$$|\tilde{\mathbf{E}}_\ell^\sigma(\mathbf{k})\rangle_{\rm H} = |e^\sigma\rangle|\tilde{E}_\ell(\mathbf{k})\rangle, \quad |\tilde{E}_\ell(\mathbf{k})\rangle \propto \delta(\theta - \theta_0)e^{i\ell\phi}. \tag{8.48}$$

Despite the fact that the norm of the Bessel beams diverges, $N = \langle\tilde{E}_\ell|\tilde{E}_\ell\rangle = \infty$, the normalized expectation values (8.33) and (8.34) converge to a finite limit.

The close interrelations between the geometric and dynamical features, are clearly traced in the Bessel beams (8.48). On the one hand, the geometric distribution of constituent plane waves along the contour C (Fig. 8.2) suggests that the interference of these waves should contain the Berry-phase difference (8.19a) gained after traversing this contour:

$$\Phi_{\rm B0} = 2\pi(1 - \cos\theta_0). \tag{8.49}$$

On the other hand, we can readily calculate the SAM, OAM, and other dynamical characteristics of the Bessel beams using the modified helicity-representation operators (8.30) and (8.31). Substituting Eq. (8.48) into Eqs. (8.33) and (8.34), with the inner product implying integration over $d\phi\, d\theta \sin\theta$, we arrive at [24]

$$\langle P_z\rangle = k\cos\theta_0, \quad \langle W\rangle = \omega, \tag{8.50}$$

$$\langle L_z\rangle = \ell + \sigma\Phi\!\!\!\!-_{\rm B0}, \quad \langle S_z\rangle = \sigma(1 - \Phi\!\!\!\!-_{\rm B0}), \quad \langle J_z\rangle = \ell + \sigma, \tag{8.51}$$

where $\Phi\!\!\!\!-_{\rm B0} = \Phi_{\rm B0}/2\pi$, and all the transverse vector components vanish.

Thus, the OAM and SAM acquire nonparaxial Berry-phase corrections [cf. operators (8.31a)] as compared with the paraxial formulas (8.38). This evidences the *intrinsic SOI of light* in free space and shows that a part $\sigma\Phi\!\!\!\!-_{\rm B0}$ of the SAM is transferred to the IOAM, with the total AM being conserved. In other words, *spin-to-orbit AM conversion* takes place. In the paraxial limit $\theta_0 \ll 1$, the SOI term vanishes as $\Phi_{B0} \approx \theta_0^2/2$, whereas in the limiting non-paraxial case $\theta_0 = \pi/2$, $\Phi_{B0} = 1$ leading to the complete AM conversion: $\langle L_z \rangle = \ell + \sigma$ and $\langle S_z \rangle = 0$. The vector Bessel beams (8.48) are eigenmodes of the total AM $(\hat{J}_z)_H$ with a discrete spectrum $(\ell + \sigma)$ independent of θ_0 (see [112–114, 125].

The SAM and OAM are observable quantities which can be separately measured, e.g., via motion of small test particles in optical fields [23, 131]. Remarkably, the IOAM $\langle L_z \rangle$, Eq. (8.51), with the SOI term immediately reveals itself in the spatial *intensity distribution* of the non-paraxial field. The real-space electric field of the vector Bessel beams, calculated via transition (8.10) of Eq. (8.48) to the laboratory circular basis, $|\tilde{\mathbf{E}}(\mathbf{k}))_C = \hat{U}_H(\mathbf{k})|\tilde{\mathbf{E}}(\mathbf{k}))_H$, and the Fourier transform (8.2), yields

$$|\mathbf{E}_\ell^\sigma(\mathbf{r}))_C \propto \begin{pmatrix} \dfrac{1+\sigma}{2} J_\ell(k_\perp \rho) - \sigma b_0 e^{i(\sigma-1)\varphi} J_{\ell+\sigma-1}(k_\perp \rho) \\ \dfrac{1-\sigma}{2} J_\ell(k_\perp \rho) + \sigma b_0 e^{i(\sigma+1)\varphi} J_{\ell+\sigma+1}(k_\perp \rho) \\ -i\sigma \sqrt{2a_0 b_0} e^{i\sigma\varphi} J_{\ell+\sigma}(k_\perp \rho) \end{pmatrix} e^{i\ell\varphi + ik_z z}, \tag{8.52}$$

where (ρ, φ, z) are the cylindrical coordinates in real space, $a_0 = \cos^2(\theta_0/2)$, $b_0 = \sin^2(\theta_0/2)$, $k_z = k\cos\theta_0$, $k_\perp = k\sin\theta_0$, and $J_n(k_\perp \rho)$ are the Bessel functions of the first kind. The field (8.52) has a cylindrically symmetric intensity distribution,[4] $I_\ell^\sigma = (\mathbf{E}_\ell^\sigma|\mathbf{E}_\ell^\sigma)_C$, given by

$$I_\ell^\sigma(\rho) \propto [a_0^2 J_\ell^2(k_\perp \rho) + b_0^2 J_{\ell+2\sigma}^2(k_\perp \rho) + 2a_0 b_0 J_{\ell+\sigma}^2(k_\perp \rho)]. \tag{8.53}$$

It is seen that the intensity distribution of the non-paraxial vector Bessel beam depends on both the IOAM (vortex) and SAM (helicity) quantum numbers, ℓ and σ. It is invariant with respect to the transformation $(\ell, \sigma) \to (-\ell, -\sigma)$ but neither with respect to $(\ell, \sigma) \to (\ell, -\sigma)$ nor $(\ell, \sigma) \to (-\ell, \sigma)$, which reflects $\ell \cdot \sigma$ symmetry typical of SOI. In the paraxial limit, $b_0 \to 0$, $a_0 \to 1$, and the intensity (8.53) represents the scalar Bessel beam of the order ℓ: $I_\ell \propto J_\ell^2(k_\perp \rho)$. In the non-paraxial case, terms proportional to $b_0 = \Phi_{B0}/2$ describe coupling to the modes of the order $(\ell + \sigma)$ and $(\ell + 2\sigma)$, which appear owing to the azimuthal geometric phases in the matrix $\hat{U}_H$, Eq. (8.10).

The intensity distributions (8.53) for different quantum numbers (ℓ, σ) are shown in Fig. 8.3. One can see that the radial positions of the first intensity maxima of beams with $(\ell, \pm 1)$ exhibit the *fine SOI splitting* which grows with θ_0 (see Fig. 8.5 below). This effect is very much similar to the fine SOI splitting of energy levels in quantum electron systems [1, 6, 7]. Akin to how *classical orbits* underlie quantum-mechanical electron states, the *geometrical-optics rays and caustics* underlie the actual wave solutions [133, 134]. The

[4] Note that for pure helicity states the magnetic-field intensity in real space coincides with the electric-field intensity, so that the electric-magnetic symmetry is maintained.

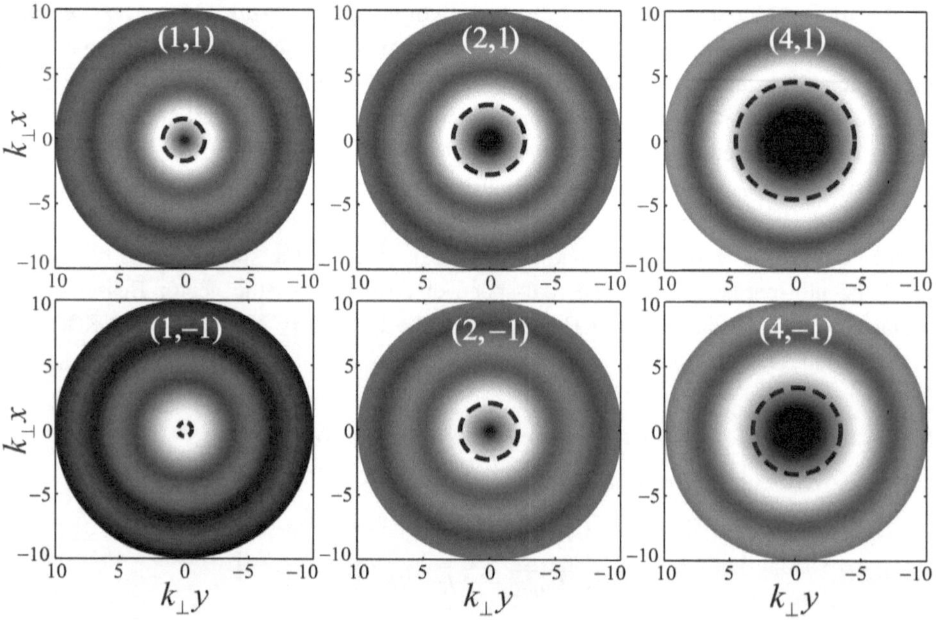

Figure 8.3 Transverse intensity distributions of the circularly polarized vector Bessel beams, I_ℓ^σ, Eq. (8.53), at $\theta_0 = 3\pi/8$, marked by quantum numbers (ℓ, σ). The dashed circles indicate position of the cylindrical caustics of radii R_ℓ^σ, Eq. (8.55), underlying the intensity maxima.

caustic of a Bessel beam represents a cylindrical surface $\{\rho = R, \varphi \in (0, 2\pi)\}$ [135], which satisfies the *quantization* (phase matching) condition similar to the semiclassical quantization of classical orbits. For a scalar Bessel beam, the quantization condition for the cylindrical caustic yields $2\pi k_\perp R_\ell = 2\pi|\ell|$, where $2\pi\ell$ represents the phase gained around the optical vortex $e^{i\ell\varphi}$ in the centre of the beam [135]. In the case of the vector Bessel beams, traversing the caustic in real space corresponds to motion across the rays represented by the contour C in momentum space, and an additional geometric Berry phase (8.49) is gained, Fig. 8.4. With the account of the Berry phase contribution, the quantization condition takes the form

$$\oint_\Gamma \mathbf{k} \cdot d\mathbf{r} = 2\pi(\ell + \sigma \Phi_{B0}), \tag{8.54}$$

where Γ is a closed orbit on the caustical surface surrounding the vortex, Fig. 8.4. Condition (8.54) yields the helicity-dependent radius of the caustic [24]:

$$k_\perp R_\ell^\sigma = |\ell + \sigma \Phi_{B0}|. \tag{8.55}$$

Remarkably, Eq. (8.55) is in exact correspondence with the OAM (8.51): $|\langle L_z \rangle| = |k_\perp R_\ell^\sigma|$. This reflects the OAM interpretation as the vector product $\mathbf{r} \times \mathbf{k}$ for the geometrical-optics

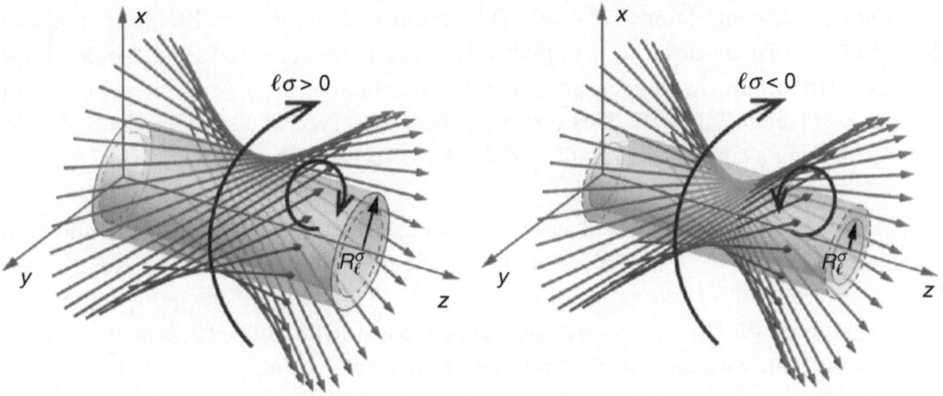

Figure 8.4 The SOI splitting of cylindrical caustics of the vector Bessel beams. Geometrical-optics rays in the real space (representing the plane waves in the beam spectrum, Fig. 8.2) are tangent to the cylindrical caustics. The co- (counter-) directed helicity and vortex result in the azimuthal Berry phase added to (subtracted from) the vortex phase, which increases (diminishes) the radius of the caustic R^σ_ℓ, Eq. (8.55). The cylinders with dashed ends indicate the scalar-beam caustic.

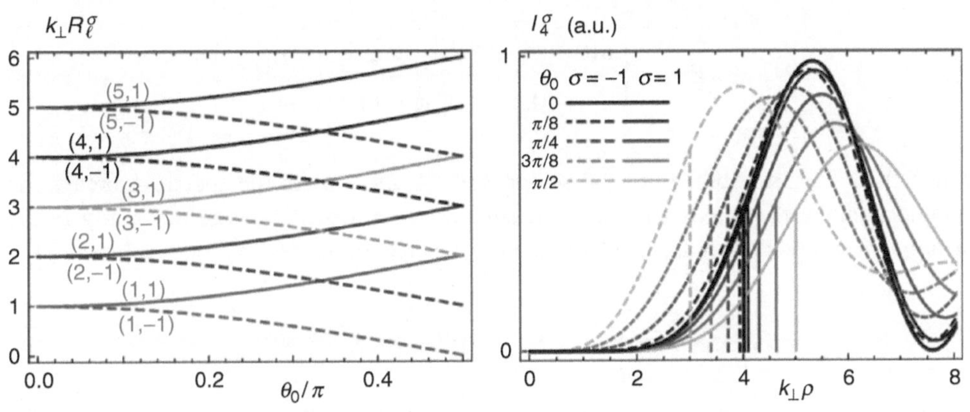

Figure 8.5 The fine structure of the caustic radii ('levels') R^σ_ℓ, Eq. (8.55), as dependent on the angle θ_0. The right-hand panel shows a comparison of the SOI splitting of the exact intensity distributions $I^\sigma_\ell(\rho)$ for $\ell = 4$ and $\sigma = \pm 1$, Eq. (8.53), with the underlying caustics R^σ_ℓ at different angles θ_0.

rays. Thus, the helicity-dependent value of the IOAM immediately reveals itself in the position of the caustic of the field.

The (ℓ, σ)-dependent structure of the caustics (8.55) as a function of θ_0 and comparison with the exact wave solutions (8.53) are shown in Fig. 8.5. This strongly resembles the SOI splitting of quantum levels. The σ-dependent splitting of caustics, $k_\perp(R^+_\ell - R^-_\ell) = 2\Phi_{B0}$, corresponds to the mutual shift of the intensity maxima and grows from 0 at $\theta_0 = 0$ to 2 at $\theta_0 = \pi/2$. Note that in the non-paraxial limit $\theta_0 = \pi/2$ the effective order of the polarized Bessel beam becomes $(\ell + \sigma)$. This results in the dramatic difference at $|\ell| = 1$, where the two polarizations $\sigma = \pm 1$ effectively generate the zero- and second-order Bessel intensity distributions, having, respectively, the maximal and minimal intensities in the centre $\rho = 0$ (see in Fig. 8.3). This effect has been observed in a spiral surface-plasmon cavity generating a circular distribution of the wave vectors, $\{\theta = \pi/2, \phi \in (0, 2\pi)\}$, with an $\ell = 1$ vortex [108, 136].

The Berry-phase effect on the quantization of orbits, Eq. (8.54), is known in numerous quantum systems [8–11, 137–140]. It underlies the fine structure of atomic energy levels [8], Landau levels of electrons in a magnetic field [11], and half-integer quantum Hall effect in graphene [9, 10]. The polarization-dependent structure of caustics in non-paraxial vortex beams represents an optical counterpart of these SOI effects.

The angular momenta (8.51) reveal the OAM-SAM coupling written in the general form in Eqs. (8.31), (8.31a), and (8.34). In turn, the spin-dependent coordinate (8.30), (8.30a), and (8.33) manifests itself in *asymmetric* non-paraxial fields with a displaced centroid: $\langle \mathbf{R}_\perp \rangle \neq \mathbf{0}$, $\langle \mathbf{P}_\perp \rangle \neq \mathbf{0}$. (Otherwise, integration of Eq. (8.30a) over $\mathrm{d}\phi$ leads to vanishing helicity-dependent terms.) To break the axial symmetry of the Bessel beam, we consider the spectrum (8.48) with a truncated azimuthal distribution:

$$|\tilde{E}_\ell(\mathbf{k})\rangle = \begin{cases} \delta(\theta - \theta_0)\mathrm{e}^{\mathrm{i}\ell\phi}, & \phi \in (-\delta, \delta) \\ 0, & \phi \in (\delta, 2\pi - \delta) \end{cases}, \tag{8.56}$$

where $0 < \delta < \pi$. Absence of the sector $\phi \in (\delta, 2\pi - \delta)$ in the spectrum breaks the symmetry along the x-axis and causes *a transverse shift* of the centre of gravity in the *orthogonal* y-direction. Indeed, calculating the expectation values (8.33), we obtain [24]

$$\langle P_x \rangle = \gamma k_\perp, \quad \langle Y \rangle = -\gamma k_\perp^{-1}(\ell + \sigma \Phi_{\mathrm{B0}}), \tag{8.57}$$

where $\gamma = (\sin\delta)/\delta$, and the longitudinal values (8.50) and (8.51) remain the same as for the symmetric beam. Naturally, the beam with the x-asymmetric spectrum acquires a non-zero x-component of the mean momentum, $\langle P_x \rangle$, which results in the tilted propagation, $\langle X \rangle = z\langle P_x \rangle / \langle P_z \rangle$. However, this is also accompanied by the transverse shift $\langle Y \rangle$ given by an expression similar to Eqs. (8.51) and (8.55): $\langle Y \rangle = -\gamma k_\perp^{-1} \langle L_z \rangle$. The ℓ- and σ-dependent parts of this shift notify the *orbital* and *spin-Hall effects* of light in free space. These Hall effects are caused, respectively, by the transverse phase gradients from the truncated azimuthal vortex and Berry phase. Figure 8.6 shows the ℓ- and σ-dependent asymmetric intensity distributions $I_\ell^\sigma(\rho, \varphi)$ calculated for the Bessel beams (8.48) with the truncated spectrum (8.56). The positions of their centres of gravity are in exact agreement with Eq. (8.57). The σ-dependent spin-Hall-effect shift of the intensity spot at $\ell = 0$ and $\theta_0 = \pi/2$ has been observed in an experiment with a semicircular surface-plasmon cavity corresponding to $\delta = \pi/2$ [47, 108].

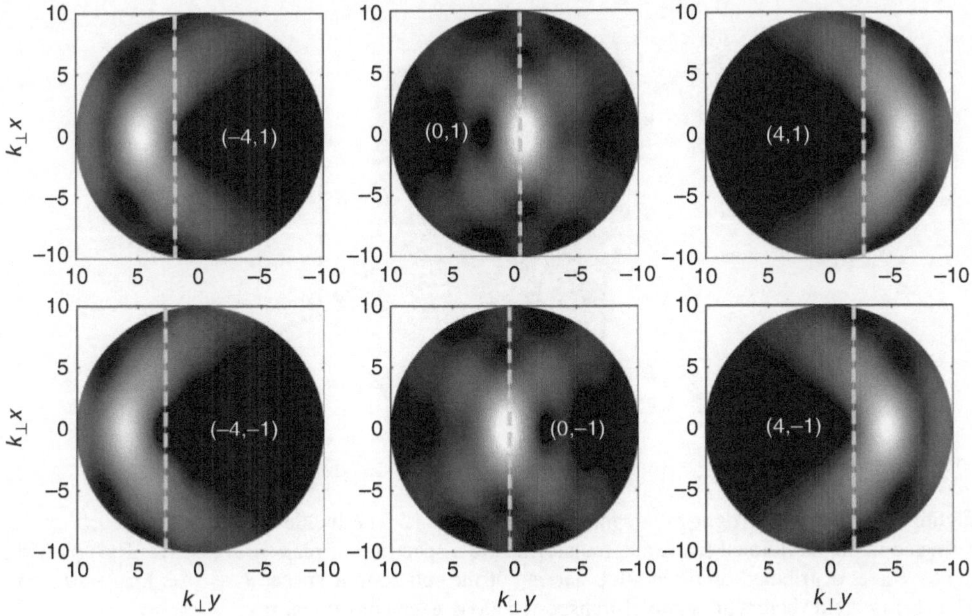

Figure 8.6 Transverse intensity distributions (at $z = 0$) of the asymmetric Bessel beam with the truncated spectrum (8.56), $\delta = \pi/2$, and $\theta_0 = 3\pi/8$, marked by quantum numbers (ℓ, σ). Dashed lines indicate the ℓ- and σ-dependent transverse shifts of the centre of gravity, $\langle Y \rangle$, Eq. (8.57), i.e., orbital and spin-Hall effects of light.

Thus, both in the symmetric and asymmetric cases, the SOI of light brings about a fine subwavelength splitting of the intensity distributions. In the symmetric case the splitting is radial and does not cause a shift of the centre of gravity. In the case of broken symmetry, there are mutually orthogonal displacements in the real and momentum spaces, $\langle Y \rangle \neq 0$ and $\langle P_x \rangle \neq 0$, which generate the longitudinal *extrinsic* OAM (8.36)

$$\langle L_z^{\rm ext} \rangle = -\langle Y \rangle \langle P_x \rangle = \gamma^2 \langle L_z \rangle, \quad \langle L_z^{\rm int} \rangle = (1 - \gamma^2) \langle L_z \rangle. \tag{8.58}$$

Hence, the Hall effects of light can be interpreted as an *intrinsic-to-extrinsic AM conversion*. It is also accompanied by a generation of a transverse component of the EOAM: $\langle L_x^{\rm ext} \rangle = \langle P_z \rangle \langle Y \rangle = -\gamma \cot\theta_0 \langle L_z \rangle$ (cf. [44]). The total conversion in Eq. (8.58) is achieved in the limit of the vanishing distribution: $\delta \to 0$, $\gamma \to 1$.

Another remarkable example of a free-space non-paraxial vector field which exhibits all the SOI features is dipole radiation [141]. While the Bessel-beam solutions provide a good model underlying properties of tightly focused fields, dipole radiation underlies the SOI phenomena in scattered fields (see Sections 8.3.2 and 8.3.3). In all these cases the spin-to-orbit AM conversion results in the helicity-dependent vortex of the Poynting energy flow in real space [30, 141, 142].

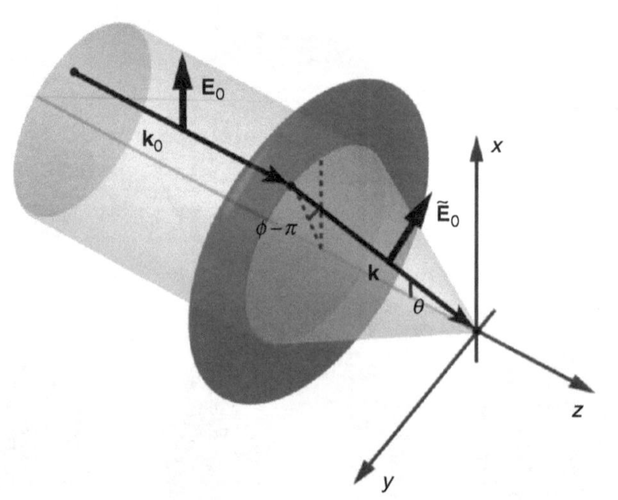

Figure 8.7 Focusing of light by a spherical high-NA lens. The incident paraxial field $\mathbf{E}_0(x, y)$ is refracted in the meridional plane and transformed to a spectrum of plane waves (rays) $\tilde{\mathbf{E}}(\theta, \phi)$ with the **k**-vectors distributed on the north polar cap of the sphere in **k**-space: $\theta \in (0, \theta_c)$, $\phi \in (0, 2\pi)$. The electric field vectors are parallel-transported along each constituent ray.

8.3.2 Focusing by a high-NA lens

Preparation of a non-paraxial beam implies a high-NA focusing. For instance, the Bessel beams considered above can be approximately produced by focusing circularly polarized paraxial beams using an axicon lens [130]. Hence, the SOI effects in free space can be attributed to interaction with the lens that transforms an incident paraxial field into a non-paraxial spectrum of plane waves. Indeed, focusing of circularly polarized light by a high-NA lens is accompanied by spin-to-orbit AM conversion [27, 28, 31, 107, 142–147], whereas breaking of the axial symmetry brings about the spin Hall effect [45–47].

Let us consider the Richards–Wolf theory of focusing with a spherical lens [148, 149], Fig. 8.7. The incident paraxial field $\mathbf{E}_0(\mathbf{r})$ is given by complex amplitudes $|\mathbf{E}_0)_C = (E_0^+, E_0^-, 0)^{\mathrm{T}}$ in the laboratory basis of circular polarizations. Entering rays with the wave vector $\mathbf{k}_0 \simeq k\mathbf{u}_z$ are marked by coordinates (x, y) at the entrance, which can be expressed via spherical angles $(\tilde{\theta}, \tilde{\phi})$ defined with respect to the origin in the focal point: $x = \mathrm{f} \sin\tilde{\theta} \cos\tilde{\phi}$, $y = \mathrm{f} \sin\tilde{\theta} \sin\tilde{\phi}$, where f is the focal distance of the lens. After refraction, the constituent rays converge at the focal point and have **k**-vectors characterized by spherical coordinates $(\theta, \phi) = (\pi - \tilde{\theta}, \pi + \tilde{\phi})$ in the momentum space, Fig. 8.7. Thus the lens performs a sort of *Fourier transform* translating the initial real-space distribution into the momentum distribution. In doing so, the angles (θ, ϕ) serve both as real-space coordinates for the incident field and momentum-space coordinates for the refracted field. The refracted rays are distributed in the range $\{\theta \in (0, \theta_c), \phi \in (0, 2\pi)\}$ (where θ_c is the aperture angle of the lens) and possess an electric field $\tilde{\mathbf{E}}(\theta, \phi)$. The Richards–Wolf theory assumes that constituent waves do *not* change their polarization state in the local *helicity basis*, and the electric fields experience pure meridional rotations together with their **k**-vectors. This is an *adiabatic approximation* which neglects the polarization dependence of the refraction coefficients (cf. [85]). As a result, $|\tilde{\mathbf{E}})_{\mathrm{H}} \propto |\tilde{\mathbf{E}}_0)_{\mathrm{H}} \simeq |\mathbf{E}_0)_{\mathrm{C}}$, and the tightly focused circularly polarized field approximately corresponds to the pure helicity state considered in Section 8.3.1. The focused field in the initial *laboratory* basis, $|\tilde{\mathbf{E}})_{\mathrm{C}}$, can be written using the inverse *helicity-basis transformation* (8.10) [31]

$$|\tilde{\mathbf{E}})_{\mathrm{C}} = \sqrt{\cos\theta}\,\hat{U}_{\mathrm{H}}(\theta,\phi)|\mathbf{E}_0)_{\mathrm{C}}, \quad \hat{U}_{\mathrm{H}}(\theta,\phi) = \begin{pmatrix} a & -b\mathrm{e}^{-2\mathrm{i}\phi} & \sqrt{2ab}\mathrm{e}^{-\mathrm{i}\phi} \\ -b\mathrm{e}^{2\mathrm{i}\phi} & a & \sqrt{2ab}\mathrm{e}^{\mathrm{i}\phi} \\ -\sqrt{2ab}\mathrm{e}^{\mathrm{i}\phi} & -\sqrt{2ab}\mathrm{e}^{-\mathrm{i}\phi} & a-b \end{pmatrix}, \tag{8.59}$$

where $a = \cos^2(\theta/2)$, $b = \sin^2(\theta/2)$, and $\sqrt{\cos\theta}$ is the apodization factor that provides the conservation of the energy flow [149]. Note that while transformation (8.10) implies the transition between different bases with the fixed electric field vector, Eq. (8.59) describes the transformation of the electric field vectors in the same basis.

The three successive rotations $\hat{R}_z(\phi)$, $\hat{R}_y(-\theta)$, and $\hat{R}_z(-\phi)$ in $\hat{U}_H = \hat{V}^\dagger \hat{R}_z(-\phi)\hat{R}_y(-\theta)\hat{R}_z(\phi)\hat{V}$ indicate, respectively: (i) the azimuthal rotation of the coordinate frame superimposing the (x, z)-plane with the local meridional plane, (ii) the refraction of the field on the angle θ therein, and (iii) the reverse azimuthal rotation compensating for the first one. Because of the non-commutativity of the rotations, this transformation is accompanied by a generation of geometrical phases which appear in the form of vortices in the off-diagonal elements of Eq. (8.59). These elements describe the effective transitions between different AM modes. For instance, if the incident wave is σ-circularly polarized, $|\mathbf{E}_0^\sigma)_{\mathrm{C}} \propto |e^\sigma)$, i.e., has only an E^σ component, the refracted wave is $|\tilde{\mathbf{E}}^\sigma)_{\mathrm{C}} \propto \hat{U}_{\mathrm{H}}|e^\sigma)$, i.e., $|\tilde{\mathbf{E}}^+)_{\mathrm{C}} \propto (a, -b\mathrm{e}^{2\mathrm{i}\phi}, -\sqrt{2ab}\mathrm{e}^{\mathrm{i}\phi})^{\mathrm{T}}$ and $|\tilde{\mathbf{E}}^-)_{\mathrm{C}} \propto (-b\mathrm{e}^{-2\mathrm{i}\phi}, a, -\sqrt{2ab}\mathrm{e}^{-\mathrm{i}\phi})^{\mathrm{T}}$. This indicates the generation of the $\tilde{E}_z$ component with the charge-1 vortex $\mathrm{e}^{\mathrm{i}\sigma\phi}$ and the oppositely-polarized component $\tilde{E}^{-\sigma}$ with the charge-2 vortex $\mathrm{e}^{2\mathrm{i}\sigma\phi}$. We emphasize once again that the actual helicities of the constituent waves remain unchanged, and these components appear because of the observation in the laboratory frame. Nonetheless, the azimuthal phases signify the real generation of the OAM in the laboratory frame, i.e., the *spin-to-orbit AM conversion*. Let the incident wave be a paraxial vortex beam with well-defined helicity σ and vortex charge ℓ, and let us denote the AM state of light with respect to the z-axis by the IOAM and SAM quantum numbers, $|\ell, \sigma\rangle$. Then, the polarization transformation (8.59) can be symbolically written as

$$|\ell,\sigma\rangle \to \sqrt{\cos\theta}\left[a|\ell,\sigma\rangle - b|\ell+2\sigma,-\sigma\rangle - \sqrt{2ab}|\ell+\sigma,0\rangle\right]. \tag{8.60}$$

Equation (8.60) exhibits the conservation of the total AM quantum number in each term, $\sigma + \ell = \mathrm{const}$, where the last term $|\ell+\sigma, 0\rangle$ corresponds to the longitudinal field $\tilde{E}_z$

which carries no SAM. Both the transverse field with the opposite polarization (b-term) and the longitudinal field ($\sqrt{2ab}$-term) contribute to the AM conversion. However, in the paraxial approximation, $\theta \ll 1$, one has $b \simeq \theta^2/4$ and $\sqrt{2ab} \simeq \theta/\sqrt{2}$, so that the main contribution is due to the z-component of the field [28].

As the focused field is smoothly distributed over the angle θ, let us calculate first the local *OAM and SAM densities*, l_z and s_z, without integration in the **k**-space. Since the field $|\tilde{\mathbf{E}})_{\mathrm{C}}$ is written in the laboratory basis, we use the AM operators (8.23) or (8.27) (alternatively, one can use the field $|\tilde{\mathbf{E}})_H \propto |\mathbf{E}_0)_C$ and the operators (8.31)). Since $\hat{L}_z = -\mathrm{i}\partial_\phi$ and $(\hat{S}_z)_{\mathrm{C}} = \hat{\sigma}$, we have

$$l_z(\theta,\phi) = -\mathrm{i}\frac{(\tilde{\mathbf{E}}|\partial_\phi|\tilde{\mathbf{E}})_{\mathrm{C}}}{(\tilde{\mathbf{E}}|\tilde{\mathbf{E}})_{\mathrm{C}}}, \quad s_z(\theta,\phi) = \frac{(\tilde{\mathbf{E}}|\hat{\sigma}|\tilde{\mathbf{E}})_{\mathrm{C}}}{(\tilde{\mathbf{E}}|\tilde{\mathbf{E}})_{\mathrm{C}}}. \tag{8.61}$$

Using the incident paraxial beam with well-defined helicity, $|\mathbf{E}^\sigma_{0\ell})_{\mathrm{C}} = |e^\sigma)E_\ell(\theta,\phi)$, $E_\ell = f(\theta)\mathrm{e}^{\mathrm{i}\ell\phi}$, the refracted field (8.59) is $|\tilde{\mathbf{E}}^\sigma_\ell)_{\mathrm{C}} = \sqrt{\cos\theta}\,\hat{U}_{\mathrm{H}}|\mathbf{e}^\sigma)E_\ell(\theta,\phi)$, and Eqs. (8.61) result in

$$l_z = \ell + \sigma\Phi_{\mathrm{B}}(\theta), \quad s_z = \sigma[1 - \Phi_{\mathrm{B}}(\theta)], \quad \Phi_{\mathrm{B}} = 1 - \cos\theta, \tag{8.62}$$

which is quite similar to Eqs. (8.49) and (8.51).

To calculate the OAM and SAM expectation values, $\langle L_z\rangle$ and $\langle S_z\rangle$, one has to integrate both numerators and denominators of Eqs. (8.61) over the spherical angles (θ,ϕ). In this manner, implying integration $\int_0^{2\pi}\mathrm{d}\phi\int_0^{\theta_{\mathrm{c}}}\mathrm{d}\theta\sin\theta$, the expectation values (8.25) yield

$$\langle L_z\rangle = -\mathrm{i}\frac{\langle\tilde{\mathbf{E}}|\partial_\phi|\tilde{\mathbf{E}}\rangle_{\mathrm{C}}}{\langle\tilde{\mathbf{E}}|\tilde{\mathbf{E}}\rangle_{\mathrm{C}}}, \quad \langle S_z\rangle = \frac{\langle\tilde{\mathbf{E}}|\hat{\sigma}|\tilde{\mathbf{E}}\rangle_{\mathrm{C}}}{\langle\tilde{\mathbf{E}}|\tilde{\mathbf{E}}\rangle_{\mathrm{C}}}, \tag{8.63}$$

which, for the field $|\tilde{\mathbf{E}}^\sigma_\ell)_{\mathrm{C}}$, results in

$$\langle L_z\rangle = \ell + \sigma\bar{\Phi}_{\mathrm{B}}, \quad \langle S_z\rangle = \sigma(1 - \bar{\Phi}_{\mathrm{B}}), \quad \bar{\Phi}_{\mathrm{B}} = 1 - \cos\bar{\theta}. \tag{8.64}$$

Here the averaged polar angle $\bar{\theta}$ represents a measure of the directional spread of the field and is defined as [150–152]

$$\cos\bar{\theta} = \frac{\langle P_z\rangle}{\langle W\rangle} = \frac{\int_0^{\theta_c}|f(\theta)|^2\cos^2\theta\sin\theta\mathrm{d}\theta}{\int_0^{\theta_c}|f(\theta)|^2\cos\theta\sin\theta\mathrm{d}\theta}. \tag{8.65}$$

Thus, the AM properties of the field tightly focused by a spherical lens differ from the non-paraxial Bessel beams, Eqs. (8.49) and (8.51), only by averaging over the polar angles θ. In the simplest case of the incident plane wave with $\ell = 0$ and $f(\theta) = 1$, Eq. (8.65) yields $\cos\bar{\theta} = 2(1-\cos^3\theta_{\mathrm{c}})/(3\sin^2\theta_{\mathrm{c}})$. Then, for a small aperture angle, $\theta_{\mathrm{c}} \ll 1$, $\cos\bar{\theta} \approx 1 - \theta_{\mathrm{c}}^2/4$ and $\langle L_z\rangle = \sigma\bar{\Phi}_{\mathrm{B}} \simeq \sigma\theta_{\mathrm{c}}^2/4$, which is reminiscent of the estimation for the multipole expansion of a strongly focused Gaussian beam [142]. For maximal aperture $\theta_c = \pi/2$, one has $\langle L_z \rangle = \sigma/3$, and the efficiency of the spin-to-orbit AM conversion reaches the value of $1/3$.

The actual real-space electric field near the focal point is determined by the interference of the constituent plane waves $|\tilde{\mathbf{E}}(\theta,\phi)\rangle_C$ and is given by the *Debye integral*, which is similar to a Fourier transform [148, 149]:

$$|\mathbf{E}(\mathbf{r})\rangle_C \propto \int_0^{2\pi} d\phi \int_0^{\theta_c} d\theta \sin\theta |\tilde{\mathbf{E}}(\theta,\phi)\rangle_C e^{i\Phi(\theta,\phi,\mathbf{r})}, \tag{8.66}$$

where $\Phi = k|\mathbf{r}-\mathbf{r}'| \simeq k\mathrm{f}(1-\mathbf{r}\cdot\mathbf{r}'/\mathrm{f}^2)$ is the optical phase gained along the path from the refraction point $\mathbf{r}' = -\mathrm{f}(\sin\theta\cos\phi, \sin\theta\sin\phi, \cos\theta)$ to the observation point $\mathbf{r}$ ($|\mathbf{r}| \ll \sqrt{\mathrm{f}/k}$). Calculating the Debye integral (8.66) and, then, the intensity $I_\ell^\sigma = (\mathbf{E}_\ell^\sigma|\mathbf{E}_\ell^\sigma)_C$ for the incident paraxial Laguerre–Gaussian beam, $E_\ell \propto (\sin\theta)^{|\ell|} e^{-\mathrm{f}^2\sin^2\theta/w_0^2} e^{i\ell\phi}$, ($w_0$ is the beam waist, which is supposed to be at the lens), one can derive an expression that is similar to Eq. (8.53) [146]:

$$I_\ell^\sigma(\rho,z) \propto \langle a J_\ell(\xi)\rangle^2 + \langle b J_{\ell+2\sigma}(\xi)\rangle^2 + 2\langle\sqrt{ab}\, J_{\ell+\sigma}(\xi)\rangle^2, \tag{8.67}$$

where we have $\xi = k\rho\sin\theta$ and $\langle\cdots\rangle \equiv \int_0^{\theta_c} d\theta \cdots (\sin\theta)^{|\ell|+1}\sqrt{\cos\theta}\, e^{-\mathrm{f}^2\sin^2\theta/w_0^2 + ikz\cos\theta}$.

The similarity between Eqs. (8.51), (8.53) and Eqs. (8.64), (8.67) shows that properties of the tightly focused polarized fields are quite similar to those of free-space Bessel-beam solutions constructed in Section 8.3.1. In this manner, the behaviour of the field components and caustics of strongly focused beams is quite similar to that described by Eqs. (8.52)–(8.55) and Figs. 8.3–8.5 [28, 107, 146]. In particular, it follows from Eq. (8.67) that the intensity of the beams with antiparallel spin and orbital AM, $\ell\sigma < 0$, does not vanish on the axis $\rho = 0$ for $\ell = 1, 2$. This manifestation of the SOI was experimentally verified in [153, 154]. In general, the size of the focal spot for a strongly focused beam has a fundamental lower bound that depends not only on the beam's directional spread but also on its IOAM and SAM, such that beams with antiparallel spin and orbital AM can achieve tighter focal spots than those for which these AM are parallel [152]. Asymmetric focusing with a truncated lens or distribution of the incident light brings about the ℓ and σ-dependent transverse shifts of the focal spot, i.e., orbital and spin Hall effects akin to Eqs. (8.56)–(8.58) and Fig. 8.6 [45–47].

8.3.3 Scattering by small particles

Scattering of paraxial light by a small particle located at the origin essentially represents the spherical redirection of the constituent plane waves (geometrical-optics rays) (Fig. 8.8). In the dipole approximation, the scattered spherical wave is generated by the dipole moment

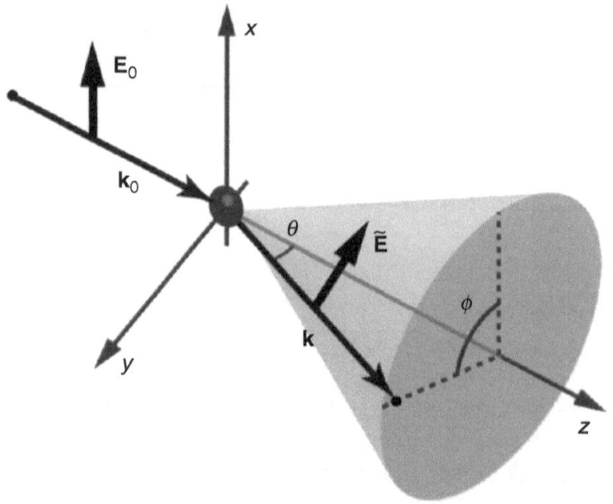

Figure 8.8 Akin to focusing (see Fig. 8.7) dipole scattering of the incident paraxial field $\mathbf{E}_0$ by a small spherical particle transforms it into a spectrum of plane waves $\tilde{\mathbf{E}}(\theta, \phi)$ with spherically distributed **k**-vectors.

$\mathfrak{S} \propto \mathbf{E}_0(\mathbf{0})$ proportional to the incident field $\mathbf{E}_0$ at the origin. Since higher-order paraxial vortex beams with $\ell \neq 0$ have zero field in the centre, $\mathbf{E}^{\sigma}_{0\ell}(\mathbf{0}) \simeq \mathbf{0}$, we consider below an incident plane wave with circular polarization: $\mathbf{E}^{\sigma}_0(\mathbf{0}) \propto \mathbf{u}^{\sigma}_0$, $|\mathbf{E}^{\sigma}_0(\mathbf{0}))_C \propto |e^{\sigma})$. The electric far field of the scattered wave is [122, 155]

$$\tilde{\mathbf{E}}(\theta, \phi) \propto -\,\widehat{\mathbf{r}} \times (\widehat{\mathbf{r}} \times \mathbf{E}_0), \tag{8.68}$$

where $\widehat{\mathbf{r}} = \mathbf{r}/r$ is the unit radial vector with spherical coordinates (θ, ϕ) of the observation point. Akin to focusing, the real-space spherical angles (θ, ϕ) serve as the coordinates in momentum space for the scattered far-field. In this manner, the $\widehat{\mathbf{r}}$-vector plays the role of the $\boldsymbol{\kappa}$-vector for scattered waves and transformation (8.68) is analogous to the vector transformation in Maxwell equations (8.3a). Therefore, Eq. (8.68) can be represented as the action of the *spherical-projection operator* (8.13), $\hat{\Pi}(\theta, \phi) = \hat{U}_{\mathrm{H}}(\theta, \phi)\hat{\mathsf{P}}_z\hat{U}^{\dagger}_{\mathrm{H}}(\theta, \phi)$, in the circular basis [31]:

$$|\tilde{\mathbf{E}})_{\mathrm{C}} \propto \hat{\Pi}(\theta, \phi)|\mathbf{E}_0)_{\mathrm{C}}, \; \hat{\Pi} = \frac{1}{2}\begin{pmatrix} 1 + a_1 & -b_1 \mathrm{e}^{-2\mathrm{i}\phi} & -\sqrt{2a_1 b_1}\mathrm{e}^{-\mathrm{i}\phi} \\ -b_1\mathrm{e}^{2\mathrm{i}\phi} & 1 + a_1 & -\sqrt{2a_1 b_1}\mathrm{e}^{\mathrm{i}\phi} \\ -\sqrt{2a_1 b_1}\mathrm{e}^{\mathrm{i}\phi} & -\sqrt{2a_1 b_1}\mathrm{e}^{-\mathrm{i}\phi} & 2b_1^2 \end{pmatrix}, \tag{8.69}$$

where $a_1 = \cos^2\theta$ and $b_1 = \sin^2\theta$. Comparing Eq. (8.69) with Eq. (8.59), we see that the field transformations upon scattering and focusing exhibit quite similar features related to the off-diagonal geometric-phase elements of matrices $\hat{U}_{\mathrm{H}}$ and $\hat{\Pi}$. In particular, the dipole scattering produces *spin-to-orbit AM conversion* [cf. Eq. (8.60)]:

$$|0,\sigma\rangle \to \frac{1}{2}\left[(1+a_1)|0,\sigma\rangle - b_1|2\sigma,-\sigma\rangle - \sqrt{2a_1b_1}|\sigma,0\rangle\right], \tag{8.70}$$

which was analyzed in several papers [29–31, 62, 156–158].

Taking an incident field with a well-defined helicity, $|\mathbf{E}_0^\sigma)_{\mathrm{C}} = |e^\sigma)$, we determine the scattered field (8.69) and calculate the OAM and SAM densities using Eqs. (8.61):

$$l_z = \sigma\frac{1-\cos^2\theta}{1+\cos^2\theta}, \quad s_z = \sigma\frac{2\cos^2\theta}{1+\cos^2\theta}. \tag{8.71}$$

Similar spherical distributions were obtained in [29]. Equations (8.71) yield $l_z \approx \theta^2\sigma$ for the paraxial angles $\theta \ll 1$ and the total conversion $l_z = \sigma$ at $\theta = \pi/2$. The integral IOAM and SAM values for the scattered field are determined via Eqs. (8.63) with the spherical integration $\int_0^{2\pi} \mathrm{d}\phi \int_0^{\pi} \mathrm{d}\theta \sin\theta$, which results in

$$\langle L_z\rangle = \sigma\frac{\int_0^{\pi}(1-\cos^2\theta)\sin\theta \mathrm{d}\theta}{\int_0^{\pi}(1+\cos^2\theta)\sin\theta \mathrm{d}\theta} = \frac{\sigma}{2}, \quad \langle S_z\rangle = \sigma\frac{2\int_0^{\pi}\sin\theta\cos^2\theta \mathrm{d}\theta}{\int_0^{\pi}\sin\theta(1+\cos^2\theta)\mathrm{d}\theta} = \frac{\sigma}{2}. \tag{8.72}$$

Thus, the spin-to-orbit AM conversion has the efficiency 1/2 [29, 156].

8.3.4 Imaging and microscopy

Strongly focused or scattered fields are essentially non-paraxial, which is the main source of the SOI phenomena. Accordingly, SOI effects can be detected via various near-field methods e.g., tracing motion of test particles [28, 144] or using near-field probes [30, 47, 108]. At the same time, traditional paraxial-optics detectors are unable to measure adequately the SOI phenomena in nonparaxial fields with strong longitudinal field component. Is there a way to transfer the SOI effects into a *paraxial* far-field? There were several experiments [31, 46, 53, 62] observing the SOI of light using standard imaging (microscopy) scheme: (i) focusing of the incident paraxial light with a high-NA lens; (ii) scattering by a small specimen; (iii) collection of the scattered light by another high-NA lens transforming it to the outgoing paraxial light, Fig. 8.9. In those cases the observed effects were ascribed either to focusing or to scattering processes, although all three elements of the imaging system contributed to the effect. The 'lens–scatterer–lens' system represents the basis for optical microscopy and it is of a great importance to describe the SOI effects in such an imaging system, taking into account all its elements [31].

Since both the input and output fields in the imaging system are paraxial, the transformation of the field by the system can be described by an effective *Jones matrix* which gives the angle-resolved polarization state of the output field. Using successive applications of the geometric transformations of the first lens, Eq. (8.59), dipole-scattering transformation (8.69), Debye integration over all incoming rays (8.66), and the inverse transformation

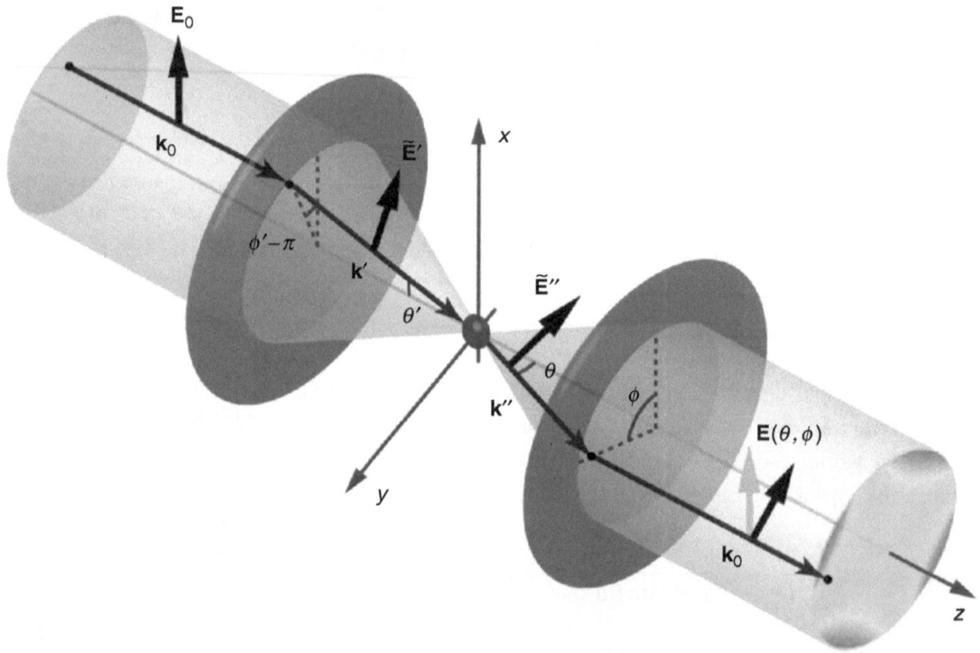

Figure 8.9 Scheme of the imaging (microscope) system 'lens–scatterer–lens'. The incident paraxial light $\mathbf{E}_0$ is focused by a high-NA lens, a small specimen in the focus scatters the non-paraxial focused field, and then the scattered light is collected by the second high-NA lens. The output paraxial field $\mathbf{E}(\theta, \phi)$ has a space-variant polarization distribution and bears information about the SOI processes inside the system. This offers an efficient tool to retrieve fine subwavelength information about the specimen.

(8.59) of the collector lens, one can obtain the three-dimensional transformation operator of the system [31]:

$$|\mathbf{E}(\theta, \phi))_{\mathrm{C}} \propto \hat{T}(\theta, \phi)|\mathbf{E}_0(\theta', \phi'))_{\mathrm{C}}, \tag{8.73}$$

$$\hat{T} = \frac{1}{\sqrt{\cos\theta}} \hat{\mathsf{P}}_z \hat{U}_{\mathrm{H}}^{\dagger}(\theta, \phi) \mathrm{e}^{\mathrm{i}\Phi(\theta,\phi,\mathbf{r}_s)} \int_0^{2\pi} \mathrm{d}\phi' \int_0^{\theta_c} d\theta' \sin\theta' \sqrt{\cos\theta'} \hat{U}_{\mathrm{H}}(\theta', \phi') \mathrm{e}^{\mathrm{i}\Phi(\theta',\phi',\mathbf{r}_s)}, \tag{8.74}$$

where the angles (θ', ϕ') and (θ, ϕ) mark the incoming and outgoing rays, respectively, (see Fig. 8.9), $\mathbf{r}_s$ is the position of the scattering particle near the common focus of the two lenses, and the projector $\hat{\mathsf{P}}_z$ ensures transversality of the output field. The transverse (upper left) 2×2 sector of the operator (8.74), $\hat{T}_{\perp}(\theta, \phi)$, provides the effective space-variant Jones matrix connecting the (x, y) components of the input and output fields in the basis of circular polarizations. Note that even if the incident light represents a homogeneous plane wave $|\mathbf{E}_0)_{\mathrm{C}}$, the output field will have nonuniform polarization $|\mathbf{E}(\theta, \phi))_{\mathrm{C}}$ bearing information about the SOI in the system and specimen properties.

First, let us consider the symmetric case when the scatterer is located precisely in the focus, where $\mathbf{r}_s = 0$. If the incident field is a homogeneous plane wave $|\mathbf{E}_0)_C$ (which implies $\ell = 0$), one can evaluate the integral (8.74) analytically. The resulting Jones operator is [31]

$$\hat{T}_{\perp}^{(0)} = \frac{A(\theta_c)}{\sqrt{\cos\theta}} \begin{pmatrix} a & -be^{-2i\phi} \\ -be^{2i\phi} & a \end{pmatrix}, \tag{8.75}$$

where, as before, $a = \cos^2(\theta/2)$, $b = \sin^2(\theta/2)$, and the aperture-dependent coefficient is $A(\theta_c) = \frac{2\pi}{15}[8 - (\cos\theta_c)^{3/2}(5 + 3\cos\theta_c)] \approx \pi\theta_c^2$ at $\theta_c \ll 1$. The Jones matrix (8.75) is proportional to the transverse part of $\hat{U}_H$, Eq. (8.59), and describes the *spin-to-orbit AM conversion* between two *paraxial* states of light:

$$|\ell, \sigma\rangle \to a|\ell, \sigma\rangle - b|\ell \pm 2, \sigma \mp 2\rangle. \tag{8.76}$$

This strongly resembles paraxial spin-to-orbit convertors based on locally anisotropic axially symmetric structures [55, 58, 59, 61, 66]. In our case, the ϕ-dependent helical phases arise from purely geometrical three-dimensional transformations of the field inside the imaging system, which demonstrate the effective birefringence of the radially (TM) and azimuthally (TE) polarized modes. Assuming σ-circularly polarized incident wave, $|\mathbf{E}_0^\sigma)_{\perp C} \propto |e^\sigma)_\perp$, the output field is $|\mathbf{E}^\sigma)_{\perp C} \propto \hat{T}_\perp^{(0)}|e^\sigma)_\perp$, i.e., $|\mathbf{E}^+)_{\perp C} \propto (a, -be^{2i\phi})^T/\sqrt{\cos\theta}$ and $|\mathbf{E}^-)_{\perp C} \propto (-be^{-2i\phi}, a)^T/\sqrt{\cos\theta}$. Using Eqs. (8.61), we determine the local OAM and SAM densities at the exit pupil:[5]

$$l_z = \sigma\left(1 - \frac{2\cos\theta}{1 + \cos^2\theta}\right), \quad s_z = \sigma\frac{2\cos\theta}{1 + \cos^2\theta}. \tag{8.77}$$

Note that, in this paraxial case, the SAM density coincides with the mean helicity and the Stokes-vector component $\mathfrak{S}_3$, Eq. (8.40). From Eqs. (8.63) for the field $|\mathbf{E}^\sigma)_{\perp C}$, implying integration over the exit pupil $\int_0^{2\pi} d\phi \int_0^{\theta_c} d\theta \sin\theta\cos\theta$, one can obtain the integral OAM and SAM values:

$$\langle L_z\rangle = \frac{1 - 3\cos\theta_c + 3\cos^2\theta_c - \cos^3\theta_c}{4 - 3\cos\theta_c - \cos^3\theta_c}, \quad \langle S_z\rangle = \frac{3 - 3\cos^2\theta_c}{4 - 3\cos\theta_c - \cos^3\theta_c},$$
$$\langle L_z\rangle + \langle S_z\rangle = 1. \tag{8.78}$$

The efficiency of the spin-to-orbit AM conversion is tiny in an imaging system with small aperture, $\langle L_z\rangle \approx \theta_c^4/24$ at $\theta_c \ll 1$, and reaches the value $\langle L_z\rangle = 1/4$ at $\theta_c = \pi/2$.

[5] Here and below, the OAM, SAM, and other quantities are determined for *real-space* fields $|\mathbf{E})_{\perp C}$. Accordingly, the angles (θ, ϕ) serve as real-space coordinates. Nonetheless, the OAM and SAM operators still have the form $\hat{L}_z = -i\partial_\phi$ and $(\hat{S}_z)_C = \hat{\sigma}$ in the coordinate representation, whereas the coordinate operator is given by $\hat{\mathbf{r}} = \mathbf{r} = \mathrm{f}(\sin\theta\cos\phi, \sin\theta\sin\phi, \cos\theta)$.

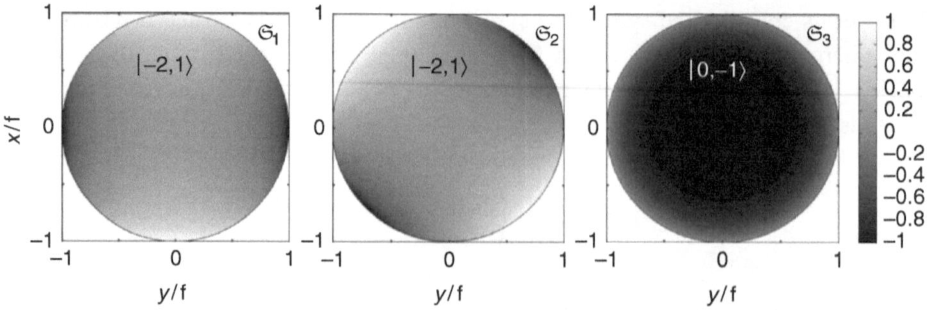

Figure 8.10 Distributions of the Stokes parameters $\vec{\mathfrak{S}} = (\mathfrak{S}_1, \mathfrak{S}_2, \mathfrak{S}_3)$, Eq. (8.79), in the exit pupil of the high-NA ($\theta_c = \pi/2$) optical microscope in the case of the on-axis location of the specimen, $\mathbf{r}_s = 0$, and left-hand circular polarization of the incident light. The four-fold patterns in the $\mathfrak{S}_1$ and $\mathfrak{S}_2$ distributions are the signature of the generation of the right-hand polarized component with optical vortex $e^{-2i\phi}$, i.e., the spin-to-orbit AM conversion (8.76).

Equations (8.75) and (8.76) show that if the incident light is circularly polarized, the output light contains the oppositely polarized component with the charge-2 optical vortex. Such superposition of the two components results in inhomogeneous polarization (radially or azimuthally oriented ellipses) which reveals itself in the typical four-fold symmetric patterns of the *Stokes parameters* [29, 31, 62]. Calculating the normalized Stokes vector (8.39) for the output light $|\mathbf{E}^\sigma)_{\perp C}$, we obtain

$$\vec{\mathfrak{S}}(\theta, \phi) = \frac{1}{a^2 + b^2}[-2ab\cos 2\phi, -2ab\sin 2\phi, \sigma(a^2 - b^2)]. \tag{8.79}$$

Distributions of the Stokes parameters (8.79) are shown in Fig. 8.10 [31].

Now we consider a small displacement of the specimen in the (x, y)-plane, $\mathbf{r}_s = (x_s, y_s, 0)$, which breaks the rotational symmetry of the system and brings about the Hall effects. As before, we restrict our consideration to the case of the uniformly polarized incident wave without OAM. In the approximation of the small subwavelength displacement, $k|\mathbf{r}_s| \ll 1$, the effective Jones matrix can be evaluated analytically from Eq. (8.74) as a correction to Eq. (8.75): $\hat{T}_\perp \simeq \hat{T}_\perp^{(0)} + \hat{T}_\perp^{(1)}$,

$$\hat{T}_\perp^{(1)} = -ik\frac{B(\theta_c)\sin\theta}{2\sqrt{\cos\theta}}\begin{pmatrix} \rho_s e^{-i\phi} & \rho_s^* e^{-i\phi} \\ \rho_s e^{i\phi} & \rho_s^* e^{i\phi} \end{pmatrix}, \tag{8.80}$$

where $\rho_s = x_s + iy_s$ and the aperture-dependent coefficient is $B(\theta_c) = \frac{\pi}{21}[8 - (\cos\theta_c)^{3/2}(11 - 3\cos 2\theta_c)] \approx \pi\theta_c^4/4$ at $\theta_c \ll 1$. The Jones matrix (8.80) reveals coupling between the position of the scatterer and polarization of the light, which results in the *spin Hall effect*. Taking the incident circularly polarized plane wave $|\mathbf{E}_0)_{\perp C} \propto |e^\sigma)_\perp$, we assume $\rho_s = x_s$, determine the output field $|\mathbf{E}^\sigma)_{\perp C} \propto \hat{T}_\perp|e^\sigma)_\perp$, and calculate the

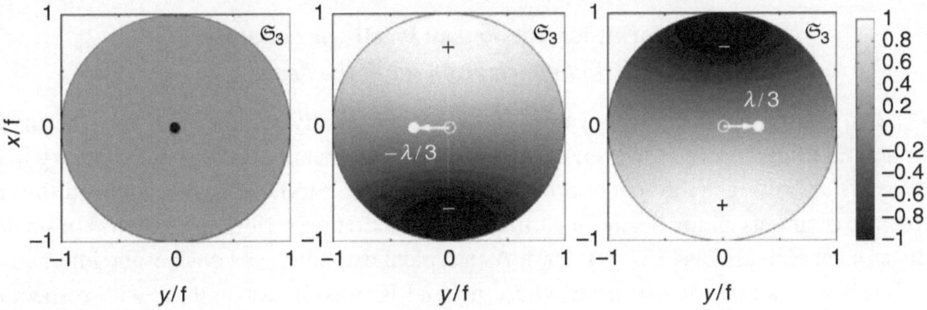

Figure 8.11 Distributions of the Stokes-vector component $\mathfrak{S}_3$, i.e., the SAM density (8.83), in the exit pupil of the high-NA ($\theta_c = \pi/2$) imaging system at different subwavelength displacements of the scattering particle: $x_s = 0, -\lambda/3, \lambda/3$. The particle's displacement induces splitting of the incident x-linearly polarized light into right-hand and left-hand circularly polarized components strongly separated along the y-axis as a result of the spin Hall effect of light.

transverse position of the centre of gravity of the field. In the coordinate representation it is determined as

$$\langle \mathbf{R}_\perp \rangle = \frac{\langle \mathbf{E}^\sigma | \mathbf{r}_\perp | \mathbf{E}^\sigma \rangle_{\perp C}}{\langle \mathbf{E}^\sigma | \mathbf{E}^\sigma \rangle_{\perp C}}, \tag{8.81}$$

with $\mathbf{r}_\perp = \mathfrak{f}(\sin\theta\cos\phi, \sin\theta\sin\phi)$, which yields

$$\langle X \rangle = 0, \quad \langle Y \rangle = -\sigma \mathfrak{f} k x_s \frac{3B\sin^4\theta_c}{2A(4 - 3\cos\theta_c - \cos^3\theta_c)}. \tag{8.82}$$

The σ-dependent shift, which is orthogonal to the displacement of the specimen, represents the spin Hall effect of light. It behaves as $\langle Y \rangle \approx -\sigma \mathfrak{f} k x_s \theta_c^4/8$ at $\theta_c \ll 1$ and reaches the value $\langle Y \rangle \simeq -0.13\sigma \mathfrak{f} k x_s$ at $\theta_c = \pi/2$. Thus, for high-NA systems and subwavelength displacements of the particle, $kx_s \sim 1$, the shift of the centre of gravity is of the order of a fraction of the focal length $\mathfrak{f} \gg \lambda$. In other words, a high-NA optical microscope dramatically *magnifies* the spin Hall effect from the typical subwavelength scale to the scale of the exit pupil [31, 159]. Similar ℓ-dependent deformations of intensity in the imaging system, i.e., the *orbital Hall effect*, were observed in [53].

Another striking manifestation of the spin Hall effect is a strong *separation* of the local SAM densities in the incident *linearly* polarized light. Assuming the incident plane wave to be linearly polarized along the x-axis, $|\mathbf{E}_0^x\rangle_{\perp C} \propto (1, 1)^{\mathrm{T}}/\sqrt{2}$, we calculate the output field $|\mathbf{E}^x\rangle_{\perp C} \propto \hat{T}_\perp |\mathbf{E}_0^x\rangle_{\perp C}$ and the SAM density s_z, Eq. (8.61), which coincides with the Stokes-vector component $\mathfrak{S}_3$:

$$s_z = \mathfrak{S}_3 \simeq -kx_s \frac{B\sin\theta\sin\phi}{A(a^2 + b^2 - 2ab\cos 2\phi)}. \tag{8.83}$$

Distribution (8.83) shows strong asymmetry along the y-axis ($y = \mathfrak{f}\sin\theta\cos\phi$) induced by the subwavelength x-displacement of the particle, (see Fig. 8.11) [31].

8.4 Spin-orbit interactions in locally paraxial fields

8.4.1 Hall effects from tilt of the beam

As shown above, the SOI effects vanish in free-space axially symmetric optical beams in the paraxial limit $\theta \to 0$. However, even at $\theta \ll 1$ a significant breaking of symmetry may induce SOI, resulting in the spin or orbital Hall effect, i.e., a transverse AM-induced shift of the centre of gravity of the beam. In particular, diffraction on various asymmetric obstacles induces such Hall effects [52, 160–163]. As a typical example, we consider the interaction of a beam with an *oblique surface*, whose normal is tilted by an angle ϑ with respect to the beam axis [44, 52, 163]. If the boundary conditions are set in the laboratory coordinate frame attached to the surface, the locally paraxial beam is tilted and no longer paraxial with respect to this frame (see Fig. 8.12). Of course, the observed effects essentially depend on the character of the interaction with the surface [25], but there are some common features, which reflect a universal behaviour coming from the geometry and the AM properties of the problem.

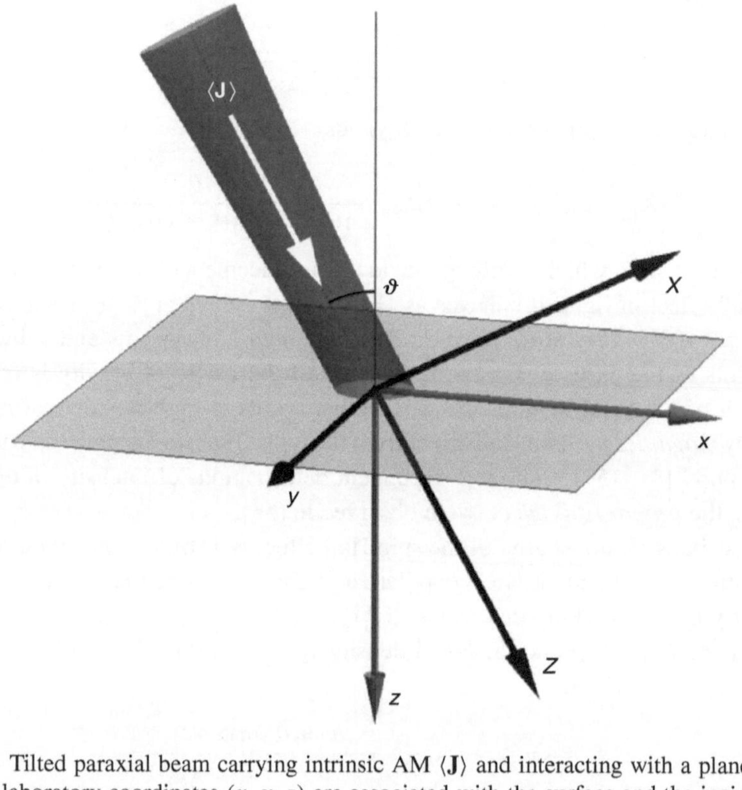

Figure 8.12 Tilted paraxial beam carrying intrinsic AM $\langle \mathbf{J} \rangle$ and interacting with a plane surface at $z = 0$. The laboratory coordinates (x, y, z) are associated with the surface and the incidence plane $y = 0$, whereas the beam coordinate frame (X, y, Z) is attached to the beam propagation direction Z tilted by the angle ϑ.

Let us consider a paraxial beam propagating in the (x, z) plane at angle ϑ with respect to the normal z-axis and impinging upon the observation plane $z = 0$ (see Fig. 8.12). We introduce the *beam coordinate frame* (X, y, Z) with the Z-axis directed along the beam axis and assume that the beam possesses a well-defined linear momentum and an intrinsic AM (8.38):

$$\langle \mathbf{P} \rangle = k\mathbf{u}_Z, \quad \langle \mathbf{J} \rangle = J\mathbf{u}_Z, \tag{8.84}$$

where $J = \ell + \sigma$ in the general case of a circularly polarized vortex beam carrying IOAM and SAM. Characterizing the internal beam structure with the real-space Poynting flow distribution (see Section 8.2.3 and Eqs. (8.45), (8.46)), the momentum and AM components can be written as

$$\langle P_{X,y} \rangle = \int \pi_{X,y} \mathrm{d}^2\mathbf{r}_\perp = 0, \quad \langle P_Z \rangle = \int \pi_Z \mathrm{d}^2\mathbf{r}_\perp \simeq \int w \mathrm{d}^2\mathbf{r}_\perp = \langle W \rangle, \tag{8.85}$$

$$\langle J_Z \rangle = \int (\mathbf{r} \times \boldsymbol{\pi})_Z \mathrm{d}^2\mathbf{r}_\perp = \int X\pi_y \mathrm{d}^2\mathbf{r}_\perp - \int y\pi_X \mathrm{d}^2\mathbf{r}_\perp \equiv J_1 + J_2, \tag{8.86}$$

where $\mathrm{d}^2\mathbf{r}_\perp = \mathrm{d}X\mathrm{d}y$ and, from the symmetry of the beam about the Z-axis, it follows that the two summands in Eq. (8.86) make equal contributions: $J_1 = J_2 = J/2$.

Now, we consider the momentum and AM of the beam in the laboratory frame (x, y, z), i.e., in the oblique cross-section $z = 0$. Since the beam and laboratory coordinates are related via a rotation by the angle ϑ about the y-axis, then

$$\begin{pmatrix} \mathbf{u}_X \\ \mathbf{u}_y \\ \mathbf{u}_Z \end{pmatrix} = \hat{R}_y(\vartheta) \begin{pmatrix} \mathbf{u}_x \\ \mathbf{u}_y \\ \mathbf{u}_z \end{pmatrix}, \quad \hat{R}_y(\vartheta) = \begin{pmatrix} \cos\vartheta & 0 & -\sin\vartheta \\ 0 & 1 & 0 \\ \sin\vartheta & 0 & \cos\vartheta \end{pmatrix}, \tag{8.87}$$

and the components of the vectors are transformed in a similar way. Applying this to the quantities (8.84), we obtain[6]

$$\langle \mathbf{P} \rangle = k \sin\vartheta \mathbf{u}_x + k \cos\vartheta \mathbf{u}_z, \quad \langle \mathbf{J} \rangle = J \sin\vartheta \mathbf{u}_x + J \cos\vartheta \mathbf{u}_z. \tag{8.88}$$

In turn, transformation of the Poynting vector in the beam yields

$$\boldsymbol{\pi} = (\pi_X \cos\vartheta + \pi_Z \sin\vartheta)\mathbf{u}_x + \pi_y \mathbf{u}_y + (-\pi_X \sin\vartheta + \pi_Z \cos\vartheta)\mathbf{u}_z. \tag{8.89}$$

[6] In actual fact, the vectors $\langle \mathbf{P} \rangle$ and $\langle \mathbf{J} \rangle$ are not invariant with respect to the rotation of the coordinate frame. In the two coordinate frames they represent the integral quantities per unit z- and Z-lengths, respectively. In the paraxial limit, in the laboratory frame, they are augmented by a factor of $1/\cos\vartheta$ as compared with the beam coordinate frame, because $\delta Z = \delta z/\cos\vartheta$ or $\mathrm{d}X\mathrm{d}y = \mathrm{d}x\mathrm{d}y \cos\vartheta$. This scaling is inessential for us and can be compensated by renormalization of the field amplitude, $\mathbf{E} \to \sqrt{\cos\vartheta}\mathbf{E}$, so that our consideration implies $\mathrm{d}X\mathrm{d}y = \mathrm{d}x\mathrm{d}y$.

Integration of Eq. (8.89) over $\mathrm{d}^2\mathbf{r}'_\perp = \mathrm{d}x\mathrm{d}y$ taking into account Eq. (8.85) immediately yields the momentum $\langle \mathbf{P} \rangle$ in Eq. (8.88). At the same time, calculations of the AM are more sophisticated. Similar to Eq. (8.86) we write

$$\langle J_z \rangle = \int x\pi_y \mathrm{d}^2\mathbf{r}'_\perp - \int y\pi_x \mathrm{d}^2\mathbf{r}'_\perp \equiv J'_1 + J'_2. \tag{8.90}$$

Substituting the Poynting-vector components (8.89) into Eq. (8.90) and using Eqs. (8.86) together with $x = X/\cos\vartheta$, one obtains $J'_1 = J_1/\cos\vartheta$ and $J'_2 = J_2\cos\vartheta - \int y\pi_Z \sin\vartheta \mathrm{d}^2\mathbf{r}'_\perp$. Using Eq. (8.85) and the definition of the centre of gravity (8.47), we arrive at

$$\langle J_z \rangle = \frac{J}{2}\left(\cos\vartheta + \frac{1}{\cos\vartheta}\right) - \langle Y \rangle k \sin\vartheta. \tag{8.91}$$

Obviously, the first term in Eq. (8.91) indicates the *intrinsic* AM with respect to the z-axis, whereas the second term describes the *extrinsic* contribution to the AM (EOAM), which appears in the laboratory frame tilted with respect to the beam. Equalling expression (8.91) and $\langle J_z \rangle = J\cos\vartheta$ from Eq. (8.88) yields

$$\langle Y \rangle = \frac{J}{2k}\tan\vartheta. \tag{8.92}$$

Thus, transformations of the Poynting flow and AM components in the tilted coordinate frame inevitably result in the subwavelength (of the order of k^{-1}) *transverse shift* of the centre of gravity of the beam, which is proportional to the intrinsic AM J and diverges as $\tan\vartheta$ at $\vartheta \to \pi/2$. This constitutes the *Hall effect of light* upon tilt of the beam. The shifts caused by the SAM and IOAM parts of the AM are associated, respectively, with the spin [44] and orbital [52, 163] Hall effects. The transverse shift (8.92) owes its origin to the non-trivial transformation of the intrinsic AM upon *elliptical deformation* of the beam. Indeed, scaling of the beam with the factor $1/\cos\vartheta$ along the x-axis brings about the transformation (8.91) of the intrinsic AM with the factor of $(\cos\vartheta + 1/\cos\vartheta)/2$ (because x and π_x are transformed in the inverse way). To be consistent with the rotational AM transformation with the factor of $\cos\vartheta$, Eq. (8.88), the extrinsic AM, and the shift of the centre of gravity, Eqs. (8.91) and (8.92), are necessary.

Of course, measurements of the AM and centre of gravity of the beam in the tilted coordinate frame essentially depend on the properties of the detector located at $z = 0$ and interacting with the field. First, the shift (8.92) can be measured by a totally absorbing detector which guarantees that the fields in the different points x do not influence each other [25, 52]. In the following subsection we will see that a vortex beam refracted at a dielectric interface undergoes a transverse shift quite similar to Eq. (8.92). However, a detector which does not perturb the free-space field cannot change the intensity distribution and, hence, the free-space centre of gravity of the beam, $\langle Y \rangle = 0$. Strikingly, even in this

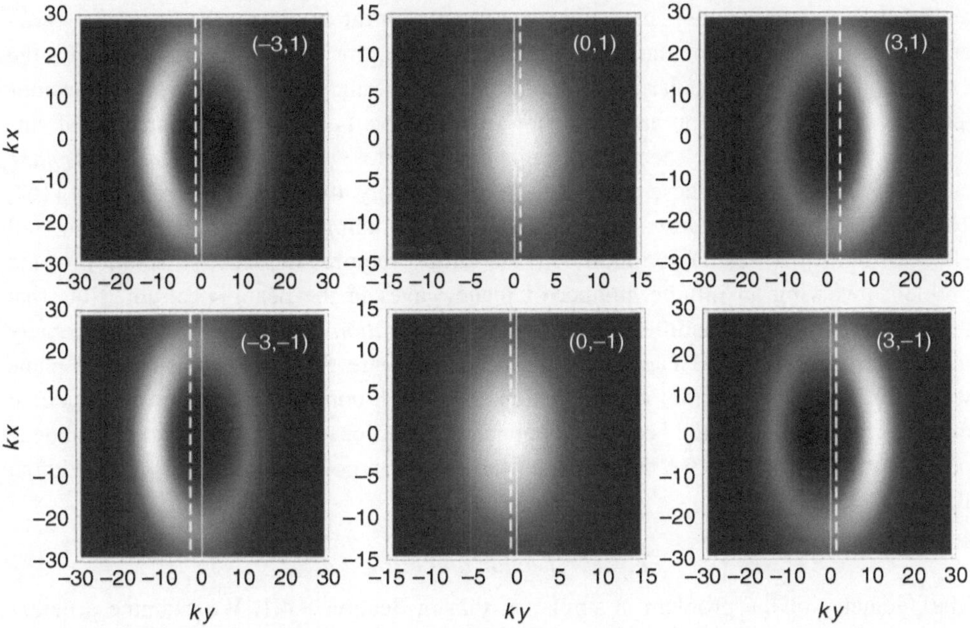

Figure 8.13 Distributions of the Poynting–vector normal component, $\pi_z(x, y)$, in free-space circularly polarized Laguerre–Gaussian beams tilted with respect to the z-axis. Although the intensity of the beam is not changed upon the tilt, the normal component of the Poynting-vector demonstrates the spin- and orbital-dependent transverse shift of the centre of gravity (8.93) (indicated by dashed lines). The beams are marked by their AM numbers (ℓ, σ) (the radial number $\mathrm{p} = 0$). The angle of incidence is $\vartheta = 1$, the aperture angle of the beam is $\theta_a \equiv 2/(kw_0) = 0.2$ (w_0 is the beam waist).

case the first moment of the Poynting energy flow π_z experiences a completely equivalent shift [44]:

$$\langle Y_P \rangle \equiv \frac{1}{P_z} \int y \pi_z \mathrm{d}^2\mathbf{r}'_\perp = \frac{J}{2k} \tan\vartheta. \tag{8.93}$$

This effect was called the *geometric Hall effect of light*. Distributions of the z-component of the Poynting flow in the free-space tilted Laguerre–Gaussian beams are shown in Fig. 8.13. They exhibit σ- and ℓ-dependent deformations and shifts typical for the spin and orbital Hall effects of light. It was also shown that the shift (8.93) is intimately related to the *transverse* AM component $\langle J_x \rangle$ in the tilted coordinate frame [44]. Transformation of the angular momentum upon coordinate rotations and the accompanying Hall effects were recently described in the general relativistic context in [247].

8.4.2 Beam shifts upon reflection and refraction

A common example of beam interaction with an oblique surface is reflection and refraction at a plane dielectric interface. This problem occurs in almost every optical system and is well studied for the plane-wave interaction with the interface: the Snell's law governs the transformation of the wave vectors, whereas the Fresnel formulas provide the connection between the polarization amplitudes of the incident, reflected, and transmitted waves [122]. Reflection and refraction of real wave beams unveils a number of surprises: it turns out that the secondary beams undergo tiny polarization-dependent *in-plane* and *out-of-plane* shifts. These are known as, respectively, the *Goos–Hänchen* effect [164, 165] and the *Imbert–Fedorov* shift [68–70], the latter being an example of the spin Hall effect of light [40–42]. Description of these effects requires an accurate solution of the problem involving an infinite number of plane waves in the beam spectrum. Note that the Fresnel boundary conditions describe the *polarization* phenomena (they are largely independent of the OAM carried by the beam) and are formulated for a single plane wave, i.e., in the *momentum* representation. This is in contrast to the above-considered oblique cross-section of the beam with boundary conditions in *real* space. To describe the polarization phenomena, we first consider the case of a polarized paraxial beam without IOAM: $\ell = 0$.

Fresnel reflection and refraction

The geometry of the problem is similar to that in Section 8.4.1. We examine a paraxial optical beam, which propagates in the (x, z) plane at an angle ϑ to the z-axis and undergoes partial reflection and refraction at a plane interface, $z = 0$, separating two dielectric media without absorption (Fig. 8.14). Below, we follow the approaches developed in [42, 166–168].

In the *geometrical-optics* description, the beams are associated with their *central plane waves* having wave vectors $\mathbf{k}_c^a$. Hereafter the index $a = \mathrm{i, r, t}$ denotes the incident, reflected, and transmitted beams, respectively, and the index i is omitted in the explicit expressions: $\mathbf{k}_c^{\mathrm{i}} \equiv \mathbf{k}_c$, etc. The second medium has the relative refractive index n, so that the wave numbers are $k^{\mathrm{r}} = k^{\mathrm{i}} \equiv k$ and $k^{\mathrm{t}} = nk$. From Snell's law, it follows that the central wave vectors of all three beams lie in the same (x, z) plane: $\mathbf{k}_c^a = k^a(\sin\vartheta^a, 0, \cos\vartheta^a)$ and form the following angles [122] with the z-axis:

$$\vartheta^{\mathrm{i}} = \vartheta, \quad \vartheta^{\mathrm{r}} = \pi - \vartheta, \quad \vartheta^{\mathrm{t}} = \sin^{-1}(n^{-1}\sin\vartheta) \equiv \vartheta'. \tag{8.94}$$

It is natural to introduce coordinate frames of individual beams (X^a, y, Z^a) with the Z^a-axes attached to their directions of propagation: $\mathbf{k}_c^a = k^a\mathbf{u}_Z^a$ (Fig. 8.14). The beam frames are obtained via rotation of the laboratory coordinate frame (x, y, z) by the angle ϑ^a about the y-axis:

$$(\mathbf{u}_X^a, \mathbf{u}_y, \mathbf{u}_Z^a)^{\mathrm{T}} = \hat{R}_y(\vartheta^a)(\mathbf{u}_x, \mathbf{u}_y, \mathbf{u}_z)^{\mathrm{T}}. \tag{8.95}$$

The electric field of the incident central plane wave can be written as $\mathbf{E}_\perp = E_X\mathbf{u}_X + E_y\mathbf{u}_y$, and we assume normalization $|E_X|^2 + |E_y|^2 = 1$. The field components form the effective normalized *Jones vector* in the beam frame, $|\mathbf{E})_{\perp\mathrm{B}} = (E_X, E_y)^{\mathrm{T}}$, and for the central waves the X–y basis coincides with the basis of TM (p) and TE (s) modes with respect to the interface. The fields of the secondary beams are determined by the

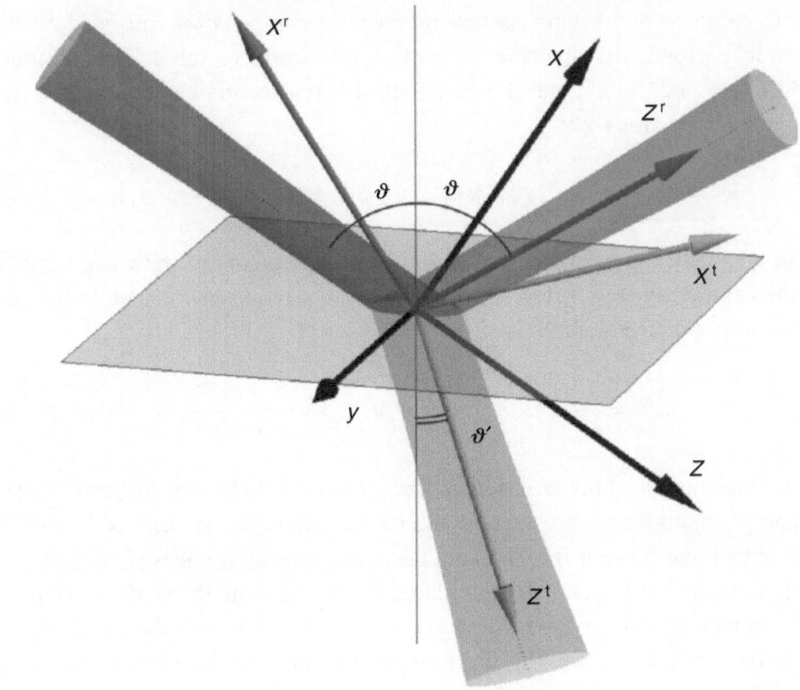

Figure 8.14 Scheme of reflection and refraction of a paraxial beam at a plane dielectric interface. The beam coordinate frames (X^a, y, Z^a) attached to the incident, reflected, and transmitted beams are shown. The laboratory frame is associated with the interface $z = 0$ and plane of incidence $y = 0$, as in Fig. 8.12.

Fresnel reflection and transmission coefficients, $R_{p,s}(\vartheta)$ and $T_{p,s}(\vartheta)$ for the p and s modes [122]. Namely, implying natural transitions between the corresponding beam coordinate frames $(\mathbf{u}_X^a, \mathbf{u}_y, \mathbf{u}_Z^a)$, the transverse electric fields of the secondary beams are obtained by application of the effective *Fresnel Jones matrix* $\hat{F}^a$:

$$|\mathbf{E}^a)_{\perp B} = \hat{F}^a |\mathbf{E})_{\perp B}, \quad \hat{F}^a = \begin{pmatrix} f_p^a & 0 \\ 0 & f_s^a \end{pmatrix}, \tag{8.96}$$

where $f^{\mathrm{i,r,t}} \equiv 1, R, T$. The normalized Jones vectors for the beams take the form:

$$|\mathbf{e}^a)_{\perp B} = Q^{a^{-1}} |\mathbf{E}^a)_{\perp B}, \quad Q^{a^2} = (\mathbf{E}^a|\mathbf{E}^a)_{\perp B}. \tag{8.97}$$

Here Q^a are the amplitude coefficients of the fields, whereas their energy coefficients, satisfying the conservation law, are given by [122]

$$\mathcal{Q}^a = \eta^a \left| \frac{\cos\vartheta^a}{\cos\vartheta} \right| Q^{a^2}, \text{ where } \mathcal{Q}^{\mathrm{r}} + \mathcal{Q}^{\mathrm{t}} = \mathcal{Q} = 1. \tag{8.98}$$

Here η^a is the relative impedance of the medium ($\eta^{\mathrm{i}} = \eta^{\mathrm{r}} = 1$).

Now, let us examine real beams possessing *finite Fourier spectra* consisted of multiple wave vectors $\mathbf{k}^a$ narrowly distributed around the central wave vectors $\mathbf{k}_c^a$. For a monochromatic field, $|\mathbf{k}^a| = |\mathbf{k}_c^a| = k^a$, and the non-central wave vectors can be characterized by small orthogonal deflections χ^a:

$$\mathbf{k}^a = \mathbf{k}_c^a + \chi^a, \quad |\chi^a| \ll k^a, \quad \chi^a \simeq k^a(\mu^a \mathbf{u}_{X^a} + \nu^a \mathbf{u}_y). \tag{8.99}$$

Thus, μ and ν specify the in-plane and out-of-plane deflections of the wave vectors with respect to the central ray (Fig. 8.15). First, Snell's law provides a connection between the components of the wave vectors (8.99):

$$\mu^a = \frac{k}{k^a}\gamma^a \mu, \quad \nu^a = \frac{k}{k^a}\nu, \quad \gamma^a = \frac{\cos\vartheta}{\cos\vartheta^a}, \tag{8.100}$$

where the coefficients γ^a of the *elliptical deformations* of the beams are given by $\gamma^{\mathrm{i}} = 1$, $\gamma^{\mathrm{r}} = -1$, and $\gamma^{\mathrm{t}} = \cos\vartheta / \cos\vartheta'$. Second, the field components are related by the Fresnel coefficients, which are written for TM and TE plane waves, i.e., waves with the electric field being parallel and orthogonal to their planes of propagation, respectively. Importantly, the plane of incidence of a wave with $\nu \neq 0$ does *not* coincide with the (x, z) plane, and, hence, the basis of p and s modes differs from the natural beam X–y basis (see Fig. 8.15). Indeed, the TM and TE modes with respect to the interface $z = 0$ are associated with polar and azimuthal polarizations along the basic vectors $\mathbf{u}_\theta$ and $\mathbf{u}_\phi$ of the global *spherical coordinate frame*, Eq. (8.7) and (8.8) (cf. [141]). Thus, the deceptively simple problem of the reflection or refraction of a wave beam essentially involves three types of coordinate frames for its adequate description: (i) the laboratory frame, (ii) the beam frames, and (iii) the spherical frame of the TM and TE modes.

Let the beam spectra be given in the (X^a, y, Z^a) coordinates as $|\tilde{\mathbf{E}}^a(\mathbf{k}^a)\rangle_{\mathrm{B}}$. Transition to the laboratory frame (x, y, z) is realized by the rotations (8.95) $\hat{R}_y(-\vartheta^a)$, $|\tilde{\mathbf{E}}^a\rangle_{\mathrm{L}} = \hat{R}_y(-\vartheta^a)|\tilde{\mathbf{E}}^a\rangle_{\mathrm{B}}$, whereas the next transitions to the global spherical coordinates are accomplished with the help of the transformation $\hat{U}_{\mathrm{S}}^\dagger(\mathbf{k}^a) = \hat{R}_y(\theta^a)\hat{R}_z(\phi^a)$, Eq. (8.8): $|\tilde{\mathbf{E}}^a\rangle_{\mathrm{S}} = \hat{U}_{\mathrm{S}}^\dagger(\mathbf{k}^a)|\tilde{\mathbf{E}}^a\rangle_{\mathrm{L}}$. Here the spherical angles (θ^a, ϕ^a) determine the direction of $\mathbf{k}^a$. The resulting rotational transformation to the basis of p and s modes is

$$|\tilde{\mathbf{E}}^a\rangle_{\mathrm{S}} = \hat{U}_\vartheta^\dagger(\vartheta^a, \mathbf{k}^a)|\tilde{\mathbf{E}}^a(\mathbf{k}^a)\rangle_{\mathrm{B}}, \quad \hat{U}_\vartheta^\dagger(\vartheta^a, \mathbf{k}^a) = \hat{R}_y(\theta^a)\hat{R}_z(\phi^a)\hat{R}_y(-\vartheta^a). \tag{8.101}$$

For the central plane wave $\mathbf{k}^a = \mathbf{k}_c^a$, we have $(\theta^a, \phi^a) = (\vartheta^a, 0)$. For a constituent wave (8.99), small wave-vector deflections μ^a and ν^a induce changes in θ^a and ϕ^a, respectively:

$$\theta^a \simeq \vartheta^a + \mu^a, \quad \phi^a \simeq \frac{\nu^a}{\sin\vartheta^a}. \tag{8.102}$$

Thus, the in-plane deflection μ changes the *angle of incidence*, whereas the out-of-plane deflection ν turns the *plane of incidence* by the angle $\nu/\sin\vartheta$ about the z-axis, Fig. 8.15.

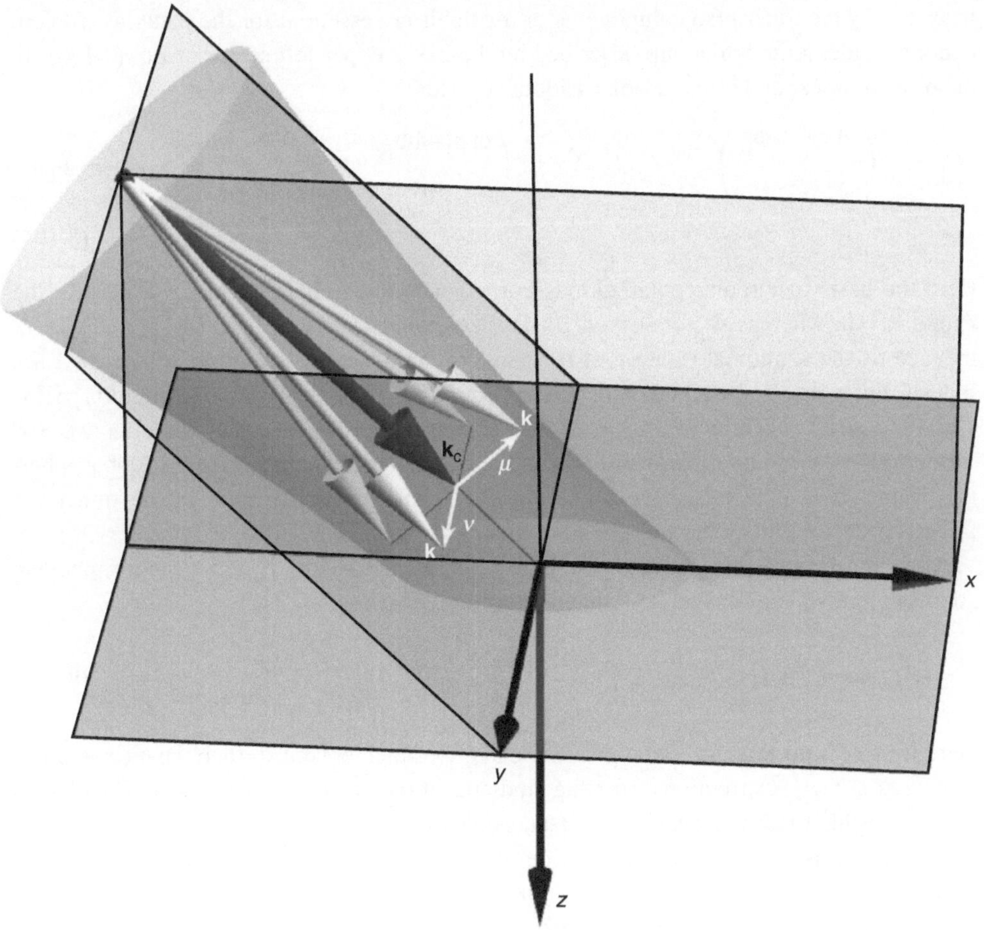

Figure 8.15 Central wave vector $\mathbf{k}_c$ (black arrow) and non-central wave vectors $\mathbf{k}$ (white arrows) in the incident beam. Small in-plane and out-of-plane deflections, μ and ν, vary the angle of incidence and the plane of incidence, respectively (see Eq. (8.99) and (8.102)).

Substituting Eq. (8.102) into Eq. (8.101), we obtain the rotation matrix $\hat{U}_\vartheta$ in the linear approximation in $\boldsymbol{\chi}^a$:

$$\hat{U}_\vartheta^\dagger(\vartheta^a, \boldsymbol{\chi}^a) \simeq \hat{R}_y(\vartheta^a + \mu^a)\hat{R}_z(\nu^a/\sin\vartheta^a)\hat{R}_y(-\vartheta^a)$$

$$\simeq \begin{pmatrix} 1 & \nu^a \cot\vartheta^a & -\mu^a \\ -\nu^a \cot\vartheta^a & 1 & -\nu^a \\ \mu^a & \nu^a & 1 \end{pmatrix}. \quad (8.103)$$

In fact, only the transverse components of the fields are essential for the paraxial problem under consideration, which are described by the 2×2 upper-left sector of Eq. (8.103). In the basis of linear and circular polarizations it yields

$$\hat{U}^{\dagger}_{\vartheta\perp} = \begin{pmatrix} 1 & -\Phi^a_{\rm B} \\ \Phi^a_{\rm B} & 1 \end{pmatrix}, \quad \hat{V}^{\dagger}_{\perp}\hat{U}^{\dagger}_{\vartheta\perp}\hat{V}_{\perp} \simeq \begin{pmatrix} \exp(-{\rm i}\Phi^a_{\rm B}) & 0 \\ 0 & \exp({\rm i}\Phi^a_{\rm B}) \end{pmatrix} = \exp(-{\rm i}\hat{\sigma}_3\Phi^a_{\rm B}). \tag{8.104}$$

Here the basis of circular polarizations corresponds to the helicity basis (8.9) with the gauge $m = 0$, whereas $\Phi^a_{\rm B} = -\nu^a \cot\vartheta^a$ is the *geometric phase*, Eqs. (8.17) and (8.18), induced by the azimuthal rotation of the plane of incidence by the angle ϕ^a, Eq. (8.102). Indeed, this phase can be understood from Eq. (8.21), where the angular rotation yields $\int \boldsymbol{\Omega}_\zeta {\rm d}\zeta = \phi^a \mathbf{u}_z$, whereas $\boldsymbol{\kappa}^a \cdot \mathbf{u}_z = \cos\vartheta^a$. Equation (8.104) describes the effective *SOI Jones matrix* caused by the transition from the beam coordinate frame to the global spherical frame. It is this free-space rotational transformation that results in the transverse Imbert–Fedorov shifts, i.e., the spin Hall effect of light.

After transforming the fields to the global spherical basis of p and s modes, one can connect them via the Fresnel boundary conditions (8.96):

$$|\tilde{\mathbf{E}}^a\rangle_{\perp{\rm S}} = \hat{F}^a|\tilde{\mathbf{E}}\rangle_{\perp{\rm S}}, \quad \hat{F}^a = \begin{pmatrix} f^a_p(1+\mu\partial_\vartheta \ln f^a_p) & 0 \\ 0 & f^a_s(1+\mu\partial_\vartheta \ln f^a_s) \end{pmatrix}. \tag{8.105}$$

Here we took into account variations in the angle of incidence, Eq. (8.102): $f(\vartheta+\mu) \simeq f(\vartheta) + \mu\partial_\vartheta f(\vartheta)$. Corrections from the gradients of the Fresnel coefficients in Eq. (8.105) are responsible for the in-plane Goos–Hänchen shifts of the beams.

Together, Eqs. (8.103)–(8.105) provide the complete transformation and Jones matrix relating the incident and secondary fields in the beam coordinate frames: $\hat{T}^a_{\perp} = \hat{U}_{\vartheta\perp}(\vartheta^a,\chi^a)\hat{F}^a(\vartheta,\chi)\hat{U}^{\dagger}_{\vartheta\perp}(\vartheta,\chi)$,

$$|\tilde{\mathbf{E}}^a\rangle_{\perp{\rm B}} = \hat{T}^a_{\perp}|\tilde{\mathbf{E}}\rangle_{\perp{\rm B}}, \quad \hat{T}^a_{\perp} = \begin{pmatrix} f^a_p(1+\mu\mathcal{X}^a_p) & f^a_p\nu\mathcal{Y}^a_p \\ -f^a_s\nu\mathcal{Y}^a_s & f^a_s(1+\mu\mathcal{X}^a_s) \end{pmatrix}, \tag{8.106}$$

where

$$\mathcal{X}^a_{p,s} = \partial_\vartheta \ln f^a_{p,s}, \quad \mathcal{Y}^a_{p,s} = \left(1 - \gamma^{a^{-1}}\frac{f^a_{s,p}}{f^a_{p,s}}\right)\cot\vartheta. \tag{8.107}$$

Equations (8.106) and (8.107) provide the complete description of the reflection and refraction of a paraxial optical beam at a plane dielectric interface. While the 'SOI transformation' (8.104) is diagonal in the basis of circular polarizations, the Fresnel boundary conditions (8.105) are naturally diagonal in the basis of linear polarizations. Therefore, there is *no* global polarization basis that would diagonalize the whole transformation (8.106).

To calculate the shifts explicitly, let us take an incident Gaussian beam at its waist $Z = 0$: $|\tilde{\mathbf{E}}\rangle_{\perp{\rm B}} = (E_X, E_y)^{\rm T} G(\mu,\nu)$, where $(E_X, E_y)^{\rm T}$ is the Jones vector of the central plane wave and $G(\mu, \nu) = \frac{w_0^2}{2\pi} \exp\left[-(kw_0)^2 \frac{\mu^2+\nu^2}{4}\right]$ is the normalized Gaussian envelope with waist w_0. By applying the Jones matrix, Eqs. (8.106) and (8.107), we obtain the fields of the secondary beams and calculate the expectation values of their transverse momenta and coordinates using paraxial versions of Eqs. (8.24):

$$\langle X^a, Y^a \rangle = Q^{a^{-2}} \langle \tilde{\mathbf{E}}^a | i\partial_{k^a_{X,y}} | \tilde{\mathbf{E}}^a \rangle_{\mathrm{B}\perp}, \quad \langle P^a_{X,y} \rangle = Q^{a^{-2}} \langle \tilde{\mathbf{E}}^a | k^a_{X,y} | \tilde{\mathbf{E}}^a \rangle_{\mathrm{B}\perp}. \tag{8.108}$$

Here the integration is taken over the transverse momentum components $k^a_X = \gamma^a k\mu$ and $k^a_y = k\nu$, and we take into account that the norms are equal to $\langle \tilde{\mathbf{E}}^a_\perp | \tilde{\mathbf{E}}^a_\perp \rangle = Q^{a^2}$. In the case of *partial* reflection and transmission, $\sin\vartheta < n$, the Fresnel coefficients $f^a_{p,s}$ are real, and Eqs. (8.106)–(8.108) at $Z^a = 0$ result in

$$\langle X^a \rangle = 0, \quad \langle P^a_X \rangle = \frac{\gamma^a}{kw_0^2} \frac{\mathrm{d}\ln Q^a}{\mathrm{d}\theta}, \tag{8.109}$$

$$\langle Y^a \rangle = -\frac{\bar{\sigma}}{2k} \frac{f_p^{a^2} \mathcal{Y}^a_p + f_s^{a^2} \mathcal{Y}^a_s}{Q^{a^2}}, \quad \langle P^a_y \rangle = \frac{\bar{\rho}}{kw_0^2} \frac{f_p^{a^2} \mathcal{Y}^a_p - f_s^{a^2} \mathcal{Y}^a_s}{Q^{a^2}}, \tag{8.110}$$

where $\bar{\sigma} = (\mathbf{e}|\hat{\sigma}_2|\mathbf{e})_{\perp B} = 2\mathrm{Im}(E_X^* E_y)$ is the average helicity of the incident beam (the Stokes parameter $\mathfrak{S}_3$) and $\bar{\rho} = (\mathbf{e}|\hat{\sigma}_1|\mathbf{e})_{\perp B} = 2\mathrm{Re}(E_X^* E_y)$ is the degree of linear polarization inclined at the angle $\pi/4$ (the Stokes parameter $\mathfrak{S}_2$).[7] In the case of total internal reflection, $\sin\vartheta > n$, there is no transmitted propagating wave, $f^{\mathrm{t}}_{p,s} = 0$, while the reflection coefficients are complex: $f^{\mathrm{r}}_{p,s} = \exp(\mathrm{i}\delta_{p,s})$, with real phases $\delta_{p,s}$. In this case Eqs. (8.106)–(8.108) bring about

$$\langle X^{\mathrm{tot\,r}} \rangle = \frac{1}{k}(|E_X|^2 \mathrm{Im}\mathcal{X}^{\mathrm{r}}_p + |E_y|^2 \mathrm{Im}\mathcal{X}^{\mathrm{r}}_s), \quad \langle P^{\mathrm{tot\,r}}_X \rangle = 0, \tag{8.111}$$

$$\langle Y^{\mathrm{tot\,r}} \rangle = -\frac{1}{2k}[\bar{\sigma}\mathrm{Re}(\mathcal{Y}^{\mathrm{r}}_p + \mathcal{Y}^{\mathrm{r}}_s) + \bar{\rho}\mathrm{Im}(\mathcal{Y}^{\mathrm{r}}_p - \mathcal{Y}^{\mathrm{r}}_s)], \quad \langle P^{\mathrm{tot\,r}}_y \rangle = 0. \tag{8.112}$$

The centroid displacement $\langle X^a \rangle$ and momentum deflection $\langle P^a_X \rangle$ in Eqs. (8.109)–(8.112) represent the *spatial and angular Goos–Hänchen shifts*, whereas the displacement $\langle Y^a \rangle$ and deflection $\langle P^a_y \rangle$–the *spatial and angular Imbert–Fedorov shifts*, respectively (see Fig. 8.16). The angular shifts depend on the beam waist and vanish in the paraxial limit $w_0 \to \infty$, whereas the spatial shifts are independent of the beam profile. Nonetheless, both spatial and angular shifts have a common origin: the *in-plane momentum gradients of the Fresnel coefficients* $\mathcal{X}^a_{p,s}$ for the Goos–Hänchen effects (8.109) and (8.111) and the *geometric SOI terms* $\mathcal{Y}^a_{p,s}$ for the Imbert–Fedorov effects (8.110) and (8.112). The latter can be associated with the *transverse momentum gradients of the Berry phases* across the beams: $\partial_{\nu^a}\Phi^a_{\mathrm{B}} = -\cot\vartheta^a$. The Imbert–Fedorov effects show divergence at $\vartheta \to 0$ because the approximate transition to the spherical coordinates, Eqs. (8.102) and (8.103), becomes singular at normal incidence; in reality, of course, the transverse shift vanishes at $\vartheta = 0$.

[7] The Pauli matrices $(\hat{\sigma}_1, \hat{\sigma}_2, \hat{\sigma}_3)$ underlie the Stokes-vector components $(\mathfrak{S}_1, \mathfrak{S}_2, \mathfrak{S}_3)$, Eq. (8.39), in the basis of *circular* polarizations. In the basis of *linear* polarizations, the permuted Pauli matrices $(\hat{\sigma}_3, \hat{\sigma}_1, \hat{\sigma}_2)$ correspond to the same Stokes parameters.

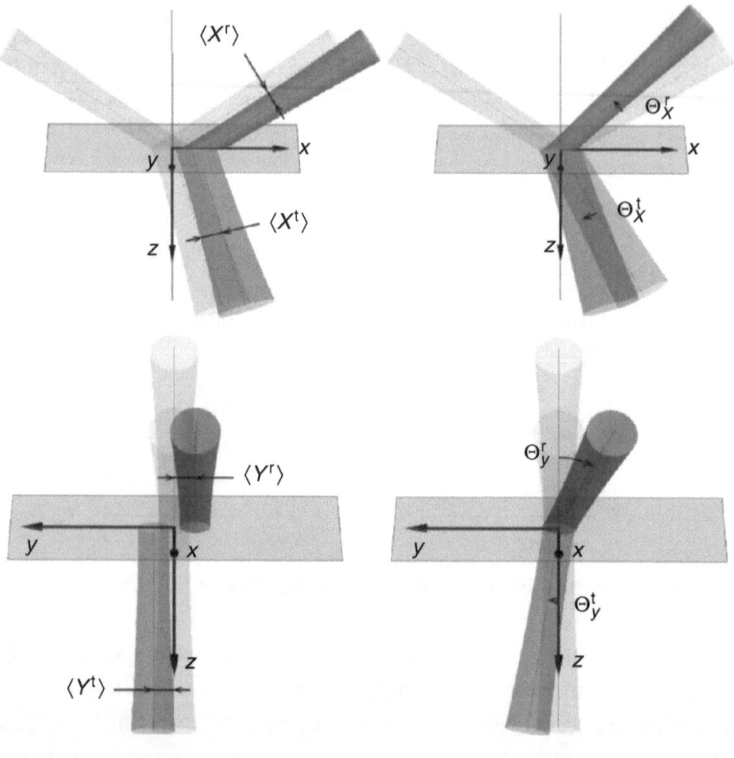

Figure 8.16 Schematic picture of the in-plane and out-of-plane, the spatial and angular shifts of the reflected and transmitted beams. The displacements $\langle X^a \rangle$ and $\langle Y^a \rangle$ represent the spatial Goos–Hänchen and Imbert–Fedorov shifts, respectively. The deflections $\Theta^a_X = \langle P^a_X \rangle / k^a$ and $\Theta^a_y = \langle P^a_y \rangle / k^a$ are associated with the angular Goos–Hänchen and Imbert–Fedorov shifts, respectively.

It is important to note that, at $Z^a \neq 0$, the real-space shifts grow according to the angular deviation: $\langle X^a \rangle (Z^a) = \langle X^a \rangle + \langle P^a_X \rangle Z^a / k^a$, $\langle Y^a \rangle (Z^a) = \langle Y^a \rangle + \langle P^a_y \rangle Z^a / k^a$ (Fig. 8.16) which is used to measure small deviation effects [42, 167–169].

The widely known spatial Goos–Hänchen shift $\langle X^a \rangle$ is caused by the angular gradient of the *phase* of complex reflection or refraction coefficients and appears, e.g, upon the total internal reflection or interaction with complex dispersive interfaces [164, 165, 170–183]. In turn, the angular Goos–Hänchen shift $\langle P^a_X \rangle$ is caused by the angular gradients of the *amplitude* of the Fresnel coefficients and can be observed in the case of partial reflection and refraction with real coefficients [51, 167, 169, 184–186]. Obviously, both Goos–Hänchen effects are intimately related to the *spatial dispersion* of the scattering coefficients. While $\langle X^a \rangle$ is a spatial analogue of the Wigner time delay in scattering [187], the effect of the angular shift $\langle P^a_X \rangle$ is also known as 'Fresnel filtering' [188, 189]. The eigenmodes of the Goos–Hänchen effects are p and s linearly polarized waves, but polarization is not crucial for the existence of these phenomena, as they can also occur for scalar waves.

The Imbert–Fedorov shifts arise essentially owing to the intrinsic polarization properties of light. In the partial reflection/transmission case, the eigenmodes of the spatial shift $\langle Y^a \rangle$ are circularly polarized waves $\bar{\sigma} = \pm 1$, the shift is proportional to the helicity of the incident wave, and, thus, represents the *spin Hall effect of light* (Fig. 8.17). This transverse shift has been discussed and studied for many years, both theoretically and experimentally [38, 68–73, 190–207]. The most important here are the early paper by Schilling [69], which contains the first reasonable spectral description of the effect and a formula equivalent to Eq. (8.112), and the papers [71, 72], which reveal the close relation of the transverse shift to the balance of the AM of light. However, for the most part, various sophisticated theories and experimental techniques gave rise to numerous controversies and were not able to give a clear self-consistent description of the effect. Interest in the transverse shift caught a second wind in 2004 when this effect was attributed to the spin Hall effect and SOI of light [40–42, 51, 54, 166–168, 208–210]. Strikingly, while the clear theoretical explanation of the Goos–Hänchen effect was given by Artmann the next year after its discovery [165], the accurate formulas (8.110) for the Imbert–Fedorov shift have been derived and verified experimentally only recently [41, 42, 166]. This could be explained by the fact that while the Goos–Hänchen effect can be fully explained within the two-dimensional geometry of the plane of incidence, the Imbert–Fedorov shift essentially involves a three-dimensional description and requires an accurate characterization of the polarization of the incident field. The latter has been a subject of recent discussion [166, 211–213]. The angular Imbert–Fedorov shift $\langle P_y^a \rangle$ was described only recently [42, 51, 166, 167, 214]. Despite the fact that the eigenmodes of this effect are linearly polarized waves with $\bar{\rho} = \pm 1$, it originates from the same SOI terms as the spatial transverse shift.

Importantly, the spatial Imbert–Fedorov shift is intimately related to the balance and *conservation of the total AM of light* [40, 41, 71, 72, 166]. Indeed, the rotational symmetry of the medium about the z-axis implies that the z-component of the total AM must be conserved upon reflection and refraction. Using equation (8.38) for paraxial beams, the SAM per photon in the three beams can be written as $\langle \mathbf{S}^a \rangle = \bar{\sigma}^a \mathbf{u}_Z^a$, $\langle S_z^a \rangle = \bar{\sigma}^a \cos\vartheta^a$. Using the Fresnel boundary conditions (8.96) and (8.97), one can determine the mean helicities $\bar{\sigma}^a = (\mathbf{e}^a | \hat{\sigma}_2 | \mathbf{e}^a)_{\perp \mathrm{B}}$, which yield the SAM for the cases of partial reflection/transmission and for total internal reflection:

$$\langle S_z^a \rangle = \bar{\sigma} \cos\vartheta^a \frac{f_p^a f_s^a}{Q^{a2}}, \quad \langle S_z^{\mathrm{tot\,r}} \rangle = -(\bar{\sigma} \cos\delta + \bar{\rho} \sin\delta) \cos\vartheta, \tag{8.113}$$

where $\delta = \delta_s - \delta_p$. Akin to Eq. (8.91), the transverse shifts (8.110) and (8.112) generate the EOAM (8.36):

$$\langle L_z^{\mathrm{ext}\,a} \rangle = [\langle \mathbf{R}^a \rangle \times \langle \mathbf{P}^a \rangle]_z = -\langle Y^a \rangle k \sin\vartheta. \tag{8.114}$$

Since the number of photons in the beams is proportional to the energy Q^a, we write the conservation of the z-component of the total AM in the scattering as

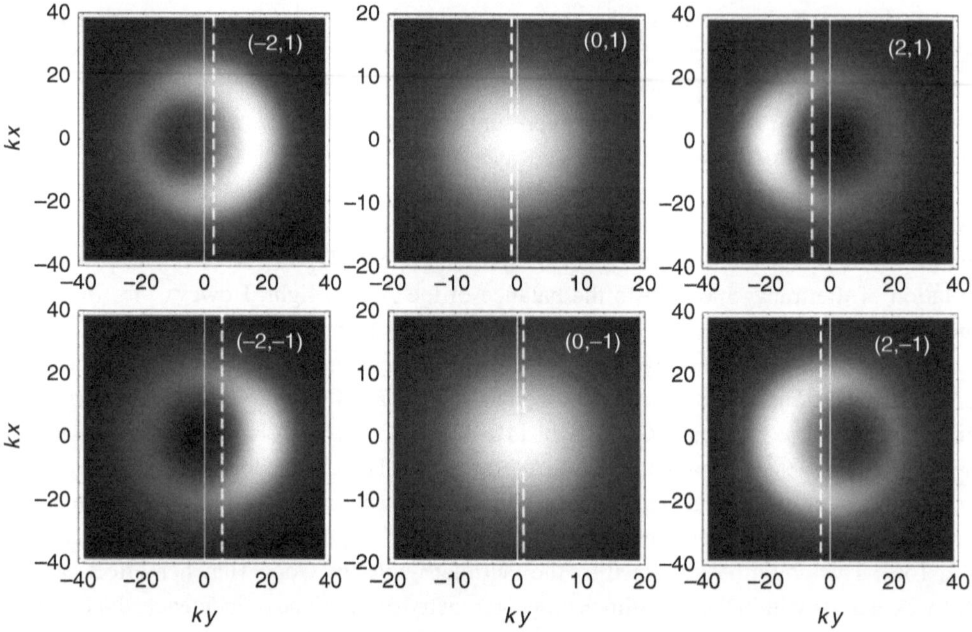

Figure 8.17 Transverse intensity distributions (at $Z^{\mathrm{r}} = 0$) in the Laguerre–Gaussian beams reflected from the air–glass interface ($n = 1.5$). As before, the beams are marked by their AM numbers (ℓ, σ) (the radial number $\mathrm{p} = 0$). The angle of incidence $\vartheta = \pi/3$, whereas the aperture angle is $\theta_a \equiv 2/(kw_0) = 0.1$. The dashed lines indicate the Imbert–Fedorov shifts of the centroids given by Eqs. (8.110) and (8.121). Akin to Fig. 8.13, the spin-dependent shifts represent perfect translations of the whole beam, while the vortex-dependent shifts are accompanied by deformations of the intensity distributions.

$$Q^{\mathrm{r}}[\langle S_z^{\mathrm{r}}\rangle - \langle Y^{\mathrm{r}}\rangle k \sin\vartheta] + Q^{\mathrm{t}}[\langle S_z^{\mathrm{t}}\rangle - \langle Y^{\mathrm{t}}\rangle k \sin\vartheta] = \langle S_z\rangle, \tag{8.115}$$

Here we took into account that $\langle L_z^{\mathrm{ext}}\rangle = \langle Y\rangle = 0$ and $Q = 1$ for the incident beam. Substituting Eqs. (8.110), (8.112), and (8.113) into Eq. (8.115) and using relations (8.94)–(8.100) and (8.107), one can verify that Eq. (8.115) is fulfilled identically. Thus, the non-zero transverse shifts of the reflected and refracted beams ensure the conservation of the total AM in the problem. In the case of total internal reflection, there is no transmitted beam, $Q^{\mathrm{t}} = 0$, $Q^{\mathrm{r}} = 1$, and the shift can be derived *solely* from the AM conservation:

$$\langle Y^{\mathrm{tot\,r}}\rangle = \frac{\langle S_z^{\mathrm{tot\,r}}\rangle - \langle S_z\rangle}{k \sin\vartheta}, \tag{8.116}$$

which, together with Eq. (8.113), immediately yields Eq. (8.112).

In a similar manner, one can verify that the angular shifts obey a constraint following from the *conservation of the linear momentum* [51, 214]. The balance of the tangential components of the beam momenta yields

$$-Q^{\mathrm{r}}\langle P_X^{\mathrm{r}}\rangle \cos\vartheta + Q^{\mathrm{t}}\langle P_X^{\mathrm{t}}\rangle \cos\vartheta' = 0, \quad Q^{\mathrm{r}}\langle P_y^{\mathrm{r}}\rangle + Q^{\mathrm{t}}\langle P_y^{\mathrm{t}}\rangle = 0, \tag{8.117}$$

where the x-component of the momentum is $\langle P_x^a\rangle = \langle P_X^a\rangle \cos\vartheta^a + \langle P_Z^a\rangle \sin\vartheta^a$ and the balance of $\langle P_Z^a\rangle \sin\vartheta^a$ is guaranteed by Snell's law (8.94). Using Eqs. (8.94)–(8.100) and (8.107), one can readily show that Eqs. (8.109) and (8.110) fulfil Eqs. (8.117) identically.

Thus, the 'microscopic' plane-wave interference resulting in the shifts (8.109)–(8.112) and 'macroscopic' arguments of the conservation laws (8.115)–(8.117) are in complete agreement with each other. This demonstrates once again that 'geometric' and 'dynamical' arguments in the description of the SOI phenomena are intimately related.

The shifts from OAM

Let us consider now the effect of the IOAM in the case of the incident vortex beam with $\ell \neq 0$. Similar to the Fresnel boundary conditions (8.96) and (8.97) that determine the changes in the SAM at the boundary, Eqs. (8.113), one can determine the boundary conditions for the change of the IOAM at the interface. The IOAM of the incident beam is $\langle \mathbf{L}\rangle = \ell \mathbf{u}_Z$. The changes of the IOAM in the secondary beams arise from *elliptical deformations* of the beams, γ^a, Eq. (8.100). Akin to changes of the intrinsic AM upon an oblique cross-section, Eq. (8.91), the IOAM of the beams can be written as [48, 51, 215]

$$\langle \mathbf{L}^{\text{int}\,a}\rangle = \frac{\ell}{2}\left(\gamma^a + \frac{1}{\gamma^a}\right)\mathbf{u}_Z^a,\ \text{i.e.,}\ \langle \mathbf{L}^{\text{int}\,\mathrm{r}}\rangle = -\ell\mathbf{u}_Z^{\mathrm{r}},\ \langle \mathbf{L}^{\text{int}\,\mathrm{t}}\rangle = \frac{\ell}{2}\left(\frac{\cos\vartheta}{\cos\vartheta'} + \frac{\cos\vartheta'}{\cos\vartheta}\right)\mathbf{u}_Z^{\mathrm{t}}. \tag{8.118}$$

As we already know, changes in the z-components of the IOAM must be compensated by transverse shifts. It turns out that the IOAM-generated transverse shifts can be obtained separately from the AM conservation between incident and reflected or incident and transmitted beams. As a result, these shifts read [cf. Eq. (8.116)] [48, 51, 215]

$$\langle Y^a\rangle = \frac{\langle L_z^{\text{int}\,a}\rangle - \langle L_z^{\text{int}}\rangle}{k\sin\vartheta},\ \text{i.e.,}\ \langle Y^{\mathrm{r}}\rangle = 0,\ \langle Y^{\mathrm{t}}\rangle = \frac{\ell}{2k}\tan\vartheta(1-n^{-2}), \tag{8.119}$$

and only the transmitted beam experiences this shift.

The ℓ-dependent transverse shift (8.119) represents an example of the *orbital Hall effect* of light caused by the orbit–orbit interaction between IOAM and EOAM. Note that the derivation in Eqs. (8.118) and (8.119) is independent of polarization and, hence, can be regarded as the shift of a *scalar* wave beam at a planar interface. The dependence on the angle of incidence ϑ is of purely geometrical origin (cf. Section 8.4.1) and does not imply any spatial dispersion of the interface. Yet, the $\langle Y\rangle \propto k^{-1}\tan\vartheta$ dependence in Eq. (8.119) is evidence that this shift is similar to the shift (8.92) that originates from the tilt of the beams, but not of the same type as the Imbert–Fedorov shift (8.110) and (8.112), which behaves as $\langle Y\rangle \propto k^{-1}\cot\vartheta$.

In addition to the 'scalar' ℓ-dependent Hall effect (8.119), there are also ℓ-dependent phenomena caused by the polarization properties of the vortex beams interacting with an interface. Specifically, the angular deviations of the Goos–Hänchen and Imbert–Fedorov effects become coupled to the complex vortex structure and induce additional shifts that depend on both polarization and vortex charge. To show this, we note that the angular shifts in Eqs. (8.109) and (8.110) can be considered as *imaginary* spatial shifts in the Gaussian envelopes of the beams, so that the resulting complex shifts can be written as [111, 167]

$$\delta X^a = \langle X^a \rangle - \mathrm{i}\gamma^{a^{-2}} \frac{w_0^2}{2} \langle P_X^a \rangle, \quad \delta Y^a = \langle Y^a \rangle - \mathrm{i}\frac{w_0^2}{2} \langle P_y^a \rangle. \tag{8.120}$$

At the same time, an imaginary shift in the vortex distribution induces the *orthogonal real shift*, proportional to the vortex charge. This is clearly seen if we write the vortex structure as $(\gamma^a X^a + \mathrm{i}(\operatorname{sgn}\ell)y)^{|\ell|}$. Obviously, here the imaginary shift along X^a is equivalent to the real shift along y, which is magnified by the power ℓ, and vice versa. After straightforward calculations, we obtain that the *angular* Goos–Hänchen and Imbert–Fedorov shifts induce the *spatial* ℓ-dependent shifts in the vortex beams (Fig. 8.17) [51, 54]:

$$\langle X^a \rangle = \ell\gamma^{a^{-1}} \frac{w_0^2}{2} \langle P_y^a \rangle|_{\ell=0}, \quad \langle Y^a \rangle = -\ell\gamma^{a^{-1}} \frac{w_0^2}{2} \langle P_X^a \rangle|_{\ell=0}. \tag{8.121}$$

In turn, the angular shifts are also modified by the vortex structure, but these changes depend on the particular shape of the beam envelope [216]. In the case of the Laguerre–Gaussian incident beam, the vortex magnifies the angular shifts by a factor of $(1 + |\ell|)$, so that the additional contribution is [51, 54]

$$\langle P_X^a \rangle = |\ell| \langle P_X^a \rangle|_{\ell=0}, \quad \langle P_y^a \rangle = |\ell| \langle P_y^a \rangle|_{\ell=0}. \tag{8.122}$$

The ℓ-dependent shifts (8.121) and (8.122) were measured experimentally in [49, 54]. They rather represent an interplay between the polarization-dependent Goos–Hänchen or Imbert–Fedorov shifts and the complex vortex structure. These shifts do not affect the AM and momentum balance, because the original angular shifts already satisfied Eq. (8.117).

In total, the spatial and angular shifts for polarized vortex beams reflected and refracted at a dielectric interface are given by the sum of *three contributions*: (i) Goos–Hänchen and Imbert–Fedorov shifts (8.109)–(8.112), (ii) the 'scalar' orbital Hall effect (8.119) related to the tilt of the beams, and (iii) additional shifts (8.121) and (8.122) due to the coupling between the polarization-dependent angular shifts and the complex vortex structure. The Imbert–Fedorov shifts (i) and (iii) are exemplified in Fig. 8.17 displaying the intensity distributions and centroids of the Laguerre–Gaussian beams reflected from the air–glass interface. Note that, while the spin-dependent shifts manifest themselves in a perfect translations of paraxial beams, the ℓ-dependent shifts are accompanied by significant *deformations* of the beam profiles [217–219]. It is worth noticing that there are no ℓ-dependent shifts in the case of total internal reflection: the z-component of the IOAM is conserved due to the effective flip of ℓ in the reflected beam, Eq. (8.118), and there are no angular shifts to be coupled with the vortex, Eqs. (8.111) and (8.112).

8.4.3 Geometrodynamics of light in a gradient-index medium

In this final section we analyse the propagation of locally paraxial light beams in inhomogeneous gradient-index media. This problem is a complete analogue of the propagation of quantum particles in an external potential field. In this spirit, a space-variant refractive index of the medium, $n(\mathbf{r})$, is a counterpart of the external scalar potential, whereas its gradient, $\nabla n(\mathbf{r})$, is an analogue of the external field. Propagation of light in gradient-index media is described by canonical *equations of motion* within the geometrical optics approach [220].

Polarization effects

Geometrical optics of inhomogeneous media represents a sort of semiclassical approximation with respect to the small parameter of the wavelength, $\lambda/2\pi = k^{-1}$. The beam is assumed to be paraxial with respect to its local direction of propagation (its central wave vector), which can vary significantly but *smoothly* along the trajectory of propagation. Below we basically follow the approach developed in [39, 40, 97, 120, 221, 222]. We start from the Maxwell equations (8.3) in a non-magnetic gradient-index medium:

$$\hat{\mathbf{p}} \times (\hat{\mathbf{p}} \times \mathbf{E}) + \omega^2 n^2(\hat{\mathbf{r}})\mathbf{E} = 0, \tag{8.123}$$

where $\hat{\mathbf{p}} = -\mathrm{i}\partial_{\mathbf{r}}$ and $\hat{\mathbf{r}} = \mathbf{r}$ are the momentum and position operators in the coordinate representation. In the momentum representation, using laboratory circular basis and Eqs. (8.3a) and (8.13), Eq. (8.123) can be written as

$$[\hat{p}^2 \hat{\Pi}(\hat{\mathbf{p}}) - \omega^2 n^2(\hat{\mathbf{r}})]|\tilde{\mathbf{E}})_C = 0, \tag{8.124}$$

where $\hat{\mathbf{p}} = \mathbf{k}$ and $\hat{\mathbf{r}} = i\partial_{\mathbf{k}}$, whereas the *spherical-projection* matrix operator $k^2\hat{\Pi}(\mathbf{k}) = k^2\hat{U}_{\mathrm{H}}(\mathbf{k})\mathsf{P}_z\hat{U}_{\mathrm{H}}^{\dagger}(\mathbf{k})$ describes the double vector product $-\mathbf{k} \times (\mathbf{k} \times \tilde{\mathbf{E}})$. Equation (8.124) resembles a relativistic wave equation with $\hat{p}^2\hat{\Pi}(\hat{\mathbf{p}})$ being a non-diagonal kinetic-energy operator and $\omega^2 n^2(\hat{\mathbf{r}})$ being the potential-energy term. As it was shown above, the operator $\hat{\Pi}$ is diagonalized via transition to the *helicity basis*, $|\tilde{\mathbf{E}})_{\mathrm{C}} = \hat{U}_{\mathrm{H}}(\mathbf{k})|\tilde{\mathbf{E}})_{\mathrm{H}}$, Eq. (8.14), which yields the kinetic term $\hat{p}^2\hat{\mathsf{P}}_z = \hat{p}^2\mathrm{diag}(1, 1, 0)$. At the same time, the momentum-dependent transformation $\hat{U}_H(\mathbf{k})$ induces non-trivial changes in the coordinate-dependent term $n^2(\hat{\mathbf{r}})$:

$$\hat{U}_{\mathrm{H}}^{\dagger}(\hat{\mathbf{p}})n^2(\hat{\mathbf{r}})\hat{U}_{\mathrm{H}}(\hat{\mathbf{p}}) = n^2(\hat{\mathbf{r}} - \hat{\mathbf{A}}), \tag{8.125}$$

where $\hat{\mathbf{A}}(\hat{\mathbf{p}}) = -\mathrm{i}\hat{U}^{\dagger}\partial_{\mathbf{p}}\hat{U}$ is the induced pure gauge potential, cf. Eqs. (8.18a) and (8.32). Using the smallness of the wavelength as compared with the scale of the variations of $n^2(\mathbf{r})$, one can write

$$n^2(\hat{\mathbf{r}} - \hat{\mathbf{A}}) \simeq n^2(\hat{\mathbf{r}}) - \hat{\mathbf{A}} \cdot \partial_{\mathbf{r}} n^2(\hat{\mathbf{r}}). \tag{8.126}$$

After that, owing to the transversality constraint (8.4), $\tilde{E}_k = 0$, the whole operator (8.124) can be projected onto the two-dimensional transversal subspace, which results in

$$[\hat{p}^2 - \omega^2 n^2(\hat{\mathbf{r}}) + \hat{\sigma}_3 \mathbf{A}_{\mathrm{B}}(\hat{\mathbf{p}}) \cdot \partial_{\mathbf{r}}\omega^2 n^2(\hat{\mathbf{r}})]|\tilde{\mathbf{E}})_{\perp\mathrm{H}} = 0, \tag{8.127}$$

where $|\tilde{\mathbf{E}})_{\perp\mathrm{H}} = (\tilde{E}^+, \tilde{E}^-)^{\mathrm{T}}$ and we used the fact that the transverse part of $\hat{\mathbf{A}}$ yields the Berry connection $\hat{\sigma}_3\mathbf{A}_{\mathrm{B}}$, Eq. (8.18a), where $\hat{\sigma}_3$ appears as the transverse part of the helicity $\hat{\sigma}$.

In the zero-wavelength limit $k^{-1} \to 0$, $\mathbf{A}_{\rm B} \to 0$, and operator $\hat{p}^2 - \omega^2 n^2(\hat{\mathbf{r}})$ in Eq. (8.127) corresponds to the geometrical optics of a scalar wave equation. The two polarization (helicity) states are *degenerated* in this approximation [220]. The matrix Berry-connection term in Eq. (8.127) lifts this degeneracy and describes the *SOI of light* [38–40, 43, 97, 120, 221, 222]. It mixes *polarization* and *spatial* degrees of freedom, which are described by *matrix* and *differential* operators, respectively. It is possible to represent the SOI term in the form of a Coriolis-type coupling (8.21) between the SAM carried by light and local rotations of the helicity coordinate frame attached to the beam trajectory [97]. Equation (8.127) can be considered as a Schrödinger equation for a two-level system, where the polarization degrees of freedom are described by the Pauli matrices, $\hat{\vec{\sigma}}$ [119–121]. Using a Taylor expansion similar to that in Eq. (8.126), one can represent Eq. (8.127) in terms of the modified coordinate operator $\hat{r} = \hat{\mathbf{r}} - \hat{\sigma}_3 \mathbf{A}_B$, Eq. (8.30):

$$\hat{\mathcal{H}}(\hat{\mathbf{p}}, \hat{r})|\tilde{\mathbf{E}})_{\perp \rm H} \equiv [\hat{p} - \omega n(\hat{r})]|\tilde{\mathbf{E}})_{\perp \rm H} = 0, \tag{8.128}$$

where we formally divided the operator $[\hat{p}^2 - \omega^2 n^2(\hat{r})]$ by $\hat{p} + \omega n(\hat{r})$ for future convenience.[8] The operator $\hat{\mathcal{H}}(\hat{\mathbf{p}}, \hat{r})$ represents an effective wave Hamiltonian, whereas the operators of orbital and polarization degrees of freedom satisfy the following commutation relations [cf. Eq. (8.28a)]:

$$[\hat{r}_i, \hat{p}_j] = \mathrm{i}\delta_{ij}, \quad [\hat{r}_i, \hat{r}_j] = -\mathrm{i}\varepsilon_{ijl}\hat{\sigma}_3 F_{B_l}, \quad [\hat{\sigma}_i, \hat{r}_j] = -2\mathrm{i}\varepsilon_{3il} A_{B_j}\hat{\sigma}_l, \tag{8.129}$$

where $\mathbf{F}_B(\hat{\mathbf{p}})$ is the Berry curvature (8.20), and we used $[\hat{\sigma}_i, \hat{\sigma}_j] = 2\mathrm{i}\varepsilon_{ijl}\hat{\sigma}_l$.

The Hamiltonian (8.128) and commutation relations (8.129) completely characterize the problem. Equations of motion for both the orbital and polarization degrees of freedom can be obtained from the Heisenberg picture of evolution, i.e., the canonical equations

$$\frac{\mathrm{d}\hat{\mathbf{p}}}{\mathrm{d}\tau} = -\mathrm{i}[\hat{\mathbf{p}}, \hat{\mathcal{H}}], \quad \frac{\mathrm{d}\hat{r}}{\mathrm{d}\tau} = -\mathrm{i}[\hat{r}, \hat{\mathcal{H}}], \quad \frac{\mathrm{d}\hat{\vec{\sigma}}}{\mathrm{d}\tau} = -\mathrm{i}[\hat{\vec{\sigma}}, \hat{\mathcal{H}}], \tag{8.130}$$

where parameter τ is determined below. After evaluating the commutators for Eqs. (8.128) and (8.129), and replacing the operators $\hat{\mathbf{p}}$, $\hat{r}$, $\hat{\vec{\sigma}}$ with their expectation values $\langle \mathbf{P} \rangle$, $\langle \mathbf{R} \rangle$, $\vec{\mathfrak{S}}$ according to the Ehrenfest theorem, we arrive at [38–40, 120, 221]

$$\frac{\mathrm{d}\langle \mathbf{P} \rangle}{\mathrm{d}\tau} = \partial_{\mathbf{r}} \omega n(\langle \mathbf{R} \rangle), \quad \frac{\mathrm{d}\langle \mathbf{R} \rangle}{\mathrm{d}\tau} = \frac{\langle \mathbf{P} \rangle}{\langle P \rangle} - \mathfrak{S}_3 \mathbf{F}_{\rm B}(\langle \mathbf{P} \rangle) \times \frac{\mathrm{d}\langle \mathbf{P} \rangle}{\mathrm{d}\tau}, \tag{8.131}$$

$$\frac{\mathrm{d}\vec{\mathfrak{S}}}{\mathrm{d}\tau} = 2\vec{\alpha} \times \vec{\mathfrak{S}}, \quad \vec{\alpha} = \left[\mathbf{A}_{\rm B}(\langle \mathbf{P} \rangle) \cdot \frac{\mathrm{d}\langle \mathbf{P} \rangle}{\mathrm{d}\tau} \right] \vec{\mathfrak{S}}_3, \tag{8.132}$$

where $\langle P \rangle = |\langle \mathbf{P} \rangle| = k = n\omega$ for paraxial light, $\mathbf{F}_{\rm B}(\langle \mathbf{P} \rangle) = \langle \mathbf{P} \rangle / \langle P \rangle^3$, and $\vec{\mathfrak{S}}_3$ is the unit vector pointing to the north pole of the Poincaré sphere. Equations (8.131) and (8.132) describe the evolution of the beam parameters along a curved *trajectory of propagation* in

[8] This can be done because we consider forward-propagating paraxial waves corresponding to the root $k = \omega n$ of the dispersion relation and neglect backward waves corresponding to the root $k = -\omega n$ [220].

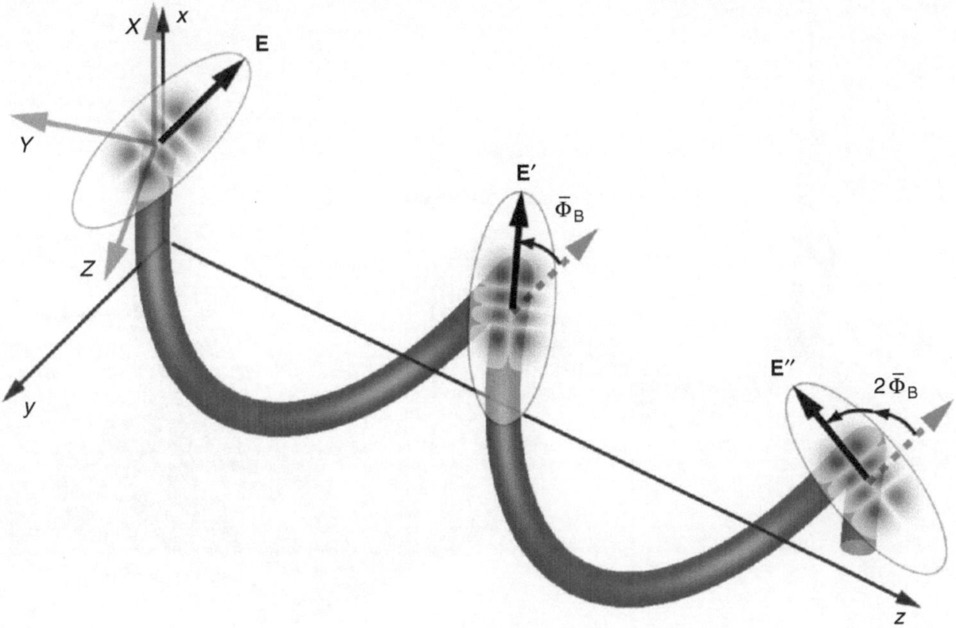

Figure 8.18 Evolution of the polarization of light along a helical trajectory in a gradient-index medium. The polarization ellipse obeys the parallel transport law, cf. Fig. 8.2. This implies rotation with respect to the global coordinates by the Berry-phase angle $\bar{\Phi}_B = 2\pi(1 - \cos\vartheta)$ after each coil of the trajectory (ϑ is the angle between the mean momentum $\langle \mathbf{P} \rangle$ and the z-axis), Eq. (8.136). For higher-order beams (e.g., Hermite–Gaussian), the transverse intensity pattern obeys similar parallel transport in the diffractionless approximation, Eq. (8.143).

an inhomogeneous medium [220]. In this manner, the parameter τ coincides with the length of the trajectory: $|\mathrm{d}\langle \mathbf{R} \rangle / \mathrm{d}\tau| \simeq 1$, whereas evolution of the polarization is represented by the precession of the Stokes vector on the Poincaré sphere. The orbital Eqs. (8.131) and polarization Eq. (8.132) are coupled through the Berry-connection and curvature terms, which describe the mutual influence of the *polarization* and *trajectory*, i.e., the *SOI of light*, Figs. 8.18 and 8.19.

Prior to the analysis of these effects, we note that polarization in Eqs. (8.131) and (8.132) is given with respect to a central-wave helicity basis which is associated with the local beam coordinate frame (X, Y, Z) undergoing a non-trivial three-dimensional evolution along the ray trajectory (see Fig. 8.18). To define such a frame, one can use the spherical basic vectors (8.7) attached to the mean momentum of the beam. Namely, if $\langle \mathbf{P} \rangle$ is given by spherical coordinates (ϑ, ψ, P) with respect to the z-axis (see Fig. 8.20 below), then $\mathbf{u}_X = \mathbf{u}_\vartheta$, $\mathbf{u}_Y = \mathbf{u}_\psi$, $\mathbf{u}_Z = \mathbf{u}_P$, where

$$\mathbf{u}_\vartheta(\langle \mathbf{P} \rangle) = \mathbf{u}_\psi \times \mathbf{K}, \quad \mathbf{u}_\psi(\langle \mathbf{P} \rangle) = \frac{\mathbf{u}_z \times \mathbf{K}}{|\mathbf{u}_z \times \mathbf{K}|}, \quad \mathbf{u}_P(\langle \mathbf{P} \rangle) = \frac{\langle \mathbf{P} \rangle}{\langle P \rangle} \equiv \mathbf{K}. \tag{8.133}$$

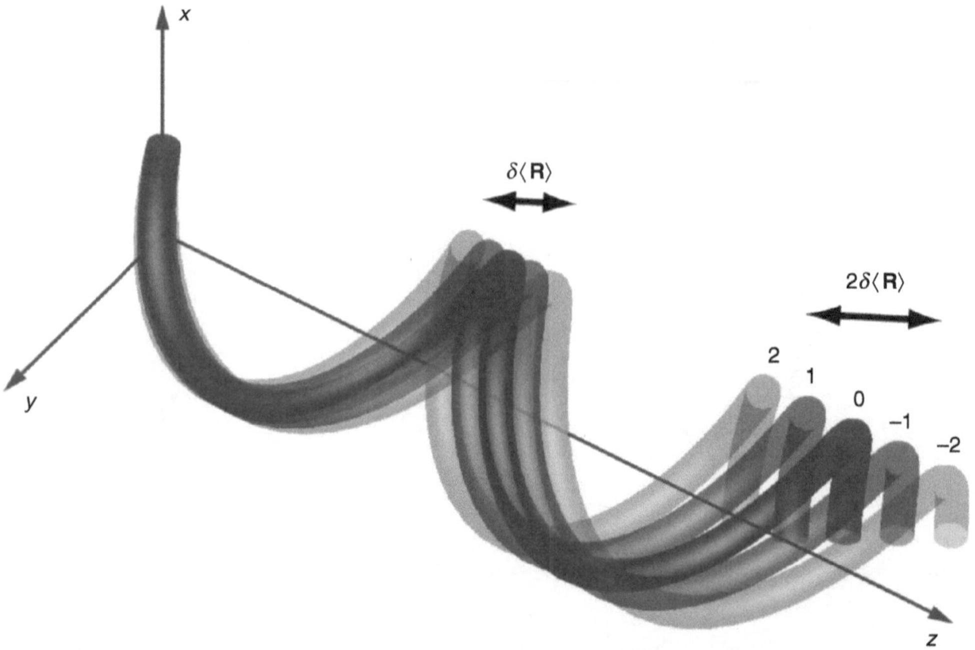

Figure 8.19 Spin and orbital Hall effects of light in a gradient-index medium. The zero-approximation helical trajectory, shown in Fig. 8.18, experiences transverse deflection proportional to the intrinsic AM carried by the paraxial beam, Eqs. (8.131), (8.138), and (8.144). The circularly polarized vortex beams are marked by the values of $(\ell + \sigma)$.

Accordingly the helicity basis of circular polarizations, Eq. (8.9), attached to $\langle \mathbf{P} \rangle$ is

$$\mathbf{u}^{\sigma}(\langle \mathbf{P} \rangle) = \frac{\mathbf{u}_{\vartheta} + i\sigma \mathbf{u}_{\psi}}{\sqrt{2}} e^{im\sigma\psi}, \tag{8.134}$$

and, as before, we usually assume $m = 1$. Similar to the discussions in Section 8.2.1, the evolution of the beam polarization can be mapped onto the unit **K**-sphere of the beam directions, cf. Fig. 8.2 and Fig. 8.20 below. Thus, in this problem, two levels of geometric Berry phase appear. First, on the 'macro'-level, the evolution of the *central* wave vector $\langle \mathbf{P} \rangle$ in the beam results in the accumulation of the Berry phase for the whole beam *along* the ray (Fig. 8.18). This effect provides the parallel transport of the beam polarization along the ray trajectory and is described by the precession of the Stokes vector, Eq. (8.132). Second, on the 'micro'-level, similar to the discussion on pp. 213–14, interference of non-central plane waves *across* the beam involves geometric phases related to the transverse distribution of the wave vectors. Akin to the Imbert–Fedorov shift, here the transverse Berry-phase gradient brings about the spin-dependent transverse shift of the beam described by the Berry-curvature term in Eq. (8.131), (Fig. 8.19).

Evolution of polarization along the trajectory, Eq. (8.132), resembles precession of the spin on the Bloch sphere, but in this case the *Stokes vector* $\vec{\mathfrak{S}}$ represents a 'pseudo-spin', as it was noted on p.188. Precession of the Stokes vector about the vertical axis on the Poincaré sphere implies rotation of the polarization ellipse about the direction of propagation of light with the helicity being conserved, $d\mathfrak{S}_3/d\tau = 0$.[9] This rotation, originally described by Rytov and Vladimirskii [223, 224], represents the *parallel transport* of the field along the curvilinear trajectory, (Fig. 8.18). It can equally be represented on the **K**-sphere of directions, Fig. 8.2, and described by the evolution of the *Jones vector*. Indeed, Eq. (8.132) is equivalent to [40, 97, 120]

$$\frac{d}{d\tau}|\mathbf{e}\rangle_{\perp B} = -i\hat{\sigma}_3\left[\mathbf{A}_B(\langle\mathbf{P}\rangle)\cdot\frac{d\langle\mathbf{P}\rangle}{d\tau}\right]|\mathbf{e}\rangle_{\perp B}, \tag{8.135}$$

where the normalized Jones vector $|\mathbf{e}\rangle_{\perp B} = (e^+, e^-)^T$ is written in the basis (8.134) of circular polarizations in the beam frame. Integration of this equation immediately yields the *Berry phase* (8.17):

$$|\mathbf{e}(\tau)\rangle_{\perp B} = \exp[-i\hat{\sigma}_3\bar{\Phi}_B(C)]|\mathbf{e}(0)\rangle_{\perp B}, \quad \bar{\Phi}_B(C) = \int_C \mathbf{A}_B(\langle\mathbf{P}\rangle)\cdot d\langle\mathbf{P}\rangle. \tag{8.136}$$

The Berry phase is associated with the evolution of the mean momentum (central plane wave) in the beam and is accumulated along the beam trajectory. In particular, propagation of light along a helical trajectory in real space (Fig. 8.18) corresponds to the circular contour C on the **K**-sphere of directions in momentum space (Fig. 8.20). The Berry phase depends only on the geometry of the trajectory, and nontrivial contours C imply non-planar trajectories with *torsion*. In the Frenet frame attached to the normal and binormal to the trajectory, the Berry phase is given by the integral of the torsion along the ray [81, 85, 91, 92, 97, 220, 223, 224]. One can show that the 'absolute' (i.e., frame-independent) complex polarization vector, $\mathbf{e}_\perp = e^+\mathbf{u}^+ + e^-\mathbf{u}^-$, obeys the parallel transport equation [92, 223, 224]:

$$\frac{d\mathbf{e}_\perp}{d\tau} = -\left(\mathbf{e}_\perp\cdot\frac{d\mathbf{K}}{d\tau}\right)\mathbf{K}. \tag{8.137}$$

This means that the electric field remains *locally non-rotating* about the ray direction **K**, but *globally* it can turn after a cycle of evolution due to the *nonintegrability* of the Berry phase (Fig. 8.18). Equations (8.132) and (8.135)–(8.137) are different representations of the same polarization evolution, where only Eq. (8.137) is coordinate-independent. The trajectory-dependent polarization evolution related to the parallel transport and Berry phase was examined theoretically and experimentally in numerous studies (see, e.g., original papers [74, 81–88] and reviews [92–97]). Being of a purely geometrical nature, evolution of polarization is rather robust against the details of the system and instead of a real

[9] The helicity is precisely conserved for plane waves in free space, and the Berry phase (8.17) and (8.19) is exact. At the same time, the helicity of locally-paraxial wave beam is an *adiabatic invariant* of smooth variations of the propagation direction. The second-order terms, $\propto k^{-2}$, describe nonadiabatic transitions between different helicity states and polarization evolution behind the first-order Berry phase (8.136) [81, 85, 225–227].

gradient-index medium the non-planar trajectory is usually produced in experiments using a coiled optical fiber [81, 83, 85].

Thus, polarization of the beam crucially depends on its trajectory, and the Berry phase (8.136) is a large effect of zeroth order in the wavelength. At the same time, there is a back reaction of the spin onto the trajectory of propagation, described by the last term in Eq. (8.131). This term resembles a Lorentz (or Coriolis) *transverse force* from the Berry-curvature monopole (8.20) in the momentum space. It brings about a spin-dependent deflection of the trajectory of the wavelength order k^{-1}, i.e., the *spin Hall effect of light* in a gradient-index medium (Fig. 8.19). This deflection is orthogonal to both the direction of propagation $\langle \mathbf{P} \rangle$ and the gradient of the refractive index, $\mathrm{d}\langle \mathbf{P} \rangle / \mathrm{d}\tau = \partial_{\mathbf{r}} \omega n$, which plays the role of the external field. Akin to the Berry phase, the spin Hall effect shift is a non-local phenomenon that can grow without limit with the trajectory length and can be expressed as a contour integral in the momentum space:

$$\delta \langle \mathbf{R} \rangle = -\bar{\sigma} \int_C \mathbf{F}_{\mathrm{B}}(\langle \mathbf{P} \rangle) \times \mathrm{d}\langle \mathbf{P} \rangle, \tag{8.138}$$

where we used $\mathfrak{S}_3 = \bar{\sigma}$, Eq. (8.40). The spin-dependent transport of light in a gradient-index medium was first predicted by Zel'dovich and co-authors in 1992 [32, 38], and then its manifestation in optical fibers was measured. Later, Eqs. (8.131) were derived rigorously in the context of the Berry-phase effects [39, 40, 97, 120, 221, 222] and the spin-dependent shift (8.138) of a light beam propagating along a curvilinear trajectory was observed recently [43]. In contrast to the Berry phase (8.136), the shift (8.138) can be observed for planar (e.g., circular) trajectories [39, 221]. In terms of the trajectory geometry, the shift is equal to k^{-1} times the integral of the trajectory *curvature* along the ray [97]. Locally, it is directed along the binormal of the ray, i.e., orthogonal to the local plane of propagation. Importantly, Eqs. (8.131) can be regarded as fundamental geometrodynamical equations of motion of quantum spinning particles in external fields [4, 76, 228, 229]. In particular, the optical spin Hall effect is entirely analogous to the intrinsic spin Hall effect of electrons in an external electric field (see, e.g., [4, 5, 12–14]). While in condensed-matter or high-energy physics it is practically impossible to measure the trajectory of a spinning particle with subwavelength accuracy, the optical spin Hall effect provides a unique opportunity to directly measure, for the first time, the deflection from the geometric Berry-phase force [43].

As other SOI effects, the spin Hall term in the equations of motion is intimately related to the *conservation of the AM*. Indeed, in a spherically symmetric medium, $n(\mathbf{r}) = n(r)$, the ray equations possess the following integral of motion [40]:

$$\langle \mathbf{J} \rangle = \langle \mathbf{S} \rangle + \langle \mathbf{L} \rangle = \bar{\sigma} \mathbf{K} + \langle \mathbf{R} \rangle \times \langle \mathbf{P} \rangle = \text{const.} \tag{8.139}$$

Evidently, this is the total AM consisting of the SAM and EOAM. Variations in the direction of SAM $\bar{\sigma}\mathbf{K}$ caused by the wave refraction are compensated by the transverse shift of the ray (8.138), $\delta\langle \mathbf{R} \rangle$, which changes the EOAM $\langle \mathbf{R} \rangle \times \langle \mathbf{P} \rangle$, cf. Eq. (8.115). In a medium with cylindrical symmetry with respect to the z-axis, the conserved quantity is $\langle J_z \rangle$. The conservation of the total AM, Eq. (8.139), links the spin Hall effect of light to the transverse *Imbert–Fedorov shift*. To show this explicitly, let us consider the paraxial beam refraction at a plane dielectric interface with a law contrast of the refractive index, δn, $|\delta n| \ll 1$ [38]. The reflected beam can be neglected in the linear approximation in δn, whereas the Imbert–Fedorov shift of the transmitted beam, Eq. (8.110), becomes $\langle Y \rangle \simeq \frac{\bar{\sigma}}{k} \frac{\delta n}{n} \tan \vartheta$. Taking into account that in the interface geometry the z-axis is directed along the gradient of the refractive index, $\partial_{\mathbf{r}} n$, and the y-axis is directed orthogonal to both $\partial_{\mathbf{r}} n$ and the direction of propagation $\langle \mathbf{P} \rangle$, this shift can written as $\delta \langle \mathbf{R} \rangle \simeq \mathfrak{S}_3 \frac{\partial_{\mathbf{r}} n}{n} \times \frac{\langle \mathbf{P} \rangle}{\langle P \rangle^2} \frac{\delta z}{\cos \vartheta} = \mathfrak{S}_3 \left[\partial_{\mathbf{r}} (\omega n) \times \frac{\langle \mathbf{P} \rangle}{\langle P \rangle^3} \right] \delta Z$. Evidently, taking the differential limit $\delta \langle \mathbf{R} \rangle / \delta Z \to \mathrm{d} \langle \mathbf{R} \rangle / \mathrm{d}\tau$, we obtain the spin-dependent term in Eq. (8.131). Thus, the Imbert–Fedorov shift and spin-dependent transport in a gradient-index medium represent two different limits of the same spin Hall effect phenomenon related to the AM conservation and transverse gradient of the Berry phase. Note that the spin Hall effect along a smooth trajectory can be regarded as a continuous sequence of the infinitesimal spatial *shifts*, which results in the local *deflection* of the beam, i.e., a change in its direction of propagation. Accordingly, the spin-dependent term in Eq. (8.131) is also called the *anomalous velocity* (non-collinear with the momentum $\langle \mathbf{P} \rangle$).

To summarize, Eqs. (8.131) and (8.132) describe the mutual influence and coupling between the internal (polarization) and external (trajectory) degrees of freedom. These can also be regarded as 'fast' and 'slow' degrees of freedom within a sort of adiabatic approximation [230, 231]. The SOI coupling terms are proportional to the Berry connection and curvature. Geometrical (Berry phase) and dynamical (AM) aspects are closely entangled in the SOI of light in a gradient-index medium and together form a *geometrodynamics* of spinning light, similar to the evolution of quantum spinning particles in external fields. Remarkably, in Lagrangian formalism, both the Berry-phase effect on polarization and spin Hall effect in the trajectory of light are described by a single SOI Lagrangian: $\mathcal{L}_{\mathrm{SOI}} = -\hat{\sigma}_3 \mathbf{A}_{\mathrm{B}}(\langle \mathbf{P} \rangle) \cdot \mathrm{d}\langle \mathbf{P} \rangle / \mathrm{d}\tau$ [43, 97].

Evolution of higher-order beams with IOAM

We have considered coupling between the polarization and the trajectory of propagation. In a similar manner, one can consider propagation of higher-order vortex beams carrying IOAM in a gradient-index medium. It turns out that, within the *geometrical-optics* consideration, the effect of the IOAM is entirely analogous to what has been discussed for the SAM.

Neglecting the diffraction phenomena in a gradient-index medium, we assume that the beam propagates *locally maintaining its internal structure* in the beam coordinate frame (X, Y, Z). In a manner similar to Eq. (8.99), we introduce deflections of constituent wave vectors in the beam with respect to the mean momentum:

$$\mathbf{k} = \langle \mathbf{P} \rangle + \boldsymbol{\chi}, \quad \boldsymbol{\chi} = k_\vartheta \mathbf{u}_\vartheta + k_\psi \mathbf{u}_\psi, \quad |\boldsymbol{\chi}| \ll k, \tag{8.140}$$

where we used spherical basis (8.133). Using the helicity basis (8.134), the Fourier representation of circularly polarized Laguerre–Gaussian beams, locally paraxial with respect to $\langle \mathbf{P} \rangle$, can be written as [cf. Eq. (8.37)]

$$\tilde{\mathbf{E}}^{\sigma}_{\ell\perp}(\boldsymbol{\chi}, \langle \mathbf{P} \rangle) = \mathbf{u}^{\sigma}(\langle \mathbf{P} \rangle)\tilde{E}_{\ell}(\boldsymbol{\chi}) \propto (\mathbf{u}_{\vartheta} + i\sigma \mathbf{u}_{\psi})e^{i\sigma\psi}(k_{\vartheta} + i(\operatorname{sgn}\ell)k_{\psi})^{|\ell|}e^{i\operatorname{sgn}\ell\psi}G(\tilde{\theta}), \tag{8.141}$$

where $G(\tilde{\theta}) \propto \exp[-(kw_0)^2\tilde{\theta}^2/4]$ is the Gaussian envelope function for the polar angle $\tilde{\theta}$ taken with respect to $\langle \mathbf{P} \rangle$: $\tilde{\theta}^2 \simeq (k_{\vartheta}^2 + k_{\psi}^2)/k^2$, whereas the factors $e^{i\sigma\psi}$ and $e^{i\operatorname{sgn}\ell\psi}$ compensate the azimuthal rotation of the spherical basis $(\mathbf{u}_{\vartheta}, \mathbf{u}_{\psi})$ to provide non-singular transition $\vartheta \to 0$. Thus, in the basic geometrical-optics approximation (8.141), the trajectory and local field properties are independent from each other, and the mean momentum $\langle \mathbf{P} \rangle$ appears as an independent parameter. This is similar to separation of 'slow', $\langle \mathbf{P} \rangle$, and 'fast', $\boldsymbol{\chi}$, variables in adiabatic approximation.

The modes (8.141) represent a *basic set* in the corresponding Hilbert space with the scalar product implying integration over the transverse momentum components

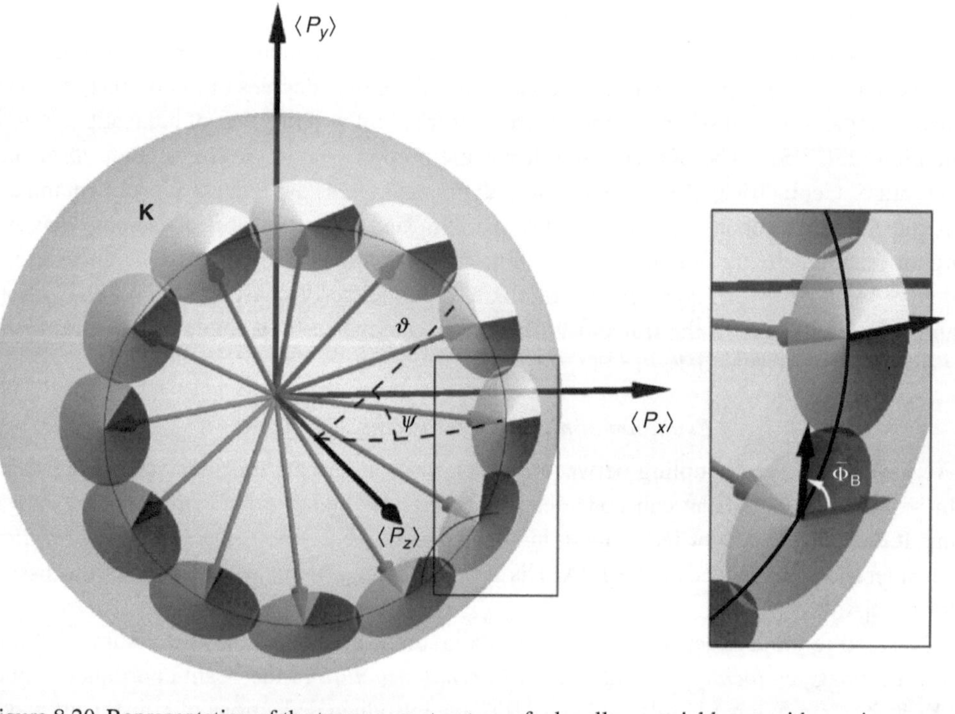

Figure 8.20 Representation of the transverse structure of a locally paraxial beam with varying mean momentum $\langle \mathbf{P} \rangle$ on the $\mathbf{K}$-sphere of directions. The circular contour of evolution C corresponds to a helical trajectory (Fig. 8.18). The optical vortex that obeys the parallel transport law and rotates by the angle numerically equal to the Berry phase $\bar{\Phi}_B(C)$ is shown. Correspondingly, the complex amplitudes of the field in the global coordinate frame acquire the Berry-phase factor $\exp[-i(\sigma + \ell)\bar{\Phi}_B]$, Eq. (8.143).

$\chi = (k_\vartheta, k_\psi)$: $\langle \tilde{\mathbf{E}}_\ell^\sigma | \tilde{\mathbf{E}}_{\ell'}^{\sigma'} \rangle_\perp = \delta_{\sigma\sigma'}\delta_{\ell\ell'}$.[10] Any paraxial field $\tilde{\mathbf{E}}_\perp(\chi, \langle\mathbf{P}\rangle)$ propagating along the ray can be represented as a superposition of the locally paraxial Laguerre–Gaussian modes (8.141): $\tilde{\mathbf{E}}_\perp(\chi, \langle\mathbf{P}\rangle) = \sum_{\sigma,\ell} \aleph_\ell^\sigma(\langle\mathbf{P}\rangle)\tilde{\mathbf{E}}_{\ell\perp}^\sigma(\chi, \langle\mathbf{P}\rangle)$. Here the amplitudes $\aleph_\ell^\sigma$ can be regarded as a *generalized Jones vector* of the field components in the basis of $\tilde{\mathbf{E}}_{\ell\perp}^\sigma$ [232], and their dependence on $\langle\mathbf{P}\rangle$ takes into account coupling between the trajectory and local fields. Since both the *polarization* $\mathbf{u}^\sigma \propto (\mathbf{u}_\vartheta + i\sigma\mathbf{u}_\psi)e^{i\sigma\psi}$ and *vortex structure* $(k_\vartheta + \mathrm{i}(\operatorname{sgn}\ell)k_\psi)^{|\ell|}\mathrm{e}^{\mathrm{i}(\operatorname{sgn}\ell)\psi}$ are *locally orthogonal to the mean momentum* $\langle\mathbf{P}\rangle$, they are naturally mapped onto the unit $\mathbf{K}$-sphere in the parameter $\langle\mathbf{P}\rangle$-space with the same *parallel transport* law (Fig. 8.20). Considering modes (8.141) as a $\langle\mathbf{P}\rangle$-dependent basis in both polarization and Hilbert space, we calculate the Berry connection similar to Eq. (8.18) [50]:

$$-\mathrm{i}\langle \tilde{\mathbf{E}}_\ell^\sigma | \partial_{\langle\mathbf{P}\rangle} | \tilde{\mathbf{E}}_\ell^\sigma \rangle_\perp = (\sigma + \ell)\mathbf{A}_\mathrm{B}. \tag{8.142}$$

Equation (8.142) shows that the geometrical aspects of the evolution of the vortex structure along the ray trajectory is similar to the evolution of the polarization state with the effective helicity ℓ. Hence, one can expect that the SOI effects for higher-order beams are multiplied by the factor of $(\sigma + \ell)$ instead of σ. In fact, the ℓ-dependent Berry-connection terms describe the *orbit-orbit interaction* between IOAM and EOAM.

First, evolution of amplitudes $\aleph_\ell^\sigma$ along the beam trajectory obeys the *Berry-phase* equation of motion analogous to Eqs. (8.135) and (8.136) [50]:

$$\frac{\mathrm{d}\aleph_\ell^\sigma}{\mathrm{d}\tau} = -\mathrm{i}(\sigma + \ell)\mathbf{A}_\mathrm{B}(\langle\mathbf{P}\rangle) \cdot \frac{\mathrm{d}\langle\mathbf{P}\rangle}{\mathrm{d}\tau}\aleph_\ell^\sigma, \quad \aleph_\ell^\sigma(\tau) = \exp[-\mathrm{i}(\sigma + \ell)\bar{\Phi}_\mathrm{B}(C)]\aleph_\ell^\sigma(0). \tag{8.143}$$

In combination with the vortex structure $\mathrm{e}^{\mathrm{i}\ell\tilde{\phi}}$ ($\tilde{\phi}$ is the azimuthal angle with respect to $\langle\mathbf{P}\rangle$) of the basic modes (8.141), the additional Berry phase $\mathrm{e}^{-\mathrm{i}\ell\bar{\Phi}_\mathrm{B}}$ results in a rotation of the transverse beam structure by an angle $\bar{\Phi}_\mathrm{B}$. Thus, in an arbitrary superposition of basic modes, both the polarization and *transverse spatial structure* of light are parallel-transported along the ray (Fig. 8.18). This phenomenon was described recently in several papers [50, 219, 233, 234] and experimentally observed almost 15 years prior to this via rotation of an image along a non-planar trajectory of the beam [89, 235, 236]. Comparing the transverse intensity-phase structure of the higher-order beam with polarization degrees of freedom, the Laguerre–Gaussian modes with circular intensity distributions and Hermite–Gaussian modes with intensity patterns oriented along the (X, Y)-axes represent counterparts of the circular and linear polarizations [232, 237].

Second, the trajectory equations of motion (8.131) for the higher-order beams take the form [50]

$$\frac{\mathrm{d}\langle\mathbf{P}\rangle}{\mathrm{d}\tau} = \partial_\mathbf{r}\omega n(\langle\mathbf{R}\rangle), \quad \frac{\mathrm{d}\langle\mathbf{R}\rangle}{\mathrm{d}\tau} = \frac{\langle\mathbf{P}\rangle}{\langle P\rangle} - (\bar{\sigma} + \bar{\ell})\mathbf{F}_\mathrm{B}(\langle\mathbf{P}\rangle) \times \frac{\mathrm{d}\langle\mathbf{P}\rangle}{\mathrm{d}\tau}, \tag{8.144}$$

[10] The radial quantum number p of the Laguerre–Gaussian modes is omitted here, because it does not make any contribution to the SOI effects [50].

where the mean SAM and IOAM helicities are defined as $\bar{\sigma} = \sum_{\sigma,\ell} \sigma |\aleph_\ell^\sigma|^2$ and $\bar{\ell} = \sum_{\sigma,\ell} \ell |\aleph_\ell^\sigma|^2$, assuming normalization $\sum_{\sigma,\ell} |\aleph_\ell^\sigma|^2 = 1$. Thus, the Coriolis-type force on the IOAM carried by the light causes the *orbital Hall effect of light*, i.e., a transverse ℓ-dependent shift of the beam trajectory in a gradient-index medium. This effect is obviously related to the *AM conservation* – akin to Eq. (8.139), in a spherically symmetric medium, Eqs. (8.144) possess an integral of motion that represents the sum of SAM, IOAM, and EOAM:

$$\langle \mathbf{J} \rangle = \langle \mathbf{S} \rangle + \langle \mathbf{L}^{\text{int}} \rangle + \langle \mathbf{L}^{\text{ext}} \rangle = (\bar{\sigma} + \bar{\ell})\mathbf{K} + \langle \mathbf{R} \rangle \times \langle \mathbf{P} \rangle = \text{const.} \tag{8.145}$$

It is also possible to derive the orbital Hall effect in a gradient-index medium from the IOAM-induced *Imbert–Fedorov shift*. From Eq. (8.119) the ℓ-dependent shift of a vortex beam refracted at a plane dielectric interface with law contrast δn is given by $\langle Y \rangle \simeq \frac{\ell}{k} \frac{\delta n}{n} \tan \vartheta$, which in a differential limit is equivalent to Eqs. (8.144) [50].

Thus, in the geometrical-optics evolution of higher-order beams along a curvilinear trajectory a number of IOAM-dependent effects appear which can be described by the simple rule $\bar{\sigma} \to (\bar{\sigma} + \bar{\ell})$ as compared to the case of zero-order Gaussian beams. In the above consideration we essentially used the assumption that the vortex structure does not vary upon propagation in the medium. In actual fact, it is affected by complex *diffraction processes* in the inhomogeneous medium and the vortex is *deformed* during evolution along the ray [219]. Moreover, the mean IOAM value $\bar{\ell}$ also varies due to diffraction processes. Nonetheless, the ℓ-dependent Berry phase is clearly observed via geometric parallel transport of the intensity structure in multiple reflections or optical-fiber propagation. Yet, the orbital Hall effect in gradient-index media can still be detected, because, unlike $|\ell|$-dependent diffraction effects, it produces opposite centroid shifts for beams with opposite ℓ. Observation of this effect is still an open experimental challenge.

8.5 Conclusion

We have considered the fundamental origins and basic manifestations of the *spin-orbit interaction of light* in locally isotropic media. This intrinsic phenomenon owes its origin to the transversality of electromagnetic waves and the angular momentum carried by light. The SOI effects cause conversions between different AM degrees of freedom, and can be observed as fine spin- and vortex-dependent variations in the field distribution, with the typical scale of a wavelength. Because of this smallness, the SOI effects were usually neglected in classical optics, but nowadays they have become highly important for all micro- and nano-optical systems. As we have demonstrated, the SOI reveals itself in most basic linear optical processes: propagation, diffraction, focusing, scattering, reflection, and refraction. It is noteworthy that the small SOI phenomena are usually robust against the details and random aberrations of the system, because they are intimately related to the fundamental AM conservation laws and global geometrical aspects of wave evolution. Furthermore, in certain cases these effects are not that small and can cause radical changes in the field distributions. This offers an efficient tool for the control and manipulation of light at the nano-scale by using the internal AM degrees of freedom.

In this chapter we aimed to construct a general theoretical approach for a diversity of optical processes involving the SOI of light. We have introduced the basic concepts that are employed for adequate interpretation of the SOI effects in various physical settings. These are: (a) *geometric Berry phases*; (b) *AM, momentum, and coordinate operators*; (c) *Poynting energy flows*. With the aid of this comprehensive formalism, we have examined and interpreted the following basic problems: (i) generic AM and coordinate operators for photons; (ii) non-paraxial AM eigenmodes (Bessel-beams) in free space; (iii) tight focusing of light with a high-NA lens; (iv) dipole scattering of light by a small nanoparticle; (v) imaging in a high-NA microscope; (vi) interaction of a paraxial beam with an oblique surface; (vii) refraction and reflection of paraxial beams at a plane dielectric interface; (viii) propagation of paraxial light in a gradient-index medium. The basic manifestations of the SOI and their classification are summarized in Table 8.1. Specifically, the *spin-to-orbit AM conversion* (the SAM–IOAM interaction) occurs in axially symmetric non-paraxial systems. This can be observed in spin- and vortex-dependent azimuthal Poynting flows and radial intensity distributions. Breakdown of the symmetry, either in the field or in optical elements, causes the spin- and vortex- dependent transverse shifts of the intensity distribution in the orthogonal direction. These are the *spin and orbital Hall effects of light* (SAM–EOAM and IOAM–EOAM interactions). Interaction of locally paraxial light beams with inhomogeneous media and surfaces also brings about the Hall effects of light, which shift the centre of gravity of light out of the plane of propagation.

Although the particular problems considered in this chapter have been discussed and analyzed with ad hoc methods in many original papers during the past decade, here, for the first time to our knowledge, they are presented within a consistent and unifying approach that unveils deep interrelations between the various manifestations of the SOI. In spite of their differences, all the SOI effects owe their origin to the same *geometrical transformations* of the fields, related to the spherical coordinates in momentum space, and all are described by the same fundamental *operators* and corresponding *expectation values*. In axially symmetric non-paraxial fields, the *azimuthal geometric phases* produce a spin-dependent vortex component which reveals itself in the OAM and orbital Poynting flow. Yet, the Berry-phase contribution to the *quantization of caustics* explains the radial shifts of the intensity distribution that underlies the spin-to-orbit AM conversion. In asymmetric fields, the *transverse Berry-phase gradient* or *transverse Poynting flow* produces a net current in the orthogonal direction which shifts the centre of gravity of the field. In all cases, when the medium possesses a rotational symmetry, the expectation values of the SAM, OAM, and centre of gravity satisfy the corresponding *AM conservation laws*.

One can notice that SOI phenomena always have a few seemingly different aspects that are equivalent on a deeper level. On the one hand, all the field deformations and redistributions arise from interference of the constituent plane waves that have mutual geometric phases because of the spread in their directions of propagation. This determines the *geometric*, or *plane-wave*, level of the description. On the other hand, the SOI implies coupling between different types of the AM of light. The AM represents an integral *dynamical* characteristic of *localized states* of light-wave packets or beams. While such integral characteristics are independent of the representation of the field, the plane-wave spectrum represents the local features of the field in *momentum space*. Alternatively, the local properties of the fields in *real space* are given by the Poynting momentum densities. On a 'micro'-level of local quantities the electromagnetic field is considered as a superposition of plane waves (momentum representation) or as a 'hydrodynamical' distribution of the Poynting flows (coordinate representation). On the 'macro'-level of the integral quantities, the localized states of light are regarded similarly to quantum particles evolving along the trajectories of propagation and obeying conservation laws. One can say that the plane-wave, Poynting-flow, and AM interpretations of the SOI reveal its *wave*, *hydrodynamical*, and *particle* aspects.

Importantly, the optical SOI can be considered within the fundamental context of the evolution of *quantum particles with spin and orbital AM*. Indeed, all the main features of the SOI of light are quite similar to those of electrons in condensed-matter physics or high-energy spinning particles in external fields. In this spirit, optics offers a unique opportunity to explore a number of fine phenomena related to the geometrical, wave, and relativistic intrinsic properties of quantum particles. In contrast to most quantum systems, one can easily scan the finest details of the field distribution *inside* the optical beam or wave packet. For instance, the intrinsic spin Hall effect and topological monopoles that have attracted enormous attention in *condensed matter physics* [12–14, 98] are detected only by means of the output collective currents strongly affected by a number of scattering processes inside the sample. In *high-energy physics*, topological monopoles, space noncommutativity, and geometrodynamical equations of motion of spinning particles were subjects of purely theoretical studies, without any experimental realizations [116, 238–240], (see also [4, 75, 76]). In sharp contrast, very simple and clean optical systems enabled one to observe the spin-Hall effect and motion of spinning light affected by topological monopole directly, by looking into fine polarization features of the field across a single-wave beam [42, 43]. Moreover, optical AM ideas stimulated the development of new research directions in quantum physics. A clear understanding of the internal structure and key features of the wave AM inspired prediction and generation of the electron vortex beams carrying an intrinsic OAM [241–245]. In addition, the enigma of the electron spin seems to be resolved now in a quite straightforward 'optical' manner, namely, by constructing a wave-packet solution and calculating the internal current distribution, quite similar to the Poynting-vector flow [246].

Thus, the physics of the SOI and AM of light represents an important and integral part of modern optics, which is both of practical interest for applications and of fundamental importance for basic physical research. We hope that this chapter will aid further progress in this rapidly advancing area of physics by forming an effective theoretical framework for the future analysis of SOI in optical and quantum systems.

Acknowledgements

We are indebted to A. Y. Bekshaev, M. V. Berry, M. R. Dennis, A. Dogariu, V. G. Fedoseyev, Y. Li, and B. Y. Zel'dovich for fruitful discussions and to E. A. Ostrovskaya for her immense help in the preparation of this review. K. Y. B. is also grateful to the Max Planck Institute for the Science of Light (Prof. Gerd Leuch's division) for hospitality. This work was supported by the European Commission (Marie Curie Action), Science Foundation Ireland (Grant No. 07/IN.1/I906), and von Humboldt Foundation.

References

[1] Berestetskii, V. B., Lifshitz, E. M., and Pitaevskii, L. P. (1982). *Quantum electrodynamics*, 2nd edn. Oxford: Pergamon Press.

[2] Mathur, H. (1991). Thomas precession, spin-orbit interaction, and Berry's phase. *Phys. Rev. Lett.*, **67**, 3325–7.

[3] Bolte, J. and Keppeler, S. (1999). A semiclassical approach to the Dirac equation. *Ann. Phys.*, **274**, 125–62.

[4] Bérard, A. and Mohrbach, H. (2006). Spin Hall effect and Berry phase of spinning particles. *Phys. Lett. A*, **352**, 190–5.

[5] Rashba, E. I. (2006). Spin-orbit coupling and spin transport. *Physica E*, **34**, 31–5.

[6] Rashba, E. I. and Sherman, E. Y. (1988). Spin-orbital band splitting in symmetric quantum wells. *Phys. Lett. A*, **129**, 175–9.

[7] Shen, S. Q., Ma, M., Xie, X. C., and Zhang, F. C. (2004). Resonant spin Hall conductance in two-dimensional electron systems with a Rashba interaction in a perpendicular magnetic field. *Phys. Rev. Lett.*, **92**, 256603.

[8] Keppeler, S. (2003). Semiclassical quantization rules for the Dirac and Pauli equations. *Ann. Phys.*, **304**, 40–71.

[9] Novoselov, K. S., Geim, A. K., Morozov, S. V. *et al.* (2005). Two-dimensional gas of massless Dirac fermions in graphene. *Nature*, **438**, 197–200.

[10] Zhang, Y. B., Tan Y. W., Stormer, H. L., and Kim, P. (2005). Experimental observation of the quantum Hall effect and Berry's phase in graphene. *Nature*, **438**, 201–4.

[11] Gosselin, P., Boumrar, H., and Mohrbach, H. (2008). Semiclassical quantization of electrons in magnetic fields: The generalized Peierls substitution. *EPL* **84**, 50002.

[12] Murakami, S., Nagaosa, N., and Zhang, S.-C. (2003). Dissipationless quantum spin current at room temperature. *Science*, **301**, 1348–51.

[13] Sinova, J., Culcer, D., Niu, Q., Sinitsyn, N. A., Jungwirth, T., and MacDonald, A. H. (2004). Universal intrinsic spin Hall effect. *Phys. Rev. Lett.*, **92**, 126603.

[14] Xiao, D., Chang, M.-C., and Niu, Q. (2010). Berry phase effects on electronic properties. *Rev. Mod. Phys.*, **82**, 1959–2007.

[15] Serebrennikov, A. Y. (2006). Coriolis force, geometric phase, and spin-electric coupling in semiconductors. *Phys. Rev. B*, **73**, 195317.

[16] Serebrennikov, A. Y. (2008). Coupling of electron spin with its rotation in semiconductors. *Phys. Lett. A*, **372**, 1106–8.

[17] Allen, L., Beijersbergen, M. W., Spreeuw R. J. C., and Woerdman, J. P. (1992). Orbital angular-momentum of light and the transformation of Laguerre–Gaussian laser modes. *Phys. Rev. A*, **45**, 8185–9.

[18] Allen, L., Padgett, M. J., and Babiker, M. (1999). The orbital angular momentum of light. *Prog. Opt.*, **39**, 291–372.

[19] Allen, L., Barnett, S. M., and Padgett, M. J. (2003). *Optical Angular Momentum.* London: Taylor & Francis.

[20] Bekshaev, A., Soskin, M., and Vasnetsov, M. (2008). *Paraxial Light Beams with Angular Momentum.* New York: Nova Science Publishers.

[21] Franke-Arnold, S., Allen, L., and Padgett, M. J. (2008). Advances in optical angular momentum. *Laser & Photon. Rev.*, **2**, 299–313.

[22] Berry, M. V. (1997). Paraxial beams of spinning light. *Proc. SPIE*, **3487**, 6–11.

[23] O'Neil, A. T., MacVicar, I., Allen, L., and Padgett, M. J. (2002). Intrinsic and extrinsic nature of the orbital angular momentum of a light beam. *Phys. Rev. Lett.*, **88**, 053601.

[24] Bliokh, K. Y., Alonso, M. A., Ostrovskaya, E. A., and Aiello, A. (2010). Angular momenta and spin-orbit interaction of nonparaxial light in free space. *Phys. Rev A*, **82**, 063825.

[25] Bekshaev, A., Bliokh, K. Y., and Soskin, M. (2011). Internal flows and energy circulation in light beams. *J. Opt.*, **13**, 053001.

[26] Aiello, A. and Bliokh K. Y. (2012). Intrinsic and extrinsic orbital angular momentum of light. (in preparation).

[27] Bomzon, Z., Gu, M., and Shamir, J. (2006). Angular momentum and geometric phases in tightly-focused circularly polarized plane waves. *Appl. Phys. Lett.*, **89**, 241104.

[28] Zhao, Y., Edgar, J. S., Jeffries, G. D. M., McGloin, D., and Chiu, D. T. (2007). Spin-to-orbital angular momentum conversion in a strongly focused optical beam. *Phys. Rev. Lett.*, **99**, 073901.

[29] Schwartz, C. and Dogariu, A. (2006). Conservation of angular momentum of light in single scattering. *Opt. Express*, **14**, 8425–33.

[30] Haefner, D., Sukhov, S., and Dogariu, A. (2009). Spin Hall Effect of light in spherical geometry. *Phys. Rev. Lett.*, **102**, 123903.

[31] Rodríguez-Herrera, O. G., Lara, D., Bliokh, K. Y., Ostrovskaya, E. A., and Dainty, C. (2010). Optical nanoprobing via spin-orbit interaction of light. *Phys. Rev. Lett.*, **104**, 253601.

[32] Dooghin, A. V., Kundikova, N. D., Liberman V. S., and Zel'dovich, B. Y. (1992). Optical Magnus effect. *Phys. Rev. A*, **45**, 8204.

[33] Darsht, M. Y., Zel'dovich, B. Y., Kataevskaya, I. V., and Kundikova, N. D. (1995). Formation of an isolated wavefront dislocation. *JETP*, **80**, 817–821 [*Zh. Eksp. Theor. Phys.*, **107**, 1464–72].

[34] Volyar, A. V., Fadeeva, T. A., and Groshenko, N. A. (1997). Angular momentum of the field of a few-mode waveguide: conversion of the angular momentum. *Tech. Phys. Lett.*, **23**, 883–6.

[35] Volyar, A. V., Zhilaitis, V. Z., and Shvedov, V. G. (1998). Spin-orbit interaction in the field of an optical vortex of a few-mode fiber. *Tech. Phys. Lett.*, **24**, 826–8.

[36] Leary, C. C., Raymer, M. G., and van Enk, S. J. (2009). Spin and orbital rotations of electrons and photons via spin-orbit interaction. *Phys. Rev. A*, **80**, 061804(R).

[37] Bliokh, K. Y. and Frolov, D. Y. (2005). Spin-orbit interaction of photons and fine splitting of levels in ring dielectric resonator. *Opt. Commun.*, **250**, 321–7.

[38] Liberman, V. S. and Zel'dovich, B. Y. (1992). Spin-orbit interaction of a photon in an inhomogeneous medium. *Phys. Rev. A*, **46**, 5199–207.

[39] Bliokh, K. Y. and Bliokh Y. P. (2004). Topological spin transport of photons: the optical Magnus effect and Berry phase. *Phys. Lett. A*, **333**, 181–6; physics/0402110.

[40] Onoda, M., Murakami, S., and Nagaosa, N. (2004). Hall effect of light. *Phys. Rev. Lett.*, **93**, 083901.

[41] Bliokh, K. Y. and Bliokh, Y. P. (2006). Conservation of angular momentum, transverse shift, and spin Hall effect in reflection and refraction of an electromagnetic wave packet *Phys. Rev. Lett.*, **96**, 073903.

[42] Hosten, O. and Kwiat, P. (2008). Observation of the spin Hall effect of light via weak measurements. *Science*, **319**, 787–90.

[43] Bliokh, K. Y., Niv, A., Kleiner, V., and Hasman, E. (2008). Geometrodynamics of spinning light. *Nature Photon.*, **2**, 748–53.

[44] Aiello, A., Lindlein, N., Marquardt, C., and Leuchs, G. (2009). Transverse angular momentum and geometric spin Hall effect of light. *Phys. Rev. Lett.*, **103**, 100401.

[45] Baranova, N. B., Savchenko, A. Y., and Zel'dovich, B. Y. (1994). Transverse shift of a focal spot due to switching of the sign of circular-polarization. *JETP Lett.*, **59**, 232–4.

[46] Zel'dovich, B. Y., Kundikova, N. D., and Rogacheva, L. F. (1994). Observed transverse shift of a focal spot upon a change in the sign of circular polarization. *JETP Lett.*, **59**, 766–9.

[47] Bliokh, K. Y., Gorodetski, Y., Kleiner, V., and Hasman, E. (2008). Coriolis effect in optics: unified geometric phase and spin-Hall effect. *Phys. Rev. Lett.*, **101**, 030404.

[48] Fedoseyev, V. G. (2001). Spin-independent transverse shift of the centre of gravity of a reflected and of a refracted light beam. *Opt. Commun.*, **193**, 9–18.

[49] Dasgupta, R. and Gupta, P. K. (2006). Experimental observation of spin-independent transverse shift of the centre of gravity of a reflected Laguerre–Gaussian light beam. *Opt. Commun.*, **257**, 91–6.

[50] Bliokh, K. Y. (2006). Geometrical optics of beams with vortices: Berry phase and orbital angular momentum Hall effect. *Phys. Rev. Lett.*, **97**, 043901.

[51] Bliokh, K. Y., Shadrivov, I. V., and Kivshar, Y. S. (2009). Goos–Hanchen and Imbert–Fedorov shifts of polarized vortex beams. *Opt. Lett.*, **34**, 389–91.

[52] Bekshaev, A. Y. (2009). Oblique section of a paraxial light beam: criteria for azimuthal energy flow and orbital angular momentum. *J. Opt. A: Pure Appl. Opt.*, **11**, 094003.

[53] Garbin, V., Volpe, G., Ferrari, E., Versluis, M., Cojoc, D., and Petrov, D. (2009). Mie scattering distinguishes the topological charge of an optical vortex: a homage to Gustav Mie. *New J. Phys.*, **11**, 013046.

[54] Merano, M., Hermosa, N., Woerdman, J. P., and Aiello, A. (2010). How orbital angular momentum affects beam shifts in optical reflection. *Phys. Rev. A*, **82**, 023817.

[55] Biener, G., Niv, A., Kleiner, V., and Hasman, E. (2002). Formation of helical beams by use of Pancharatnam–Berry phase optical elements. *Opt. Lett.*, **27**, 1875–77.

[56] Bomzon, Z., Biener, G., Kleiner, V., and Hasman, E. (2002). Space-variant Pancharatnam–Berry phase optical elements with computer-generated subwavelength gratings. *Opt. Lett.*, **27**, 1141–3.

[57] Berry, M. V., Jeffrey, M. R., and Mansuripur, M. (2005). Orbital and spin angular momentum in conical diffraction. *J. Opt. A: Pure Appl. Opt.*, **7**, 685–90.

[58] Marrucci, L., Manzo, C., and Paparo, D. (2006). Optical spin-to-orbit angular momentum conversion in inhomogeneous anisotropic media. *Phys. Rev. Lett.*, **96**, 163905.
[59] Calvo, G. F. and Picón, A. (2007). Spin-induced angular momentum switching. *Opt. Lett.*, **32**, 838–40.
[60] Nagali, E., Sciarrino, F., De Martini, F. *et al.* (2009). Quantum information transfer from spin to orbital angular momentum of photons. *Phys. Rev. Lett.*, **103**, 013601.
[61] Brasselet, E., Izdebskaya, Y., Shvedov, V., Desyatnikov, A., Krolikowski, W., and Kivshar, Y. S. (2009). Dynamics of optical spin-orbit coupling in uniaxial crystals. *Opt. Lett.*, **34**, 1021–3.
[62] Brasselet, E., Murazawa, N., Misawa, H., and Juodkazis, S. (2009). Optical vortices from liquid crystal droplets. *Phys. Rev. Lett.*, **103**, 103903.
[63] Fadeyeva, T. A., Rubass, A. F., and Volyar, A. V. (2009). Transverse shift of a higher-order paraxial vortex-beam induced by a homogeneous anisotropic medium. *Phys. Rev. A*, **79**, 053815.
[64] Brasselet, E. and Loussert, C. (2011). Electrically controlled topological defects in liquid crystals as tunable spin-orbit encoders for photons. *Opt. Lett.*, **36**, 719–21.
[65] Shitrit, N., Bretner, I., Gorodetski, Y., Kleiner, V., and Hasman, E. (2011). Optical spin Hall effects in plasmonic chains. *Nano Lett.*, **11**, 2038–42.
[66] Hasman, E., Biener, G., Niv, A., and Kleiner, V. (2005). Space-variant polarization manipulation. *Prog. Opt.*, **47**, 215–89.
[67] Marrucci, L., Karimi, E., Slussarenko, S. *et al.* (2011). Spin-to-orbit conversion of the angular momentum of light and its classical and quantum applications. *J. Opt.*, **13**, 064001.
[68] Fedorov, F. I. (1955). K teorii polnovo otrazenija. *Dokl. Akad. Nauk SSSR*, **105**, 465–7. English translation available at http://master.basnet.by/congress2011/symposium/spbi.pdf.
[69] Schilling, H. (1965). Die strahlversetzung bei der reflexion linear oder elliptisch polarisierter ebener wellen an der trennebene zwischen absorbierenden medien. *Ann. Phys. (Berlin)*, **16**, 122–34.
[70] Imbert, C. (1972). Calculation and experimental proof of the transverse shift induced by total internal reflection of a circularly polarized light beam. *Phys. Rev. D*, **5**, 787–96.
[71] Player, M. A. (1987). Angular momentum balance and transverse shift on reflection of light. *J. Phys. A: Math. Gen.*, **20**, 3667–78.
[72] Fedoseyev, V. G. (1988). Conservation laws and transverse motion of energy on reflection and transmission of electromagnetic waves. *J. Phys. A: Math. Gen.*, **21**, 2045–59.
[73] Dutriaux, L., Le Floch, A., and Bretenaker, F. (1993). Measurement of the transverse displacement at total reflection by helicoidal laser eigenstates. *Europhys. Lett.*, **24**, 345–9.
[74] Bialynicki-Birula, I. and Bialynicka-Birula, Z. (1987). Berry phase in the relativistic theory of spinning particles. *Phys. Rev. D*, **35**, 2383–7.
[75] Skagerstam, B.-S. K. (1992). Localization of massless spinning particles and the Berry phase. arXiv:hep-th/9210054.
[76] Duval, C., Horváth, Z., and Horváthy, P. A. (2006). Fermat principle for spinning light. *Phys. Rev. D*, **74**, 021701(R).
[77] Hawton, M. and Baylis, W. E. (2005). Angular momentum and the geometrical gauge of localized photon states. *Phys. Rev. A*, **71**, 033816.

[78] Mashhoon, B. (1988). Neutron interferometry in a rotating frame of reference. *Phys. Rev. Lett.*, **61**, 2639–42.
[79] Mashhoon, B. (1989). Electrodynamics in a rotating frame of reference. *Phys. Lett. A*, **139**, 103–8.
[80] Mashhoon, B. (2009). Optics of rotating systems. *Phys. Rev. A*, **79**, 062111.
[81] Ross, J. N. (1984). The rotation of the polarization in low birefringence monomode optical fibers due to geometric effects. *Opt. Quantum Electron.*, **16**, 455–61.
[82] Chiao, R. Y. and Wu, Y. S. (1986). Manifestations of Berry topological phase for the photon, *Phys. Rev. Lett.*, **57** 933–6.
[83] Tomita, A. and Chiao, R. Y. (1986). Observation of Berry topological phase by use of an optical fiber. *Phys. Rev. Lett.*, **57**, 937–40.
[84] Haldane, F. D. M. (1986). Path dependence of the geometric rotation of polarization in optical fibers. *Opt. Lett.*, **11**, 730–2.
[85] Berry, M. V. (1987). Interpreting the anholonomy of coiled light. *Nature*, **326**, 277–8.
[86] Segert, J. (1987). Photon Berry phase as a classical topological effect. *Phys. Rev. A*, **36**, 10–15.
[87] Jordan, T. F. (1987). Direct calculation of the Berry phase for spins and helicities. *J. Math. Phys.*, **28**, 1759–60.
[88] Lipson, S. G. (1990). Berry's phase in optical interferometry – a simple derivation. *Opt. Lett.*, **15**, 154–5.
[89] Galvez, E. J. and Holmes, C. D. (1999). Geometric phase of optical rotators. *J. Opt. Soc. Am. A*, **16**, 1981–5.
[90] Shapere, A. and Wilczek, F. (1989). *Geometric Phases in Physics*. Singapore: World Scientific.
[91] Markovski, B. and Vinitsky, S. I. (1989). *Topological Phases in Quantum Theory*. Singapore: World Scientific.
[92] Vinitskii, S. I., Derbov, V. L., Dubovik, V. M., Markovski, B. L., and Stepanovskii, Y. P. (1990). Topological phases in quantum-mechanics and polarization optics *Usp. Fiz. Nauk*, **160**(6) 1–49 [*Sov. Phys. Usp.*, **33**, 403].
[93] Bhandari, R. (1997). Polarization of light and topological phases. *Phys. Rep.*, **281**, 2–64.
[94] Galvez, E. J. (2002). Applications of geometric phase in optics. In *Recent Research Developments in Optics 2*. Kerala: Research Signpost, pp. 165–82.
[95] Ben-Aryeh, Y. (2004). Berry and Pancharatnam topological phases of atomic and optical systems. *J. Opt. B: Quantum Semiclass.*, **6**, R1–18.
[96] Malykin, G. B. and Pozdnyakova, V. I. (2004). Geometric phases in singlemode fiber lightguides and fiber ring interferometers. *Physics–Uspekhi*, **47**, 289–308.
[97] Bliokh, K. Y. (2009). Geometrodynamics of polarized light: Berry phase and spin Hall effect in a gradient-index medium. *J. Opt. A: Pure Appl. Opt.*, **11**, 094009.
[98] Fang, Z., Nagaosa, N., Takahashi, K. S. *et al.* (2003) The anomalous Hall effect and magnetic monopoles in momentum space. *Science*, **302**, 92–5.
[99] Hannay, J. H. (1998). The Majorana representation of polarization, and the Berry phase of light. *J. Mod. Opt.*, **45**, 1001–8.
[100] Opat, G. I. (1991). The precession of a Foucault pendulum viewed as a beat phenomenon of a conical pendulum subject to a Coriolis force. *Am. J. Phys.*, **59**, 822–3.
[101] Khein, A. and Nelson, D. F. (1992). Hannay angle study of the Foucault pendulum in action-angle variables. *Am. J. Phys.* **61**, 170–4.

[102] Garetz, B. A. and Arnold, S. (1979). Variable frequency-shifting of circularly polarized laser-radiation via a rotating half-wave retardation plate. *Opt. Commun.*, **31**, 1–3.

[103] Garetz, B. A. (1981). Angular Doppler-effect. *J. Opt. Soc. Am.*, **71**, 609–11.

[104] Allen, L., Babiker, M., and Power, W. L. (1994). Azimuthal Doppler shift shift in light beams with orbital angular momentum. *Opt. Commun.*, **112**, 141–4.

[105] Bialynicki-Birula, I. and Bialynicka-Birula, Z. (1997). Rotational frequency shift. *Phys. Rev. Lett.*, **78**, 2539–42.

[106] Courtial, J., Dholakia, K., Robertson, D. A., Allen, L., and Padgett, M. J. (1998). Rotational frequency shift of a light beam. *Phys. Rev. Lett.*, **81**, 4828–30.

[107] Bomzon, Z. and Gu, M. (2007). Space-variant geometrical phases in focused cylindrical light beams. *Opt. Lett.*, **32**, 3017–9.

[108] Gorodetski, Y., Niv., A., Kleiner, V., and Hasman, E. (2008). Observation of the spin-based plasmonic effect in nanoscale structures. *Phys. Rev. Lett.*, **101**, 043903.

[109] Akhiezer, A. I. and Berestetskii, V. B. (1965). *Quantum Electrodynamics*. New York: Interscience Publishers.

[110] van Enk, S. J. and Nienhuis, G. (1992). Eigenfunction description of laser beams and orbital angular momentum of light. *Opt. Commun.*, **94**, 147–58.

[111] Aiello, A., Merano, M., and Woerdman, J. P. (2009). Duality between spatial and angular shift in optical reflection. *Phys. Rev. A*, **80**, 061801(R).

[112] Barnett, S. M. and Allen, L. (1994). Orbital angular-momentum and nonparaxial light-beams. *Opt. Commun.*, **110**, 670–8.

[113] van Enk, S. J. and Nienhuis, G. (1994). Spin and orbital angular-momentum of photons. *Europhys. Lett.*, **25**, 497–501.

[114] van Enk, S. J. and Nienhuis, G. (1994). Commutation rules and eigenvalues of spin and orbital angular momentum of radiation fields. *J. Mod. Opt.*, **41**, 963–77.

[115] Barnett, S. M. (2010). Rotation of electromagnetic fields and the nature of optical angular momentum. *J. Mod. Opt.*, **57**, 1339–43.

[116] Pryce, M. H. L. (1948). The mass-centre in the restricted theory of relativity and its connexion with the quantum theory of elementary particles. *Proc. R. Soc. London, Ser. A*, **195**.

[117] Hawton, M. and Baylis, W. E. (2001). Photon position operators and localized bases. *Phys. Rev. A*, **64**, 012101.

[118] Oemrawsingh, S. S. R., Eliel, E. R., Nienhuis, G., and Woerdman, J. P. (2004). Intrinsic orbital angular momentum of paraxial beams with off-axis imprinted vortices. *J. Opt. Soc. Am. A*, **21**, 2089–96.

[119] Kuratsuji, H. and Kakigi, S. (1988). Maxwell-Schrodinger equation for polarized light and evolution of Stokes parameters. *Phys. Rev. Lett.*, **80**, 1888–91.

[120] Bliokh, K. Y., Frolov, D. Y., and Kravtsov, Y. A. (2007). Non-Abelian evolution of electromagnetic waves in a weakly anisotropic inhomogeneous medium. *Phys. Rev. A*, **75**, 053821.

[121] Kravtsov, Y. A., Bieg, B., and Bliokh, K. Y. (2007). Stokes-vector evolution in a weakly anisotropic inhomogeneous medium. *J. Opt. Soc. Am. A*, **24**, 3388–96.

[122] Born, M. and Wolf, E. (2005). *Principles of Optics*, 7th edn. London: Pergamon.

[123] Berry, M. V. (2009). Optical currents. *J. Opt. A: Pure Appl. Opt.*, **11**, 094001.

[124] Bekshaev, A. Y. and Soskin, M. S. (2007). Transverse energy flows in vectorial fields of paraxial beams with singularities. *Opt. Commun.*, **271**, 332–48.

[125] Li, C.-F. (2009). Spin and orbital angular momentum of a class of nonparaxial light beams having a globally defined polarization. *Phys. Rev. A*, **80**, 063814.

[126] Bialynicki-Birula, I. (1996). Photon wave function. *Prog. Opt.*, **36**, 245–94.

[127] Bergmann, J. E. S., Mohammadi, S. M., Daldorff, L. K. S. *et al.* (2008). Conservation laws in generalized Riemann–Silberstein electrodynamics. arXiv:0803.2383v6.

[128] Durnin, J. (1987). Exact solutions for nondiffracting beams. I. The scalar theory. *J. Opt. Soc. Am. A*, **4**, 651–4.

[129] Bouchal, Z. and Olivik, M. (1995). Non-diffractive vector Bessel beams. *J. Mod. Opt.*, **42**, 1555–66.

[130] Arlt, J. and Dholakia, K. (2000). Generation of high-order Bessel beams by use of an axicon. *Opt. Commun.*, **177**, 297–301.

[131] Volke-Sepulveda, K., Garcés-Chavez, V., Chávez-Cerda, S., Arlt, J., and Dholakia, K. (2002). Orbital angular momentum of a high-order Bessel light beam. *J. Opt. B: Quantum Semiclass. Opt.*, **4**, S82–9.

[132] Jáuregui, R. and Hacyan, S. (2005). Quantum-mechanical properties of Bessel beams. *Phys. Rev. A*, **71**, 033411.

[133] Berry, M. V. and Mount, K. E. (1972). Semiclassical approximations in wave mechanics. *Rep. Prog. Phys.*, **35**, 315–97.

[134] Kravtsov, Y. A. and Orlov, Y. I. (1999). *Caustics, Catastrophes, and Wave Fields*, 2nd edn. Berlin: Springer-Verlag.

[135] Berry, M. V. and McDonald, K. T. (2008). Exact and geometrical optics energy trajectories in twisted beams. *J. Opt. A: Pure Appl. Opt.*, **10**, 035005.

[136] Yang, S., Chen, W., Nelson, R. L., and Zhan, Q. (2009). Miniature circular polarization analyzer with spiral plasmonic lens. *Opt. Lett.*, **34**, 3047–9.

[137] Wilkinson, M. (1984). An example of phase holonomy in WKB theory. *J. Phys. A: Math. Gen.*, **17**, 3459–76.

[138] Kuratsuji, H. and Iida, S. (1985). Semiclassical quantization with a quantum adiabatic phase. *Phys. Lett. A*, **111**, 220–2.

[139] Littlejohn, R. G. and Flynn, W. G. (1991). Geometrical phases and the Bohr-Sommerfeld quantization of multicomponent wave fields. *Phys. Rev. Lett.*, **66**, 2839–42.

[140] Mikitik, G. P. and Sharlai, Y. V. (1999). Manifestation of Berry's phase in metal physics. *Phys. Rev. Lett.*, **82**, 2147–50.

[141] Berry, M. V. (2011). Lateral and transverse shifts in reflected dipole radiation. *Proc. R. Soc. A*, **467**, 2500–19.

[142] Nieminen, T. A., Stilgoe, A. B., Heckenberg, N. R., and Rubinsztein-Dunlop, H. (2008). Angular momentum of a strongly focused Gaussian beam. *J. Opt. A: Pure Appl. Opt.*, **10**, 115005.

[143] Zhan, Q. (2006). Properties of circularly polarized vortex beams. *Opt. Lett.*, **31**, 867–9.

[144] Zhao, Y., Shapiro, D., Mcgloin, D., Chiu, D. T., and Marchesini, S. (2009). Direct observation of the transfer of orbital angular momentum to metal particles from a focused circularly polarized Gaussian beam. *Opt. Express*, **17**, 23316–22.

[145] Chen, B. and Pu, J. (2009). Tight focusing of elliptically polarized vortex beams. *Appl. Opt.*, **48**, 1288–94.

[146] Monteiro, P. B., Neto, P. A. M., and Nussenzveig, H. M. (2009). Angular momentum of focused beams: beyond the paraxial approximation. *Phys. Rev. A*, **79**, 033830.

[147] Bekshaev, A. Y. (2010). A simple analytical model of the angular momentum transformation in strongly focused light beams. *Cent. Eur. J. Phys.*, **8**, 947–60.

[148] Wolf, E. (1959). Electromagnetic diffraction in optical systems. I. An integral representation of the image field. *Proc. R. Soc. London. Ser. A*, **253**, 349–57.

[149] Richards, B. and Wolf, E. (1959). Electromagnetic diffraction in optical systems. II. Structure of the image field in an aplanatic system. *Proc. R. Soc. London. Ser. A*, **253**, 358–79.

[150] Alonso, M. A. and Forbes, G. W. (2000). Uncertainty products for nonparaxial wave fields. *J. Opt. Soc. Am. A*, **17**, 2391–402.

[151] Alonso, M. A., Borghi, R., and Santarsiero, M. (2006). Nonparaxial fields with maximum joint spatial-directional localization. II. Vectorial case. *J. Opt. Soc. Am. A*, **23**, 701–12.

[152] Alonso, M. A. (2011). The effect of orbital angular momentum and helicity in the uncertainty-type relations between focal spot size and angular spread. *J. Opt.*, **13**, 064016.

[153] Bokor, N., Iketaki, Y., Watanabe, T., and Fujii, M. (2005). Investigation of polarization effects for high-numerical-aperture first-order Laguerre–Gaussian beams by 2D scaning with a single fluorescent microbead. *Opt. Express*, **13**, 10440–7.

[154] Iketaki, Y., Watanabe, T., Bokor, N., and Fujii, M. (2007). Investigation of the centre intensity of first- and second-order Laguerre–Gaussian beams with linear and circular polarization. *Opt. Lett.*, **32**, 2357–9.

[155] Jackson, J. D. (1998). *Classical Electrodynamics*, 3rd edn. New York: Wiley & Sons.

[156] Moe, G. and Happer, W. (1977). Conservation of angular momentum for light propagating in a transparent anisotropic medium. *J. Phys. B: Atom. Molec. Phys.*, **10**, 1191–208.

[157] Schwartz, C. and Dogariu, A. (2006). Backscattered polarization patterns, optical vortices, and the angular momentum of light. *Opt. Lett.*, **31**, 1121–3.

[158] Vuong, L. T., Adam, A. J. L., Brok, J. M., Planken, P. C. M., and Urbach, H. P. (2010). Electromagnetic spin-orbit interactions via scattering of subwavelength apertures. *Phys. Rev. Lett.*, **104**, 083903.

[159] Rodríguez-Herrera, O. G., Lara, D., and Dainty, C. (2010). Far-field polarization-based sensitivity to sub-resolution displacements of a sub-resolution scatterer in tightly focused fields. *Opt. Express*, **18**, 5609–28.

[160] Vasnetsov, M. V., Marienko, I. G., and Soskin, M. S. (2000). Self-reconstruction of an optical vortex. *JETP Lett.*, **71**, 130–3.

[161] Gorshkov, V. N., Kononenko, A. N., and Soskin, M. S. (2001). Diffraction and self-restoration of a severely screened vortex beam. *SPIE Proc.*, **4403**, 127–37.

[162] Anan'ev, Y. A. and Bekshaev, A. Y. (1996). Changes in the light beam structure induced by transmission through dispersive elements. II. Special cases. *Opt. Spectrosc.*, **80**, 445–52.

[163] Bekshaev, A. Y. and Popov, A. Y. (2002). Method of light beam orbital angular momentum evaluation by means of space-angle intensity moments. *Ukr. J. Phys. Opt.*, **3**, 249–57.

[164] Goos, F. and Hänchen, H. (1947). Ein neuer und fundamentaler Versuch zur Totalreflexion. *Ann. Phys.*, **1**, 333–46.

[165] Artmann, K. (1948). Berechnung der Seitenversetzung des totalreflectierten Strahles. *Ann. Phys.*, **2**, 87–102.

[166] Bliokh, K. Y. and Bliokh, Y. P. (2007). Polarization, transverse shifts, and angular momentum conservation laws in partial reflection and refraction of an electromagnetic wave packet. *Phys. Rev. E*, **75**, 066609.
[167] Aiello, A. and Woerdman, J. P. (2008). Role of beam propagation in Goos–Hänchen and Imbert–Fedorov shifts. *Opt. Lett.*, **33**, 1437–9.
[168] Qin, Y., Li, Y., Feng, X., Xiao, Y.-F., Yang, H., and Gong, Q. (2011). Observation of the in-plane spin separation of light. *Opt. Express*, **19**, 9636–45.
[169] Merano, M., Aiello, A., van Exter, M. P., Eliel, E. R., and Woerdman, J. P. (2009). Observing angular deviations in the specular reflection of a light beam. *Nature Photon.*, **3**, 337–40.
[170] Lotsch, H. K. V. (1970). Beam displacement at total reflection: the Goos-Hänchen shift. *Optik* **32**, 116–37, 189–204, 299–319, 553–69.
[171] McGuirk, M. and Carniglia, C. K. (1976). An angular spectrum representation approach to the Goos–Hänchen shift. *J. Opt. Soc. Am.*, **67**, 103–7.
[172] Wild, W. J. and Giles, C. L. (1982). Goos–Hänchen shifts from absorbing media. *Phys. Rev. A*, **25**, 2099–101.
[173] Bretenaker, F., Le Floch, A., and Dutriaux, L. (1992). Direct measurement of the optical Goos–Hänchen effect in lasers. *Phys. Rev. Lett.*, **17**, 931–3.
[174] Pfleghaar, E., Marseille, A., and Weis, A. (1993). Quantitative investigation of the effect of resonant absorbers on the Goos–Hänchen shift. *Phys. Rev. Lett.*, **70**, 2281–4.
[175] Emile, O., Galstyan, T., Le Floch, A., and Bretenaker, F. (1995). Measurements of the nonlinear Goos–Hänchen effect for Gaussian optical beams. *Phys. Rev. Lett.*, **75**, 1511–3.
[176] Bonnet, C., Chauvat, D., Emile, O., Bretenaker, F. and Le Floch, A. (2001). Measurement of positive and negative Goos–Hänchen effects for metallic gratings near Wood anomalies. *Opt. Lett.*, **26**, 666–8.
[177] Berman, P. R. (2002). Goos–Hänchen shift in negatively refractive media. *Phys. Rev. E*, **66**, 067603.
[178] Li, C.-F. (2003). Negative lateral shift of a light beam transmitted through a dielectric slab and interaction of boundary effects. *Phys. Rev. Lett.*, **91**, 133903.
[179] Fan, J., Dogariu, A., and Wang, L. J. (2003). Amplified total internal reflection. *Opt. Express*, **11**, 299–308.
[180] Shadrivov, I. V., Zharov, A. A., and Kivshar, Y. S. (2003). Giant Goos–Hänchen effect at the reflection from left-handed metamaterials. *Appl. Phys. Lett.*, **83**, 2713–5.
[181] Felbacq, D., Moreau, A., and Smaâli, R. (2003). Goos–Hänchen effect in the gaps of photonic crystals. *Opt. Lett.*, **28**, 1633–5.
[182] Yin, X., Hesselink, L., Liu, Z., Fang, N., and Zhang, X. (2004). Large positive and negative lateral optical beam displacements due to surface plasmon resonance. *Appl. Phys. Lett.*, **85**, 372–4.
[183] Merano, M., Aiello, A., 't Hooft, G. W., van Exter, M. P., Eliel, E. R., and Woerdman, J. P. (2007). Observation of Goos–Hänchen shifts in metallic reflection. *Opt. Express*, **15**, 15928–34.
[184] Ra, J. W., Bertoni, H. L., and Felsen, L. B. (1973). Reflection and transmission of beams at a dielectric interface. *SIAM J. Appl. Math.*, **24**, 396–413.
[185] Antar, Y. M. and Boerner, W. M. (1974). Gaussian beam interaction with a planar dielectric interface. *Can. J. Phys.*, **52**, 962–72.

[186] Chan, C. C. and Tamir, C. (1985). Angular shift of a Gaussian beam reflected near the Brewster angle. *Opt. Lett.*, **10**, 378–80.

[187] Chiu, K. W. and Quinn, J. J. (1972). On the Goos–Hänchen effect: a simple example of a time delay scattering process. *Am. J. Phys.*, **40**, 1847–51.

[188] Tureci, H. E. and Stone, A. D. (2002). Deviation from Snell's law for beams transmitted near the critical angle: application to microcavity lasers. *Opt. Lett.*, **27**, 7–9.

[189] Schomerus, H. and Hentschel, M. (2006). Correcting ray optics at curved dielectric microresonator interfaces: phase-space unification of Fresnel filtering and the Goos–Hänchen shift. *Phys. Rev. Lett.*, **96**, 243903.

[190] Costa de Beauregard, O. (1965). Translational inertial spin effect with photons. *Phys. Rev.*, **139**, B1443–6.

[191] Boulware, D. G. (1973). Phase-shift analysis of the translation of totally reflected beams. *Phys. Rev. D*, **7**, 2375–82.

[192] Ashby, N. and Miller, S. C., Jr. (1973). Shifts of light beams due to total internal reflection. *Phys. Rev. D*, **7**, 2383–9.

[193] Julia, B. and Neveu, A. (1973). On birefringence phenomena associated with total reflection. *J. Phys. (Paris)*, **34**, 335–40.

[194] Ricard, J. (1974). Calcul du déplacement transversal par réflexion d'un faisceau lumineux limité latéralement. *Nouv. Rev. Opt.*, **5**, 7–24.

[195] Hugonin, J. P. and Petit, R. (1977). Étude générale des déplacements a la réflexion totale. *J. Opt.*, **8**, 73–87.

[196] Cowan, J. J. and Aničin, B. (1977). Longitudinal and transverse displacements of a bounded microwave beam at total internal reflection. *J. Opt. Soc. Am.*, **67**, 1307–14.

[197] Costa de Beauregard, O., Imbert, C., and Levy, Y. (1977). Observation of shifts in total reflection of a light beam by a multilayered structure. *Phys. Rev. D*, **15**, 3553–62.

[198] Turner, R. G. 1980. Shifts of coherent light beams on reflection at plane interfaces between isotropic media. *Aust. J. Phys.*, **33**, 319–35.

[199] Pun'ko, N. N. and Filippov, V. V. (1984). Beam splitting due to the finite size of the medium during total reflection. *JETP Lett.*, **39**, 20–3.

[200] Fedoseev, V. G. (1985). Lateral shift of a refracted light beam. *Opt. Spectrosk.*, **58**, 491–3.

[201] Fedoseev, V. G. (1991). Lateral displacement of light under reflection and refraction. 1. General results. *Opt. Spectrosk.*, **71**, 829–34.

[202] Fedoseev, V. G. (1991). Lateral displacement of light under reflection and refraction. 2. Displacement ratings. *Opt. Spectrosk.*, **71**, 992–7.

[203] Nasalski, W. (1996). Longitudinal and transverse effects on nonspecular reflection. *J. Opt. Soc. Am. A*, **13**, 172–81.

[204] Baida, F. I., Labeke, D. V., and Vigoureux, J.-M. (2000). Numerical study of the displacement of a three-dimensional Gaussian beam transmitted at total internal reflection: near-field applications. *J. Opt. Soc. Am. A*, **17**, 858–66.

[205] Pillon, F., Gilles, H., and Girard, S. (2004). Experimental observation of the Imbert-Fedorov transverse displacement after a single total reflection. *Appl. Opt.*, **43**, 1863–9.

[206] Pillon, F., Gilles, H., Girard, S., and Laroche, M. (2005). Goos–Hänchen and Imbert–Fedorov shifts for leaky guided modes. *J. Opt. Soc. Am. B*, **22**, 1290–9.

[207] Pillon, F., Gilles, H., Girard, S., Laroche, M., and Emile, O. (2005). Transverse displacement at a total reflection near the grazing angle: a way to discriminate between theories. *Appl. Phys. B*, **80**, 355–7.

[208] Qin, Y., Li, Y., He, H., and Gong, Q. (2009). Measurements of spin Hall effect of reflected light. *Opt. Lett.*, **34**, 2551–3.

[209] Ménard, J.-M., Mattacchione, A. E., Betz, M., and van Driel, H. M. (2009). Imaging the spin Hall effect of light inside semiconductors via absorption. *Opt. Lett.*, **34**, 2312–4.

[210] Ménard, J.-M., Mattacchione, A. E., van Driel, H. M., Hautmann, C., and Betz, M., (2010). Ultrafast optical imaging of the spin Hall effect of light in semiconductors. *Phys. Rev. B*, **82**, 045303.

[211] Onoda, M., Murakami, S., and Nagaosa, N. (2006). Geometrical aspects in optical wave-packet dynamics. *Phys. Rev. E*, **74**, 066610.

[212] Liu, B.-Y. and Li, C.-F. (2008). The Imbert–Fedorov shift of paraxial light beams. *Opt. Commun.*, **281**, 3427–33.

[213] Bliokh, K. Y. (2009). Comment on "The Imbert–Fedorov shift of a paraxial light beams." *Opt. Commun.*, **282**, 1521–2.

[214] Fedoseyev, V. G. (2009). Conservation laws and angular transverse shifts of the reflected and transmitted light beams. *Opt. Commun.*, **282**, 1247–51.

[215] Fedoseyev, V. G. (2008). Transformation of the orbital angular momentum at the reflection and transmission of a light beam on a plane interface. *J. Phys. A: Math. Theor.*, **41**, 505202.

[216] Aiello, A. and Woerdman, J. P. (2011). Goos–Hänchen and Imbert–Fedorov shifts of a nondiffracting Bessel beam. *Opt. Lett.*, **36**, 543–5.

[217] Okuda, H. and Sasada, H. (2006). Huge transverse deformation in nonspecular reflection of a light beam possessing orbital angular momentum near critical incidence. *Opt. Express*, **14**, 8393–402.

[218] Okuda, H. and Sasada, H. (2008). Significant deformations and propagation variations of Laguerre–Gaussian beams reflected and transmitted at a dielectric interface. *J. Opt. Soc. Am. A*, **25**, 881–90.

[219] Bliokh, K. Y. and Desyatnikov, A. S. (2009). Spin and orbital Hall effects for diffracting optical beams in gradient-index media. *Phys. Rev. A*, **79**, 011807(R).

[220] Kravtsov, Y. A. and Orlov, Y. I. (1990). *Geometrical Optics of Inhomogeneous Medium*. Berlin: Springer-Verlag.

[221] Bliokh, K. Y. and Bliokh, Y. P. (2004). Modified geometrical optics of a smoothly inhomogeneous isotropic medium: The anisotropy, Berry phase, and the optical Magnus effect. *Phys. Rev. E*, **70**, 026605.

[222] Bliokh, K. Y. and Freilikher, V. D. (2005). Topological spin transport of photons: magnetic monopole gauge field in Maxwell's equations and polarization splitting of rays in periodically inhomogeneous media. *Phys. Rev. B*, **72**, 035108.

[223] Rytov, S. M. (1938). On transition from wave to geometrical optics. *Dokl. Akad. Nauk. SSSR*, **18**, 263–7 (reprinted in [91]).

[224] Vladimirskii, V. V. (1941). The rotation of polarization plane for curved light ray. *Dokl. Akad. Nauk. SSSR*, **31**, 222–6 (reprinted in [91]).

[225] Liberman, V. S. and Zel'dovich, B. Y. (1994). Birefringence by a smoothly inhomogeneous locally isotropic medium. *Phys. Rev. E*, **49**, 2389–96.

[226] Savchenko, A. Y. and Zel'dovich, B. Y. (1994). Birefringence by a smoothly inhomogeneous locally isotropic medium: three-dimensional case. *Phys. Rev. E*, **50**, 2287–92.

[227] Bliokh, K. Y. and Stepanovskii, Y. P. (2003). On the change in electromagnetic wave polarization in a smooth one-dimensionally inhomogeneous medium. *J. Exp. Theor. Phys.*, **97**, 479–84.

[228] Duval, C., Horváth, Z., and Horváthy, P. A. (2007). Geometrical spinoptics and the optical Hall effect. *J. Geom. Phys.*, **57**, 925–41.

[229] Gosselin, P., Bérard, A., and Mohrbach, H. (2007). Spin Hall effect of photons in a static gravitational field. *Phys. Rev. D*, **75**, 084035.

[230] Berry, M. V. (1989). The quantum phase, five years after. In *Geometric Phases in Phyics*, ed. A. Shapere and F. Wilczek. Singapore: World Scientific.

[231] Nori, F. (2008). The dynamics of spinning light. *Nature Photon.*, **2**, 716–7.

[232] Allen, L., Courtial, J., and Padgett, M. J. (1999). Matrix formulation for the propagation of light beams with orbital and spin angular momenta. *Phys. Rev. E*, **60**, 7497–503.

[233] Alexeyev, C. N. and Yavorsky, M. A. (2006). Topological phase evolving from the orbital angular momentum of 'coiled' quantum vortices. *J. Opt. A: Pure Appl. Opt.*, **8**, 752–8.

[234] Alexeyev, C. N. and Yavorsky, M. A. (2007). Berry's phase from optical vortices in coiled optical fibers. *J. Opt. A: Pure Appl. Opt.*, **9**, 6–14.

[235] Segev, M., Solomon, R., and Yariv, A. (1992). Manifestation of Berry's phase in image-bearing optical beams. *Phys. Rev. Lett.*, **69**, 590–3.

[236] Kataevskaya, I. V. and Kundikova, N. D. (1995). Influence of the helical shape of a fiber waveguide on the propagation of light. *Quantum Electron.*, **25**, 927–8.

[237] Padgett, M. J. and Courtial, J. (1999). Poincaré-sphere equivalent for light beams containing orbital angular momentum. *Opt. Lett.*, **24**, 430–2.

[238] Papapetrou, A. (1951). Spinning test-particles in general relativity. *Proc. R. Soc. London Ser. A*, **209**, 248–58.

[239] Dixon, W. G. (1965). On a classical theory of charged particles with spin and the classical limit of the Dirac equation, *Nuovo Cimento*, **38**, 1616–43.

[240] Skagerstam, B. S. and Stern, A. (1981). Lagrangian descriptions of classical charged particles with spin. *Physica Scripta*, **24**, 493–7.

[241] Bliokh, K. Y., Bliokh, Y. P., Savel'ev, S., and Nori, F. (2007). Semiclassical dynamics of electron wave packet states with phase vortices. *Phys. Rev. Lett.*, **99**, 190404.

[242] Uchida, M. and Tonomura, A. (2010). Generation of electron beams carrying orbital angular momentum. *Nature*, **464**, 737–9.

[243] Verbeeck, J., Tian, H., and Schattschneider, P. (2010). Production and application of electron vortex beams. *Nature*, **467**, 301–4.

[244] McMorran, B. J., Agrawal, A., Anderson, I. M. *et al.* (2011). Electron vortex beams with high quanta of orbital angular momentum. *Science*, **331**, 192–5.

[245] Bliokh, K. Y., Dennis, M. R., and Nori, F. (2011). Relativistic electron vortex beams: angular momentum and spin-orbit interaction. *Phys. Rev. Lett.*, **107**, 174802.

[246] Chuu, C.-P., Chang, M.-C., and Niu, Q. (2010). Semiclassical dynamics and transport of the Dirac spin. *Soild State Commun.*, **150**, 533–7.

[247] Bliokh, K. Y. and Nori, F. (2012) Relativistic Hall effect. *Phys. Rev. Lett.*, **108**, 120403.

9

Quantum electrodynamics, angular momentum and chirality

DAVID L. ANDREWS AND MOHAMED BABIKER

9.1 Introduction

When light engages with matter, the interactions that take place at the photon level are subject to the operation of powerful underlying symmetry laws. Such principles underpin the physics of even the simplest photon interactions, as for instance in the familiar Planck–Einstein relation $E = h\nu$ for the absorption or emission of radiation. As emerged from Noether's work [1], this manifestation of overall conservation of energy is a direct consequence of a system invariance under temporal translation [2]. In connection with specific atomic photophysics, the term 'selection rule' is often used in connection with other space or time symmetries, these frequently being manifest as constraints over the conservation of quantized angular momentum. Obvious examples are the rules that govern the 'allowed' and 'forbidden' lines in atomic spectra, where the associated conditions over the geometric disposition and flow of charge emerge in the form of transition multipoles [3, 4].

The angular momentum attributes of light are most familiar in connection with the integer spin of the photon. Circularly polarized states have well-defined spin angular momentum along the direction of propagation [5], and numerous chiral or gyrotropic interactions exploit the differences in behaviour – observed in a material that is itself chirally constituted – between light beams of left- and right-handedness [6–8]. The principles are well known, and their applications have a surprisingly wide compass. Magnetic circular dichroism can characterize electronic transitions between spin states in a wide range of transition metal structures [9]; optical rotation distinguishes mirror-image compounds of remarkably different chemical action, and circular differential Raman scattering can reveal particular forms of protein folding [10]; surface reflections of circularly polarized specificity can reveal [11] or exploit [12, 13] microstructures such as gratings and chiral thin films [14].

The connectivity of angular momentum and chirality is clear in the case of circularly polarized light. It is well understood how gyrotropic phenomena can occur even in media that, whilst being macroscopically isotropic, comprise constituents of a specific

---

The Angular Momentum of Light, ed. D. L. Andrews and M. Babiker. Published by Cambridge University Press.

handedness, such as chiral molecules. However, when attention is turned to interactions that engage other forms of optical angular momentum whose form was largely unsuspected prior to the emergence of phase-structured radiation [15], it quickly emerges that few of the familiar precepts translate directly into this new field. There have until recently been many unanswered questions, and a significant number remain; nonetheless, many important issues have been resolved and the principles identified. This chapter aims to set them forth.

9.2 Quantum fields

The following analysis is based on a quantum electrodynamical (QED) representation of the light and its interactions. Although it is not so widely deployed in connection with structured light, this is a formalism that enables underlying symmetry issues to be addressed in a uniquely well suited representation, honouring both the quantum character of material transitions and the photonic properties of the radiation. Indeed it has been argued that it is only in such a formulation that the significance of spin and orbital angular momentum components can be unambiguously understood.

It is appropriate to begin by reference to the Hamiltonian operator for conventional, unstructured radiation, for which a suitable basis set comprises plane wave optical modes:

$$H_{\text{rad}} = \sum_{\boldsymbol{k},\eta} \left(a^{(\eta)\dagger}(\boldsymbol{k}) a^{(\eta)}(\boldsymbol{k}) + \frac{1}{2} \right) \hbar c k, \tag{9.1}$$

where $\boldsymbol{k}$ denotes the wave-vector, and η is a polarization label, which together represent four independent degrees of freedom; $a^{(\eta)}(\boldsymbol{k})$ is the photon annihilation operator, and its Hermitian conjugate $(h.c.)$ $a^{(\eta)\dagger}(\boldsymbol{k})$ is the corresponding operator for photon creation. The eigenstates of the Hamiltonian given by equation (9.1) are photon number states; the formalism readily extends to states with a statistical distribution of photon numbers. In terms of the same operators, the quantized mode expansion for the transverse vector potential $\boldsymbol{a}^{\perp}(\boldsymbol{r})$ of the radiation, at a position $\boldsymbol{r}$ within a quantization volume V, is represented by[1]

$$\boldsymbol{a}^{\perp}(\boldsymbol{r}) = \sum_{\boldsymbol{k},\eta} \left\{ \left(\frac{\hbar}{2\varepsilon_0 V c k} \right)^{\frac{1}{2}} \boldsymbol{\varepsilon}^{(\eta)}(\boldsymbol{k}) a^{(\eta)}(\boldsymbol{k}) \exp(i\boldsymbol{k}\cdot\boldsymbol{r}) + h.c. \right\}. \tag{9.2}$$

The expansion (9.2) establishes in quantum form a discrete basis Fourier transform of the vector potential, cast in terms of unit polarization vectors $\boldsymbol{\varepsilon}^{(\eta)}(\boldsymbol{k})$ transverse to the $\boldsymbol{k}$-vectors. Since the following analysis is to be based on a multipolar form of interaction Hamiltonian, we proceed to secure corresponding mode expansions for the electric and magnetic fields, from

$$\boldsymbol{b}(\boldsymbol{r}) = \nabla \times \boldsymbol{a}(\boldsymbol{r});\ \boldsymbol{e}(\boldsymbol{r}) = -\dot{\boldsymbol{a}}(\boldsymbol{r}). \tag{9.3}$$

[1] Here we follow a standard choice for the symbols [17]; one must take care to distinguish $\boldsymbol{a}$ from a, but the difference in superscripts and context always remove ambiguity.

To this end the requisite time derivative of the vector potential in its operator formulation can then be recovered using the familiar Heisenberg commutator relationship [16], leading to

$$\mathrm{i}\hbar\dot{a}^{(\eta)}(\boldsymbol{k}) = [a^{(\eta)}(\boldsymbol{k}), H_{\mathrm{rad}}] \Rightarrow \dot{a}^{(\eta)}(\boldsymbol{k}) = -\mathrm{i}cka^{(\eta)}(\boldsymbol{k}); \quad \dot{a}^{(\eta)\dagger}(\boldsymbol{k}) = \mathrm{i}cka^{(\eta)\dagger}(\boldsymbol{k}), \tag{9.4}$$

and hence

$$\boldsymbol{e}^{\perp}(\boldsymbol{r}) = \sum_{\boldsymbol{k},\eta}\left\{\mathrm{i}\left(\frac{\hbar ck}{2\varepsilon_0 V}\right)^{\frac{1}{2}} \boldsymbol{\varepsilon}^{(\eta)}(\boldsymbol{k})a^{(\eta)}(\boldsymbol{k})\exp(\mathrm{i}\boldsymbol{k}\cdot\boldsymbol{r}) + h.c.\right\}, \tag{9.5}$$

$$\boldsymbol{b}(\boldsymbol{r}) = \sum_{\boldsymbol{k},\eta}\left\{\mathrm{i}\left(\frac{\hbar k}{2\varepsilon_0 cV}\right)^{\frac{1}{2}} (\hat{\boldsymbol{k}}\times\boldsymbol{\varepsilon}^{(\eta)}(\boldsymbol{k}))a^{(\eta)}(\boldsymbol{k})\exp(\mathrm{i}\boldsymbol{k}\cdot\boldsymbol{r}) + h.c.\right\}. \tag{9.6}$$

Both fields are transverse – the magnetic field necessarily so because it is, by the above definition, the curl of a vector field. In the classical interpretation, the terms represented explicitly in equations (9.5) and (9.6) signify *positive frequency* analytic signals; the corresponding *negative frequency* components are conveyed by the Hermitian conjugate terms. Using these equations, it is readily verified that the radiation Hamiltonian, cast as an integrated energy density,

$$H_{\mathrm{rad}} = \frac{\varepsilon_0}{2}\int\{\boldsymbol{e}^{\perp 2}(\boldsymbol{r}) + c^2\boldsymbol{b}^2(\boldsymbol{r})\}\mathrm{d}^3\boldsymbol{r} \tag{9.7}$$

delivers the operator construct given in equation (9.1). These results are all perfectly general, although for some angular momentum issues to be pursued below a different basis set will be employed, better suited to the representation of structured light.

To approach the angular momentum issues we begin with some vector identities that hold irrespective of the basis. As demonstrated by Mandel and Wolf [5] the total angular momentum of a radiating beam, $\boldsymbol{J}_{\mathrm{rad}}$, derived from the volume integral of a cross-product between the position vector and Poynting vector (linear momentum density)

$$\boldsymbol{J}_{\mathrm{rad}} = \varepsilon_0\int\boldsymbol{r}\times\{\boldsymbol{e}(\boldsymbol{r})\times\boldsymbol{b}(\boldsymbol{r})\}\mathrm{d}^3\boldsymbol{r}, \tag{9.8}$$

can be partitioned into two terms, spin and orbital, respectively as follows:

$$\boldsymbol{S}_{\mathrm{rad}} = \varepsilon_0\int\{\boldsymbol{e}(\boldsymbol{r})\times\boldsymbol{a}(\boldsymbol{r})\}\mathrm{d}^3\boldsymbol{r}. \tag{9.9}$$

$$\boldsymbol{L}_{\mathrm{rad}} = \varepsilon_0\int e_i(\boldsymbol{r})\{\boldsymbol{r}\times\nabla\}a_i(\boldsymbol{r})\mathrm{d}^3\boldsymbol{r}. \tag{9.10}$$

Indeed, Bekshaev and Soskin have proposed that the corresponding energy flux can be similarly partitioned [18]. A division into spin and orbital parts is usually considered physically meaningful only in the case of paraxial beams – a commonly imposed condition that is assumed in the following theory – with non-paraxial beams requiring a significantly different treatment [19–23]. Bialynicki-Birula *et al.* have observed that, in contrast to energy, momentum, and the total angular momentum, the integrands in the above expressions for spin and orbital parts cannot in general be considered local densities, as they represent intrinsically non-local objects [24]. They have also shown that it is possible to cast the partition in another form due to Darwin [25], removing the manifest gauge dependence. With a suitably non-local form of vector potential, a quantum framework directly delivers results consistent with the Darwin formulation.

In the case of plane waves, serving as a point of reference for later discussing the distinctive features of more complex beams, it is readily shown that the spin component of angular momentum, (9.9), gives non-vanishing results, and the optical spin angular momentum operator $\boldsymbol{S}_{\mathrm{rad}}$ is expressible in a similar form to the Hamiltonian (9.1), through the photon mode expansion

$$\boldsymbol{S}_{\mathrm{rad}} = \mathrm{i}\hbar \sum_{\boldsymbol{k}} \sum_{\eta,\eta'} \left(a^{(\eta')\dagger}(\boldsymbol{k}) a^{(\eta)}(\boldsymbol{k}) + \frac{1}{2}\delta_{\eta,\eta'} \right) (\boldsymbol{e}^{(\eta)}(\boldsymbol{k}) \times \bar{\boldsymbol{e}}^{(\eta')}(\boldsymbol{k})), \tag{9.11}$$

where the overbar on the second polarization vector denotes complex conjugation. The congruence in form between this operator and the Hamiltonian has an obvious consequence, in that the two operators support a set of common eigenstates – the two circular polarization states (for each mode wave-vector $\boldsymbol{k}$), with sharp values for spin:

$$\boldsymbol{e}^{(\mathrm{L/R})}(\boldsymbol{k}) = \frac{1}{\sqrt{2}}(\hat{\boldsymbol{i}} \pm \mathrm{i}\hat{\boldsymbol{j}}), \tag{9.12}$$

where the positive sign superscript denotes the left-circular polarization, L, a minus signifying the right, R–a common convention, though at odds with [5]. It is useful to note that $\boldsymbol{e}^{(\mathrm{L/R})}(\boldsymbol{k}) = \bar{\boldsymbol{e}}^{(\mathrm{R/L})}(\boldsymbol{k})$ and $\boldsymbol{e}^{(\mathrm{L/R})}(-\boldsymbol{k}) = -\bar{\boldsymbol{e}}^{(\mathrm{R/L})}(\boldsymbol{k})$; the wave-vector $\boldsymbol{k}$, $\boldsymbol{e}^{(\mathrm{L})}(\boldsymbol{k})$ and $\boldsymbol{e}^{(\mathrm{R})}(\boldsymbol{k})$ together form a right-handed orthogonal triad. It emerges that the spin angular momentum operator reduces to

$$\boldsymbol{S}_{\mathrm{rad}} = \hbar \sum_{\boldsymbol{k}} \hat{\boldsymbol{k}} (a^{(\mathrm{L})\dagger}(\boldsymbol{k}) a^{(\mathrm{L})}(\boldsymbol{k}) - a^{(\mathrm{R})\dagger}(\boldsymbol{k}) a^{(\mathrm{R})}(\boldsymbol{k})), \tag{9.13}$$

which delivers $\hbar$ for each quantum difference between the occupation numbers n_{L} and n_{R} of the left- and right-handedly polarized modes. It is here that the picture of each circular photon conveying unit spin, its sign determined by the sense of rotation, emerges.

For the generation and description of structured radiation endowed with orbital angular momentum, electromagnetic fields are most often cast in terms of Laguerre–Gaussian (LG) modes, these being azimuthal phase-bearing paraxial solutions of the source-free Maxwell equations. The four degrees of freedom for such modes, propagating along a specified direction z, are a wave-vector magnitude k, polarization η, and integers l and p that together designate the Laguerre polynomial [26]. As such, the radiation Hamiltonian is explicitly cast in the following second-quantized form:

$$H_{\text{rad}} = \sum_{k,\eta,l,p} \left(a_{l,p}^{(\eta)\dagger}(\boldsymbol{k}) a_{l,p}^{(\eta)}(\boldsymbol{k}) + \frac{1}{2} \right) \hbar ck. \tag{9.14}$$

The electric and magnetic field expansions emerge in the following form as functions of cylindrical coordinates comprising the off-axis radial distance r, axial position z and azimuthal angle ϕ [27]:

$$\boldsymbol{e}^{\perp}(\boldsymbol{r}) = \sum_{k,\eta,l,p} \left\{ \mathrm{i} \left(\frac{\hbar ck}{\varepsilon_0 V} \right)^{1/2} \boldsymbol{\varepsilon}_{l,p}{}^{(\eta)}(\boldsymbol{k})\, a_{l,p}^{(\eta)}(\boldsymbol{k})\, f_{l,p}(r) \exp(\mathrm{i}kz + \mathrm{i}l\phi) + h.c. \right\}; \tag{9.15}$$

$$\boldsymbol{b}(\boldsymbol{r}) = \sum_{k,\eta,l,p} \left\{ \mathrm{i} \left(\frac{\hbar k}{\varepsilon_0 c V} \right)^{1/2} (\hat{\boldsymbol{k}} \times \boldsymbol{\varepsilon}_{l,p}{}^{(\eta)}(\boldsymbol{k}))\, a_{l,p}^{(\eta)}(\boldsymbol{k})\, f_{l,p}(r) \exp(\mathrm{i}kz + \mathrm{i}l\phi) + h.c. \right\}. \tag{9.16}$$

For a paraxial beam of waist w_0, the radial distribution function is given by

$$f_{l,p}(r) = \frac{C_p^{|l|}}{w_0} \left[\frac{\sqrt{2}r}{w_0} \right]^{|l|} \exp\left(\frac{-r^2}{w_0{}^2} \right) L_p^{|l|} \left(\frac{2r^2}{w_0{}^2} \right). \tag{9.17}$$

To determine the angular momentum attributes of the LG modes, we observe that the radial functions $f_{l,p}(r)$ in equation (9.17) depend only on the modulus of l; thus all information on wavefront handedness is contained within the phase factor $\exp(\mathrm{i}l\phi)$, hence being determined by the *sign* of l. The last factor in (9.17) is the Laguerre polynomial. For all such modes, the operator for *orbital* angular momentum is expressible as

$$\boldsymbol{L}_{\text{rad}} = \hbar \sum_{k,\eta,l,p} l\hat{\boldsymbol{k}} \left[a_{l,p}^{(\eta)\dagger}(\boldsymbol{k}) a_{l,p}^{(\eta)}(\boldsymbol{k}) + \frac{1}{2} \right]. \tag{9.18}$$

Again, the form of this operator manifests its commutation with the corresponding Hamiltonian (9.14), as does the spin operator (9.13), exhibiting the commensurability of orbital angular momentum with photon energy and spin. Since the sum of l in equation (9.18) (accommodating both positive and negative values) is itself zero, it emerges that an (l, p) LG mode with an occupation number n carries an angular momentum of $n\hbar l$, equivalent to $\hbar l$ per photon. Thus, l acquires the status of an orbital angular momentum quantum number, for which a positive sign again denotes left helicity and a negative sign, right. By writing equation (9.18) explicitly in terms of the $\eta = \text{L, R}$ components, it is therefore possible to express the total angular momentum of a freely propagating paraxial Laguerre–Gaussian beam as follows:

$$\boldsymbol{J}_{\text{rad}} = \boldsymbol{L}_{\text{rad}} + \boldsymbol{S}_{\text{rad}} = \sum_{k,l,p} \hbar\hat{\boldsymbol{k}} \left[(l+1) a_{l,p}^{(\text{L})\dagger}(\boldsymbol{k}) a_{l,p}^{(\text{L})}(\boldsymbol{k}) + (l-1) a_{l,p}^{(\text{R})\dagger}(\boldsymbol{k}) a_{l,p}^{(\text{R})}(\boldsymbol{k}) \right]. \tag{9.19}$$

Equation (9.19) makes plain the arithmetic summing of orbital and spin angular momentum for each species of photon circularity. The result, here cast explicitly in terms of a mode expansion, is consistent with the familiar axial ratio of optical angular momentum and energy [15, 28].

The self-evident commutability of the operators $\boldsymbol{J}_{\text{rad}}$ and H_{rad} deserves further comment. In particular, it is to be contrasted with the familiar commutation of $\boldsymbol{J}^2$ and H operators in the context of three-dimensional rotations, as for example arises in connection with the electronic states of atoms. One feature of this difference is that the operation of $\boldsymbol{J}^2_{\text{rad}}$ on its eigenstates delivers the modulus square of the $\boldsymbol{J}_{\text{rad}}$ eigenvalue, not a '$j(j+1)$' form of result. For example, in the case of a beam with only spin angular momentum, the eigenvalue of $\boldsymbol{J}^2_{\text{rad}}$ is simply $\hbar^2(n_{\text{L}} - n_{\text{R}})^2$ It is also significant that in the atomic case the direction of the $\boldsymbol{J}$ vector cannot be ascertained, only its projections on an imposed axis–whereas here, in the case of paraxial radiation, the direction of the angular momentum vector is unequivocal.

The separate identification of integer spin and orbital angular momentum, evident in the case of Laguerre–Gaussian beams, extends to Bessel beams [29], and Mathieu beams–though the orbital angular momentum content of the latter is less straightforward to interpret on a photon basis [30]. However, with the use of a suitably patterned birefringent plate it is possible to engineer a structured beam that entangles the orbital angular momentum with a non-uniform polarization state [31], undermining any such separation. Moreover, it is also possible to create multi-vortex structured beams that exhibit a non-integer angular momentum per photon [32]. The quantum optical interpretation of any such state, which is not an eigenstate of either of the conventionally defined spin and orbital angular momentum operators, hinges on the action of angular momentum measurement effecting wave-packet collapse, and thereby delivering one of the permissible integer values.

9.3 Interactions with matter

To describe the interactions of angular momentum-bearing light with matter, the formalism to be employed is largely based on the Power–Zienau–Woolley multipolar representation, as set forth in numerous works and reviews [17, 33–35]. For an assembly of N optically distinct centres differentiated by a label ξ, the QED Hamiltonian comprises the unperturbed operators for the radiation, those for the matter components, and the corresponding light–matter interaction operators

$$H = H_{\text{rad}} + \sum_{\xi}^{N} H_{\text{mat}}(\xi) + \sum_{\xi}^{N} H_{\text{int}}(\xi). \tag{9.20}$$

As befits energy operators, each of the Hamiltonian terms is of even parity with respect to both space inversion and time reversal. The first two terms of (9.20), which together represent an 'unperturbed' Hamiltonian H_0, determine the basis in which the different states of the system can be described, i.e. direct product eigenstates of the radiation and the matter

Hamiltonians. The interaction terms are expressed in terms of electric and magnetic fields. Alternative formulations cast in *minimal coupling* form, i.e. in terms of the vector potential of the radiation field, lead to identical results for energy-conserving processes [36, 37]. Woolley [34] has powerfully argued that the deployment of a multipolar representation–even the simplest electric dipole coupling–is ultimately consistent only with a fully fledged QED cast of the theory.

The multipolar interaction Hamiltonian comprises linear couplings of the molecular polarization field $\boldsymbol{p}(\boldsymbol{r})$ (accommodating all electric multipoles En) with the transverse electric field $\boldsymbol{e}^{\perp}(\boldsymbol{r})$ of the radiation, and of the molecular magnetization field $\boldsymbol{m}(\boldsymbol{r})$ (constituent magnetic multipoles Mn) with the magnetic radiation field $\boldsymbol{b}(\boldsymbol{r})$

$$H_{\text{int}} = -\int \boldsymbol{p}(\boldsymbol{r}) \cdot \boldsymbol{e}^{\perp}(\boldsymbol{r}) d^3\boldsymbol{r} - \int \boldsymbol{m}(\boldsymbol{r}) \cdot \boldsymbol{b}(\boldsymbol{r}) \mathrm{d}^3\boldsymbol{r}. \tag{9.21}$$

Equation (9.21) can be rendered exact by the addition of a diamagnetizatino term, quadratic in the magnetic field. However diamagnetization, associated with variations of current density, is typically responsible for couplings an order of magnitude smaller than those that relate to the two linear terms exhibited in (9.21) (by a factor typically the square of the fine structure constant); there has been no suggestion of the former being involved in angular momentum exchange, and for present purposes it warrants no further consideration. The usual Taylor series expansion of the linear terms separates the coupling into multipolar orders, whose leading contributions are E1, E2 and M1, cast in terms of the quantum operators: $\boldsymbol{\mu}(\xi)$ (electric dipole); $\boldsymbol{Q}(\xi)$ (electric-quadrupole), and $\boldsymbol{m}(\xi)$ (magnetic-dipole) respectively; the latter pair represent couplings of broadly similar magnitude, deriving from the same level of expansion in the minimal coupling representation. The leading terms of (9.21) are thus cast as follows, with $\boldsymbol{R}_\xi$ denoting the position of centre ξ and implied summation over Cartesian subscripts:

$$H_{int} = -\sum_{\xi} \mu_i(\xi) e_i^{\perp}(\boldsymbol{R}_\xi) - \sum_{\xi} Q_{ij}(\xi) \nabla_j e_i^{\perp}(\boldsymbol{R}_\xi) - \cdots - \sum_{\xi} m_i(\xi) b_i(\boldsymbol{R}_\xi) - \cdots. \tag{9.22}$$

Issues of angular momentum transfer can now be addressed by constructing the quantum amplitudes for optical interactions of arbitrary nature. When the general results are subsequently developed for specific processes of interest, the prospective final states can be cast in a form that exhibits specific angular momentum properties; it will then be possible to identify the symmetry principles that determine the collapse or emergence of finite amplitudes under particular conditions. Such a procedure proves considerably more expedient and straightforward than attempting to interrogate the time-evolved radiation state using the quantum angular momentum operators. For a molecule ξ we therefore begin from the usual development of the quantum amplitude M_{fi}^{ξ}:

$$M_{fi}^{\xi} = \left\langle f \left| \sum_{n=1}^{\infty} \left(H_{\text{int}} \frac{1}{(H_0 - E_i)} \right)^{n-1} H_{\text{int}} \right| i \right\rangle \equiv \sum_{n=1}^{\infty} M_{fi}^{\xi(n)}. \tag{9.23}$$

Since the interaction Hamiltonian H_{int} (9.22), is linear in the electromagnetic fields–and since the corresponding field expansions such as equations (9.5), (9.6), (9.15), (9.16) are themselves linear in photon creation and annihilation operators–the nth term in the above expansion delivers the leading contribution for any process involving n photons–indeed this is usually the only significant contributor to the corresponding quantum amplitude. Moreover, it emerges from the operator structure of those electromagnetic fields that each contribution to the quantum amplitude, for a component ξ located at $\boldsymbol{R}_\xi \equiv (r_\xi, \phi_\xi)$, will entail a factor $\exp\{-\mathrm{i}(\Delta\boldsymbol{k}.\boldsymbol{R}_\xi + \phi_\xi \Delta l)\}$, where $\Delta\boldsymbol{k}$ is the final minus the initial wave-vector sum of the photons involved in the process (reflecting the number of photon creations and annihilations), and Δl signifies any corresponding change in orbital angular momentum.

The quantum amplitude $M_{fi}^{\xi(n)}$ for a specific n-photon interaction thus emerges in the form of a linear combination of scalar terms, each one the inner product of a radiation tensor (comprising in the general case a product of components of the mode polarization vectors and wave-vectors) and a molecular tensor in the form of a generalized non-linear optical susceptibility. With a view to the present focus on symmetry and chirality, we assume a fixed centre of mass positioning of the individual optical centres, ignoring any forces or torques associated with response to the linear and angular momentum of the radiation. Where condensed phase matter is concerned, this is consistent with the physical influence of surrounding matter acting as a bath. Even with this simplification, considerable complexity still ensues in the general case; to make further progress it is expedient to focus on the leading multipole terms, explicitly given in equation (9.22), from which principles of general applicability can be resolved. We therefore write

$$M_{fi}^{\xi} \sim \exp\{-\mathrm{i}(\Delta\boldsymbol{k}.\boldsymbol{R}_\xi + \phi_\xi \Delta l)\} \sum_{e=0}^{n} \sum_{m=n-e}^{n} \boldsymbol{S}_{e;m;n-e-m}^{(r)} \otimes \boldsymbol{T}_{e;m;n-e-m}^{(r)(\xi)}. \tag{9.24}$$

In equation (9.24), each summand on the right is a scalar contraction of two tensors having the same rank: the result, of course, depends on the relative orientations of the properties by which they are defined. The radiation tensor $\boldsymbol{S}^{(r)} \equiv S_{i_1 i_2 \ldots i_r}$ comprises an outer product of radiation components (specifically, a product of components of the electric field and magnetic field), and the corresponding molecular tensor $\boldsymbol{T}^{(r)} = T_{i_1 i_2 \ldots i_r}$ can be written in a position-independent form that entails a product of n molecular transition integrals. Both tensors are distinguished by labels (e, m, q) corresponding to the number of E1, M1, and E2 interactions, respectively; the sum of these equals the number of photon interactions involved in the process, $n = e + m + q$, whilst the tensor rank r is given by $r = e + m + 2q$. We can illustrate the symmetry analysis with the simple case of single-photon absorption, where just one interaction directly couples the initial and final states. In order for any such transition integral not to vanish identically, the product of the symmetry representations for these two states must span one or more irreducible representations for which components of the transition moment themselves transform, under the full set of symmetry operations determined by the molecular point group [38].

Similar considerations arise in the realm of angular momentum, and the rules that operate in connection with photon spin in particular are mostly well known, particularly with regard to the well-known angular momentum selection rules for conventional single photon and multiphoton atomic transitions. The same principles operate with equal rigour for absorption and emission by molecules or within solids, though mapping the irreducible representations of the full three-dimensional rotation-inversion group O(3) onto a lower symmetry sub-group does in many cases permit transitions to occur between states of more than one symmetry class [39]. The requirements for invariance under a more restricted set of translational, reflectional or rotational operations lead to criteria that can reflect specific point or space group symmetries, affording important information on molecular structure and crystal morphology.

It is as well to be aware that there are numerous pitfalls if such principles, which are nearly always applied with particular reference to electric dipole transitions, are extended without due caution to the angular momentum associated with multipolar transitions. The detection of any individual photon at a specific point in space serves to demolish the directional uncertainty in its emission, and the price is a loss of angular momentum information [40–42]. Andrews [43] has shown that any photon emitted by multipolar decay, with the sole exception of dipole emission, conveys away from its source an angular momentum that acquires as it propagates beyond the near-zone a progressively broader spectrum of integer values. At the quantum level each order of multipole can be associated with a distinctive spatial distribution (as well as distance-dependence) that serves to identify the relative probabilities of photons propagating in different directions. However, the designation of a specific multipolar character to any individually emitted photon, on the assumption of a direct mapping from the character of a material transition, contravenes the limits of quantum uncertainty. In such a connection, literature references to terms such as 'quadrupole photon' generally reflect no more than an inference unsupported by rigorous analysis, and should be discouraged. There are, moreover, numerous cases where an atomic decay transition is allowed by more than one form of multipole–such as the M1-E2 interference that is observable in Pb [44].

9.4 Parametric and non-parametric processes

The overall conservation of energy in light–matter interactions is, in certain cases, satisfied individually within the material and radiation components of the system. Such optical processes, lacking energy exchange, are commonly termed 'elastic'. Momentum considerations afford the basis for a secondary classification; once again, there are specific processes for which momentum is separately conserved by the radiation and by the material. Processes that satisfy both criteria, usually with regard to linear momentum alone, are termed 'parametric'. Where beams that lack orbital angular momentum are concerned, the condition $\Delta \boldsymbol{k} = 0$ provides for the quantum amplitudes for spatially different sites to carry the same overall phase and hence interfere constructively, as is apparent from (9.24); this is the origin of the high conversion efficiency for harmonic generation in materials of suitable symmetry [45].

The connection of photon spin with optically parametric interactions is simple, though it is less widely familiar; in such processes there is an additional requirement for angular momentum to be conserved separately, again both by the radiation field and by the material system. This constraint imposes particular conditions where circular polarization states are concerned. It is for this reason that forward Rayleigh scattering invariably preserves the helicity of a circularly polarized beam, for example. This is also why, in isotropic media, odd harmonic generation is forbidden with a circularly polarized input (even order harmonics being precluded for other reasons); it is not possible to conserve angular momentum when the emergent photons can each convey only one unit–whilst three or more units, from photons of the fundamental mode, must be spent in generating each harmonic photon [46]. Again, the potential involvement of multipole coupling broadens the scope for consideration, especially when local symmetry is undermined by the presence of a surface or interface [47].

A fresh tier of considerations arises for light that conveys orbital angular momentum. For parametric processes in a paraxial beam it has been verified, both by theoretical analysis and by experiment, that orbital angular momentum is separately conserved in the radiation field [48, 49]. The principle can be seen to operate as follows. When Δk and Δl are zero–in other words when there is both wave-vector matching and orbital angular momentum matching between the optical input and output–the quantum amplitudes (9.24) for spatially different sites ξ carry the same phase factor, and in consequence interfere constructively. This is evident from the fact that an interaction in a system comprising N molecules has a rate observable Γ delivered by the modulus square of the summed quantum amplitudes for each individual component:

$$\Gamma \sim \left| \sum_{\xi}^{N} M_{fi}^{\xi} \right|^2 . \tag{9.25}$$

In a system of randomly ordered units, such as gases, unstructured fluids and randomly disordered solids, the only term in M_{fi}^{ξ} that contains ξ-specific information is the phase factor $\exp\{-\mathrm{i}(\Delta \boldsymbol{k}.\boldsymbol{R}_{\xi} + \phi_{\xi}\Delta l)\}$. With $\Delta \boldsymbol{k} = 0$ and $\Delta l = 0$, this phase is the same for all components ξ, so that the rate becomes quadratically dependent on N, giving the high efficiency that is indeed a characteristic of parametric processes. As a significant corollary, the condition $\Delta l = 0$ also serves to ensure that there is no dependence on the *sign* of the orbital angular momentum, in the sense that the same process rate would be observed if every photon involved were to have its orbital angular momentum reversed. Parametric processes therefore do not provide grounds for chiral discrimination on the basis of orbital angular momentum. In every sense, the conservation of orbital angular momentum is consistent with separate operation of the other conservation laws, for the material and optical parts of the system.

For non-parametric processes, more common amongst the wealth of optical interactions, the phase factor from equation (9.24) generally differs from one component to another, and the interference terms within the squared modulus in (9.25) average to zero. For simplicity we can assume all the components ξ are identical, so that the general result (9.26) is colligative, i.e. linearly proportional to N. Nonetheless the same principle applies even in systems where there is, for example, a statistical distribution of fluctuating orientations. By virtue of the ergodic theorem, it may be assumed that the time-averaged response from each component contributes identically to the ensemble rate; thus, using angular brackets to denote an orientational average, we can write

$$\Gamma \sim N\langle |M_{fi}^{\xi}|^2 \rangle. \tag{9.26}$$

This result, for non-parametric interactions, does not lend itself to the same transparency of physical interpretation, on issues of parity and chiral discrimination; to proceed further, some more detailed analysis is required. We turn first to the broad issues of parity.

9.5 Parity issues

A link between observations of material chirality and the angular momentum of light can correctly be anticipated on the basis that any form of chiral discrimination has to require a chiral probe. The propagative nature of electromagnetic radiation means that when light conveys angular as well as linear momentum, there are finite and collinear attributes of both pseudo-vector (space inversion parity $+1$) and regular vector (parity -1) character. Thus, any process that engages both of these properties cannot support parity as a good quantum number. To begin addressing the symmetry issues that arise in the interactions of twisted beams of light, it is therefore useful to first review the fundamental symmetries of the salient fields under space inversion and time reversal, P and T. Consideration of the third fundamental parity, charge reversal C, does not deliver any additional information. The space and time parity signatures of the vector potential, electric and magnetic fields, multipolar coupling terms and angular momenta are given in Table 9.1. It is worth noting that the generic electromagnetic field symmetry assignments for spatial parity [50] apply to the mode expansions only in their summed entirety; they do not generally apply to individual structured modes. To recognize the parity under time reversal in particular it is expedient to make use of the relation $\bar{\boldsymbol{\varepsilon}}^{(\eta)}(-\boldsymbol{k}) = -\boldsymbol{\varepsilon}^{(\eta)}(\boldsymbol{k})$ [51], since in each of the mode expansions the sums over wave-vector $\boldsymbol{k}$ extend in all directions.

To first resolve the possible dependence of any light–matter interaction on the *sign* of its net angular momentum content requires considerations of optical chirality. In systems that display macroscopic isotropy, manifestations of chiral response can arise only in media comprising molecular units that are non-superimposable upon their mirror images, which means that they must lack S_n rotation-inversion symmetry axes [8]. In such systems, where parity is not a good quantum number, the spatial transformation properties (under symmetry operations of the molecular point group) for electric and magnetic counterpart

Table 9.1 *Space and time parity of electromagnetic quantities.*

Operator	Time parity	Space parity
Hamiltonian *H*	+1	+1
Vector potential *a*	−1	−1
Electric field *e*	+1	−1
Electric field of LG mode (k, λ, l, p)	+1	$(-1)^{l+1}$
Electric multipole E*n*	+1	$(-1)^n$
Magnetic field *b*	−1	+1
Magnetic field of LG mode (k, λ, l, p)	−1	$(-1)^l$
Magnetic multipole M*n*	−1	$(-1)^{n-1}$
Angular momentum *J*, *L*, *S*	−1	+1

quantities become the same–permitting optical transitions to be simultaneously allowed by both electric and magnetic dipoles, E1 and M1, for example.

For a non-parametric interaction, the transition rate of a single centre ξ is determined by the modulus square of its quantum amplitude M_{fi}^{ξ} – the latter, as exhibited in equation (9.24), comprising a sum of contributions that entail electric and magnetic multipoles. The distinctive feature of a chiral centre is that, included in the sum of terms that arise in the square, are quantum interferences of a particular class, i.e. products that are odd in spatial parity, both for a specific matter tensor $\boldsymbol{T}^{(r)}$ and the corresponding radiation tensor $\boldsymbol{S}^{(r)}$, even though the product $\boldsymbol{S}^{(r)} \otimes \boldsymbol{T}^{(r)}$ is necessarily of even parity. Circular dichroism is a classic example. This absorption process is inelastic and thus non-parametric, and since it entails only one photon interaction we can develop equation (9.26) as

$$\begin{aligned} \Gamma \sim N\langle |M_{fi}^{\xi}|^2 \rangle &= N\langle |M_{fi}^{\xi(\mathrm{E1})} + M_{fi}^{\xi(\mathrm{M1})} + \cdots|^2 \rangle \\ &= N\left[\left\langle |M_{fi}^{\xi(\mathrm{E1})}|^2 \right\rangle + 2\mathrm{Re}\left\langle M_{fi}^{\xi(\mathrm{E1})} \overline{M_{fi}^{\xi(\mathrm{M1})}} \right\rangle + \cdots \right]. \end{aligned} \tag{9.27}$$

The leading $(\mathrm{E1})^2$ term delivers the same result for either left- or right-handed circularly polarized radiation, whereas the E1-M1 interference correction changes sign if the helicity of the radiation is reversed. Hence there emerges the differential rate measured as circular dichroism

$$\Gamma_{\mathrm{L}} - \Gamma_{\mathrm{R}} \sim 4\mathrm{Re}\left\langle M_{fi}^{\xi}(E1)\overline{M_{fi}^{\xi}(M1)} \right\rangle. \tag{9.28}$$

Terms such as the one on the right in equation (9.28), odd in parity for the radiation and for the molecule, can only arise in the case of chiral systems. In consequence, handedness is in fact apparent in two respects. When the operation of space inversion is performed on each of the molecular multipoles (but not the radiation–signifying a change of enantiomer, but retention of optical handedness), these interference terms change sign. The same is true if the radiation changes handedness on space inversion, but the molecule retains a specific enantiomeric form: again, these interference terms will change sign. It is such behaviour that provides the usual basis for chiral discrimination, generally involving circular polarizations and the associated optical spin.

Before turning attention to the implications of the helicity associated with light conveying orbital angular momentum [52], it should be clear that there are two scenarios to consider. On the one hand, if the optical process is parametric then $\Delta l = 0$ and, as observed in the last section, any rate-dependence on the sign of the orbital angular momentum of the radiation is immediately lost. On the other hand, for a non-parametric process where $\Delta l \neq 0$, the incoherent character means that the interaction rate is proportional to the sum of the square moduli of each quantum amplitude, M^{ξ}_{fi}, in which the azimuthal phase factor disappears in every summand; again, the observable transition rate proves not to involve the orbital angular momentum or its sign. Hence the wave-front helicity of the beam is immaterial; chiral discrimination depends only on circular polarization. So we can conclude that it is only the spin angular momentum of light, not any orbital angular momentum, that is associated with chiral discrimination. First proved theoretically [53], this conclusion has been amply supported by subsequent experiment [54].

9.6 Chirality

The conclusions that have been established on parity grounds lead us to expect no difference between the behaviour of right and left forms of any twisted optical mode, at least if the photons it comprises are plane polarized (no spin angular momentum) and the material they interact with is achiral. In symmetry terms this relatively trivial condition appears equivalent to the case of a circularly polarized beam with a planar wave-front (e.g. Hermite–Gaussian mode), interacting with an achiral material. However, there are several differences between LG photons and those associated with planar wave-fronts. In the present context, one significant difference is that a plane wave basis imposes no restriction on the direction of photon propagation, whereas the nature of LG photons designates propagation in one specified direction, z. This indicates that mirror inversion along a plane containing the z-axis becomes the relevant symmetry element for determining chiral differentiation, as opposed to space inversion. Arguing symmetry on these grounds is sufficient to establish that there is no differentiation between LG beams of opposite handedness when achiral material is interrogated, since a left-handed optical vortex transforms to a right-handed one under the specified mirror reflection. It is against this context that we can now approach the interaction of a twisted beam with chiral matter [52, 53].

Consider first the interactions of a chiral material with a beam of twisted wave-front comprising plane polarized photons. Superficially the symmetry aspects appear analogous to the case of circular polarizations in a plane wave. Mirror inversion of one part of the system alone (molecule or radiation) will give a system different from the original, and a corresponding change in signal signifying chiral specificity can be anticipated. Representing in brackets the circularity (in one form or other) of the radiation and the matter respectively, both cases impose identities between the results of (L, L) and (R, R) experiments, and so too between (L, R) and (R, L). There is no symmetry-based reason for an identity between all four cases. Succinctly, denoting either handedness L/R by a circularity c and its reverse by $\bar{c}$, we have $(c, c) = (\bar{c}, \bar{c})$, while the latter may or may not be equal to $(c, \bar{c})$. It thus remains to ascertain whether there is a means by which signals that are permissibly different in this fashion may in fact differ. In Section 9.4 it was shown that for a parametric optical process, since $\Delta l = 0$ there is accordingly no dependence on the sign of the orbital angular momentum. However, for a non-parametric process where $\Delta l \neq 0$, with incoherent character, the rate is given by equation (9.26); the phase factor $\exp\{-\mathrm{i}(\Delta \boldsymbol{k}.\boldsymbol{R}_\xi + \phi_\xi \Delta l)\}$ within the quantum amplitude (9.24) cancels for each centre ξ, and again the observable transition rate ceases to involve the orbital angular momentum or its sign. Hence the circularity of the beam is again immaterial.

In the most intricate case, where chiral molecules interact with twisted light comprising circular photons, then in view of the observations made above, the problem clearly reduces to exactly that which applies to other beams of circular polarization–but now there is a mechanism for the associated signals actually to differ. Again the helicity of the wave-front is immaterial so far as chiral specificity is concerned. Consider for example an experiment in which an LG beam interacts with a system of molecules, where the beam wave-front, the polarization state, and the chiral molecules themselves all have the same sense of circularity c. It is expedient to address again the simplest optical process in which chirality can be manifest, namely circular dichroism–photon absorption whose rate is sensitive to the handedness of the radiation. Here, the quantum interference terms responsible for chiral discrimination, principally a product of E1 and M1 amplitudes (see Section 9.5) change sign if the circularity of either the photons or the isomeric form of the molecules is reversed. No component of the rate changes if, however, the input is replaced by an LG beam of circularity $\bar{c}$, leaving the circularity of the photons and molecular handedness the same as before. The general conclusion is that there is no circumstance in which the interactions of an optical vortex beam, with orientationally disordered matter, can be influenced by the wave-front handedness, i.e. the sign of the associated orbital angular momentum. It is not yet clear whether such a rule will extend to solid crystalline materials, for which the point group symmetry is necessarily only a sub-group of O(3). Intriguingly, recent measurements of Raman optical activity by Alfano's group [55] do appear to indicate that there is some involvement of wave-front helicity in chiral discrimination; resolution of the mechanism for any particular crystal will necessitate an analysis based on the symmetry constraints imposed by the specific lattice structure.

9.7 Conclusion

The connectivity of photon angular momentum with chirality is central to a wide range of issues in the interactions of light with matter. The development of structured beams has enormously expanded both the scope and relevance of recent studies, which extend from fundamentals through to assessments of potential for numerous device and analytical applications.

The capacity of photons to convey individually a greater quota of angular momentum than was previously thought conceivable has prompted wide-ranging propositions for quantum communication and data handling applications [56–60]. Some questions concerning the effects of propagation through scattering media have only just begun to be addressed, concerning issues of whether the transmission of information at low intensity levels might be undermined by fundamental limitations at the single-photon level. Tackling the problem by the quantum electrodynamical methods described in this chapter, identifying the relevant components of the electromagnetic scattering and coupling tensors, a recent analysis by Andrews [61] broadly supports the fidelity of quantum information communication in such an application; the transmission of the photon through a forward scattering event conserves orbital angular momentum, just as it also conserves energy and linear momentum–provided that the incident photon itself conveys the necessary modal information.

One other important and very different area of application lies in the controlled delivery of angular momentum to trapped particles, as is described in detail in other chapters. For example in several recent studies we have shown how surface plasmons generated by counterpropagating, internally reflected Laguerre–Gaussian beams can produce locally confined plasmon vortex fields capable of manipulating and rotating trapped atoms [62–64]. The full extent of the possibilities for nanostructure control and fabrication using such a setup are only now becoming apparent–see Chapter 11. Employing the same principles with trapped molecules instead of atoms should afford opportunities for creating rotation upon a molecular axis as well as around the beam–with the enticing possibility of producing differing results for chiral molecules of opposite enantiomeric form.

Acknowledgement

We are pleased to acknowledge support from the UK Engineering and Physical Sciences Research Council for this work.

References

[1] E. Noether, Invariante Variationsprobleme, *Nachr. v. d. Ges. d. Wiss. zu Göttingen, Math-phys. Klasse*, 235–57 (1918).
[2] J. Hanc, S. Tuleja and M. Hancova, Symmetries and conservation laws: consequences of Noether's theorem, *Amer. J. Phys.* **72**, 428–35 (2004).
[3] B. W. Shore and D. H. Menzel *Principles of Atomic Spectra* (New York: Wiley, 1968), p. 290 ff.
[4] V. B. Berestetskii, E. M. Lifshitz and L. P. Pitaevskii, *Quantum Electrodynamics*, 2nd edn. (Oxford: Butterworth, 1982), p. 18 ff.
[5] L. Mandel and E. Wolf, *Optical Coherence and Quantum Optics* (Cambridge: Cambridge University press 1995), p. 490.
[6] S. F. Mason, *Molecular Optical Activity and the Chiral Discriminations* (Cambridge: Cambridge University Press 1982).

[7] N. Berova, K. Nakanishi and R. W. Woody, *Circular Dichroism: Principles and Applications* (Weinheim: Wiley-VCH, 2000).

[8] L. D. Barron, *Molecular Light Scattering and Optical Activity* 2nd edn. (Cambridge: Cambridge University Press 2004).

[9] W. R. Mason, *A Practical Guide to Magnetic Circular Dichroism Spectroscopy* (Hoboken, NJ: Wiley, 2007).

[10] L. D. Barron, L. Hecht, I. H. McColl and E. W. Blanch, Raman optical activity comes of age, *Mol. Phys.* **102**, 731–44 (2004).

[11] M. J. Huttunen, M. Virkki, M. Erkintalo, *et al.* Absolute probe of surface chirality based on focused circularly polarized light, *J. Phys. Chem. Lett.* **1**, 1826–29 (2010).

[12] P. C. Deguzman and G. P. Nordin, Stacked subwavelength gratings as circular polarization filters, *Appl. Opt.* **40**, 5731–7 (2001).

[13] Z. Bomzon, G. Biener, V. Kleiner and E. Hasman, Radially and azimuthally polarized beams generated by space-variant dielectric subwavelength gratings, *Opt. Lett.* **27**, 285–7 (2002).

[14] A. Lakhtakia and M. McCall, Sculptured thin films as ultranarrow-bandpass circular-polarization filters, *Opt. Commun.* **168**, 457–65 (1999).

[15] L. Allen, M. W. Beijersbergen, R. J. C. Spreeuw and J. P. Woerdman, Orbital angular momentum of light and the transformation of Laguerre-Gaussian laser modes, *Phys. Rev. A* **45**, 8185–9 (1992).

[16] W. H. Louisell, *Quantum Statistical Properties of Radiation* (New York: Wiley, 1973).

[17] D. P. Craig and T. Thirunamachandran, *Molecular Quantum Electrodynamics* (New York: Dover, 1998).

[18] A. Bekshaev and M. Soskin, Transverse energy flows in vectorial fields of paraxial beams, *Proc. SPIE* **6729**, 67290G (2007).

[19] S. M. Barnett and L. Allen, Orbital angular momentum and nonparaxial light beams, *Opt. Commun.* **110**, 670–8 (1994)

[20] S. J. van Enk and G. Nienhuis, Spin and orbital angular momentum of photons, *Europhys. Lett.* **25**, 497–501 (1994).

[21] E. Santamato, Photon orbital angular momentum: problems and perspectives, *Fortschr. Phys.* **52**, 1141–53 (2004).

[22] K. Y. Bliokh, M. A. Alonso, E. A. Ostrovskaya and A. Aiello, Angular momenta and spin-orbit interaction of nonparaxial light in free space, Phys. Rev. A **82**, 063825 (2010).

[23] A. Ya. Bekshaev, A simple analytical model of the angular momentum transformation in strongly focused light beams, *Cent. Eur. J. Phys.* **8**, 947–60 (2010).

[24] I. Bialynicki-Birula and Z. Bialynicka-Birula, Canonical separation of angular momentum of light into its orbital and spin parts, *J. Opt.* **13**, 064014 (2011).

[25] C. G. Darwin, Notes on the theory of radiation, *Proc. R. Soc. Lond. A* **136**, 36–52 (1932).

[26] L. Allen, M. J. Padgett and M. Babiker, The orbital angular momentum of light, *Prog. Opt.* **39**, 291–372 (1999).

[27] G. Nienhuis, Angular momentum and vortices in optics, in *Structured Light and Its Applications*, ed. D. L. Andrews (Boston, MA: Academic Press, 2008), Chapter 2.

[28] L. Allen, V. E. Lembessis and M. Babiker, Spin-orbit coupling in free-space Laguerre-Gaussian light beams, *Phys. Rev. A* **53**, R2937–9 (1996).

[29] L. Allen and M. Padgett, Introduction to phase-structured electromagnetic waves, in *Structured Light and Its Applications*, ed. D. L. Andrews (Boston, MA: Academic Press, 2008), Chapter 1.

[30] S. Chávez-Cerda, M. J. Padgett, I. Allison, *et al.* Holographic generation and orbital angular momentum of high-order Mathieu beams, *J. Opt. B: Quantum Semiclass. Opt.* **4**, S52–7 (2002).

[31] E. Nagali, F. Sciarrino, F. De Martini, *et al.* Quantum information transfer from spin to orbital angular momentum of photons, *Phys. Rev. Lett.* **103**, 013601 (2009).

[32] J. Leach, E. Yao and M. J. Padgett, Observation of the vortex structure of a non-integer vortex beam, *New J. Phys.* **6**, 71 (2004).

[33] R. G. Woolley, Reformulation of molecular quantum electrodynamics, *J. Phys. B: At. Mol. Opt. Phys.* **7**, 488–99 (1974).

[34] R. G. Woolley, Charged particles, gauge invariance, and molecular electrodynamics, *Int. J. Quantum. Chem.* **74**, 531–45 (1999).

[35] C. Cohen-Tannoudji, J. Dupont-Roc and G. Grynberg, *Photons and Atoms: Introduction to Quantum Electrodynamics* (New York: Wiley-Interscience, 1989).

[36] G. Compagno and E. A. Power, Alternative effective-Hamiltonians in non-relativistic quantum electrodynamics, *Phys. Rev. A* **38**, 4340–3 (1988).

[37] E. A. Power and T. Thirunamachandran, Maxwell's equations and the multipolar Hamiltonian, *Phys. Rev. A* **26**, 1800–1 (1982).

[38] D. L. Andrews, Symmetry characterisation in molecular multiphoton spectroscopy, *Spectrochim. Acta* **46A**, 871–85 (1990).

[39] M. Hamermesh, *Group Theory and Its Application to Physical Problems*, (Mineola, NY: Dover, 1989).

[40] S. Franke and S. M. Barnett, Angular momentum in spontaneous emission, *J. Phys. B: At. Mol. Opt. Phys.* **29**, 2141–50 (1996).

[41] S. Franke-Arnold, S. M. Barnett, E. Yao, J. Leach, J. Courtial J and M. Padgett, Uncertainty principle for angular position and angular momentum, *New. J. Phys.* **6**, 103 (2004).

[42] B. Jack, P. Aursand, S. Franke-Arnold, *et al.* Demonstration of the angular uncertainty principle for single photons, *J. Opt.* **13**, 064017 (2011).

[43] D. L. Andrews, Optical angular momentum: multipole transitions and photonics, *Phys. Rev. A* **81**, 033825 (2010).

[44] S. Werbowy and J. Kwela, M1-E2 interference in the Zeeman spectra of PbI and PbII, *J. Phys. B: At. Mol. Opt. Phys.* **42**, 065002 (2009).

[45] D. L. Andrews and P. Allcock, *Optical Harmonics in Molecular Systems* (Weinheim: Wiley-VCH, 2002), Chapter 5.

[46] D. L. Andrews, Harmonic generation in free molecules, *J. Phys. B: Atom. Mol. Phys.* **13**, 4091–9 (1980).

[47] D. L. Andrews, The role of longitudinal polarization in surface second harmonic generation, *J. Mod. Opt.* **40**, 939–46 (1993).

[48] J. Courtial, K. Dholakia, L. Allen and M. J. Padgett, Second-harmonic generation and the conservation of orbital angular momentum with high-order Laguerre-Gaussian modes, *Phys.Rev. A* **56** 4193–6 (1997).

[49] L. C. Dávila Romero, D. L. Andrews and M. Babiker, A quantum electrodynamics framework for the nonlinear optics of twisted beams, *J. Opt. B: Quantum Semiclass. Opt.* **4**, S66–72 (2002).

[50] S. Naguleswaran and G. E. Stedman, Time reversal selection rules and gauge invariance in nonlinear optics, *J. Phys. B: At. Mol. Opt. Phys.* **29**, 4027–40 (1996).

[51] B. J. McKenzie and G. E. Stedman, Virtual phonon exchange between Kramers ions in a field theoretic formalism, *J. Phys. C: Solid State Phys.* **12**, 5061–75 (1979).
[52] D. L. Andrews, L. C. Dávila Romero and M. Babiker, Twisted laser beams and their optical interactions with chiral matter, in *Trends in Chemical Physics Research*, ed. A. N. Linke (Hauppauge, NY: Nova Science, 2006), pp. 155–76.
[53] D. L. Andrews, L. C. Dávila Romero and M. Babiker, On optical vortex interactions with chiral matter, *Opt. Commun.* **237**, 133–9 (2004).
[54] F. Araoka, T. Verbiest, K. Clays and A. Persoons, Interactions of twisted light with chiral molecules: an experimental investigation, *Phys. Rev. A* **71**, 055401 (2005).
[55] G. Milione, J. Secor, G. Michel, S. Evans and R. R. Alfano, Raman optical activity by light with spin and orbital angular momentum, *Proc. SPIE* **7950**, 79500H (2011).
[56] S. M. Barnett, Optical angular-momentum flux, *J. Opt. B: Quantum Semiclass. Opt.* **4**, S7–16 (2002).
[57] G. Molina-Terriza, J. P. Torres and L. Torner, Management of the angular momentum of light: preparation of photons in multidimensional vector states of angular momentum, *Phys. Rev. Lett.* **88**, 013601 (2002).
[58] G. Gibson, J. Courtial, M. J. Padgett, *et al.* Free-space information transfer using light beams carrying orbital angular momentum, *Opt. Express* **12**, 5448–56 (2004).
[59] S. Franke-Arnold and J. Jeffers, Orbital angular momentum in quantum communication and information, *Structured Light and Its Applications* ed. D. L. Andrews, (Burlington, MA: Academic Press, 2008), pp. 271–93.
[60] J. C. Garcia-Escartin and P. Chamorro-Posada, Quantum multiplexing with the orbital angular momentum of light, *Phys. Rev. A* **78**, 062320 (2008).
[61] D. L. Andrews, The effect of scattering on single photon transmission of optical angular momentum, *J. Opt.* **13**, 064003 (2011).
[62] V. E. Lembessis, M. Babiker and D. L. Andrews, Surface optical vortices, *Phys. Rev. A* **79**, 011806(R) (2009).
[63] D. L. Andrews, M. Babiker, V. E. Lembessis and S. Al-Awfi, Surface plasmons with phase singularities and their effects on matter, *Phys. Status Solidi RRL* **4**, 241–3 (2010).
[64] V. E. Lembessis, S. Al-Awfi, M. Babiker and D. L. Andrews, Surface plasmon optical vortices and their influence on atoms, *J. Opt.* **13**, 064002 (2011).

10

Trapping of charged particles by Bessel beams

I. BIALYNICKI-BIRULA, Z. BIALYNICKA-BIRULA AND N. DROZD

10.1 Introduction

There exist two well-established methods to trap charged particles: the Penning trap [1] and the Paul trap [2]. In the Penning trap the particle is confined in space by a combination of static magnetic and electric fields. In the Paul trap the trapping is caused by a high frequency electric quadrupole field. The subject of this article is to present a third mechanism for trapping charged particles – trapping by beams of electromagnetic radiation. It was discovered some time ago [3–6] that a properly prepared beam acts as a "waveguide" for particles, confining their motion in the transverse directions. Similar phenomena occur in RF fields [7]. In all these cases, the essential role is played by the electric field configuration in the plane perpendicular to the beam axis (for nonrelativistic electrons, the magnetic field is less important). Particles are confined to the vicinity of the minimum-energy points. In particular, for beams of electromagnetic radiation carrying orbital angular momentum such points lie on the beam axis. One beam may confine particles only in the transverse direction. Two or three crossing beams may fully confine particles, acting as a substitute for the Paul trap.

Trapping of charged particles by beams of electromagnetic radiation is based on the same general principles as the Paul trap. In the Paul trap the quadrupole radio-frequency field is produced by a system of electrodes. In our case, the field is that of freely propagating electromagnetic beams. In both cases the essential role is played by two factors: the proper shape of the force field and fast oscillation or rotation. These features are clearly seen when we invoke the notion of the ponderomotive potential. This notion is applicable when the the motion of a charged particle can be unambiguously split into a slow drift and a fast quiver motion. Such a splitting is possible when a nonrelativistic particle is moving in a rapidly oscillating field. We shall restrict ourselves in this paper to monochromatic beams since they are especially well suited for trapping particles. The best example of such beams are Bessel beams which will be analyzed in detail.

The Angular Momentum of Light, ed. D. L. Andrews and M. Babiker. Published by Cambridge University Press.

10.2 Electric and magnetic fields of Bessel beams

Every monochromatic electromagnetic field can be represented by the following general expressions:

$$\boldsymbol{E}(\boldsymbol{r},t) = \boldsymbol{E}_c(\boldsymbol{r})\cos(\omega t) + \boldsymbol{E}_s(\boldsymbol{r})\sin(\omega t), \tag{10.1a}$$

$$\boldsymbol{B}(\boldsymbol{r},t) = \boldsymbol{B}_c(\boldsymbol{r})\cos(\omega t) + \boldsymbol{B}_s(\boldsymbol{r})\sin(\omega t). \tag{10.1b}$$

According to Maxwell's equations the electric and magnetic field vectors are interdependent,

$$\omega\boldsymbol{B}_c(\boldsymbol{r}) = \nabla\times\boldsymbol{E}_s(\boldsymbol{r}), \quad \omega\boldsymbol{B}_s(\boldsymbol{r}) = -\nabla\times\boldsymbol{E}_c(\boldsymbol{r}). \tag{10.2}$$

Generically, the electric and magnetic field vectors at each point trace two ellipses spanned by the pairs of vectors $\{\boldsymbol{E}_c, \boldsymbol{E}_s\}$ and $\{\boldsymbol{B}_c, \boldsymbol{B}_s\}$.

Bessel beams are most easily analyzed with the use of the Riemann–Silberstein (RS) vector [8], defined as a complex combination of the electric and magnetic fields,

$$\boldsymbol{F}(\boldsymbol{r},t) = \boldsymbol{E}(\boldsymbol{r},t) + ic\boldsymbol{B}(\boldsymbol{r},t). \tag{10.3}$$

This vector obeys the following equations:

$$\partial_t\boldsymbol{F}(\boldsymbol{r},t) = c\nabla\times\boldsymbol{F}(\boldsymbol{r},t), \quad \nabla\cdot\boldsymbol{F}(\boldsymbol{r},t) = 0. \tag{10.4}$$

For monochromatic fields we can write

$$\boldsymbol{F}(\boldsymbol{r},t) = e^{-i\omega t}\boldsymbol{F}_+(\boldsymbol{r}) + e^{i\omega t}\boldsymbol{F}_-(\boldsymbol{r}), \tag{10.5}$$

and the Maxwell equations reduce to the two equations for Trkalian fields [9]

$$c\nabla\times\boldsymbol{F}_\pm(\boldsymbol{r}) = \pm\omega\boldsymbol{F}_\pm(\boldsymbol{r}). \tag{10.6}$$

The two parts in (10.5), as we show below, correspond to two helicities. By comparing the formulas (10.1a) and (10.5), we find

$$\boldsymbol{E}_c = \mathrm{Re}\boldsymbol{F}_+ + \mathrm{Re}\boldsymbol{F}_-, \quad \boldsymbol{E}_s = \mathrm{Im}\boldsymbol{F}_+ - \mathrm{Im}\boldsymbol{F}_-, \tag{10.7}$$

$$\boldsymbol{B}_c = \mathrm{Re}\boldsymbol{F}_+ + \mathrm{Im}\boldsymbol{F}_-, \quad \boldsymbol{B}_s = -\mathrm{Im}\boldsymbol{F}_+ + \mathrm{Re}\boldsymbol{F}_-. \tag{10.8}$$

It follows from these formulas that for fields of definite helicity the ellipses traced by the electric and magnetic field vector coincide.

Following [10] (with some small changes in notation), we shall describe the electromagnetic field of the Bessel beam in cylindrical coordinates, with the z axis in the direction of propagation, in terms of the RS vector in the following form:

$$\begin{pmatrix} F_x(\rho,\phi,z,t) \\ F_y(\rho,\phi,z,t) \\ F_z(\rho,\phi,z,t) \end{pmatrix} = E_0 e^{-i\chi\omega t}\, e^{ik_\parallel z} \times \begin{pmatrix} \kappa_- e^{i(M+1)\phi} J_{M+1}(k_\perp\rho) + \kappa_+ e^{i(M-1)\phi} J_{M-1}(k_\perp\rho) \\ -i\kappa_- e^{i(M+1)\phi} J_{M+1}(k_\perp\rho) + i\kappa_+ e^{i(M-1)\phi} J_{M-1}(k_\perp\rho) \\ -2ie^{iM\phi} J_M(k_\perp\rho) \end{pmatrix}. \tag{10.9}$$

where

$$\kappa_\pm = \frac{\chi\sqrt{k_\parallel^2 + k_\perp^2} \pm k_\parallel}{k_\perp}. \tag{10.10}$$

We shall consider only positive values of M since Bessel beams with negative values can be obtained (up to an overall sign) by the following transformations of spacetime variables and helicity:

$$M \to -M, \quad \chi \to -\chi \quad \Longrightarrow \quad \phi \to -\phi, \quad t \to -t. \tag{10.11a}$$

The formulation in terms of the RS vector enables one to introduce, by analogy with quantum mechanics, the classification of the solutions of Maxwell's equations with the use of quantum numbers [10]. One may view these quantum numbers as characteristics of the photons making the beam. In the case of Bessel beams these quantum numbers are the eigenvalues of the following four operators (we dropped $\hbar$ in the definitions): the component of momentum in the direction of propagation $\hat{k}_\parallel$, the length of momentum in the transverse direction $\hat{k}_\perp$, the projection of the total angular momentum (orbital plus spin) on the direction of propagation $\hat{M}_z$, and the helicity operator (the projection of the total angular momentum on the direction of momentum) $\hat{\chi}$,

$$\hat{k}_\parallel = -i\partial_z, \quad \hat{k}_\perp = \sqrt{-\partial_x^2 - \partial_y^2}, \tag{10.12}$$

$$\hat{M}_z = -i(x\partial_y - y\partial_x) + \begin{bmatrix} 0 & -i & 0 \\ i & 0 & 0 \\ 0 & 0 & 0 \end{bmatrix}, \tag{10.13}$$

$$\hat{\chi} = \frac{1}{2\pi^2}\int d^3r' \frac{1}{|\boldsymbol{r} - \boldsymbol{r}'|^2} \begin{bmatrix} 0 & -\partial_{z'} & \partial_{y'} \\ \partial_{z'} & 0 & -\partial_{x'} \\ -\partial_{y'} & \partial_{x'} & 0 \end{bmatrix}. \tag{10.14}$$

The 3×3 matrix in Eq. (10.13) represents the z component of the photon spin in the vector representation. The representation of the helicity operator in the form (10.14) was given in [11]. The factors $\exp[i(M \pm 1)\phi]$ and $\exp[iM\phi]$ in the definition of Bessel beams reflect the decomposition of the total angular momentum M into the components with three different orbital angular momenta.

In classical physics, quantum numbers are just convenient labels attached to the solutions of Maxwell equations. Thus, each Bessel beam (10.9) is characterized by the following four parameters: a real parameter $k_\parallel$, a positive parameter $k_\perp$, an integer M, and χ (± 1). Every Bessel beam is monochromatic with the frequency $\omega = c\sqrt{k_\parallel^2 + k_\perp^2}$. The amplitude E_0 in Eq. (10.9) measures the strength of the beam in volt per meter (1 V/m corresponds to the beam intensity 2.65×10^{-3} W/m^2). The dimensionless electric and magnetic field vectors for the Bessel beam are obtained by taking the real and imaginary parts in Eq. (10.9) and dropping the factor E_0,

$$\begin{pmatrix} \mathcal{E}_x \\ \mathcal{E}_y \\ \mathcal{E}_z \end{pmatrix} = \begin{pmatrix} \kappa_- c_{M+1} + \kappa_+ c_{M-1} \\ \kappa_- s_{M+1} - \kappa_+ s_{M-1} \\ 2 s_M \end{pmatrix}, \tag{10.15a}$$

$$\begin{pmatrix} \mathcal{B}_x \\ \mathcal{B}_y \\ \mathcal{B}_z \end{pmatrix} = \begin{pmatrix} \kappa_- s_{M+1} + \kappa_+ s_{M-1} \\ -\kappa_- c_{M+1} + \kappa_+ c_{M-1} \\ -2 c_M \end{pmatrix}, \tag{10.15b}$$

where

$$c_M = \cos(k_\parallel z - \chi\omega t + M\phi) J_M(k_\perp \rho), \tag{10.16a}$$

$$s_M = \sin(k_\parallel z - \chi\omega t + M\phi) J_M(k_\perp \rho). \tag{10.16b}$$

In Fig. 10.1 we show the electric field configuration in the transverse plane near the beam axis for different values of M. Note that for $M = 2$ the configuration of electric forces is the same as of the gravitational forces on the saddle surface in the mechanical model of the Paul trap [2]. Therefore, at least in this case we can expect trapping of particles in the transverse plane. To exhibit rotation of field vectors in time, we plotted these vectors in Fig. 10.2 one quarter of the cycle later. In Fig. 10.3 we show the rotation of the field vectors for $M = 2$.

10.3 Motion of charged particles in a Bessel beam

All field components of the Bessel beam with $|M| > 1$ vanish along the z axis since in these cases all Bessel functions in Eq. (10.9) vanish at the origin. The motion of charged particles in Bessel beams in the vicinity of this line was thoroughly studied in [4] by solving numerically the standard equations of motion

$$m\frac{d^2 x^\mu(s)}{ds^2} = e f^{\mu\nu}(x(s))\frac{dx_\nu(s)}{ds}, \tag{10.17}$$

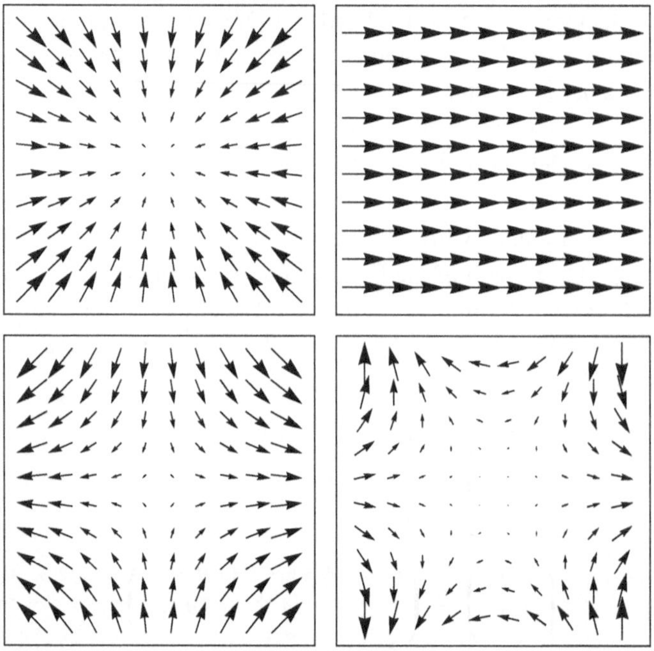

Figure 10.1 Electric field vectors in the xy plane near the beam axis for $\chi = 1$ and for $M = 0, 1, 2, 3$. This field configuration corresponds to $z = 0$ and $t = 0$. For other values the vectors would be rotated by the angle $(k_\parallel z - \chi\omega t)$. The top configurations correspond to $M = 0$ and $M = 1$. The bottom configurations correspond to $M = 2$ and $M = 3$.

where s is the proper time. These equations can be rewritten in the following dimensionless form

$$\frac{\mathrm{d}^2\xi^\mu(\tau)}{\mathrm{d}\tau^2} = \gamma\,\varphi^{\mu\nu}(\xi(\tau))\frac{\mathrm{d}\xi_\nu(\tau)}{\mathrm{d}\tau}, \tag{10.18}$$

where the particle coordinates ξ^μ are measured in units of the longitudinal wave length $2\pi/k_\parallel$ and the dimensionless proper time τ is measured in units of the inverse circular frequency $\tau = \omega s$. The dimensionless effective coupling constant γ is

$$\gamma = \frac{eE_0k_\parallel}{m\omega^2} = \frac{eE_0/k_\parallel}{mc^2(1 + (k_\perp/k_\parallel)^2)}. \tag{10.19}$$

The dimensionless electromagnetic field tensor $\varphi^{\mu\nu}$ is composed of the dimensionless electric and magnetic components given by Eqs. (10.15). The parameter γ is usually very small. It is approximately equal to the ratio of the energy gain (i.e. the energy that the particle acquires moving under the influence of the electric field E_0 along the distance of $1/k_\parallel$) to the particle rest energy mc^2. The particle trajectories scaled by the wavelength with initial velocities scaled by the speed of light will look the same for different situations provided the value of γ remains the same. The smaller the mass, the more effective is the action of

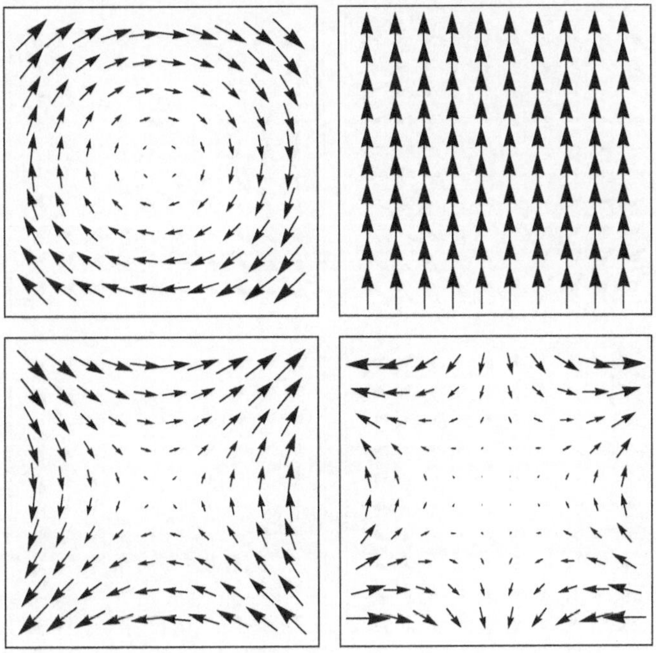

Figure 10.2 Electric field vectors in the xy plane as in Fig. 10.1 but a quarter of the cycle later. Each field vector is rotated counterclockwise by $\pi/2$.

the electromagnetic wave. Therefore, in all our specific examples we shall consider electrons. In what follows we shall consider nonrelativistic electrons whose motion is governed mainly by the electric field. One beam may confine the motion of an electron only in the transverse directions – the electron is guided along the direction of propagation. For full trapping we will need at least two beams.

Typical trajectories of an electron guided by the Bessel beam are shown in Fig. 10.4. For Bessel beams describing the propagation of light, the effective transverse dimension $2\pi/k_\perp$ is much larger than the longitudinal wave length. This means that the dimensionless parameter $k_\perp/k_\parallel$ is usually very small. The trajectories shown in Fig. 10.4 are obtained for $k_\perp/k_\parallel = 0.01$ and $\gamma = 0.00001$. For visible light, this value of γ corresponds to the beam intensity of the order of $10^{13}\,\mathrm{W/m^2}$. This high intensity resulted from our choice of rather high transverse velocities that were taken for illustrative purposes. The three trajectories guided by the beam start at the same point near the beam axis and with the same initial longitudinal velocity $\dot{z}(0) = 0.001c$ but they have different initial transverse velocities $\dot{x}(0) = 0.0002c$, $0.0001c$, and $0.00005c$. Trapping of electrons with initial transverse velocity exceeding the value $0.001c$ will require the increase of the beam intensity. For moderate beam intensities, relativistic electrons will not be trapped.

In Fig. 10.5 we plotted the trajectories trapped near the beam center for different M obtained when $\dot{x}(0) = 0.0001c$ and with the values of the remaining parameters the same

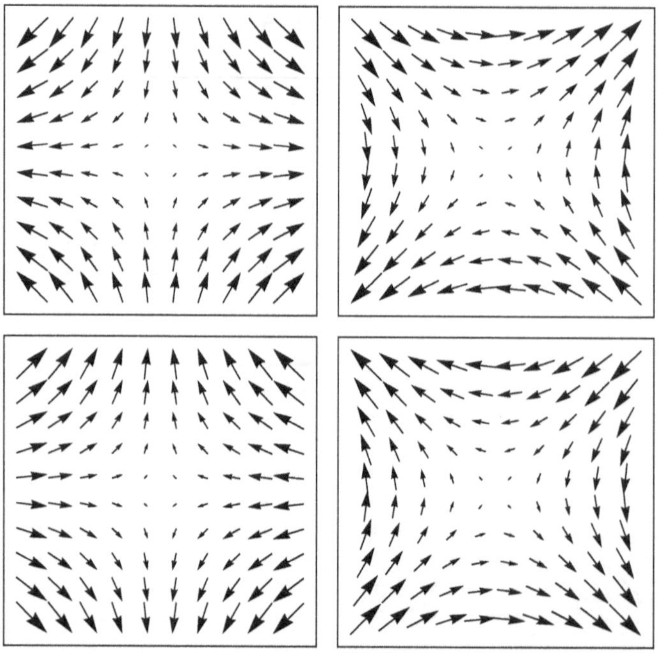

Figure 10.3 Snapshots of the electric field vector in the transverse plane for $M = 2$ and $\chi = 1$, taken at $t = 0,\ T/4,\ T/2,\ 3T/4$, where T is the period of the wave.

as in Fig. 10.4. We can easily explain qualitatively these results, with the help of the formulas (10.15) and Fig. 10.1. First, when $k_\perp/k_\parallel$ is small, only the Bessel function $J_{M-\chi}$ contributes significantly to the transverse components of the fields. This observation underscores the role of the *orbital angular momentum.* For realistic Bessel beams only one orbital angular momentum component, with $M - \chi$, plays the dominant role. Therefore, for $\chi = 1$, the trapping will occur only for $M = 0$, 2 and 3 but not for $M = 1$. This is so because in the last case the dominant part comes from the Bessel function J_0 and near the beam center the field configuration is that of a circularly polarized plane wave. Trapping is the best for $M = 0$ and $M = 2$ and for $\chi = 1$ because only in these two cases the dominant role in the formula (10.9) is played by the Bessel function J_1. Therefore, the field strength increases linearly with the distance from the beam axis, as in the Paul's mechanical model. For $M = 3$ the trapping is weak, because the Bessel function J_2 increases more slowly near the beam center, so that the radius of the effective potential well is larger. With the increase of M, the trapping will become weaker and weaker.

The problem of an electron motion near the Bessel beam axis was solved explicitly in [3] in classical and in quantum mechanics for the fields described in the paraxial approximation. The analytic solution in this case was possible due to the fact that the field configuration close to the center of the beam with $M = 2$ and $\chi = 1$ is particularly simple– the field increases linearly with the distance from the beam axis. The main feature of the

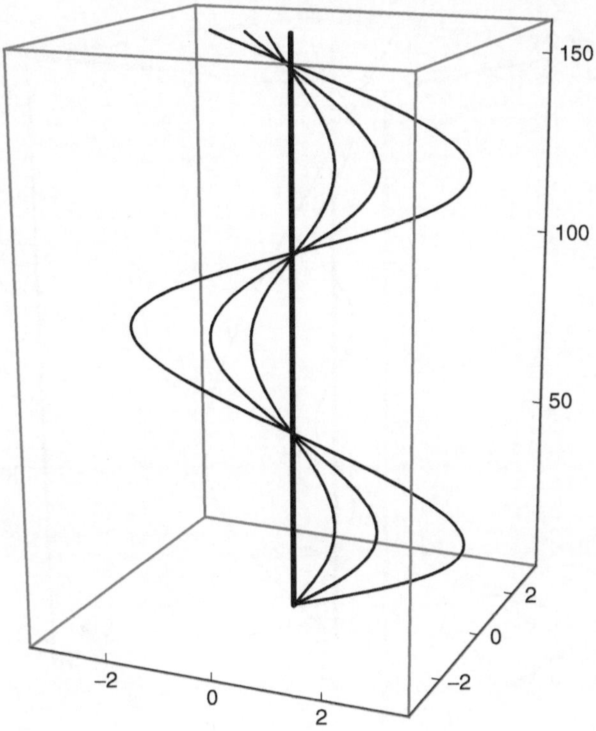

Figure 10.4 Three trajectories of an electron guided by the Bessel beam with $M = 2$ and $\chi = 1$ obtained for different initial transverse velocities. All distances are measured in units of the longitudinal wavelength. Thick line represents the beam axis. Note a large difference in the scaling of transverse and longitudinal distances.

analytic solutions is the separation of the longitudinal motion from the motion in the transverse plane. The motion in the transverse plane is bounded whereas the longitudinal motion decomposes into oscillations and a uniform motion with velocity determined by the initial conditions. The analytic solutions of the equations of motion (10.17) in this simplified model fully confirm our numerical results.

Bessel functions vanish not only at the center of the beam but they also have a succession of other zeros. It turns out that charged particles may be trapped in regions corresponding to these zeros. In Fig. 10.6 we show the four trajectories trapped at the first four zeros for $M = 2$ and $\chi = 1$.

10.4 Ponderomotive potential

The trajectories of particles trapped by Bessel beams in the nonrelativistic regime can be reproduced remarkably well in the ponderomotive approximation. The notion of the ponderomotive potential introduced by Gaponov and Miller [12] and independently by

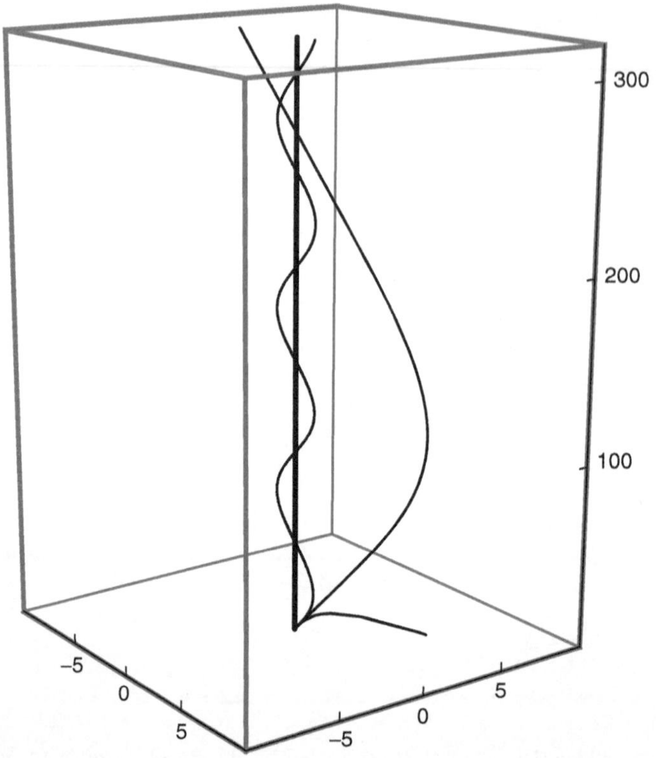

Figure 10.5 Four trajectories of an electron guided by the Bessel beams with $M = 0$, 1, 2 and 3. The tightly trapped trajectories for $M = 0$ and $M = 2$ are indistinguishable. Trapping is less pronounced for $M = 3$ (wide trajectory). There is no trapping for $M = 1$ (runaway trajectory).

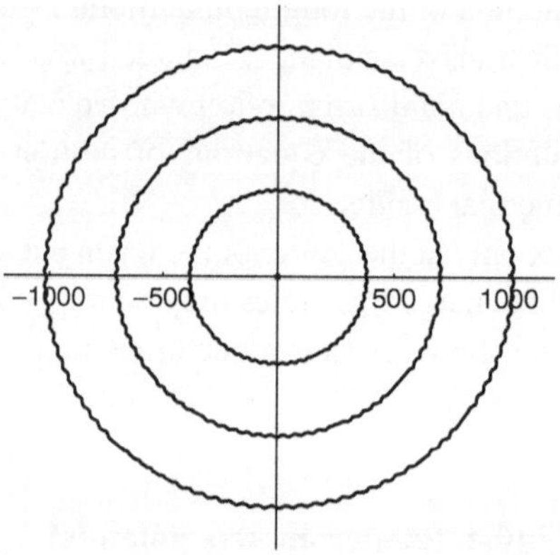

Figure 10.6 Projections on the transverse plane of the three trajectories of an electron guided by the Bessel beams with $M = 2$. The trajectories started with the same initial velocities in the vicinity of the zeros of the Bessel function.

Boot and Harvie [13] is very useful when the particle moves slowly in an inhomogeneous, rapidly oscillating electromagnetic field. The approximation is based on the decomposition of the particle trajectory into a sum $\boldsymbol{R}(t) + \boldsymbol{r}(t)$, where the first part is slowly varying and the second part describes small but fast oscillations with zero mean value. We start from the equations of motion with the Lorentz force,

$$\ddot{\boldsymbol{R}} + \ddot{\boldsymbol{r}} = \frac{e}{m}\left[\boldsymbol{E}(\boldsymbol{R}+\boldsymbol{r},t) + \left(\dot{\boldsymbol{R}} + \dot{\boldsymbol{r}}\right) \times \boldsymbol{B}(\boldsymbol{R}+\boldsymbol{r},t)\right]. \tag{10.20}$$

The necessary condition for the applicability of the ponderomotive approximation is $|\boldsymbol{v}\cdot\nabla f|/|\omega f| \ll 1$, where f is the force field. Under this assumption, one may expand the fields in the equations of motion around the slow trajectory $\boldsymbol{R}(t)$ keeping only linear terms in $\boldsymbol{r}(t)$. In the lowest order, using the notation of Eqs. (10.1a) and (10.2), we obtain

$$\ddot{\boldsymbol{r}} = \frac{e}{m}\left[\boldsymbol{E}_c(\boldsymbol{R})\cos(\omega t) + \boldsymbol{E}_s(\boldsymbol{R})\sin(\omega t)\right]. \tag{10.21}$$

After disregarding the slow variation of $\boldsymbol{R}$ with time, we obtain by a straightforward integration

$$\boldsymbol{r} = -\frac{e}{m\omega^2}\left[\boldsymbol{E}_c(\boldsymbol{R})\cos(\omega t) + \boldsymbol{E}_s(\boldsymbol{R})\sin(\omega t)\right], \tag{10.22}$$

where we dropped the integration constants to obtain zero mean values. In the next order, the equation reads

$$\begin{aligned}\ddot{\boldsymbol{R}} = \frac{e}{m}\Big\{&\Big[(\boldsymbol{r}\cdot\nabla)\boldsymbol{E}_c(\boldsymbol{R}) + \frac{1}{\omega}\dot{\boldsymbol{r}} \times (\nabla\times\boldsymbol{E}_s(\boldsymbol{R}))\Big]\cos(\omega t)\\ &+\Big[(\boldsymbol{r}\cdot\nabla)\boldsymbol{E}_s(\boldsymbol{R}) - \frac{1}{\omega}\dot{\boldsymbol{r}} \times (\nabla\times\boldsymbol{E}_c(\boldsymbol{R}))\Big]\sin(\omega t)\Big\}.\end{aligned} \tag{10.23}$$

In the final step, after inserting the solutions of Eq. (10.21) for $\boldsymbol{r}$ and $\dot{\boldsymbol{r}}$, we perform time averaging of the right-hand side over one period of fast oscillations $2\pi/\omega$ and we obtain

$$\begin{aligned}\ddot{\boldsymbol{R}} &= -\frac{e^2}{2m^2\omega^2}\big[(\boldsymbol{E}_c(\boldsymbol{R})\cdot\nabla)\boldsymbol{E}_c(\boldsymbol{R}) + \boldsymbol{E}_s(\boldsymbol{R})\times(\nabla\times\boldsymbol{E}_s(\boldsymbol{R}))\\ &\qquad + (\boldsymbol{E}_s(\boldsymbol{R})\cdot\nabla)\boldsymbol{E}_s(\boldsymbol{R}) + \boldsymbol{E}_c(\boldsymbol{R})\times(\nabla\times\boldsymbol{E}_c(\boldsymbol{R}))\big]\\ &= -\frac{e^2}{4m^2\omega^2}\nabla[\boldsymbol{E}_c(\boldsymbol{R})^2 + \boldsymbol{E}_s(\boldsymbol{R})^2] = -\nabla V_p(\boldsymbol{R}),\end{aligned} \tag{10.24}$$

where we used the vector identity

$$\boldsymbol{E}\times(\nabla\times\boldsymbol{E}) = \frac{1}{2}\nabla\boldsymbol{E}^2 - (\boldsymbol{E}\cdot\nabla)\boldsymbol{E}. \tag{10.25}$$

The ponderomotive potential $V_p(\boldsymbol{R})$ can also be expressed as the square of the electric field averaged over time

$$V_p(\boldsymbol{R}) = \frac{e^2}{2m\omega^2}\langle \boldsymbol{E}(\boldsymbol{R},t)\cdot\boldsymbol{E}(\boldsymbol{R},t)\rangle. \tag{10.26}$$

Despite the fact that only the electric field enters into this formula, the ponderomotive potential contains also a contribution from the magnetic field, as seen from our derivation. In our dimensionless variables the equations of motion have the form

$$\ddot{\boldsymbol{\xi}} = -\frac{\gamma^2}{2}\nabla\langle \boldsymbol{E}(\boldsymbol{\xi},\tau)\cdot\boldsymbol{E}(\boldsymbol{\xi},\tau)\rangle. \tag{10.27}$$

The use of ponderomotive approximation requires some care in handling the initial conditions. There is no need to adjust the initial value of the slowly varying part of the trajectory (the term $\boldsymbol{R}$ in Eq. (10.21)). We can always identify $\boldsymbol{R}(t=0)$ with the initial position of the exact trajectory because the amplitude of fast oscillations is small. However, the adjustment of the initial velocity is necessary because the velocity of fast oscillations is large and it gives a significant contribution to the true velocity. This adjustment is especially important when the trajectory starts in a region where the field is strong. According to Eq. (10.22), $\dot{\boldsymbol{r}}$ can be related to the electric field by the following formula

$$\begin{aligned}\dot{\boldsymbol{r}} &= \frac{e}{m\omega^2}[\boldsymbol{E}_c(\boldsymbol{R})\sin(\omega t) - \boldsymbol{E}_s(\boldsymbol{R})\cos(\omega t)]\\ &= -\frac{e}{m\omega^2}[\boldsymbol{E}_c(\boldsymbol{R})\cos(\omega t+\pi/2) + \boldsymbol{E}_s(\boldsymbol{R})\sin(\omega t+\pi/2)]\\ &= -\frac{e}{m\omega^2}\boldsymbol{E}(\boldsymbol{R}, t+\pi/2/\omega).\end{aligned} \tag{10.28}$$

Since the ponderomotive approximation reproduces only the slowly changing part of the trajectory, the initial velocity of the slow motion $\dot{\boldsymbol{R}}$ must be identified with the difference of the exact solution and $\dot{\boldsymbol{r}}$. The significance of the initial velocity adjustment will be illustrated with the examples given in the next section.

The ponderomotive potential for a Bessel beam is a function of the distance from the axis only,

$$V_p(\rho) = V_0[(\kappa_+ J_{M-1}(k_\perp\rho))^2 + 2(J_M(k_\perp\rho))^2 + (\kappa_- J_{M+1}(k_\perp\rho))^2], \tag{10.29}$$

where V_0 is the amplitude determined by the beam intensity; in our dimensionless variables, $V_0 = \gamma^2/2$. For realistic Bessel beams, when $k_\perp \ll k_\parallel$, only one term in this formula contributes significantly. This term represents the contribution from the component with orbital angular momentum $M-\chi$. The ponderomotive potentials for $\chi = 1,\ k_\perp/k_\parallel = 0.01$, and for different values of M are shown in Fig. 10.7. The minima correspond to the zeros of the dominant Bessel function. Only the minima at $\rho = 0$ for $M > 1$ are exactly equal zero, the remaining ones are close to zero.

The trajectories obtained by integration of exact equations, depicted in Fig. 10.4 and in Fig. 10.5 coincide with those obtained from the ponderomotive potential (10.29).

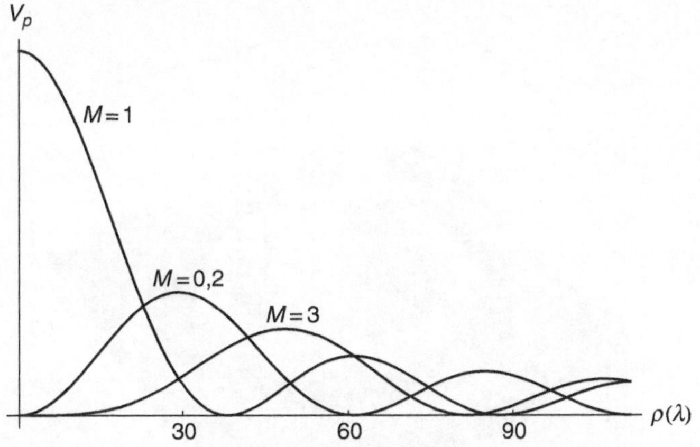

Figure 10.7 Ponderomotive potential for the Bessel beams as a function of the distance from beam axis ρ.

The vertical line in those figures is the line of the potential minima (except for $M = 1$, where it is a maximum). For $M > 1$ the beam axis is an obvious potential minimum, since the whole electromagnetic field vanishes there. For $M = 0$ the z-component of the field does not vanish on the axis but its contribution is very small when $k_\perp/k_\parallel$ is small. The notion of the ponderomotive potential enables one to understand the existence of a limiting transverse particle velocity that allows for trapping. The transverse kinetic energy cannot exceed the height of the ponderomotive potential.

The ponderomotive potential for the Bessel beams, in addition to the potential well along the axis, has a succession of potential valleys corresponding to the minima in Fig. 10.7. In Fig. 10.8 we show the potential for $M = 2$ and $\chi = 1$ as a function of x and y where the potential valleys are clearly seen. The trajectories depicted in Fig. 10.6 are trapped in those valleys and they are reproduced very well by solving the equations with the ponderomotive potential.

So far we considered trapping by beams carrying definite angular momentum. Such beams are most natural in this context since the ponderomotive potential has easily identified minima. However, as shown in the next section, trapping is found also for field configurations that do not have definite angular momentum.

10.5 Trapping of particles by superpositions of Bessel beams

10.5.1 Trapping of particles along a helix

We shall now consider the motion of an electron in the superposition of a Bessel beam and a circularly polarized plane wave. We assume that both waves propagate in the same direction and have the same frequency and the same helicity. The dimensionless electric and magnetic fields are in this case ($\chi = 1$)

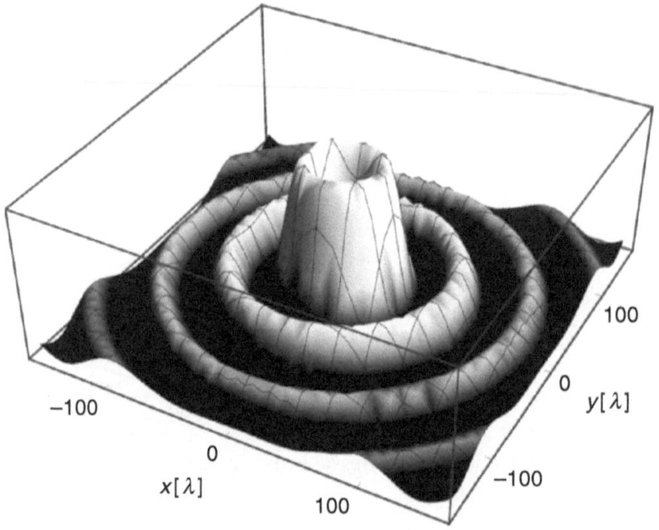

Figure 10.8 Ponderomotive potential for the Bessel beam with $M = 2$ and $\chi = 1$ as a function of x and y.

$$\begin{pmatrix} \mathcal{E}_x \\ \mathcal{E}_y \\ \mathcal{E}_z \end{pmatrix} = \begin{pmatrix} a\cos(\omega(z/c - t)) + (\kappa_- c_{M+1} + \kappa_+ c_{M-1}) \\ -a\sin(\omega(z/c - t)) + (\kappa_- s_{M+1} - \kappa_+ s_{M-1}) \\ 2s_M \end{pmatrix}, \tag{10.30a}$$

$$\begin{pmatrix} \mathcal{B}_x \\ \mathcal{B}_y \\ \mathcal{B}_z \end{pmatrix} = \begin{pmatrix} a\sin(\omega(z/c - t)) + (\kappa_- s_{M+1} + \kappa_+ s_{M-1}) \\ a\cos(\omega(z/c - t)) + (-\kappa_- c_{M+1} + \kappa_+ c_{M-1}) \\ -2c_M \end{pmatrix}, \tag{10.30b}$$

where a measures the relative strength of the plane wave as compared to the Bessel beam. The ponderomotive potential is given by the formula

$$\begin{aligned} V_p(\rho, \phi, z) = V_0\big[&(\kappa_+ J_{M-1}(k_\perp\rho))^2 + 2\,(J_M(k_\perp\rho))^2 + (\kappa_- J_{M+1}(k_\perp\rho))^2 \\ &+ a^2 + 2a\kappa_+ J_{M-1}(k_\perp\rho)\cos\big((M-1)\phi - (\omega c - k_\parallel)z\big)\big]. \end{aligned} \tag{10.31}$$

The minima of $V_p(\rho, \phi, z)$ can be easily found. The minimum in variables ϕ and z is reached when the cosine attains the lowest value -1. The minimum in ρ for this value of the cosine can be easily found numerically. These condition define the helix $(\omega/c - k_\parallel)z = (M - 1)\phi + \pi$. The line of the helix traces the minima of the ponderomotive potential and is the analog of the beam axis for a Bessel beam. The helical motion of these minima along the beam axis is visualized in Fig. 10.9. Therefore, in this case we can expect

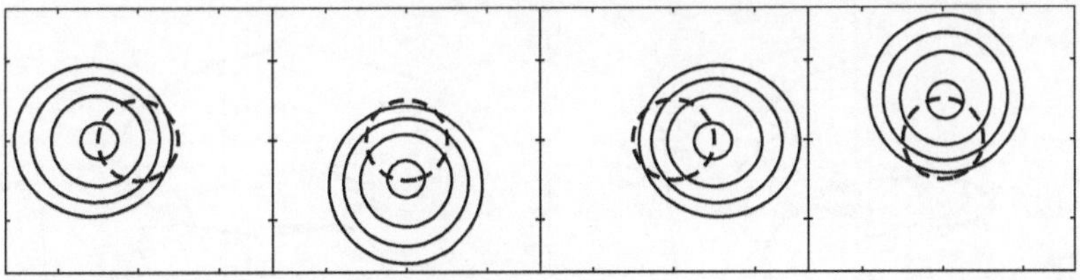

Figure 10.9 Contour lines of the ponderomotive potential in the xy plane created by a superposition of the Bessel beam with $M = 2$ and a plane wave with the same frequency ω drawn for four values of z. The dashed circle represents the projection of the helix on the xy plane.

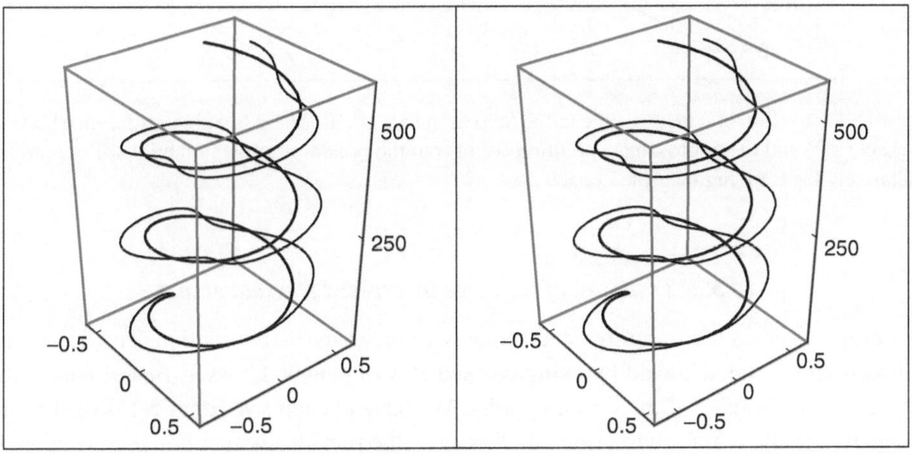

Figure 10.10 Motion of a slow particle guided by the helix (thick line) created by a superposition of the Bessel beam and a plane wave. There are no noticeable differences between the exact solution (left) and the ponderomotive approximation (right).

that the electrons will be guided along the helical line. However, this requires stronger fields. For electrons with initial transverse velocities the same as for a single Bessel beam we increased γ to the value 0.0005. In Fig. 10.10 we show the trajectories for $M = 2$, $\chi = 1$, and $a = 3$ obtained by solving exact equations and those with the ponderomotive potential. The small parameter $k_\perp/k_\parallel$ determines the radius of the helix and to make this figure more transparent we have increased this parameter tenfold from the value used in previous figures.

In order to trap more energetic particles, we need stronger fields. As an example, we have chosen the initial transverse velocity as 0.01c and we have increased the field strength to $\gamma = 0.01$. In Fig. 10.11 we show that the particle is trapped around the helix. The ponderomotive approximation is still applicable but it does not reproduce exactly the fine details of the trajectories.

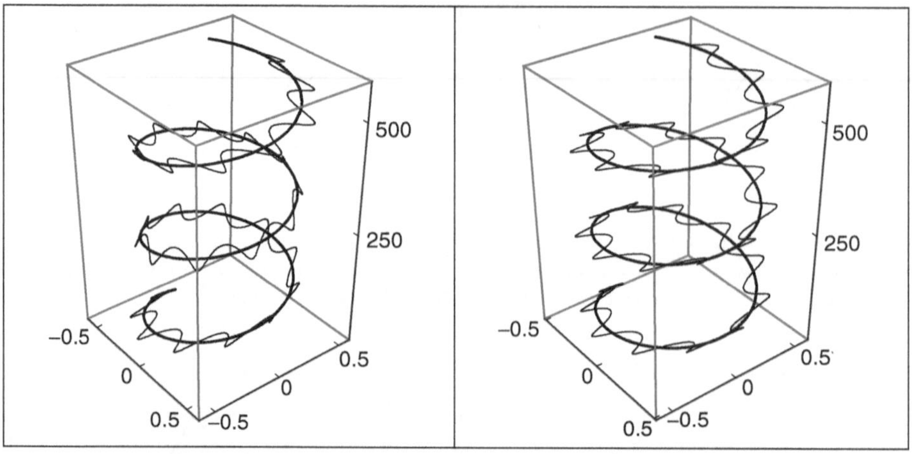

Figure 10.11 Motion of a fast particle in the field as in Fig. 10.10. There is a clear difference between the exact (left) and the approximate solution (right) but the ponderomotive potential still reproduces qualitatively the trapping near the helix.

10.5.2 Trapping of particles by crossed Bessel beams

In all previous cases the trapping occurred only in the transverse direction. The trapping in all directions can be achieved by using crossed Bessel beams. Crossed Bessel beams trap particles in the vicinity of the crossing point. We have chosen two identical Bessel beams propagating in the x and y directions. In this case, the ponderomotive potential is given by the formula:

$$\begin{aligned} V_p = V_0\big[&\kappa_+^2(J_{M-1}^2(\rho_x) + J_{M-1}^2(\rho_y)) + \kappa_-^2(J_{M+1}^2(\rho_x) + J_{M+1}^2(\rho_y)) \\ &- 2\kappa_+ J_M(\rho_x)J_{M-1}(\rho_y)\cos(\psi + \phi_y) + 2\kappa_- J_M(\rho_x)J_{M+1}(\rho_y)\cos(\psi - \phi_y) \\ &- 2\kappa_+ J_{M-1}(\rho_x)J_M(\rho_y)\sin(\psi - \phi_x) - 2\kappa_- J_{M+1}(\rho_x)J_M(\rho_y)\sin(\psi + \phi_x) \\ &- \kappa_+^2 J_{M-1}(\rho_x)J_{M-1}(\rho_y)\sin(\psi - \phi_x + \phi_y) \\ &+ \kappa_-^2 J_{M+1}(\rho_x)J_{M+1}(\rho_y)\sin(\psi + \phi_x - \phi_y) \\ &- J_{M-1}(\rho_x)J_{M+1}(\rho_y)\sin(\psi - \phi_x - \phi_y) \\ &+ J_{M+1}(\rho_x)J_{M-1}(\rho_y)\sin(\psi + \phi_x + \phi_y) \\ &+ 2(J_M^2(\rho_x) + J_M^2(\rho_y))\big], \end{aligned} \tag{10.32}$$

where we used the following abbreviations:

$$\begin{aligned} &\phi_x = \arctan(z/y), \quad \phi_y = \arctan(x/z), \quad \psi = k_\parallel(x - y) + M(\phi_x - \phi_y), \\ &\rho_x = k_\perp\sqrt{y^2 + z^2}, \quad \rho_y = k_\perp\sqrt{z^2 + x^2}. \end{aligned} \tag{10.33}$$

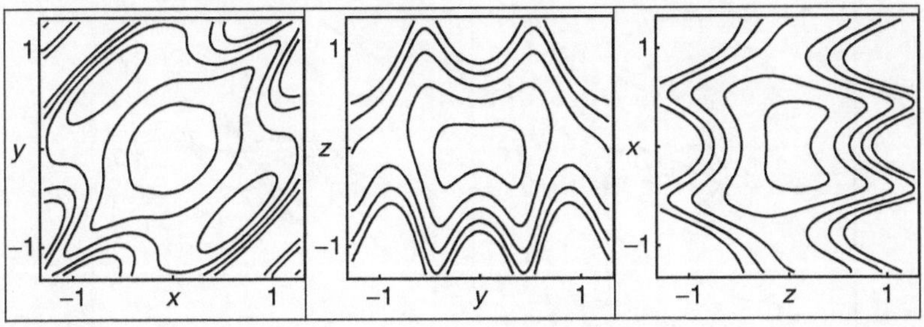

Figure 10.12 Contour lines in the xy, yz, and zx planes of the ponderomotive potential for two crossed Bessel beams with $M = 0$ propagating in the x and y directions. There is a potential minimum at the crossing point.

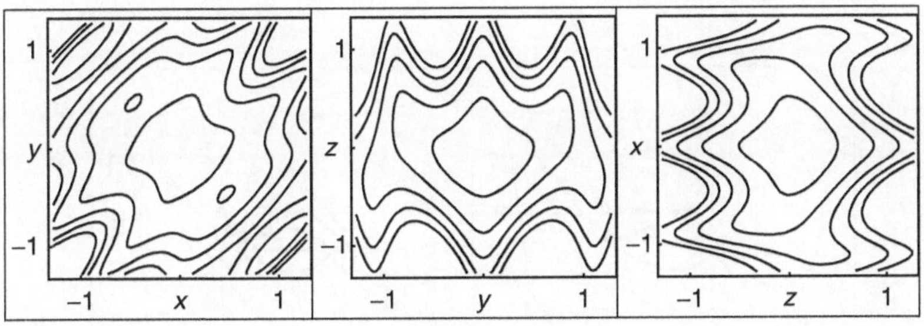

Figure 10.13 Contour lines in the xy, yz, and zx planes of the ponderomotive potential for two crossed Bessel beams with $M = 2$ propagating in the x and y directions. There is a potential minimum at the crossing point.

Again the value $M = 1$ is exceptional – it does not lead to trapping. The shapes of the potential wells for $M = 0$ and $M = 2$ are very similar, as shown in Fig. 10.12 and Fig. 10.13). These shapes clearly indicate that the potentials produced by crossed Bessel beams can trap particles in the vicinity of the beam crossing point. Solutions of the equations of motion fully confirm these predictions. For higher values of M trapping is less effective since the wells become shallower.

For crossed Bessel beams the ponderomotive approximation is less successful than for a single Bessel beam. In Fig. 10.14 and Fig. 10.15 the exact trajectory for $M = 0$ is compared with the approximate trajectories. Even though the adjustment of the initial velocity significantly improves the result, there remain substantial differences. The discrepancy is most likely due to a rather complicated shape of the potential well which may lead to a chaotic behavior–small changes produce large effects.

10.5.3 Trapping of particles by a standing Bessel wave

In all previous cases we considered trapping by running waves. In practical applications it may be easier to work with standing waves. It turns out that a standing Bessel wave can also

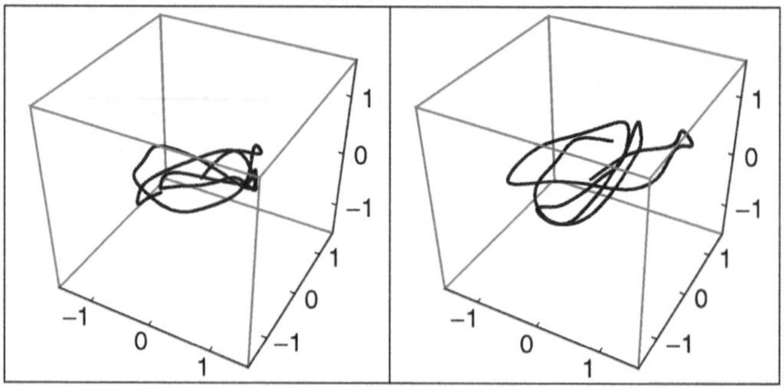

Figure 10.14 Trajectories of a particle trapped in the vicinity of the potential minimum. Exact trajectory (left) is completely different from the trajectory obtained in the ponderomotive approximation without the velocity adjustment (right).

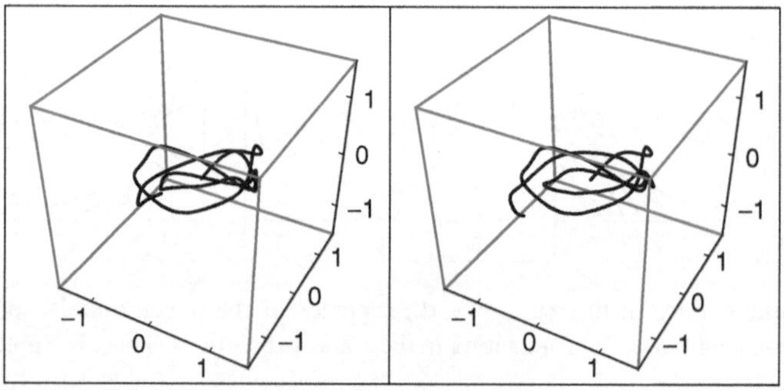

Figure 10.15 The trajectory obtained in the ponderomotive approximation with the velocity adjustment (right) is much closer to the exact trajectory (left) than in Fig. 10.14.

fully trap charged particles. A standing Bessel wave is constructed by adding two Bessel beams that differ only in the sign of $k_\parallel$. The ponderomotive potential of this superposition is given by the formula

$$V_p(\boldsymbol{r}) = 2V_0\Big[\,2k_\parallel^2/k_\perp^2\left(J_{M-1}^2(k_\perp\rho) + J_{M+1}^2(k_\perp\rho)\right) + \left(1+\cos(2k_\parallel z)\right)\left(J_{M-1}^2(k_\perp\rho) + J_{M+1}^2(k_\perp\rho) + 2J_M^2(k_\perp\rho)\right)\Big]. \qquad (10.34)$$

The ponderomotive potential is a periodic function of z with the period equal to $\pi/k_\parallel$. The amplitude of these oscillations is exceedingly small. Owing to the large value of $k_\parallel^2/k_\perp^2$, the z-independent terms dominate. The case $M=0$ is exceptional since then the z-dependent term dominates near the beam axis. In Fig. 10.16 we show contour lines of the ponderomotive potential for $M=0$ cut by the xz plane that exhibit the potential minima. The trajectories of a particle trapped in three dimensions near such a minimum are shown in Figs. 10.17–10.19. In Fig. 10.17 we have a fairly good agreement between the exact

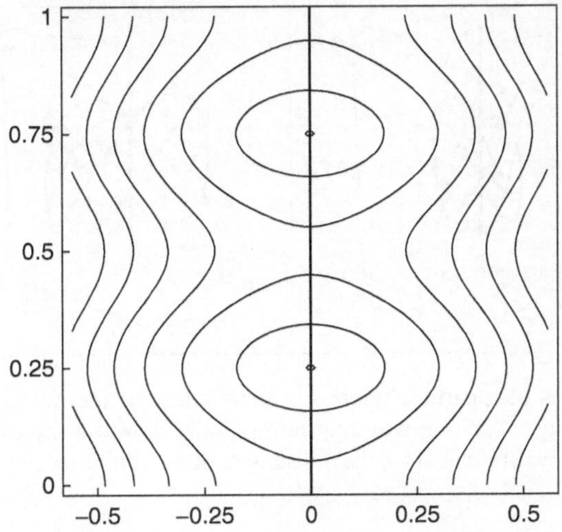

Figure 10.16 Contour lines in the xz plane of the ponderomotive potential for the standing Bessel wave ($M = 0$). Owing to cylindrical symmetry the contour lines would be the same after an arbitrary rotation around the beam axis shown as the center line. The minima of the potential are marked by the dots.

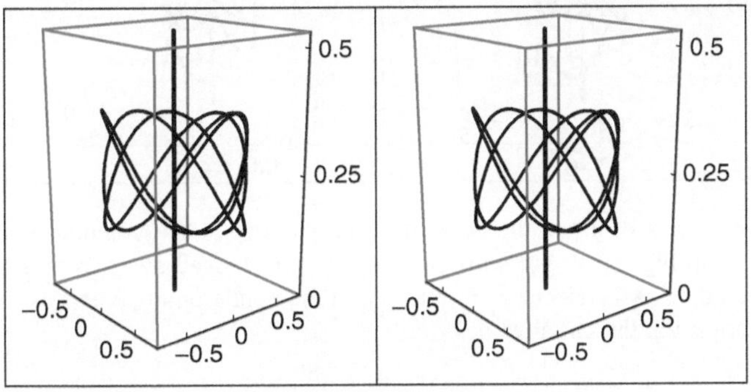

Figure 10.17 Trajectory of a particle trapped in the vicinity of the potential minimum at $z = \pi/2k_{\|}$. Thick line is the beam axis. There is no noticeable difference between the exact solution (left) and the one obtained within the ponderomotive approximation (right).

trajectory and its ponderomotive approximation because the starting point is very close to the potential minimum. In Fig. 10.18 the trajectories are completely different because the adjustment of the initial velocity has not been made. Finally, in Fig. 10.19 the adjustment has been made and the agreement has been fully restored.

10.6 Outlook

We have shown that Bessel beams might be used as a "waveguide" and also as a trap for charged particles. So far we only considered a single charged particle. The extension of

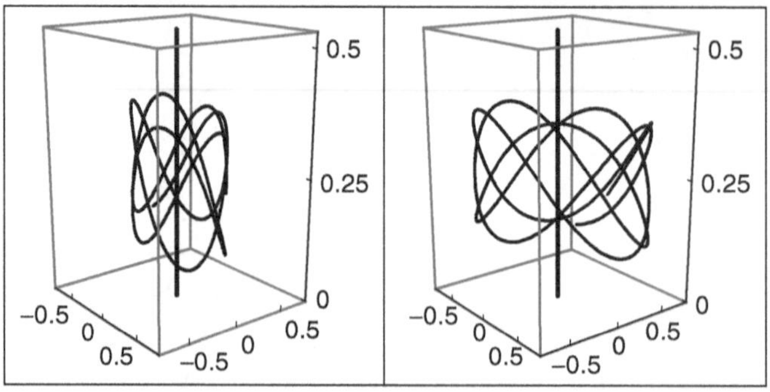

Figure 10.18 Trajectories of a particle trapped in the vicinity of the potential minimum with the same parameters as in Fig. 10.17 but with the starting point moved away from the minimum. The trajectories are completely different since the adjustment of the initial velocity has not been made.

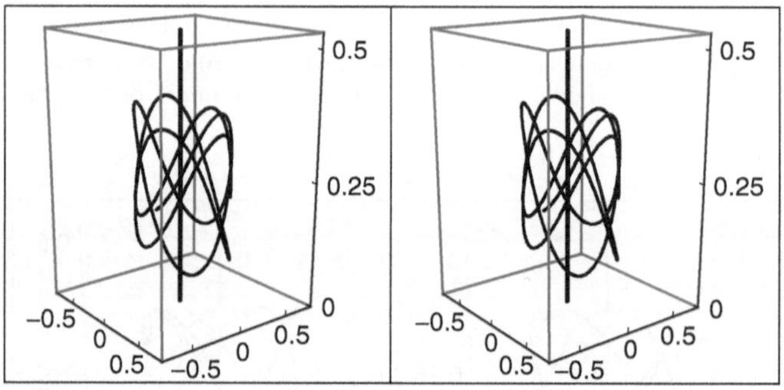

Figure 10.19 Trajectories of a particle trapped in the vicinity of the potential minimum with the same parameters as in Fig. 10.17 but with the starting point moved away from the minimum. The agreement between the exact trajectory and the one in the ponderomotive approximations has been fully restored by adjusting the initial velocity.

our results to many particles is an open problem because it would require the inclusion of the Coulomb interaction between particles. The standing Bessel wave configuration offers a possibility of trapping many particles in successive minima of the ponderomotive potential. In this way one may construct an analog of the optical lattice for charged particles. The only condition for such a construction is that the Coulomb repulsion should not overcome the ponderomotive potential barrier. This means that the intensity of the beam must satisfy the inequality $I > I_c = 4\pi m(c/\lambda)^3$. For the wavelength $\lambda = 6 \cdot 10^{-7}$ m, this gives a rather high (but not totally unrealistic) value, $I_c = 1.4 \cdot 10^{15}$ W/m^2. When the intensity obeys the condition $I < 2I_c$ then only one electron can be trapped at each potential minimum– the Coulomb repulsion will push extra electrons away. One may significantly improve the standing wave trap by adding a constant magnetic field in the direction perpendicular to the beam axis. Such a magnetic field will also quantize the spin degree of freedom of electrons, transforming each electron into a qubit. All these conditions apply to electrons at rest. In order to hold electrons that undergo motion in a trap, we need higher intensities.

Acknowledgements

We would like to thank Piotr Kucharski for useful comments. We acknowledge support from the Polish Ministry of Science and Higher Education under a grant for the years 2008–2010.

References

[1] H. Dehmelt, Experiment with an isolated subatomic particle at rest, *Rev. Mod. Phys.* **62**, 525 (1990).
[2] W. Paul, Electromagnetic traps for charged and neutral particles, *Rev. Mod. Phys.* **62**, 531 (1990).
[3] I. Bialynicki-Birula, Particle beams guided by electromagnetic vortices: new solutions of the Lorentz, Schrödinger, Klein-Gordon, and Dirac equations, *Phys. Rev. Lett.* **93**, 020402 (2004).
[4] I. Bialynicki-Birula, Z. Bialynicka-Birula, and B. Chmura, Trojan states of electrons guided by Bessel beams, *Laser Phys.* **15**, 1371 (2005).
[5] I. Bialynicki-Birula and T. Radożycki, Pinning and transport of cyclotron (Landau) orbits by electromagnetic vortices, *Phys. Rev. A* **73**, 052114 (2006).
[6] V. H. Mellado, S. Hacyan, and R. Jáuregui, Trapping and acceleration of charged particles in Bessel beams, *Laser and Particle Beams* **24**, 1 (2006).
[7] D. Gerlich, Molecular ions and nanoparticles in RF and AC traps, *Hyperfine Interactions* **146/147**, 293 (2003).
[8] I. Bialynicki-Birula, *Progress in Optics*, ed. E. Wolf (Amsterdam: Elsevier, 1996), Vol. 36; arXiv:quant-ph/0508202.
[9] V. Trkal, A note on the hydrodynamics of viscous fluids, *Czech. J. of Phys.* **44**, 97 (1994).
[10] I. Bialynicki-Birula and Z. Bialynicka-Birula, Beams of electromagnetic radiation carrying angular momentum: the Riemann–Silberstein vector and the classical-quantum correspondence, *Opt. Comm.* **263**, 342 (2006).
[11] I. Bialynicki-Birula and Z. Bialynicka-Birula, Why photons cannot be sharply localized, *Phys. Rev. A.* **79**, 032112 (2009).
[12] E. V. Gaponov and M. E. Miller, On the potential wells for charged particles in a high-frequency electromagnetic field, *Zh. Eksp. Teor. Fiz.* **34**, 242 (1958); *JETP* **7**, 68 (1958).
[13] H. A. H. Boot and R. B. R.-S. Harvie, Charged particles in a non-uniform radio-frequency field, *Nature* **180**, 1187 (1957).

11
Theory of atoms in twisted light

M. BABIKER, D. L. ANDREWS AND V. E. LEMBESSIS

11.1 Introduction

The theoretical foundations for describing the influence of twisted light on atoms, ions and molecules are essentially the same as those leading to their cooling and trapping by ordinary plane wave laser light [1–4]. A third of a century has passed since the Doppler mechanism was first put forward by Wineland and Dehmelt as a mechanism for cooling ions [5], and then by Hänsch and Schawlaw for neutral atoms [6]. The key role played by the Doppler mechanism is best illustrated by the simplest case of one-dimensional optical molasses, where a two-level atom is subject to two identical counter-propagating plane waves of a frequency near to resonance with the atomic transition frequency. In terms of a classical description, at small velocities the atom experiences a friction force $\mathbf{F} = -\alpha\mathbf{v}$, where $\mathbf{v}$ is the velocity vector of the atomic centre of mass parallel to the common axis of the light beams, and α is a friction coefficient. The friction force gradually cools the atoms to progressively smaller, albeit limited, velocities and can be made to operate in all three dimensions by use of multiple beams. The cooling technique based on the Doppler mechanism has been superseded by more effective cooling mechanisms, especially evaporative and Sisyphus cooling mechanisms, culminating, with the achievement of super-cooling, in the realisation of Bose–Einstein condensates (BECs) [7]. Nevertheless, the Doppler mechanism remains important as a mechanism for the early stages of cooling and still plays a key role in the interaction of atomic systems with laser light.

This chapter is primarily concerned with how atoms in twisted laser light experience translational and trapping forces as well as additional effects in the form of rotational forces associated with the exchange and transfer of the orbital angular momentum of the light. For simplicity, we concentrate only on effects mediated by the Doppler mechanism because, as will become clear, much new physics has been explored by theory in this context over the last two decades or so. By contrast, the effort in experimental work on atoms in twisted light has been rather modest. However, some notable experimental work has been reported [8, 9]. The material presented in this chapter is not intended to be an exhaustive

The Angular Momentum of Light, ed. D. L. Andrews and M. Babiker. Published by Cambridge University Press.

account but, rather, our aim is to present the essential features of the theory and highlight the results for a number of predicted effects that should be amenable to experimental verification.

The chapter is organised as follows. In Section 11.2 we begin with a brief overview, outlining the basics of the processes involving exchange of energy momentum and orbital angular momentum responsible for the cooling and trapping of atoms in twisted laser fields. In Section 11.3 we briefly explore the issue of transfer of orbital angular momentum to atomic systems. In Section 11.4, we discuss the novel features of twisted light interacting with atoms at near-resonance, namely the radiation forces and torques acting on the atoms in unbounded space, in addition to the translational forces that also exist for twisted light. We concentrate first on a two-level system, as the most widely discussed atomic model in this context. We briefly explore the transient regime, but move on to discuss at length the steady state regime in Section 11.5. The theory of multiple twisted light beams requires appropriate coordinate transformations, which we discuss in Section 11.6. In Section 11.7 we consider three-level atoms in counter-propagating twisted light fields and evaluate the light-induced torque in that context. In Section 11.8, we discuss the case of twisted light in evanescent form due to total internal reflection at a material interface and its effects on atomic systems in the vicinity of the interface. Section 11.9 contains our comments and final conclusions.

11.2 Overview

The realisation that orbital angular momentum carried by twisted light beams can lead to observable effects when acting on two-level atoms and molecules was first put forward by Allen *et al.* [10] in their seminal paper. A number of theoretical investigations [11–22] subsequently led to significant predictions, most notably the existence of a light-induced torque [11] acting on the atomic centre of mass about the beam axis and the azimuthal Doppler shift [12]. With adaptations of the standard optical molasses setup, studies of atomic motion in twisted light beams were carried out, leading to trajectories signifying rotational features due to processes of cooling and trapping of the atoms and ions at well defined regions of space. The character of photon spin as a natural property of light when discussed along with orbital angular momentum was clarified [14] and interesting conclusions emerged from considerations of spin-orbit coupling effects [15].

There are also recent suggestions that atom trapping is best achieved in the dark regions of the twisted light beam where there would be negligible heating effects [23]. Other work exploring the selection rules when twisted light induces transitions involving electronic and centre of mass states were carried out by by van Enk [24] and Babiker *et al.* [16], while Juzeliūnas *et al.* uncovered novel features in the interactions of Bose–Einstein condensates [25, 26] with twisted light. Other studies dealt with the channelling of atoms in material structures possessing cylindrical symmetry. Theoretical studies [27] have shown that, in such structures, the channelling of atoms involves light torques similar to those produced by free space Laguerre–Gaussian beams, which have also been employed as atom guides [28]. Various other groups have explored diverse effects of optical angular momentum on atomic and molecular systems [29–37].

As pointed out above, although most of the research to date exploring the interaction of twisted light with atoms has been theoretical in nature, there has been some notable experimental activity. The first experimental work was by Tobosa and Petrov [9] who demonstrated the transfer of orbital angular momentum from twisted light to caesium atoms. Subsequently, Anderson *et al.* [8] demonstrated persistent current flow in a superfluid sodium atom Bose–Einstein condensate (BEC). Here the BEC was forced into an optical toroidal trap due to a Laguerre–Gaussian (1, 0) (so-called doughnut) beam, together with a counter-propagating Gaussian (0, 0) beam. This situation creates a twisted optical molasses which slows the atoms axially and imparts a torque on each trapped atom, leading to a rotation of the BEC trapped in the toroidal high intensity regions about the common axis, creating the persistent current. These experiments are now regarded as significant for understanding superfluidity in atomic gas condensates and could have potential applications in the realisation of superconducting atomic devices.

Another important issue is whether the left- or right-handed property of twisted light might produce differences in interaction with chiral molecules. The coupling of light to chiral systems permits electronic transitions to be simultaneously allowed by multipoles of opposite parity: the possible implications of using light with a wave-front twist are addressed in Chapter 9.

11.3 OAM transfer

It is clearly natural to argue that, just as spin angular momentum is exchanged in a radiative atomic transition one might expect orbital angular momentum to be exchanged when twisted light interacts with an atom or a molecule. This appears to be plausible since Berry [38] showed that the orbital angular momentum is an intrinsic property of all types of azimuthal phase-bearing light. Furthermore, one can expect orbital angular momentum to be involved, not just in electric dipole interactions, but also in electric quadrupole and higher electric multipole interactions with atoms. This prospect has been analysed [16], leading to the significant prediction that the exchange of OAM does occur in the electric dipole approximation, but that it affects only the motion of the atomic centre of mass. To this leading multipolar order, no OAM transfer occurs to the internal 'electronic' motion. These conclusions suggest that no experiments can detect OAM exchange between Laguerre–Gaussian light and the internal dynamics of molecular systems through changes involving electric dipole transitions. The analysis confirms the fact that OAM effects are manifest primarily in the centre mass motion by the imposition of additional forces and associated torques. The study of these additional forces is best carried out by extending the formalism of Doppler cooling and trapping to the case of light possessing OAM, as we discuss next.

11.4 Radiation forces–cooling and trapping

It is well known that the Doppler effect, which is responsible for the broadening of atomic spectral lines, can play a key role in the process of laser cooling and trapping. When an atomic system is subject to light which is red-detuned from a transition frequency, the mechanisms of absorption and subsequent emission lead to laser-induced forces against the atomic direction of motion, culminating in the cooling, i.e. a slowing down of the atomic motion. In optical molasses the atomic motion is cooled when the atoms are subject to paired, counter-propagating light beams, leading to the slowing down in all three directions.

Clearly one expects that twisted light should influence atomic motion, but in different ways from plane wave light. The key features arising when an atomic gas is subject to twisted light are: (i) there is a light-induced torque which is capable of rotating the atoms about the beam axis. This is in addition to translational effects along the beam axis; (ii) the intensity profiles of the twisted light exhibit regions of maxima and minima which act to trap the atoms. The light-induced forces and torque are, in general, both position- and time-dependent. As we explain shortly, the position- and time-dependence, in general, exhibit a transient regime, and after a sufficiently long time has elapsed (much longer than the typical timescale of the problem) a steady state regime is established.

11.4.1 Two-level atoms in twisted light

The most widely discussed model for an atom interacting with near-resonance laser light is one incorporating the centre of mass motion, with the internal motion modelled in terms of only two energy levels. The total Hamiltonian of the atom in the presence of the laser is written as

$$H = \hbar\omega a^{\dagger}a + \frac{\mathbf{P}^2}{2M} + \hbar\omega_0\pi^{\dagger}\pi - \mathrm{i}\hbar[\tilde{\pi}^{\dagger} f(\mathbf{R}) - h.c.], \tag{11.1}$$

where $h.c.$ stands for hermitian conjugate, and $\tilde{\pi}$ and $f(\mathbf{R})$ are given by

$$\tilde{\pi} = \pi \mathrm{e}^{\mathrm{i}\omega t}; \quad f(\mathbf{R}) = (\boldsymbol{\mu}_{12}\cdot\hat{\boldsymbol{\epsilon}})\alpha\mathcal{F}_{klp}(\mathbf{R})\mathrm{e}^{\mathrm{i}\Theta_{klp}(\mathbf{R})}/\hbar. \tag{11.2}$$

Here $\mathbf{P}$ is the centre-of-mass momentum operator with M the total atomic mass, ω_0 the atomic transition frequency, while π and $\pi^{\dagger}$ are the ladder operators for the two-level system. The operators a and $a^{\dagger}$ are the annihilation and creation operators of the laser light and ω is its frequency. For a coherent beam, it is appropriate to impose the classical limit in which case the a and $a^{\dagger}$ operators become c-numbers involving the parameter α such that

$$a(t) \to \alpha\mathrm{e}^{-\mathrm{i}\omega t}; \qquad a^{\dagger}(t) \to \alpha^*\mathrm{e}^{\mathrm{i}\omega t}. \tag{11.3}$$

The final term in Eq. (11.1) represents the interaction energy between the laser light and the atom, taken as electric dipole coupling in the rotating wave approximation. Note that the electric field is evaluated at $\mathbf{R}$, the position of the atomic centre of mass, where the transition dipole moment operator $\boldsymbol{\mu}$ is located. The function $f(\mathbf{R})$ in Eq. (11.2) involves the transition dipole matrix element $\boldsymbol{\mu}_{12}$ in which the atom interacts with a Laguerre–Gaussian mode, which is assumed to be polarised along $\hat{\boldsymbol{\varepsilon}}$.

The main characteristics of the light beam are the amplitude function, $\mathcal{F}_{klp}(\mathbf{R})$ and phase $\Theta_{klp}(\mathbf{R})$. For a Laguerre–Gaussian mode these are given by

$$\mathcal{F}_{klp}(\mathbf{R}) = \mathcal{F}_{k00}\frac{N_{lp}}{(1+z^2/z_R^2)^{1/2}}\left(\frac{\sqrt{2}r}{w(z)}\right)^{|l|} L_p^{|l|}\left(\frac{2r^2}{w^2(z)}\right)\mathrm{e}^{-r^2/w^2(z)}, \tag{11.4}$$

$$\Theta_{klp}(\mathbf{R}) = \frac{kr^2z}{2(z^2+z_R^2)} + l\phi + (2p+|l|+1)\tan^{-1}(z/z_R) + kz, \tag{11.5}$$

where $N_{lp} = \sqrt{p!/(|l|+p)!}$ is a normalisation constant and $\mathcal{F}_{k00}$ is the amplitude for a plane wave propagating along the z-axis with wavevector k. The function $w(z)$ is the beam width at axial coordinate z, where $w^2(z) = 2(z^2+z_R^2)/kz_R$ and z_R is the Rayleigh range. For ease of notation we shall drop the labels klp in the following analysis, but these labels can, of course, be restored when the need arises.

Next we adopt the classical approximation only for the centre of mass motion, in which the position $\mathbf{R}$ and momentum $\mathbf{P}$ assume their average values $\mathbf{r}$ and $M\mathbf{V}$, where $\mathbf{V}$ is the velocity. The internal motion of the two-level atom, however, continues to be treated quantum mechanically. This step is justified provided that the spread in the atomic wavepacket is smaller than the wavelength of the light, and that the recoil energy is much smaller than the atomic linewidth. Under these conditions, the density matrix can then be written as

$$\rho_S = \delta(\mathbf{R}-\mathbf{r})\delta(\mathbf{P}-M\mathbf{V})\rho(t), \tag{11.6}$$

where $\rho(t)$ is the density matrix of the internal motion, which is governed by

$$\frac{\mathrm{d}\rho}{\mathrm{d}t} = -\frac{\mathrm{i}}{\hbar}[H,\rho] + \mathcal{R}\rho, \tag{11.7}$$

where $\mathcal{R}\rho$ represents the relaxation processes in the two-level system. The optical Bloch equations governing the evolution of the two-level density matrix elements are as follows:

$$\begin{bmatrix}\dot{\hat{\rho}}_{21}(t)\\ \dot{\hat{\rho}}_{12}(t)\\ \dot{\rho}_{22}(t)\end{bmatrix} = \begin{bmatrix}-(\Gamma_2-\mathrm{i}\Delta) & 0 & 2f(\mathbf{r})\\ 0 & -(\Gamma_2+\mathrm{i}\Delta) & 2f^*(\mathbf{r})\\ -f^*(\mathbf{r}) & -f(\mathbf{r}) & -\Gamma_1\end{bmatrix}\begin{bmatrix}\hat{\rho}_{21}(t)\\ \hat{\rho}_{12}(t)\\ \rho_{22}(t)\end{bmatrix} + \begin{bmatrix}-f(\mathbf{r})\\ -f^*(\mathbf{r})\\ 0\end{bmatrix}. \tag{11.8}$$

We have assumed the presence of two types of relaxation processes, namely those due to inelastic collisions with a rate Γ_1, as well as those due to elastic collisions with a rate Γ_2. The symbol Δ appearing in the OBEs stands for a velocity-dependent detuning

$$\Delta = \Delta_0 - \nabla\Theta\cdot\mathbf{V}. \tag{11.9}$$

We have also set $\hat{\rho} = \tilde{\rho}\exp(-\mathrm{i}t\mathbf{V}\cdot\nabla\Theta)$ and made use of the sum rule $\rho_{11}(t) + \rho_{22}(t) = 1$.

The average force acting on the atomic centre of mass due to interaction with the light is given by

$$\langle \mathbf{F} \rangle = -\langle tr(\rho \nabla H) \rangle . \tag{11.10}$$

This force is divisible into two distinct forces

$$\langle \mathbf{F} \rangle = \langle \mathbf{F}_{\text{diss}} \rangle + \left\langle \mathbf{F}_{\text{dipole}} \right\rangle . \tag{11.11}$$

The term $\langle \mathbf{F}_{\text{diss}} \rangle$ is referred to as the dissipative force and $\left\langle \mathbf{F}_{\text{dipole}} \right\rangle$ as the dipole force

$$\langle \mathbf{F}_{\text{diss}}(\mathbf{R}, t) \rangle = -\hbar \nabla \Theta (\hat{\rho}_{21}^* f(\mathbf{R}) + \hat{\rho}_{21} f^*(\mathbf{R})), \tag{11.12}$$

$$\langle \mathbf{F}_{\text{dipole}}(\mathbf{R}, t) \rangle = \mathrm{i}\hbar \frac{\nabla \Omega}{\Omega} (\hat{\rho}_{21}^* f(\mathbf{R}) - \hat{\rho}_{21} f^*(\mathbf{R})), \tag{11.13}$$

where $\Omega(\mathbf{R})$ is the Rabi frequency, which is a function of the centre of mass coordinate

$$\hbar \Omega(\mathbf{R}) = |(\boldsymbol{\mu}_{12} \cdot \hat{\epsilon}) \alpha \mathcal{F}(\mathbf{R})|; \quad f(\mathbf{R}) = \Omega(\mathbf{R}) \mathrm{e}^{\mathrm{i}\Theta(\mathbf{R})}. \tag{11.14}$$

In this theoretical framework, the centre of mass motion is governed by Newton's second law,

$$M \frac{\mathrm{d}^2 \mathbf{R}}{\mathrm{d}t^2} = \langle \mathbf{F}(t) \rangle . \tag{11.15}$$

As this equation is second order in time, its solution requires initial conditions in the form of values of the components of position and velocity vectors $\mathbf{R}(0)$ and $\mathbf{V}(0)$ evaluated at $t = 0$. We seek to determine the atomic trajectory $\mathbf{R}(t)$ and the corresponding evolution of the velocity $\mathbf{V}(t)$.

11.4.2 Transient forces

The dissipative and dipole forces defined above typically attain steady state forms (in which they are essentially time-independent) after a sufficiently long time has elapsed from time $t = 0$, the instant at which the light has been switched on. For atomic transitions possessing a relatively small Γ (the de-excitation rate of the upper state of the atomic transition), i.e. for which the upper state has a relatively long lifetime, the time evolution of the forces cannot be ignored. The rare-earth ion Eu^{3+} provides an example for illustration in this context. The Eu^{3+} ion has a mass $M = 25.17 \times 10^{-26}$ kg, a transition ${}^5D_0 \rightarrow {}^7D_1$ with a wavelength $\lambda = 614$ nm and an upper state decay transition rate $\Gamma = 1111\ \mathrm{s}^{-1}$, which is much smaller than the Γ values for typical atomic transitions, for example for Mg^+ ion, or rubidium or sodium atoms.

If we focus on Eu^{3+} ions interacting with the $l = 1$, $p = 0$ doughnut mode with $w_0 = 35\lambda$, we are led to a scenario in which there are three special cases: (a) exact resonance; (b) strong collisions; and (c) intense field. For (a) and (b) we may assume a moderate laser intensity $I = 10^5\ \mathrm{W\ cm}^{-2}$ while for case (c) we may assume a much higher intensity, $I = 10^8\ \mathrm{W\ cm}^{-2}$. To explore the transient regime as well as the steady state, we need a

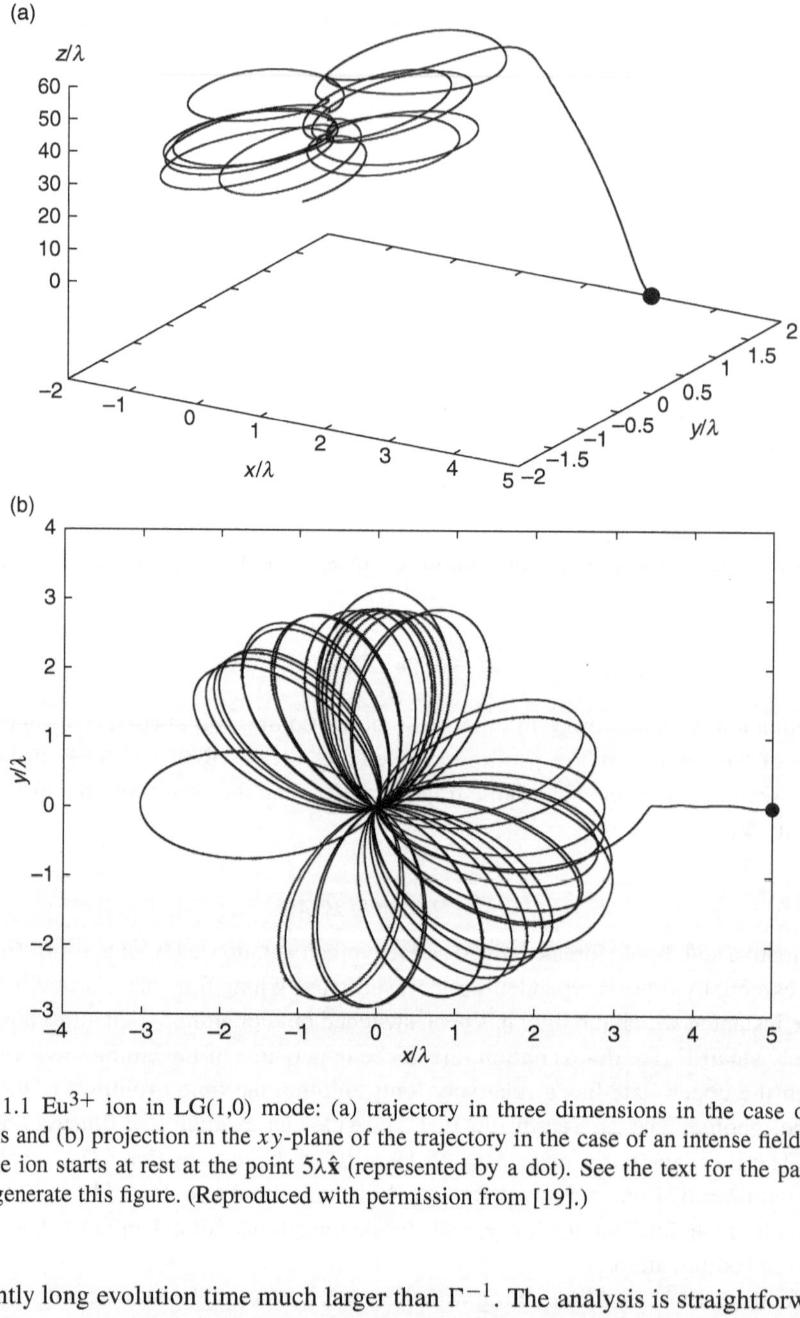

Figure 11.1 Eu^{3+} ion in LG(1,0) mode: (a) trajectory in three dimensions in the case of strong collisions and (b) projection in the xy-plane of the trajectory in the case of an intense field. In both cases, the ion starts at rest at the point $5\lambda\hat{\mathbf{x}}$ (represented by a dot). See the text for the parameters used to generate this figure. (Reproduced with permission from [19].)

sufficiently long evolution time much larger than Γ^{-1}. The analysis is straightforward but too cumbersome to present here since for most purposes, the steady state regime is applicable. The reader is referred to the published account for details [19–22] and it suffices to present here only representative results. The result shown in Fig. 11.1(a) is for the case of strong collisions. The atom follows a typical path composed of an axial motion and an in-plane motion. The in-plane motion is seen to be in the form of loops. It is in this looping trajectory in the transient regime that the effects of the light-induced transient torque are evident. A similar trajectory, this time presented as a projection on the xy-plane, is shown in Fig. 11.1(b) arising in the case of an intense external field, but here there are many more loops due to the atom gaining kinetic energy with enhanced force and torque.

11.5 Steady state forces

The steady state regime corresponds to the condition $\Gamma t \gg 1$. This signifies that the time derivatives in the optical Bloch equations can be set equal to zero. The steady state dissipative and dipole forces acting on a two-level atom due to a single beam propagating along the positive z-axis, emerging from Eqs. (11.12) and (11.13), add up to make up the total force as follows:

$$\langle \mathbf{F} \rangle_{klp} = \langle \mathbf{F}_{\text{diss}} \rangle_{klp} + \langle \mathbf{F}_{\text{dipole}} \rangle_{klp} . \tag{11.16}$$

For the dissipative force $\langle \mathbf{F}_{\text{diss}} \rangle_{klp}$ we have

$$\langle \mathbf{F}_{\text{diss}}(\mathbf{R}, \mathbf{V}) \rangle_{klp} = 2\hbar\Gamma\Omega^2_{klp}(\mathbf{R}) \left(\frac{\nabla\Theta_{klp}(\mathbf{R})}{\Delta^2_{klp}(\mathbf{R}, \mathbf{V}) + 2\Omega^2_{klp}(\mathbf{R}) + \Gamma^2} \right), \tag{11.17}$$

while for the dipole force $\langle \mathbf{F}_{\text{dipole}}(\mathbf{R}, \mathbf{V}) \rangle_{klp}$ we have

$$\langle \mathbf{F}_{\text{dipole}}(\mathbf{R}, \mathbf{V}) \rangle_{klp} = -2\hbar\Omega_{klp}(\mathbf{R})\nabla\Omega_{klp} \left(\frac{\Delta_{klp}(\mathbf{R}, \mathbf{V})}{\Delta^2_{klp}(\mathbf{R}, \mathbf{V}) + 2\Omega^2_{klp}(\mathbf{R}) + \Gamma^2} \right), \tag{11.18}$$

where we have restored the klp subscript to indicate explicit reference to a specific Laguerre–Gaussian mode. Note that the detuning $\Delta_{klp}(\mathbf{R}, \mathbf{V})$ is now both position-and velocity-dependent

$$\Delta_{klp}(\mathbf{R}, \mathbf{V}) = \Delta_0 - \mathbf{V} \cdot \nabla\Theta_{klp}(\mathbf{R}, \mathbf{V}) \tag{11.19}$$

where $\Delta_0 = \omega - \omega_0$. The above dissipative and dipole forces are twisted-beam versions of the forces in conventional laser cooling and trapping, i.e. using 'untwisted' laser light. The dissipative force here can therefore be provisionally identified in the context of optical molasses as a frictional cooling force, and the dipole force as the force responsible for the trapping of the atom in the bright or dark regions of field intensity, depending on the detuning.

11.5.1 Steady state light-induced torque

The expectation that twisted light creates a light induced torque acting on the centre of mass of the atom about the beam axis can be realised as follows. Consider for simplicity the case where $\mathbf{V}=0$; for an atomic position near the beam waist, i.e. corresponding to $z \ll z_R$, we have

$$\left\langle \mathbf{F}^0_{\mathrm{diss}}(\mathbf{R}) \right\rangle_{klp} = \frac{2\hbar\Gamma\Omega^2_{klp}(\mathbf{R})}{\Delta_0^2 + 2\Omega^2_{klp}(\mathbf{R}) + \Gamma^2} \left[k\hat{\mathbf{z}} + \frac{l}{r}\hat{\phi} \right]. \tag{11.20}$$

This force possesses an axial component as well as an azimuthal component. The azimuthal component has a non-vanishing moment about the beam axis, i.e. a torque, given by

$$\begin{aligned} \mathcal{T} &= \mathbf{r} \times \left\langle \mathbf{F}^{01}_{\mathrm{diss}} \right\rangle \\ &= \frac{2\hbar\Gamma\Omega^2_{klp}(\mathbf{R})}{\Delta_0^2 + 2\Omega^2_{klp}(\mathbf{R}) + \Gamma^2} l\hat{\mathbf{z}}. \end{aligned} \tag{11.21}$$

In the saturation limit, where $\Omega \gg \Delta_0$ and $\Omega \gg \Gamma$, we have

$$\mathcal{T} \approx \hbar l \Gamma \hat{\mathbf{z}}. \tag{11.22}$$

In this very simple and physically intuitive result the maximum torque is defined as the number of atomic transitions per second, Γ, multiplied by the angular momentum transferred $\hbar l$, in each transition. This feature of the light-induced torque was first pointed out by Babiker *et al.* [11]. It must be stressed that, in general, the light-induced torque is both position- and velocity-dependent, and therefore evolves along the trajectory.

11.5.2 Dipole potential

The velocity-independent dipole force can be derived from the dipole potential

$$\langle U(\mathbf{R})\rangle_{klp} = \frac{\hbar\Delta_0}{2} \ln\left[1 + \frac{2\Omega^2_{klp}(\mathbf{R})}{\Delta_0^2 + \Gamma^2}\right], \tag{11.23}$$

such that $\langle \mathbf{F}^0_{\mathrm{dipole}}\rangle_{klp} = -\nabla\langle U(\mathbf{R})\rangle_{klp}$. This potential would trap atoms in the high-intensity regions of the beam for $\Delta_0 < 0$ ($\omega < \omega_0$ red–detuning). For blue–detuning, $\Delta_0 > 0$, the trapping would be in the dark regions of the field. For example, consider the LG mode for which $l=1$, $p=0$. On the plane of the beam waist $z=0$, the potential minimum occurs at $r = r_0 = w_0/\sqrt{2}$. For a beam propagating along the z-axis the locus of the potential minimum is a circle in the xy-plane given by

$$x^2 + y^2 = r_0^2. \tag{11.24}$$

Expanding $\langle U(\mathbf{R})\rangle_{k10}$ about r_0 we have

$$\langle U\rangle_{k10} \approx U_0 + \frac{1}{2}\Lambda_{k10}(R - r_0)^2, \tag{11.25}$$

where U_0 is the minimum potential value with depth given by

$$|U_0| = \frac{1}{2}\hbar|\Delta_0| \ln\left[1 + \frac{2\Omega^2_{k10}(r_0)}{\Delta_0^2 + \Gamma^2}\right], \tag{11.26}$$

and Λ_{k10} is an elastic constant given by

$$\Lambda_{k10} = \frac{4\hbar|\Delta_0|}{\Delta_0^2 + 2e^{-1}\Omega_{k00}^2 + \Gamma^2}\left(\frac{e^{-1}\Omega_{k00}^2}{w_0^2}\right). \tag{11.27}$$

An atom of mass M, which would be trapped if the magnitude of its energy is less than $|U_0|$, would exhibit a vibrational motion about $r = r_0 = w_0/\sqrt{2}$ of frequency ν approximately given by $\nu = \frac{1}{2\pi}\{\Lambda_{k10}/M\}^{1/2}$.

11.5.3 Doppler shift

The force expressions in Eqs. (11.17) and (11.18) contain the effective velocity-dependent detuning Δ_{klp} defined by

$$\Delta_{klp} = \omega - \omega_0 - \nabla\Theta_{klp}.\mathbf{V}. \tag{11.28}$$

This can be written as

$$\Delta_{klp} = \omega - \omega_0 - \delta, \tag{11.29}$$

where δ is the effective Doppler shift. On substituting for $\Theta(\mathbf{R})$, using Eq. (11.5), we obtain

$$\delta = \left(\frac{krz}{z^2 + z_R^2}\right)V_r + \frac{lV_\phi}{r}$$
$$+\left\{\frac{kr^2}{2(z^2+z_R^2)}\left[1 - \frac{2z^2}{z^2+z_R^2}\right] + \frac{(2p+l+1)z_R}{z^2+z_R^2} + k\right\}V_z, \tag{11.30}$$

where V_r, V_ϕ and V_z are the velocity components of the atom centre of mass in cylindrical coordinates. The Doppler shift is composed of four types of contribution: an axial term, a Gouy term, a curvature-related term and a term associated with the azimuthal motion, such that

$$\delta = \delta_{\text{axial}} + \delta_{\text{Gouy}} + \delta_{\text{curve}} + \delta_{\text{azimuth}}. \tag{11.31}$$

The axial term coincides with a Doppler shift due to a plane wave travelling along the beam axis

$$\delta_{\text{axial}} = kV_z. \tag{11.32}$$

This would be the largest shift if the atom has a substantial axial velocity component. The Gouy shift is

$$\delta_{\text{Gouy}} = \left(\frac{(2p+|l|+1)z_{\text{R}}}{z^2+z_{\text{R}}^2}\right)V_z. \tag{11.33}$$

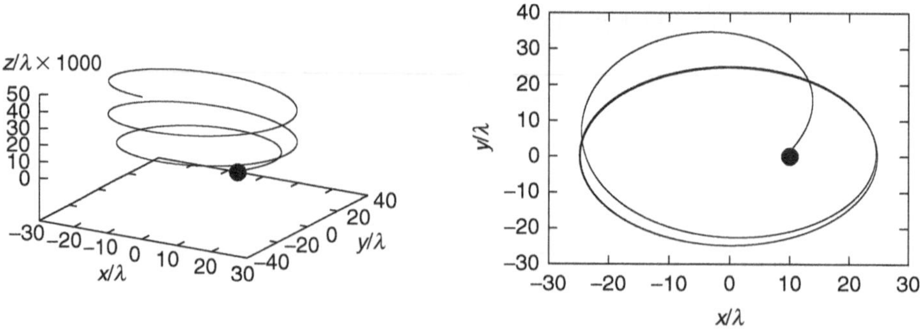

Figure 11.2 Trajectory of an Mg^+ ion in a single LG light beam with $l = 1$, $p = 0$. The ion starts from rest at $\mathbf{r}_0 = 10\lambda\hat{\mathbf{x}}$.

Since in most cases we shall assume that $z_R \gg w_0$, the Gouy shift would be negligibly small. Next, the shift due to the beam curvature is

$$\delta_{\text{curve}} = \left(\frac{krz}{z^2 + z_R^2}\right) V_r + \frac{kr^2}{2(z^2 + z_R^2)}\left[1 - \frac{2z^2}{z^2 + z_R^2}\right] V_z. \tag{11.34}$$

This is due to beam spreading in the radial direction as the beam propagates axially and arises from the curvature of the wavefront. It, too would be small if $z_R \gg w_0$. Finally, the azimuthal Doppler shift is

$$\delta_{\text{azimuth}} = \frac{lV_\phi}{r}. \tag{11.35}$$

This term increases with the winding number l of the Laguerre–Gaussian mode and so it is directly attributable to orbital angular momentum content of the light. The 'Doppler term' in Eq. (11.28) involving the gradient plays a key role in the dissipative and dipole forces and hence the atomic trajectories, as we discuss next.

11.5.4 Atom dynamics

Once the form of the forces in twisted light are known for a given set of the parameters, the motion of the atoms follows from Newton's second law subject to initial conditions. The solutions lead to the evolution of trajectory $\mathbf{R}(t)$ as well as other attributes of the motion. Unfortunately, the analytical form of $\mathbf{R}(t)$ cannot, in general, be determined and we must resort to numerical methods.

To see the influence of the torque due to twisted light on the form of the trajectory we consider two cases in which an Mg^+ ion moves in turn in the fields of two single twisted doughnut beams LG(1, 0) and LG(−1, 0), i.e differing only in the sign of l. This scenario should display two trajectories differing only in the the direction of rotation, with one turning in a sense opposite to that of the other. This situation is displayed in Figs. 11.2 and 11.3.

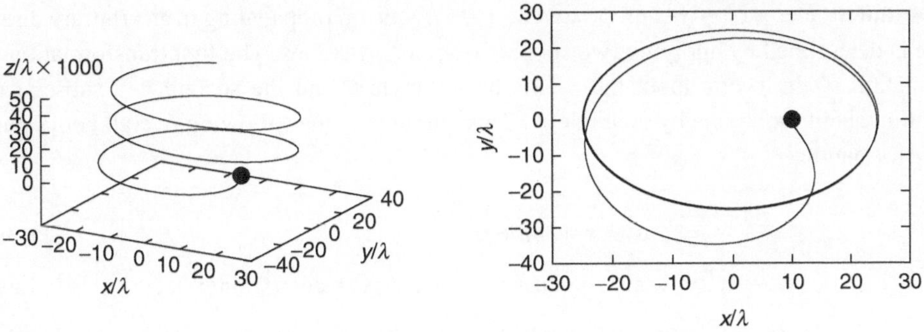

Figure 11.3 As Fig. 11.2, except that $l = -1$.

The reversal of the sense of motion between Figs. 11.2 and 11.3 is consistent with the existence of the light-induced torque.

11.6 Multiple beams

Doppler cooling is operative in optical molasses configurations in one, two and three dimensions. For beams endowed with OAM a description of optical molasses requires the specification of field distributions of individual beams referred to a common laboratory coordinate system. We therefore need to apply multiple coordinate transformations with reference to the original Cartesian axes. The total force acting on the atom is the vector sum of individual forces in a given configuration of light beams.

For a light beam of frequency ω, axial wavevector k and quantum numbers l and p coupled to an atom or an ion at a general position vector $\mathbf{R} = (r, \phi, z)$ in cylindrical coordinates, the phase $\Theta_{klp}(\mathbf{R})$ and the Rabi frequency $\Omega_{klp}(\mathbf{R})$ can be taken as follows:

$$\Theta_{klp} = l\phi + kz \tag{11.36}$$

and

$$\Omega_{klp}(\mathbf{R}) \approx \Omega_0 \left(\frac{r\sqrt{2}}{w_0}\right)^{|l|} \exp(-r^2/w_0^2) L_p^{|l|}\left(\frac{2r^2}{w_0^2}\right). \tag{11.37}$$

These expressions are applicable for a Laguerre–Gaussian beam in the limit $z \ll z_R$, where z_R is the Rayleigh range; also setting $w(z) = w_0$, i.e. ignoring all beam curvature effects.

The total forces acting on the atomic centre of mass moving with velocity $\mathbf{V} = \dot{\mathbf{R}}$ are as given in the steady state, Section 11.5, but with the approximate phase $\Theta_{klp}(\mathbf{R})$ and Rabi frequency $\Omega_{klp}(\mathbf{R})$, as in Eqs. (11.36) and (11.37). These are written in cylindrical polar coordinates, with the beam's propagation parallel to the z-axis. In order to consider multiple beams, however, it is convenient to begin by expressing the position dependence in the Rabi frequency and phase in Cartesian coordinates $\mathbf{R} = (x, y, z)$, simply by the substitutions $r = \sqrt{x^2 + y^2}$ and $\phi = \arctan(y/x)$. A beam propagating in an arbitrary direction is determined by applying two successive transformations. The first transformation is a rotation of the beam about the y-axis by an angle θ and the second is a subsequent rotation about the x-axis by an angle ψ. This amounts to the following overall coordinate transformation:

$$x \to x' = \cos(\theta)x + \sin(\theta)z, \tag{11.38}$$

$$y \to y' = -\sin(\theta)\sin(\psi)x + \cos(\psi)y + \cos(\theta)\sin(\psi)z, \tag{11.39}$$

$$z \to z' = -\sin(\theta)\cos(\psi)x - \sin(\psi)y + \cos(\theta)\cos(\psi)z. \tag{11.40}$$

By suitable choices of the angles θ and ψ we obtain the force distribution due to a twisted light beam propagating in any direction. In this manner, we are able to consider geometrical arrangements involving counter-propagating beams (especially those corresponding to one-, two- and three-dimensional optical molasses configurations), for beams possessing OAM.

We concentrate on the case of optical molasses of magnesium ions, Mg^+, which have a transition of frequency ω_0 corresponding to the transition wavelength $\lambda = 280.1$ nm and transition rate $\Gamma = 2.7 \times 10^8\ \mathrm{s}^{-1}$. The Mg^+ mass is $M = 4.0 \times 10^{-26}$ kg. We consider red-detuned light to induce trapping in areas of high intensity, $\Delta_0 = -\Gamma$ and $w_0 = 35\lambda$. The equation of motion for the Mg^+ ion is now

$$M\frac{\mathrm{d}^2}{\mathrm{d}t^2}\mathbf{R}(t) = \sum_i \langle \mathbf{F}_i \rangle, \tag{11.41}$$

where the sum is taken over individual (total) force contributions from each beam present. In the one-dimensional molasses configuration, a pair of counter-propagating beams is set up along the z-axis. The specification of the force due to the beam propagating in the negative z-direction is given by Eqs. (11.38)–(11.40) with $\theta = \pi$ and $\psi = 0$. Figure 11.4 shows the trajectory of the Mg^+ ion with $l_1 = -l_2 = 1$ and $p_1 = p_2 = 0$. The initial radial position is $r = 10\lambda$ and the initial velocity is $\mathbf{V}(0) = 5\,\hat{\mathbf{z}}\,\mathrm{m\,s^{-1}}$. The motion is analysed for a time duration equal to $2 \times 10^5\Gamma^{-1}$. It is clear that the atom is slowed down to a halt in the z-direction, while in its motion in the xy-plane it is attracted to the region of high beam intensity at approximately $r_0 > w_0/\sqrt{2}$.

Figure 11.5 indicates that once the ion is trapped axially, it continues to rotate anticlockwise about the axis, subject to a torque which, in the saturation limit, is given by $|\langle \mathcal{T} \rangle| \approx l_1\hbar\Gamma - l_2\hbar\Gamma = 2\hbar\Gamma$. The motion of the ion gives rise to an electric current equal to $e/\tau \equiv ev_s/2\pi r_0$. With v_s of about $2\,\mathrm{m\,s^{-1}}$ and $r_0 \approx w_0 = 35\lambda$ we therefore have an ionic current of the order of 1 fA (femtoamp) if one singly ionised atom is involved. It is significant that the current scales with the number of trapped ions; obviously a million or so ions can produce a current on the nA scale.

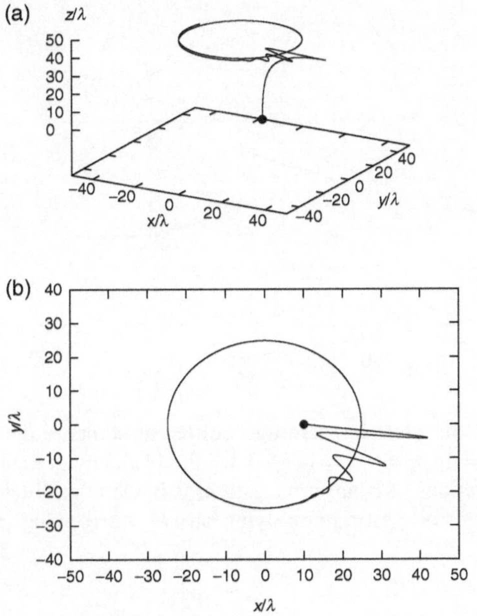

Figure 11.4 (a) Trajectory of an Mg^+ ion in two counter-propagating LG beams for which $l_1 = 1$, $p_1 = 0$; $l_2 = -1$, $p_2 = 0$. The initial conditions are such that $v_x(0) = v_y(0) = 0$; $v_z = 5\,\mathrm{m\,s^{-1}}$. (b) projection of the trajectory in the xy-plane.

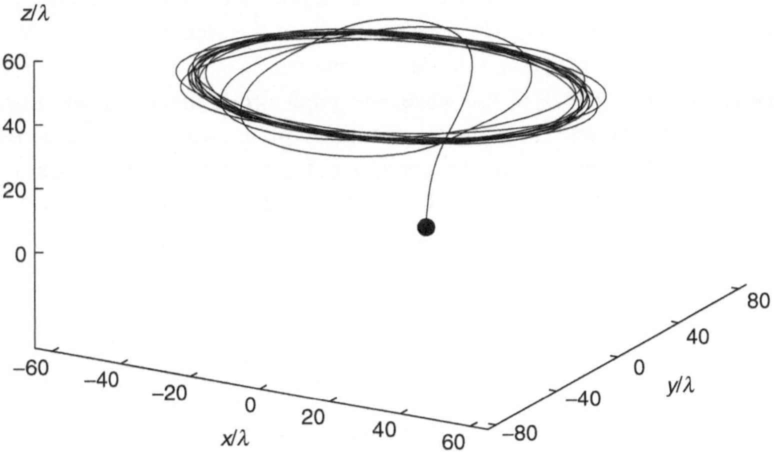

Figure 11.5 Trajectory of a Mg^+ion in one-dimensional counter-propagating twisted beams with $l_1 = -l_2 = 1$ and $p_1 = 0 = p_2$. The initial velocity is $\mathbf{v} = 5\hat{\mathbf{z}}\,\mathrm{m\,s^{-1}}$. (Reproduced with permission from [22].)

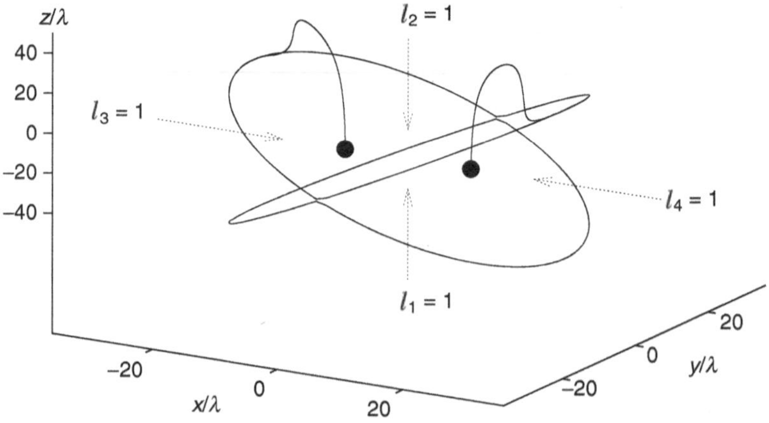

Figure 11.6 Trajectory of the Mg^+ ion starting at different points in the two-dimensional twisted optical molasses where $l_i = 1$; $p_i = 0$; $i = 1 - 4$. The initial velocity in each case is $\mathbf{v}(0) = 5\hat{\mathbf{z}}\,\mathrm{m\,s}^{-1}$. The arrows show the directions of beam propagation. It is seen that the ions end up motionless at points lying on the locus of deepest minima lying on two intersecting circles. (Reproduced with permission from [22].)

11.6.1 Optical molasses in two and three dimensions

We now introduce a second pair of counter-propagating beams along the x-axis and allow that one pair could be characterised by a different width w_0'. The total force is now the vector sum of individual forces from the four beams. The specifications of three of the beams is made with the help of the transformation equations (11.38)–(11.40). The trajectories of two Mg^+ ions positioned at different initial points, each having an initial velocity of $v_z = 5\,\mathrm{m\,s}^{-1}$, are shown in Fig. 11.6, where each of the four beams has an azimuthal index, $l = 1$, radial index, $p = 0$ and the same waist w_0.

Since by choice of l values in this case, the total torque arising from either pair of beams is zero; each ion ends up at a specific fixed point, where it remains essentially motionless. It should be noted that the deepest potential well is four times as deep as that of a single beam, with the potential minima situated along the locus of spatial points defined simultaneously by the two equations $x^2 + y^2 = w_0^2/2$ and $y^2 + z^2 = w'^2_0/2$. For $w'_0 = w_0$ these equations describe two orthogonal oblique circles representing the intersection curves of two cylinders of radii $w_0/\sqrt{2}$. Solving for x and y we have

$$x = \pm z; \;\; y = \pm\sqrt{w_0^2/2 - z^2}. \tag{11.42}$$

The locus of spatial points where the dipole potential is a minimum can be described by the parametric equations

$$x(u) = (w_0'/\sqrt{2})\cos u; \;\; y(u) = (w_0'/\sqrt{2})\sin u; \;\; z(u) = \pm\sqrt{w_0^2/2 - (w'^2_0/2)\sin^2 u}. \tag{11.43}$$

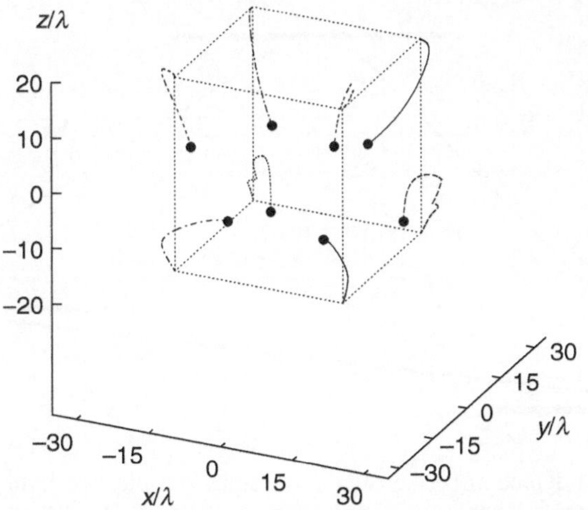

Figure 11.7 Trajectories of Mg^+ ions starting at different positions in three dimensions in the fields of three pairs of counter-propagating twisted LG beams along the three Cartesian axes, with $l_i = 1$; $p_i = 0$; $i = 1, \ldots, 6$. The initial velocity of the atom is $\mathbf{v} = 5\hat{\mathbf{z}}\,\mathrm{m\,s^{-1}}$. It is seen that each atom ends up localised at a point on a corner of a cube of side w_0. (Reproduced with permission from [22].)

All ions in the two-dimensional configuration of orthogonal counter-propagating pairs of twisted beams will be trapped at points lying on one of the two oblique circles, as determined by the initial conditions. An ensemble of Mg^+ ions with a distribution of initial positions and velocities will populate the two circles, producing two orthogonal essentially static ion loops. Associated with this system of charges would be a Coulomb field whose spatial distribution, for example, for ions uniformly distributed in the ring, can easily be evaluated. When the values of l are such that each pair of beams generates a torque, the motion becomes more complicated, but the ions will seek to congregate in the region of potential minima, while responding to the combined effects of two orthogonal torques and orthogonal axial cooling forces.

When a third pair of counter-propagating beams is added to the two-dimensional configuration, orthogonal to the plane containing the original beams, we have a three-dimensional configuration. In this case the deepest potential minima are located at eight discrete points defined by the coordinates: $x = \pm\frac{w_0}{\sqrt{2}}$, $y = \pm\frac{w_0}{\sqrt{2}}$, $z = \pm\frac{w_0}{\sqrt{2}}$. These coincide with the eight corners of a cube of side w_0, centred at the origin of coordinates. Figure 11.7 exhibits the trajectories of eight atoms which end up being trapped at the eight corners of the cube.

11.7 Three-level atoms in LG beams

In laser spectroscopy, the three-level atom model in both Λ and V configurations has exhibited a variety of novel interference effects and has revealed a number of interesting phenomena, most notably, electromagnetically induced transparency, the generation

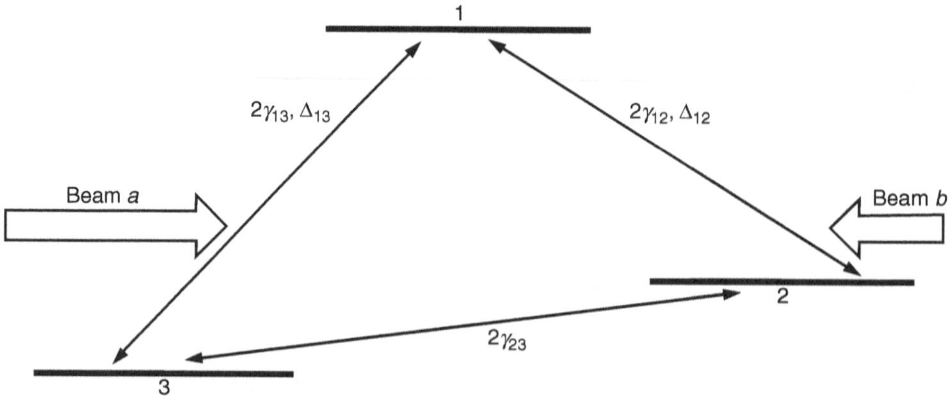

Figure 11.8 Schematic depiction of the three-level atom and the two light beams a and b. The transition frequencies are ω_{12} and ω_{13} and the corresponding upper state transition rates are γ_{12} and γ_{13}.

of slow light, and lasing without inversion. In these applications, laser light is invariably taken as ordinary (TEM_{00}) light with planar wavefronts. The three-level atom subject to interaction with light endowed with orbital angular momentum has featured in recent investigations, leading to interesting predictions involving the creation of an effective magnetic field acting on the ensemble [25, 26].

Here we are interested in the optical forces on the three-level atoms and the corresponding torque. Figure 11.8 schematically represents the three-level system and the two beams, together with relevant parameters. For simplicity, the atom is assumed to interact with two counter-propagating Laguerre–Gaussian beams for which $l = 1$, $p = 0$, i.e both 'doughnut' beams. The two counter-propagating light beams are denoted by the labels a and b, as shown in Fig. 11.8, each of which carries orbital angular momentum of the same sense and the same magnitude. We aim to derive an expression for the light-induced torque imposed by the two beams on the centre of mass of the atom about the common beam axis and, finally, to examine how the torque varies with the adjustable parameters, particularly the frequency of the light.

It is easy to see that the two beams create an optical toroidal region of high intensity and generate two physical effects, namely an optical molasses which traps the atoms axially and, by virtue of the orbital angular momentum carried by the LG beam, imparts a torque on the centre of mass of each trapped atom and so rotates the atom about the beams axis, generating an atomic current. We take the centre of mass position vector of the atom as **R** in cylindrical coordinates. The interaction of the beams with the atom can be written in terms of the quantum states in the form

$$H_{int} = -\mathrm{i}\hbar \left\{ |1\rangle \langle 3| \, f_{13} a e^{\mathrm{i}\Delta_{13} t} - |3\rangle \langle 1| \, f_{13}^{*} a^{\dagger} e^{-\mathrm{i}\Delta_{13} t} \right\}$$
$$- \mathrm{i}\hbar \left\{ |1\rangle \langle 2| \, f_{12} b e^{\mathrm{i}\Delta_{12} t} - |2\rangle \langle 1| \, f_{12}^{*} b^{\dagger} e^{-\mathrm{i}\Delta_{12} t} \right\}, \quad (11.44)$$

where the operators a, b and $a^\dagger, b^\dagger$ are annihilation and creation operators and f_{13}, f_{12} are position-dependent functions representing the interaction in the electric dipole approximation, defined as

$$f_{13} = G_{13}e^{i\Theta_a(\mathbf{R})}; \quad f_{12} = G_{12}e^{i\Theta_b(\mathbf{R})}, \tag{11.45}$$

where G_{13} and G_{12} are functions representing the Rabi frequencies and $\Theta_b(\mathbf{R})$ and $\Theta_a(\mathbf{R})$ are phase functions associated with the light beams. The derivation of the scattering force is best carried out using the density matrix formalism based on Eq. (11.44). The scattering force emerges in the form

$$\mathbf{F}_{\text{scat}} = 2\hbar\nabla\Theta_a(\mathbf{R})\left\{G_{13}\rho_{13} + G^*_{13}\rho_{31}\right\} + 2\hbar\nabla\Theta_b(\mathbf{R})\left\{G_{12}\rho_{12} + G^*_{12}\rho_{21}\right\}, \tag{11.46}$$

where ρ_{12} and ρ_{13} are density matrix elements of the three internal atomic levels. To arrive at Eq. (11.46), we have assumed that the single atom wavefunction is separable as a product of internal and centre of mass motions, and that the centre of mass motion can be described classically, while the internal dynamics is described quantum mechanically. As appears in Eq. (11.46), it turns out that the force acting on the centre of mass coordinate would then involve only density matrix elements involving the internal states.

We shall now assume that the two beams have the same spatial dependence, so we can write

$$G_{1i} = \Omega_1\mathcal{R}_i(r); \quad i = 1, 2, \tag{11.47}$$

where r is the radial cylindrical coordinate and $\mathcal{R}_1(r) = \mathcal{R}_2(r) = \mathcal{R}(r)$; with each of the two Laguerre–Gaussian beams now taken as doughnut beams we have

$$\mathcal{R}(r) = \frac{1}{(2^{|l|}|l|!)^{1/2}}\left(\frac{r\sqrt{2}}{w_0}\right)^{|l|} L_p^{|l|}\left(\frac{2r^2}{w_0^2}\right) e^{-(r^2/w_0^2)}, \tag{11.48}$$

and

$$\Theta_a = -l\phi - kz; \quad \Theta_b = -l\phi + kz. \tag{11.49}$$

We have ignored the Gouy phase and all curvature effects as we are interested in effects arising near the beam waist situated in the plane $z = 0$. Note that in the above $\mathcal{R}$ the Laguerre–Gaussian field profile depends on the radial cylindrical coordinate r, the azimuthal angle variable ϕ and the angular momentum indices l and p, along with the beam waist parameter w_0. Note also that we have assumed that $p = 0$ allowing only for doughnut modes, but the treatment can accommodate any fields belonging to the well-known family of Laguerre–Gaussian beams. We obtain for the force, using Eqs. (11.46) to (11.49),

$$\mathbf{F}_{\text{scat}} = 2\hbar k\Psi_1\mathcal{R}^2(r)\hat{\mathbf{z}} + \frac{2\hbar l}{r}\Psi_2\mathcal{R}^2(r)\hat{\boldsymbol{\phi}}, \tag{11.50}$$

with

$$\Psi_1 = -G_{12}\rho_{12} - G^*_{12}\rho_{21} + G_{13}\rho_{13} + G^*_{13}\rho_{31},$$
$$\Psi_2 = -\left\{G_{12}\rho_{12} + G^*_{12}\rho_{21} + G_{13}\rho_{13} + G^*_{13}\rho_{31}\right\}, \qquad (11.51)$$

where in Eq. (11.50) carets denote unit vectors in cylindrical coordinates. It is straightforward to show that the torque acting on the centre of mass of the atoms about the beam axis emerges in the following form:

$$\mathbf{T} = \mathbf{r} \times \mathbf{F}_{\text{scat}}$$
$$= 2\hbar l \Psi_2 \mathcal{R}^2(r)\hat{\mathbf{z}}. \qquad (11.52)$$

Note that this torque is quantised and increases with increasing winding integer number l. It also depends on the radial position and for the case $p = 0$, corresponding to doughnut modes, there is a maximum intensity region and an associated optical dipole potential trap to which the atoms will be attracted. This region is also a maximum torque region and the atoms would then experience the torque, so generating an atomic flow in the shape of a ring.

The final step in the evaluation of the torque is to find expressions for the off-diagonal density matrix elements and entering the function Ψ_2. These matrix elements emerge from a treatment of optical Bloch equations (OBEs) for a three-level system. The details of the evaluation are lengthy and are not reproduced here [39]. It suffices to say that in addition to the parameters identified above, the density matrix elements depend on the γ parameters for the different transitions, the frequency detunings Δ_{12} and Δ_{13} and the parameters Ω_{12} and Ω_{12}. For illustration we plot the function Ψ_2 in the torque expression Eq. (11.52) against the detuning Δ_{13}, choosing the following values of the parameters:

$$\Delta_{12} = -\gamma;\quad \gamma_{12} = \gamma_{13} = 0.5\gamma;\quad \Omega_{12} = 0.1\gamma;\quad \Omega_{13} = 0.6\gamma;\quad \gamma_{23} = 0.001\gamma, \quad (11.53)$$

where γ is a convenient scaling transition rate. The result is shown in Fig. 11.9 and we note from this figure that the torque reaches a maximum for some value of Δ_{13}. It also vanishes when $\Delta_{13} = -\gamma$, which corresponds to the condition of coherent population trapping (CPT) [40, 41].

The torque can be made to vanish for another choice of parameters. To show this we consider the following parameter set:

$$\Delta_{12} = 0;\quad \gamma_{12} = \gamma_{13} = 0.5\gamma;\quad \Omega_{12} = 0.1\gamma;\quad \Omega_{13} = 0.6\gamma;\quad \gamma_{23} = 0.001\gamma. \quad (11.54)$$

The results are shown in Fig. 11.10 for the torque function plotted against Δ_{13}. Clearly the torque vanishes as we achieve the CPT condition where both $\Delta_{13} = 0$ and $\Delta_{12} = 0$. This is a special case of the condition in which the transition frequency difference matches the beams frequency difference, namely

$$\omega_{13} - \omega_{12} = \omega_a - \omega_b. \qquad (11.55)$$

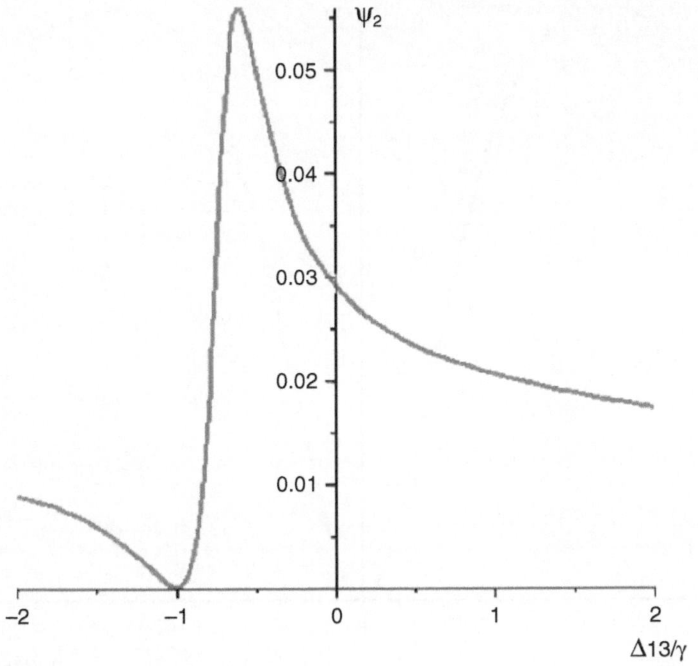

Figure 11.9 Variation of the torque function Ψ_2 with the detuning Δ_{13} (in units of γ) for the set of parameters displayed in Eq. (11.53). (Reproduced with permission from [39].)

Thus it emerges that three-level atoms also experience a light-induced torque, whose characteristics depend sensitively on the atomic parameters as well as the mode parameters. This scenario involves two counter-propagating LG beams appropriately detuned from the atomic resonances, creating a torque influencing rotational motion, as well as leading to the axial cooling of the atomic motion. We have also shown that the system is amenable to coherent population trapping [39] and in such a way that it is possible to create conditions in which the torque vanishes. This leads to the prospect of controlling rotational motion by judicious choice of the resonance conditions between the light beams and the atomic transitions.

This work is relevant to the experiments by Anderson *et al.* [8] who were first to demonstrate persistent current flow in a superfluid sodium atomic gas. The current flow was made possible by trapping the BEC in an optical toroidal trap created by a Laguerre–Gaussian (1, 0) light beam incident on the BEC, along with a counter-propagating Gaussian (0, 0) light beam. This scenario creates two physical effects, namely an optical molasses which traps the atoms axially and, by virtue of the orbital angular momentum carried by the LG beam, imparts a torque on the centre of mass of each trapped atom and so rotates the ensemble about the axis of the beams, generating the atomic current. These first experiments on superfluidity in BECs have been regarded as of fundamental significance for

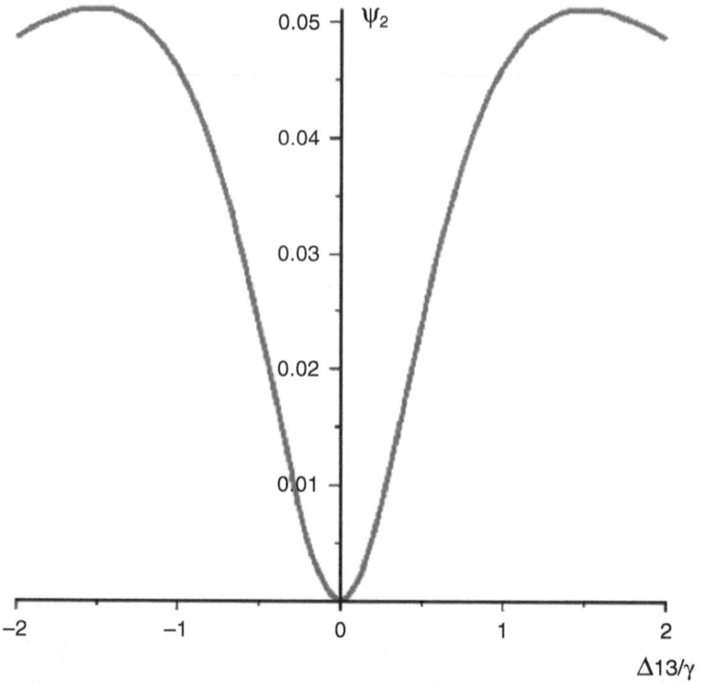

Figure 11.10 Variation of the torque function Ψ_2 (vertical axis) with the detuning Δ_{13} in units of γ for the set of parameters displayed in Eq. (11.54). (Reproduced with permission from [39].)

understanding superfluidity in atomic condensates, and they could have potential applications in the realization of superconducting atomic devices.

The arrangement we discussed here in which two counter-propagating Laguerre–Gaussian light beams create the torque should be more effective in the creation of rotational motion in a BEC, leading to persistent current flow, than the experimental arrangement of Anderson *et al.* [8], and we hope that this will spur experimental activity along these new lines.

11.8 Surface plasmonic optical vortices (SPOVs)

Finally we consider the possibility of generating optical vortices with surface-type features [42–44], and which could also be strongly plasmonic. We aim only to outline the analysis leading to the general form of the plasmonic fields, and explore their effects on atomic systems near the surface. In recent studies, Gorodetski *et al.* [43], have established that spin-based plasmonic effects can be created in nanoscale structures by the use of circularly polarised light; such phenomena depend on a geometric phase associated with a polarisation state that is locally variant in space. It has been shown that a surface spiral surrounding a nanoaperture can also confer the necessary phase conditions.

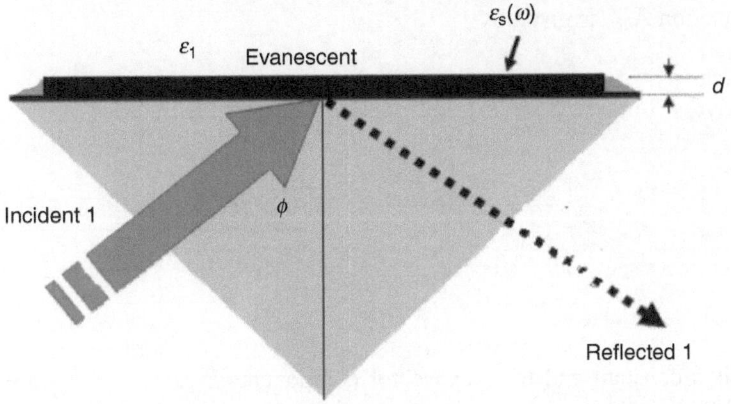

Figure 11.11 Schematic depiction of total internal reflection of a single LG beam at a planar dielectric interface with a metallic film of thickness d creating an evanescent mode. (Reproduced with permission from [45].)

Here we shall determine the behaviour associated with angular momentum whose origin is neither photon spin nor surface morphology, the analogous phase properties being primarily conferred by a helical beam structure. The present focus is on the exploitation of surface plasmonic field enhancement for the control of vicinal atoms. We first outline the procedure involved in the derivation of the plasmonic surface optical fields for the case of a single LG light beam, and the associated optical forces are identified. We then consider a system of two counter-propagating LG beams in a setup designed to laterally trap adsorbed atoms. Typical parameters are used to determine the optical forces, and to obtain the trajectories for sodium atoms.

From Maxwell's equations the electric field vector components emerge with in-plane polarisation in the three regions of a layered structure as shown in Fig. 11.11, namely a dielectric, occupying the region $z < -d$; a surface film occupying the region $-d < z < 0$, and the vacuum region occupying the space $z > 0$. These fields are then subject to boundary and phase-matching conditions and the procedure leads to the evanescent field in the vacuum region in the form

$$\mathbf{E}_{klp} = \frac{2A_{klp}}{\xi}\left(\hat{\mathbf{x}} + i\frac{k_\parallel}{k_{z1}}\hat{\mathbf{z}}\right) \mathrm{e}^{-\mathrm{i}\{\sqrt{\epsilon_1}\frac{\omega}{c}x\sin\phi + k_{z2}d\}}\mathrm{e}^{-k_{z1}z}, \tag{11.56}$$

where ξ is defined as

$$\xi = \left(1 - \mathrm{i}\frac{k_{z2}\epsilon_1}{k_{z1}\epsilon_2}\right)\cosh(k_{zs}d) + \left(\frac{k_{z2}\epsilon_1}{k_{z1}\epsilon_2} - \mathrm{i}\frac{k_{z2}\epsilon_s}{k_{zs}\epsilon_2}\right)\sinh(k_{zs}d), \tag{11.57}$$

while the function A_{klp} is given by

$$A_{klp}(x, y) = \frac{\xi_{k00} N_{lp}}{2^{p+|l|} p! |l|!} \left(\frac{[2(y^2 + (x \cos\phi + d \sin\phi)^2)]^{\frac{1}{2}}}{w_0} \right)^{|l|}$$

$$\times \, \mathrm{e}^{\frac{-[y^2 + (x\cos\phi + d\sin\phi)^2]}{w_0^2}} L^{|l|} \left\{ \frac{[y^2 + (x\cos\phi + d\sin\phi)^2]}{w_0^2} \right\}$$

$$\times \, \mathrm{e}^{-\mathrm{i}l \tan^{-1}\left[\frac{y}{(x\cos\phi - d\sin\phi}\right]}, \tag{11.58}$$

where N_{lp} is a constant and the wavevector components k_{z1}, k_{z2} and k_{zs} are in general complex, while $k_{\|}$ is the common in-plane wavevector component. The dielectric functions $\epsilon_{1,2}$ are constants determined by the constitution of the substrate and film layers, while ϵ_s is given by

$$\epsilon_s = 1 - \frac{\omega_{\mathrm{p}}^2}{\omega(\omega + \mathrm{i}\gamma)}, \tag{11.59}$$

with ω_{p} the plasma frequency and $\gamma \ll \omega_{\mathrm{p}}$ represents loss. The form of the evanescent field is now specified in full by the expressions given above.

We now consider an atom in the vacuum region with position vector $\mathbf{R(t)} = (x(t), y(t), z(t))$. The atom is again represented as two-level system, and its interaction with the surface plasmon vortex is taken in the dipole approximation, with an interaction Hamiltonian

$$H_{\mathrm{int}} = -\boldsymbol{\mu} \cdot \mathbf{E}_{klp}(\mathbf{R}). \tag{11.60}$$

This interaction is responsible for two important dynamical attributes of the atomic motion, namely the Rabi frequency $\Omega(\mathbf{R}(t))$ and the phase $\Theta(\mathbf{R}(t))$. The Rabi frequency is

$$\Omega(\mathbf{R}(t)) = \left| \frac{\boldsymbol{\mu} \cdot \mathbf{E}_{klp}(\mathbf{R}, t)}{\hbar} \right|, \tag{11.61}$$

and the phase function is

$$\Theta(\mathbf{R}(t)) = l \tan^{-1} \left\{ \frac{y}{x\cos\phi - d\sin\phi} \right\} - \frac{nx\omega \sin\phi}{c}. \tag{11.62}$$

The total steady state optical force acting on the atom at time t in the vacuum region, when its velocity vector is $\mathbf{V(t)} = \dot{\mathbf{R}}(t)$, are formally as given in the previous section. It is clear that the gradient of the second term of the phase function gives rise to a component of the dissipative force purely along the x-direction. In addition to the angular force arising from the gradient of the first term of the phase function, the atom is subsequently forced in a direction parallel to the surface. To confine the atom to an angular path in the plane parallel to the surface, it is necessary to use two internally incident beams.

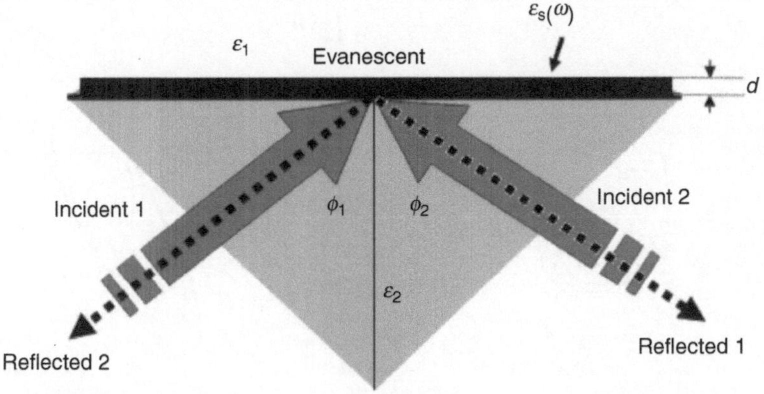

Figure 11.12 Schematic depiction of total internal reflection of two LG beams at a planar dielectric interface with a metallic film of thickness d, creating a surface plasmon using the counter-propagating evanescent modes. (Reproduced with permission from [45].)

An undesirable feature in the case of a single beam SPOV, as far as its interaction with atoms is concerned, is that there is a plane wave with a wave-vector equal to the in-plane component of the incident light travelling along the surface. The corresponding term in the phase function gives rise to a dissipative force, as for a plane wave in free space, but one that acts on the atom along the x-direction. A possible arrangement in which this undesirable motion can be eliminated is by the creation of a situation whereby the in-plane waves of the two beams counter-propagate, as shown in Fig. 11.12. Here two LG beams, labelled 1 and 2, are incident at angles ϕ_1 and ϕ_2 ; both are reflected and have field components within the film, and surface plasmonic components in the vacuum region. We assume that the two beams are identical in modal form and spatial distribution, except for their directions of propagation and the possibility of an associated change in the sign of the angular momentum quantum number l.

Once the forces acting on the atom centre of mass are specified, the motion of the atom in the vicinity of the surface can be determined. The dynamics again follows a Newtonian equation of motion, driven by a sum of the forces delivered by each beam. To illustrate the solutions of this equation leading to typical trajectories, we consider a sodium atom in two confocal, counter-propagating Laguerre–Gaussian beams for which the parameters are as follows. The beam waist is taken as $w_0 = 35\lambda$, with $\lambda = 589.0$ nm. The intensity is assumed to be $I = 2.0 \times 10^6$ W m^{-2}. The layer structure consists of a thin silver film of thickness $d = 59$ nm and electron density $n = 5.57 \times 10^{28}$m^{-3}, deposited on glass of dielectric constant $\epsilon_2 = 2.298$. The counter-propagating beams are both doughnut modes. We also take a detuning of magnitude $|\Delta_0| = 100\Gamma_0$, where $\Gamma_0 = 6.13 \times 10^7$ s^{-1}. The dipole moment is taken as $\mu = 2.6ea_B$, where a_B is the Bohr radius.

The results are displayed in Figs 11.13 and 11.14 for the same magnitude of the detuning Δ_0, but one is positive and the other is negative. As can be seen, the trajectories of

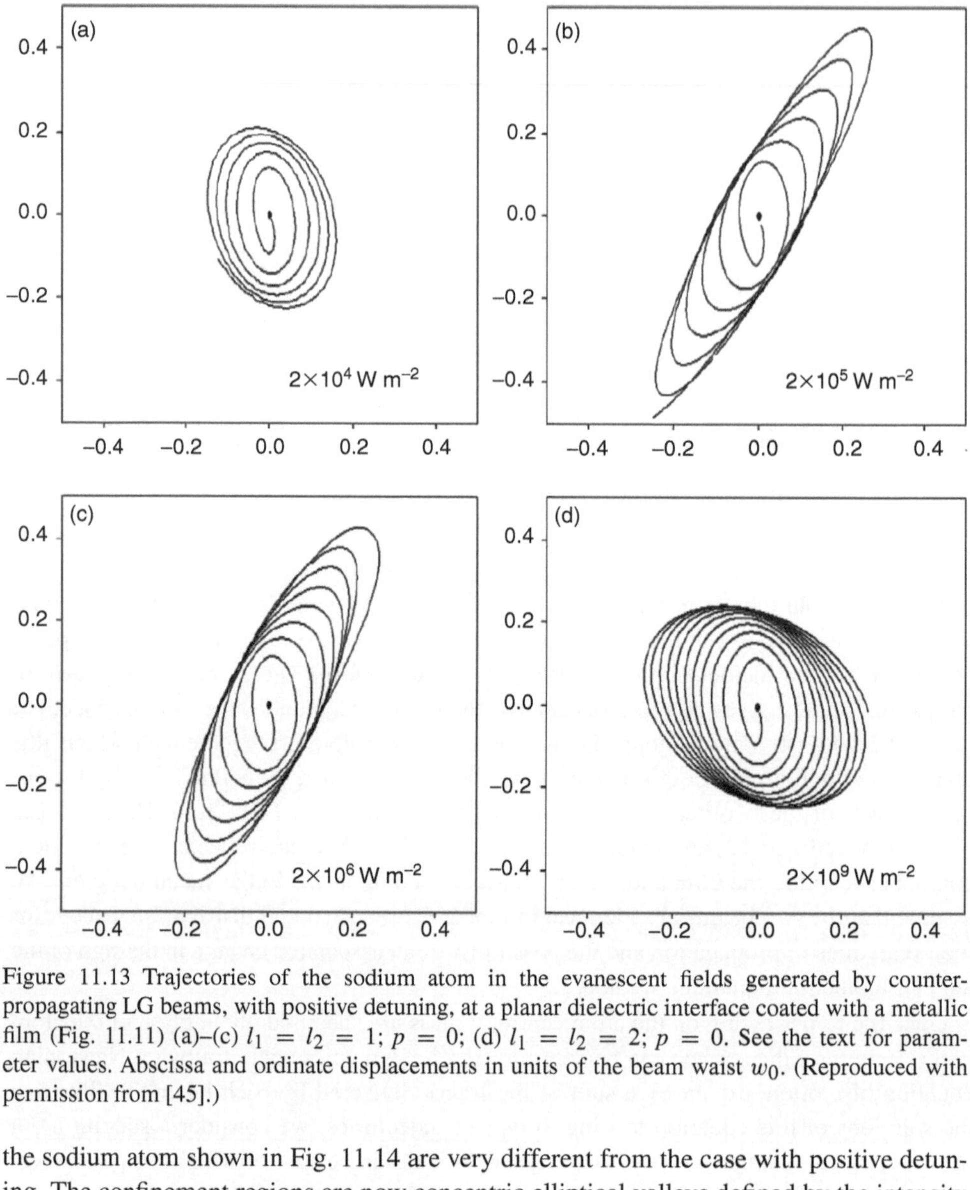

Figure 11.13 Trajectories of the sodium atom in the evanescent fields generated by counter-propagating LG beams, with positive detuning, at a planar dielectric interface coated with a metallic film (Fig. 11.11) (a)–(c) $l_1 = l_2 = 1$; $p = 0$; (d) $l_1 = l_2 = 2$; $p = 0$. See the text for parameter values. Abscissa and ordinate displacements in units of the beam waist w_0. (Reproduced with permission from [45].)

the sodium atom shown in Fig. 11.14 are very different from the case with positive detuning. The confinement regions are now concentric elliptical valleys defined by the intensity distributions, radial confinement leading to vibrational motion in a radial direction and resulting in an overall zigzag trajectory.

Our analysis has focused in particular on an evanescent mode generated by laser light with an azimuthal phase dependence and orbital angular momentum as typified by Laguerre–Gaussian beams. The fields associated with this mode can be as much as two orders of magnitude stronger than those in the absence of the metal film. In order to ensure

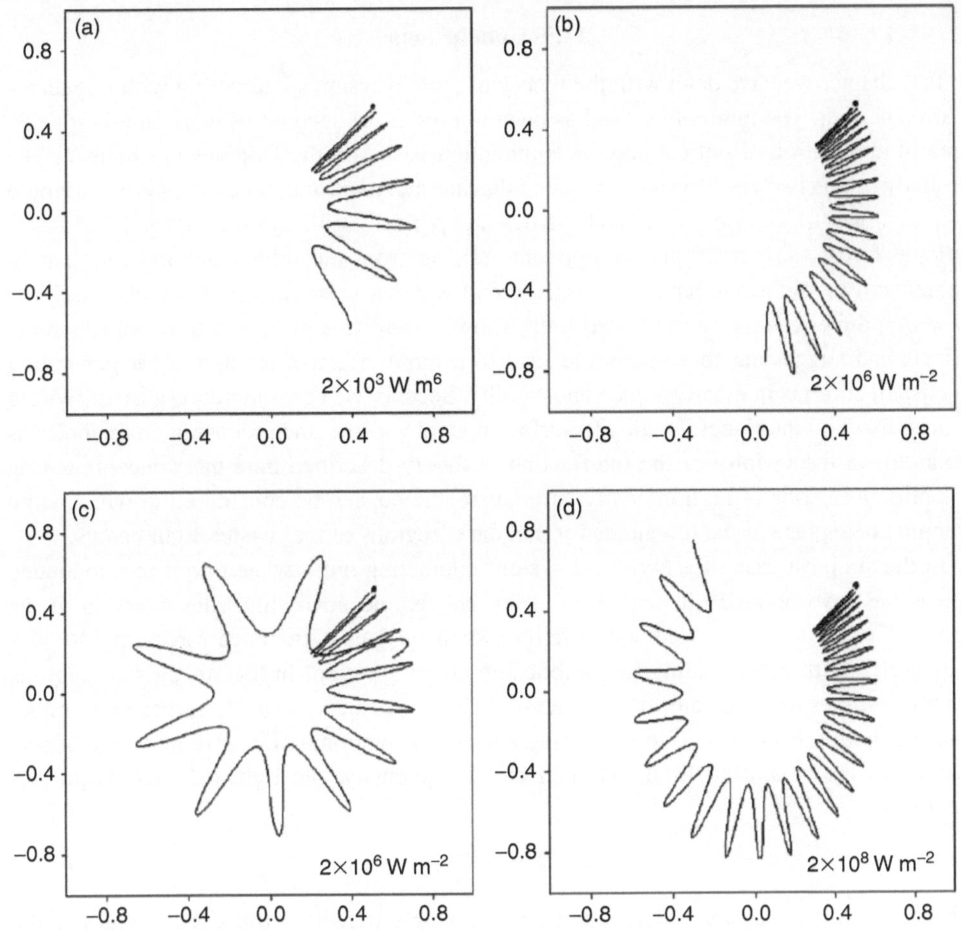

Figure 11.14 Trajectories of the sodium atom in the evanescent fields generated by counter-propagating LG beams, with negative detuning, at a planar dielectric interface coated with a metallic film: (a)–(c) $l_1 = l_2 = 1$; $p = 0$; (d) $l_1 = l_2 = 2$; $p = 0$. Abscissa and ordinate displacements in units of the beam waist w_0. See the text for parameter values. (Reproduced with permission from [45].)

atomic confinement within the high intensity regions of the evanescent light we need a second, counter-propagating light beam with the same angular momentum attributes, incident in such a way that the in-plane plane wave effects lead to a trapping of the atom, producing confinement within the high intensity regions of the evanescent fields. The propensity for any such vortices to be concentrated in a small volume, tightly bound to the planar surface, makes such modes interesting – not just in their own right as essentially two-dimensional modes with rotational features, but also for possible applications. The rich features that have been identified in the anticipated dynamics should spur on the progress of experimental activity in this area.

11.9 Conclusions

In this chapter we have dealt with the theory of atomic systems interacting with Laguerre–Gaussian light. The main objective has been to present an account of work in this specific area of application of optical angular momentum based on the Doppler mechanism. The groundwork on two-level systems revealed that atomic, ionic and molecular systems should display novel trajectories, with such particles experiencing in addition to modified translational forces, also rotational components arising from the orbital angular momentum contents of the light. We have also considered two other scenarios. First we discussed the case of a three-level atom in twisted light, showing how this gives rise to novel rotational effects and suggesting that this could lead to a more effective technique for generating persistent currents in macroscopic superfluidity. Second, we considered the case of twisted light at material interfaces, leading to surface optical vortices, then focusing on their effects on atoms in the vicinity of the interface. The theory described here has concentrated on trapping in regions of high intensity; similar treatments can be considered in which atom trapping take place in the low intensity (i.e. dark) regions of the twisted light beams.

In the simplest case of a two-level system interacting with twisted light in unbounded space, we have pointed out that in the transient regime, involving time intervals of the order of $\tau = \Gamma^{-1}$, the excited state lifetime, such systems experience forces and torques that evolve with time, leading to complicated atomic motion. In the steady state regime, involving times much greater than τ, the particles are subject to steady forces and torques and the motion leads to cooling, trapping and novel rotational effects in a variety of scenarios. In the saturation limit, attained at high intensity, the light-induced torque has the form

$$\mathbf{T} \approx \hbar l \Gamma \hat{\mathbf{z}}. \tag{11.63}$$

This result has the simple interpretation that a single transition transfers orbital angular momentum of magnitude $\hbar l$ to the atom centre of mass, and since there are Γ transitions per second, the angular momentum transfer per transition multiplied by the number of transitions per second represents a torque acting on the atomic centre of mass of magnitude $\hbar \Gamma l$ along the beam axis $\hat{\mathbf{z}}$. This is consistent with the fundamental understanding that the orbital angular momentum of light is quantised in units of $\hbar$. In addition to OAM transfer, we have also seen that a two-level atom is subject to a new (azimuthal) Doppler shift due to the rotational motion, leading to additional azimuthal components of the dissipative force.

On considering processes involved in optical molasses experiments, namely those due to pairs of counter-propagating beams in one, two and three dimensions, we have encountered different forms of steady state optical forces that can produce a variety of novel trajectories. Twisted light generates potential wells arising from the total dipole force contributions from all beams present, while the total dissipative force for a given beam is responsible for the cooling or the heating of both the azimuthal motion and the motion along the beam axis. In a given pair of counter-propagating beams, the torque is either doubled or annulled, depending on the signs of the winding number l. As a result, we have been able to show that in the case of one-dimensional molasses with the simplest LG beams, namely the (1, 0) beams, we establish an annular region in which the atoms or ions are trapped and can be made to generate a current. When two sets of (1, 0) counter-propagating beams participate in a two-dimensional molasses, the atoms or ions are trapped in regions of space in the form of two circles. In three-dimensions the atoms or ions are trapped at the eight corners of a cube.

We have also seen how a three-level system subject to two counter-propagating twisted light beams experience a torque which can lead to persistent current. Finally, we have considered a case in which twisted light is internally reflected at an interface, generating an evanescent light mode possessing vortex features, revealing the possibility for atoms in the vicinity of the surface to be laterally trapped in the light fields so generated.

It is clear that a fuller understanding of the interaction of atoms and ions with twisted light is yet to be achieved. Recent trends indicate that experimental activity in this area is gathering pace, and we should then see increasingly accurate and meaningful correlations being secured between theory and experiment.

Acknowledgements

The authors are very grateful to Professor L. Allen, Dr. W. K. Lai, Dr. C. R. Bennett, Dr. L. Dávila Romero, Dr. A. R. Carter and Professor S. Al-Awfi, with whom the authors collaborated in much of the work presented here. Thanks are also due to the EPSRC for supporting the research.

References

[1] V. S. Letokhov and V. G. Minogin, *Laser Light Pressure on Atoms* (New York: Gordon and Breach, 1987).
[2] A. P. Kazantsev, G. I. Surdutovich and V. P. Yakovlev, *Mechanical Action of Light on Atoms* (Singapore: World Scientific, 1990).
[3] C. S. Adams and E. Riis, Laser cooling and trapping of neutral atoms, *Prog. Quant. Electr.* **21**, 1 (1997).
[4] H. Metcalf and P. van der Straten, *Laser Cooling and Trapping*, (New York: Springer-Verlag, 1999).
[5] D. Wineland and H. Dehmelt, Proposed $10^{14}\delta\nu/\nu$ laser fluorescence spectroscopy on TI^+ mono-ion oscillator III (side band cooling), *Bull. Am. Phys. Soc.* **20**, 637 (1975).
[6] T. Hänsch and A. Schawlow, Cooling of gases by laser radiation, *Opt. Comm.* **13**, 68 (1975).
[7] C. J. Pethick and H. Smith, *Bose-Einistein Condensation in Dilute Gases.* (Cambridge: Cambridge University Press, 2002).
[8] M. F. Anderson, C. Ryu, P. Calde *et al.*, Quantized rotation of atoms from photons with orbital angular momentum, *Phys. Rev. Lett.* **97**, 170406 (2006).
[9] J. W. R. Tabosa and D. V. Petrov, Optical pumping of orbital angular momentum of light in cold caesium atoms, *Phys. Rev. Lett.* **83**, 4967 (1999).

[10] L. Allen, M. W. Beijersbergen, R. J. C. Spreeuw and J. P. Woerdman, Orbital angular momentum of light and the transformation of Laguerre–Gaussian laser modes, *Phys. Rev.* **A45**, 8185 (1992).

[11] M. Babiker, W. L. Power and L. Allen, Light-induced torque on moving atoms, *Phys. Rev. Lett.* **73**, 1239 (1994).

[12] L. Allen, M. Babiker and W. L. Power, Azimuthal Doppler shift in light beams with orbital angular momentum, *Opt. Comm.* **112**, 141 (1994).

[13] L. Allen, M. Babiker, W. K. Lai, and V. E. Lembessis Atom dynamics in multiple Laguerre–Gaussian beams, *Phys. Rev.* **A54**, 4259 (1996).

[14] S, M. Barnett and L. Allen, Orbital angular momentum and nonparaxial light beams, *Opt. Comm* **110**, 670 (1994).

[15] L. Allen, V. E. Lembessis and M. Babiker, Spin-orbit coupling in free-space Laguerre–Gaussian light beams, *Phys. Rev.* **A53**, R2937 (1996).

[16] M. Babiker, C. R. Bennett, D. L. Andrews, and L. C. Dávila Romero, Orbital angular momentum exchange in the interaction of twisted light with molecules, *Phys. Rev. Lett.* **89**, 143601 (2002).

[17] D. L. Andrews, A. R. Carter, M. Babiker and M. Al-Amri, Transient optical angular momentum effects and atom trapping in multiple twisted beams, *Proc. SPIE* **6131** (2006).

[18] L. C. Dávila Romero, A. R. Carter, M. Babiker and D. L. Andrews, Interaction of Laguerre–Gaussian light with liquid crystals, *Proc. SPIE* **5736**, 150–7 (2005).

[19] A. R. Carter, M. Babiker, M. Al-Amri and D. L. Andrews, Transient optical angular momentum effects in light-matter interactions, *Phys. Rev.* **A72** (2006).

[20] A. R. Carter and M. Babiker, Twisted optical molasses for all-optical cooling and trapping, *Phys. Rev.* **A77**, 043401 (2008).

[21] V. E. Lembessis, D.L. Andrews, M. Babiker and S. Al-Awfi, Optical vortex singularities and atomic circualtion in evanescent waves, *Proc. SPIE* **7950**, 101–10 (2011).

[22] A. R. Carter, M. Babiker, M. Al-Amri and D. L. Andrews, Generation of microscale current loops, atom rings and cubic clusters using twisted optical molasses, *Phys. Rev.* **A73**, 021401 (2006).

[23] E. Courtrade, O. Houde, J. F. Clément, P. Verkek and D. Hennquin, Dark optical lattice of ring traps for cold atoms, *Phys. Rev.* **A74**, 031403(R) (2006).

[24] S. J. van Enk, Selection rules and centre-of-mass motion of ultra cold atoms, *Quant. Opt.* **6**, 445 (1994).

[25] G. Juzeliūnas and P. Öhberg, Slow light in degenerate Fermi gases, *Phys. Rev. Lett.* **93**, 033602 (2004).

[26] G. Juzeliūnas, P. Öhberg, J. Ruseckas and A. Klein, Effective magnetic fields in degenerate atomic gases induced by light beams with orbital angular momenta, *Phys. Rev.* **A71**, 053614 (2005).

[27] S. Al-Awfi and M. Babiker, Field-dipole orientation mechanism and higher order modes in atom guides, *J. Mod. Opt.* **48**, 847 (2001).

[28] J. P. Dowling and J. Gea-Banacloche, Evanescent light-wave atom mirrors, resonators, waveguides, and traps, *Adv. Atom. Mol. Opt. Phy.* **37**, 1 (1997).

[29] M. A. Clifford, J. Arlt, J. Courtial and K. Dholakia, High-order Laguerre–Gaussian light beams for studies of cold atoms, *Opt. Comm.* **156**, 300 (1998).

[30] Y. Song, D. Millam and W. T. Hill III, Long, narrow all-light atom guide, *Opt. Lett.* **24**, 1805 (1999).

[31] D. M. Villeneuve, S. A. Asyev, P. Dietrich, M. Spanner, M. Yu Ivanov and P. B. Corkum, Forced molecular rotation in an optical centrifuge, *Phys. Rev. Lett.* **85**, 542 (2000).
[32] A. Muthukrishnan and C. R. Stroud Jr, Entanglement of internal and external angular momenta of a single atom, *J. Opt. Quantum Semiclass.* **4**, S73 (2002).
[33] M. S. Bigelow, P. Zernom and R. W. Boyd, Breakup of ring beams carrying orbital angular momentum in sodium vapor, *Phys. Rev. Lett.* **92**, 083902 (2004).
[34] A. V. Bezverbny, V. G. Niz'ev and A. M. Tomaikin, Dipole traps for neutral atoms formed by non-uniformly polarised Laguerre–Gaussian modes, *Quantum Electron.* **34**, 685 (2004).
[35] A. Alexanderscu, E. De Fabrizio and D. Kojoc, Electronic and centre of mass transitions driven by Laguerre–Gaussian beams, *J. Opt.* **B7**, 87 (2005).
[36] W. Zheng-Ling and Y. Jiang-Ping, Average dissipative and dipole forces on a three-level atom in a Laguerre–Gaussian beam, *Chin. Phys. Lett.* **22**, 1386 (2005).
[37] R. Jauregui, Rotational effects of twisted light on atoms beyond the paraxial approximation, *Phys. Rev.* **A70**, 033415 (2004).
[38] M. V. Berry, Paraxial beams of spinning light, in *Singular Optics*, eds. M. S. Soskin and M. V. Vasnetsov, *Proc. SPIE*, **3487**, 6 (1998).
[39] V. E. Lembessis and M. Babiker, Light-induced torque for the generation of persistent current flow in atomic gas Bose–Einstein condensates, *Phys. Rev.* **A82**, 051402 (2010).
[40] A. Aspect, E. Arimondo, R. Kaiser, N. Vansteenkiste and C. Cohen-Tannoudji, Laser cooling below the one-photon recoil energy by velocity-selective coherent population trapping: theoretical analysis, *J. Opt. Soc. Am.* **B6**, 2112–24 (1989).
[41] E. Arimondo and G. Orriols, Nonabsorbing atomic coherences by coherent two-photon transitions in a three-level optical pumping, *Lett. Al Nuovo Cimento*, **17** 333 (1976).
[42] Y. Gorodetski, A. Niv, V. Kleiner and E. Hasman, Observation of the spin-based plasmonic effect in nanoscale structures, *Phys. Rev. Lett.* **101**, 043903 (2008).
[43] Y. Gorodetski, N. Shitrit, I. Bretner, V. Kleiner and E. Hasman, Observation of optical spin symmetry breaking in nanoapertures, *Nano Lett.* **9**, 3016 (2009).
[44] V. E. Lembessis, M. Babiker and D. L. Andrews, Surface optical vortices, *Phys. Rev.* **A79**, 011806 (2009).
[45] V. E. Lembessis, S. Al-Awfi, M. Babiker and D. L. Andrews, Surface plasmon optical vortices and their influence on atoms, *J. Opt.* **13**, 064002 (2011).

12

An experimentalist's introduction to orbital angular momentum for quantum optics

J. ROMERO, D. GIOVANNINI, S. FRANKE-ARNOLD AND M. J. PADGETT

12.1 Introduction

Although the orbital angular momentum (OAM) of light is often quantified in terms of $\ell\hbar$ per photon, one should recognise that this result can be derived directly from Maxwell's equations and is most certainly not an exclusively quantum property. Following the recognition of optical OAM in 1992 [1], the immediate thrust of experimental work was to observe OAM transfer to microscopic objects. These exciting experiments by Rubinsztein-Dunlop [2, 3] and others established that the predicted angular momentum was truly mechanical, but did not provide insight on OAM at the level of single photons. Laguerre–Gaussian modes were identified as being a natural basis set for describing beams with OAM, where their phase terms describe the helical phasefronts corresponding to an OAM of $\ell\hbar$ per photon.

The polarisation of light has played an important historical role in the experimental investigation of quantum entanglement, quantum gate operations and quantum cryptography. A potential advantage of OAM is that it resides in a higher-dimensional state space which, as follows from the theory, is infinite-dimensional. One of the experimental challenges is to access a large number of these dimensions. In quantum communication, the most obvious advantage of high n-dimensional systems is the increased bandwidth [4], but qunits also lead to stronger security [5, 6]. An increase in the number of entangled quantum modes has also been shown to improve measures of nonlocality [7], to allow quantum computation with better fault tolerance [8] and to simplify quantum logic structures [9].

12.2 Conservation of OAM

The helical nature of the phasefronts corresponds to a Poynting vector colinear to the local wavevector, which is skewed with respect to the beam axis. This skewed wavevector suggests interesting behaviours for OAM-carrying beams in their interaction with nonlinear materials. Specifically, the skewed nature of the wavevector could result in unusual phase

The Angular Momentum of Light, ed. D. L. Andrews and M. Babiker. Published by Cambridge University Press.

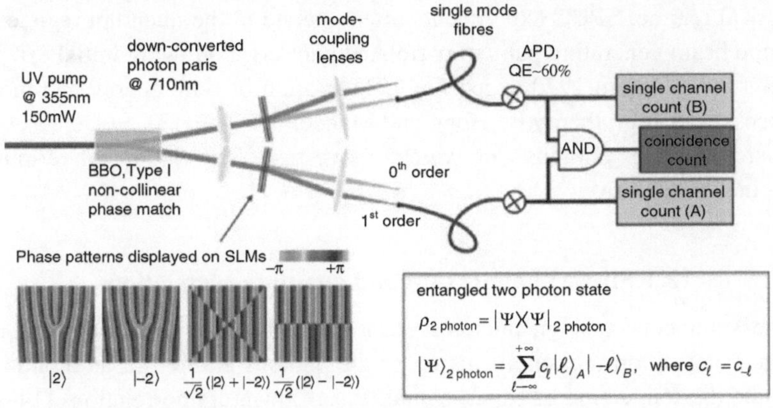

Figure 12.1 In spontaneous parametric down-conversion, a single pump photon is split into two lower-energy, entangled signal and idler photons. The photons can be projected into OAM and OAM superposition states by using holograms and a single mode fibre as mode filters. The two photon state can be written in the OAM basis, with $|c_\ell|^2$ giving the probability of a coincidence when phase masks corresponding to $|\ell\rangle$ and $-|\ell\rangle$ are encoded on the SLMs.

matching conditions, perhaps allowing access to new geometries. Initial experiments by Padgett and coworkers [10] focused on the nonlinear crystal KTP, which is frequently used in second-harmonic generation to convert an input beam at 1064 nm to an output beam at 532 nm. Efficient conversion requires both energy and momentum to be conserved. The latter is achieved by setting the transmission angle through the crystal such that the dispersion between the fundamental and second harmonic light is compensated by the birefringence of the crystal itself. Rather than revealing a new form of phase matching, these experiments showed that the conditions for frequency-doubling a Laguerre–Gaussian mode were exactly the same as for a regular Gaussian laser mode. Upon closer inspection, it was noticed that the generated mode was not quite the same, its ℓ-index being double that of the input beam. This result is in perfect agreement with a photon picture where two infrared photons combine to form a green photon of twice the energy and, it would seem, twice the angular momentum–both quantities being conserved. A full quantum electrodynamic description of this process is described in detail by Babiker and coworkers [11]. Of course, although consistent with this photon picture, frequency doubling can also be completely understood in terms of the classical electric field and the resulting polarisation of the material.

The reverse process to frequency doubling is the process of spontaneous parametric down-conversion (SPDC), which forms the basis of most current experiments on the quantum nature of OAM correlations (Fig. 12.1). Here, a high-frequency photon is incident on a nonlinear crystal, which parametrically generates two lower-frequency photons. These signal and idler photons may or may not be distinguishable from each other, depending on the crystal type and geometry. In contrast to second-harmonic generation, usually described in the classical regime, SPDC experiments often operate in the quantum regime, with an intense pump beam generating pairs of correlated photons. Following initial experimental work by Burnham [12] in 1970 supporting conservation of both energy and momentum in this process, and solid theory by Hong and Mandel in 1985 [13], SPDC has become a reliable source of single photons with which to investigate correlations of various photon properties, including OAM.

12.3 Single photon OAM and quantum correlations

The first measurements on single photon correlations focused on polarisation and position-momentum. Polarisation measurements on single photons emitted in an atomic cascade transition laid the foundation of experimental tests of quantum correlations [14–16]. The two photons from the cascade could be separated in such a way that the polarisation state of either of them was undetermined, but highly correlated (e.g. always opposite) for any given pair. These correlations survive also when the photons are spatially separated at the time of the measurement, indicating nonlocal correlations. A stringent test of quantum entanglement was provided by violating a Bell inequality [14–16], thus proving that the correlations could not be explained by any local hidden variable theory [17].

Before similar experiments could be repeated for the OAM of single photons, a 'polariser' for OAM states had to be developed. It was not until 1990, when Soskin and coworkers [18] in parallel with Heckenberg [19] and coworkers had pioneered the use of diffractive optical elements for the generation of beams with helical phasefronts, that such a device was brought about. Their teams showed that when a diffraction grating containing an ℓ-fold edge dislocation was illuminated with a Gaussian beam, the first-order diffracted beam had an annular intensity cross-section consistent with helical phasefronts (Fig. 12.2) described by a phase cross-section $\exp(i\ell\phi)$. OAM beams of any desired order can be generated through diffraction off a suitable ℓ-forked hologram. Operated in reverse, these diffraction holograms serve to detect the OAM of a laser beam. A helically phased beam incident on the hologram is converted back into a fundamental Gaussian beam that can be coupled via a single-mode fibre to a detector (Fig. 12.2). Varying the order of the dislocation on the hologram such that the light is transformed to the fundamental Gaussian beam with on-axis intensity which is coupled through the fibre allows one to determine the ℓ-index of the incident light. If the detector is sensitive to single photons then the combination of grating, fibre and detector is a mode-projector which allows the measurement of the ℓ-index of a single photon.

Equipped with this mode-projector, Zeilinger and coworkers measured in 2001 the OAM of single photons generated by SPDC [20] and investigated quantum correlations in OAM. They showed that OAM is conserved in the parametric down-conversion process, just as it had earlier been shown for second harmonic generation [10]. For a Gaussian pump beam (without angular momentum) they found strongest coincidences when the OAM of the signal photon (ℓ_s) was opposite to that of the idler photon ($\ell_i = -\ell_s$) and, more generally, for a pump beam with $\ell_p = -1, 0, +1$, that the OAM was conserved: $\ell_s + \ell_i = \ell_p$.

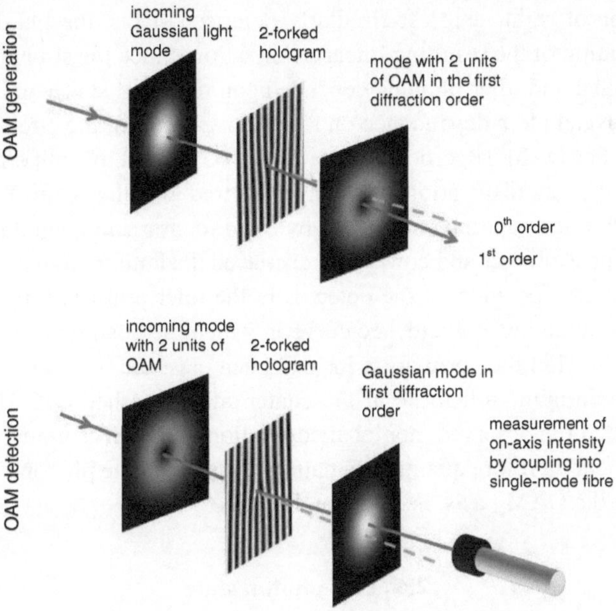

Figure 12.2 A forked diffraction grating can be used to generate a beam with a helical phasefront in the direction of the first order, with the charge of the dislocation corresponding to the OAM of the generated beam. The reverse is also true, when a helically phased beam is incident on the hologram, the beam is transformed into the fundamental Gaussian mode, which can be coupled to a single mode fibre and a photon detector.

The conservation of OAM can be understood as a direct consequence of phase-matching in SPDC [21]. It is important to appreciate that the signal and idler beams independently have low spatial coherence and both are a mixture of many different OAM states. The correlations exist only at the level of the individual photon pairs and are hidden when the measurements average over many photon pairs, hence the detection of quantum correlations requires single photon detection. Zeilinger and coworkers [20] confirmed that for a Gaussian pump beam the signal and idler photons could have a range of OAM values, but once the signal OAM was determined, the paired idler photon always had the opposite value. These results are in agreement with previous work by Monken and coworkers, who showed that the angular spectrum of the pump is transferred to the transverse correlation properties of the signal and idler photons [22]. Since then, their group has investigated the conservation of OAM in SPDC in more detail, showing that it is the transfer of this pump angular spectrum that leads to the conservation of OAM [23].

While showing that $\ell_s + \ell_i = \ell_p$ for any measurement demonstrates conservation of OAM and indicates correlations between signal and idler, it is in itself not a proof of OAM entanglement. Unlike classical correlations, quantum correlations show a phase dependence which can be tested by showing correlations for OAM states and also for their superpositions. In the case of polarisation, any linear polarisation state can be regarded as a superposition of right- and left-circularly polarised states, the phase between them setting the orientation of the resulting linear polarisation. Since the strength of the correlations between signal and idler photons depends upon the relative orientation of the linear states, this is equivalent to a dependence on the relative phase of the circular states in each individual beam. For OAM, superposition states can be created by deliberately moving the hologram so that the fork dislocation is no longer centred with the beam. The relative phase of the modes within the superposition is set by the direction and magnitude of this transverse displacement. Zeilinger and coworkers displaced the hologram in the signal arm, and scanned the transverse position of the detector in the idler arm (with the idler hologram removed). This resulted in a coincidence curve in which the position of the minimum was also displaced [20]. Had the correlation instead been classical, one would have expected the coincidence minimum still to be at the center, albeit washed out. The shifted minimum confirmed that the observed, nonlocal correlation in the measurement of the OAM is phase dependent and therefore quantum in nature–the state of the photon pair generated by SPDC, written in the OAM basis, is an entangled state.

12.4 Two-photon state

Any two-photon, entangled, pure state, $|\Psi_{2\,\text{photon}}\rangle$ can be written as

$$|\Psi_{2\,\text{photon}}\rangle = \sum_i |\psi_i\rangle_{\text{A}} |\phi_i\rangle_{\text{B}} \tag{12.1}$$

where the sets $\{|\psi_i\rangle_{\text{A}}\}$ and $\{|\phi_i\rangle_{\text{B}}\}$ form complete, orthonormal bases for photons A and B respectively [24]. The Laguerre–Gaussian set of modes, being a complete, orthonormal set, can be used as a basis for describing the two-photon OAM state, that is $\{|\psi_i\rangle_{\text{A}}\} \to \{|\ell\rangle_{\text{A}}\}$ and $\{|\phi_i\rangle_{\text{A}}\} \to \{|\ell\rangle_{\text{B}}\}$ as in Fig. 12.1. A quantitative measurement of a two-photon OAM state, and in particular the entangled OAM state anticipated for SPDC photons, requires one to measure both signal and idler photon projected into defined superpositions of OAM states. Similar measurements had previously been performed for polarisation entangled states, and the analogy between the Poincaré spheres helped to identify the complementary states required to measure the density matrix for the OAM entangled photon pairs. The first tomographic measurement of a two-photon OAM density matrix ($\rho_{2\,\text{photon}}$ in Fig. 12.1), was done by White and coworkers in 2004 using modes of order 1, corresponding to the Hermite–Gaussian modes described earlier [25]. Subsequent precise measurement of an OAM density matrix was done for chosen two-dimensional subspaces, with $|\ell|$ ranging from 1 to 30 [26] showing a fidelity with the maximally entangled Bell state of up to 0.97. The fidelity decreases for subspaces with large $|\ell|$ for which background noise eventually becomes dominant (Fig. 12.3).

12.5 Analogy with polarisation

The mathematical analogy between polarisation and OAM states was recognised by Allen, Woerdman and coworkers in their seminal studies in the early 1990s [1, 27]. Their mode

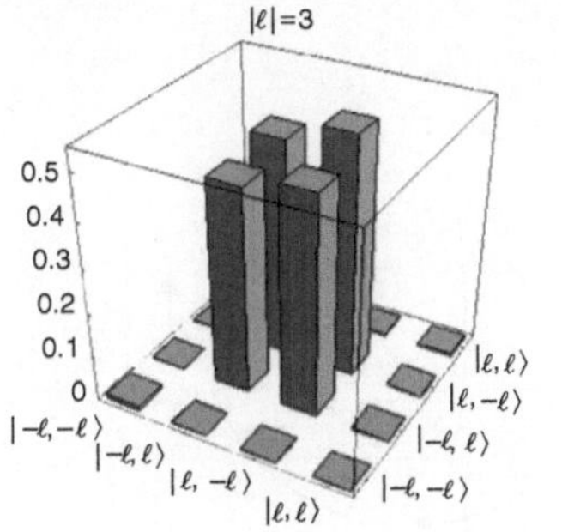

Figure 12.3 A sample density matrix measurement for a subspace of $|\pm 3\rangle$ reveals an almost maximally entangled two-photon OAM state.

converters, based on cylindrical lenses, transformed Hermite–Gaussian modes into Laguerre–Gaussian modes. Those devices were the equivalent optical components to wave-plates for polarisation states. This analogy was exploited by Padgett and Courtial [28, 29], who proposed a representation of the OAM modal state with an equivalent to the Poincaré sphere. The Poincaré sphere which describes polarisation is an example of a Bloch sphere and represents the complex superposition of two orthogonal polarisation states. Right- and left-circular polarisation states are located at the 'north' and 'south' poles of the Poincaré sphere, respectively. Their coherent superpositions result in linear polarisation states, placed around the 'equator', where the longitude corresponds to the orientation of the linear polarisation state which depends on the relative phase between the right- and left-circular polarisation components. Intermediate 'latitudes' correspond to elliptical polarisation states with the 'longitude' denoting the orientation of the major axis (Fig. 12.4).

While polarisation is a two-state system, completely described by a Poincaré sphere, OAM is more complicated. In general, transverse modes can be divided into sets of the same mode order N, where all modes of a particular mode order experience the same Gouy phase upon propagation [30]. For the Laguerre–Gaussian modes, $N = |\ell| + 2p + 1$ (p is the radial mode index describing the number of radial nodes in the field distribution) and there are $N + 1$ modes of order N. It follows that the Laguerre–Gaussian modes with $p = 0$ and $\ell = \pm 1$ represent all the modes of order 1 – a two-dimensional subspace which can be fully represented on the surface of the equivalent Poincaré sphere.

In this case the north and south poles represent Laguerre–Gaussian modes with $p = 0$ and $\ell = \pm 1$ respectively. The coherent superposition of these two modes give a Hermite–Gaussian mode (of order 1) with an orientation that depends on the relative phase between the $\ell = \pm 1$ components, which are again represented by different longitudes around the equator. Intermediate latitudes correspond to more complicated spatial modes but the analogy with polarisation states is maintained.

For Laguerre–Gaussian modes with $|\ell| > 1$, the situation is more complicated since the number of modes of the same order is greater than two, and these modes can no longer

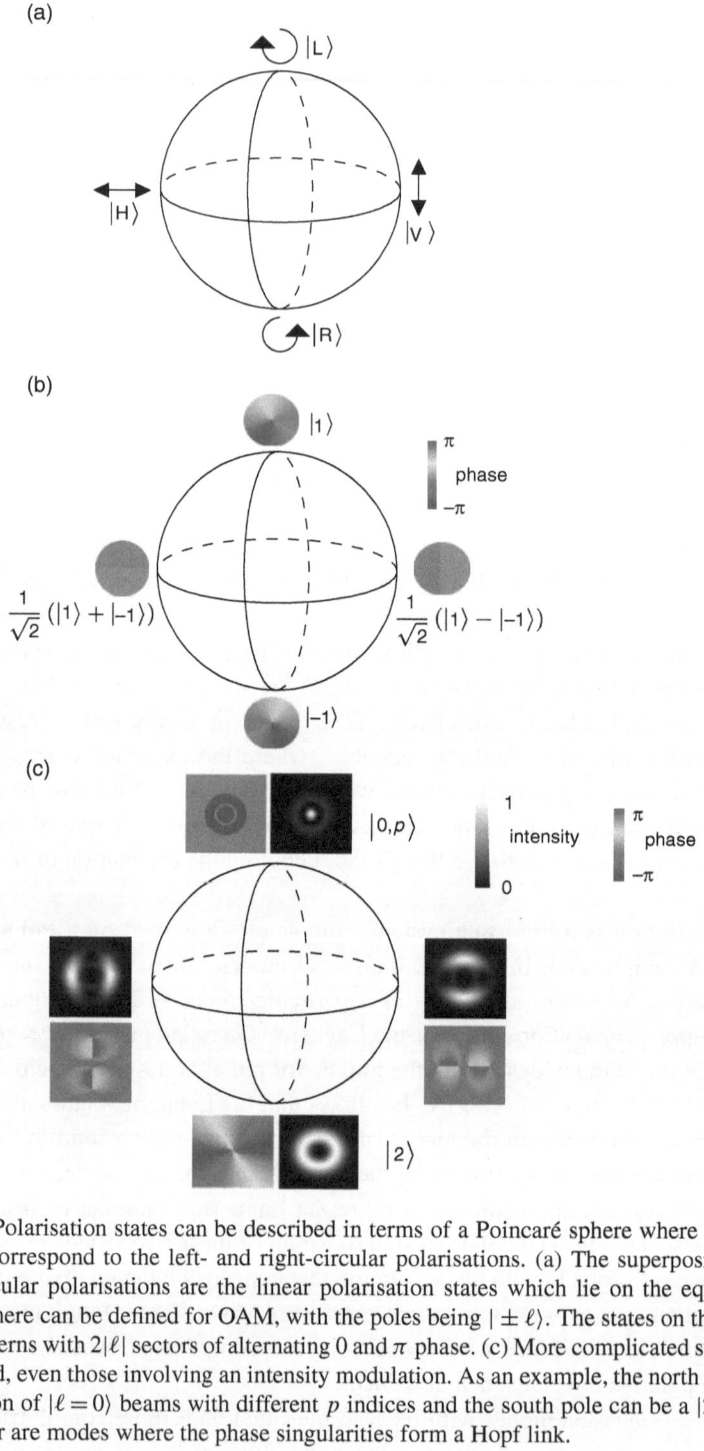

Figure 12.4 Polarisation states can be described in terms of a Poincaré sphere where the north and south poles correspond to the left- and right-circular polarisations. (a) The superpositions of left- and right-circular polarisations are the linear polarisation states which lie on the equator. (b) An analogous sphere can be defined for OAM, with the poles being $|\pm\ell\rangle$. The states on the equator are $2|\ell|$ petal patterns with $2|\ell|$ sectors of alternating 0 and π phase. (c) More complicated superpositions can be defined, even those involving an intensity modulation. As an example, the north pole could be a superposition of $|\ell=0\rangle$ beams with different p indices and the south pole can be a $|2\rangle$. The states on the equator are modes where the phase singularities form a Hopf link.

be represented on a single sphere. It is possible, however, to consider a subspace of just two modes and their superpositions, e.g. modes with $p=0$ and opposite values of ℓ. If the states $\pm\ell$ are positioned at the north and south poles, the states at the equator have intensity cross-sections consisting of a single ring of 2ℓ petals, and their orientation again varies with longitude around the equator (Fig. 12.4).

It is possible to construct Poincaré spheres for more complicated mode structures, as long as they occupy only two dimensions. An example used for the demonstration of entanglement between three-dimensional vortex structures, namely Hopf links, is a Poincaré sphere with a specific superposition of states with $\ell=0$ and various p values at the north pole and a state with $\ell=2$, $p=0$ at the south pole [31]. The equatorial states represent Hopf links (Fig. 12.4).

The Poincaré sphere equivalent for OAM has been used early on to analyse the frequency shifts that occur when beams carrying OAM are rotated around the propagation axis, where the dynamic phase shift is seen as a geometric or Berry phase [28]. The clear analogy with polarisation also lends itself to repeat experiments on quantum entanglement devised for two-dimensional systems and previously studied for the polarisation of entangled photons, which will be discussed in Section 12.6.

12.6 Tests of quantum mechanics in higher-dimensional OAM spaces

Many experiments have investigated OAM entanglement in two-dimensional subspaces, but such measurements do not exploit the high dimensionality offered by OAM. Higher-dimensional spaces can be accessed in different ways. One technique is to specifically address three or more OAM modes and test their entanglement, as previously implemented to test for qutrit entanglement [32] and for generalised Bell states [33]. Alternatively, one can take measurements in the Fourier space of OAM, which is represented by the continuous space of angle states. Simultaneous strong correlations between both angle and OAM allow demonstration of the EPR paradox [34], and therefore confirm entanglement. Operating in the Fourier space also allows the measurement of the Shannon dimensionality of OAM systems, work which has been driven forward by Eliel, Woerdman and coworkers [35, 36].

In 2002, Zeilinger and coworkers [32] generated and tested correlations between qutrits, OAM superposition states in a three-dimensional OAM space consisting of contributions from $\ell=-1,0,+1$. For each down-converted photon, these superpositions were generated by a combination of two holograms with a single fork dislocation each, that could be translated along orthogonal directions. In fact, the generated modal superpositions were more complicated, containing additional contributions from higher-order radial modes as well as higher $|\ell|$ modes: post-selection, however, reduced the system to the mentioned three-dimensional Hilbert space. In each arm, a probabilistic mode detector would identify the modal contributions of modes $\ell=-1,0,+1$ by directing the photons onto suitable hologram-fibre detectors with a probability of 1/3 respectively. Coincidences were recorded for a variety of detection settings. Collins and coworkers [37] had identified a

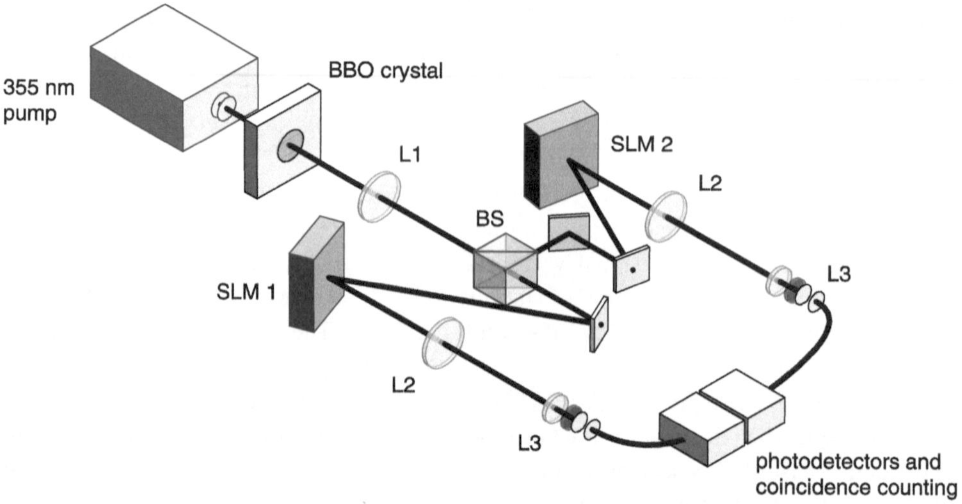

Figure 12.5 A 355 nm beam pumps a BBO crystal to generate two 710 nm photons colinearly. The photons are separated by a beamsplitter BS. The photons are projected into the desired OAM state by the combination of a forked diffraction grating and a single mode fibre. The hologram, which could in practice be any desired measurement mode (in addition to the diffraction grating), is encoded in a spatial light modulator (SLM). To facilitate alignment (i.e. position and angle are decoupled), the output face of the crystal is imaged by L1 onto the SLMs, and the SLMs are in turn imaged onto the input facets of single-mode fibres by the combination of L2 and L3.

generalised type of Clauser–Horne–Shimony–Holt (CHSH) Bell inequality that allowed tests of local hidden variable theories in state spaces of more than two dimensions. Depending on the settings of the local analysers, combinations of coincidence counts are in violation with any local hidden variable theory. Zeilinger's team analysed over 20 million combinations of analyser positions and found a maximum violation of the Bell type inequality by 18 standard deviations, clearly more than is possible in two-dimensional systems.

In addition to translating holograms, precise measurement of OAM superpositions can also be done by programming spatial light modulators (SLMs). SLMs are pixelated liquid crystal devices that can be programmed such that the light reflected by each pixel can be phase-delayed between 0 and 2π. Such a device can therefore be programmed to act as a diffractive optical element, of which a forked diffraction grating (in the case of an OAM state) is just an example. In practice, any phase and intensity modulation can be programmed (Fig. 12.5). This capability has been exploited in recent experiments [33, 34].

Violations of local hidden variable theories have been observed by programming specific superpositions of OAM states that span 12-dimensional OAM state spaces. The results demonstrated genuine 11-dimensional entanglement, when entanglement concentration was applied. The main experimental difficulty in observing higher-dimensional entanglement is caused by the fact that the efficiency of down-conversion falls for higher-order

OAM, as the overlap of the down-converted modes with the pump beam gets smaller. Maximally entangled states, however, require equal amplitude magnitudes for all the different mode contributions. The so-called Procrustean method of entanglement concentration does allow to increase the entanglement (by equalising the mode amplitudes via filters), but comes at the cost of decreased count rates. Eventually, entanglement is lost if noise dominates over the signal count rates. While filters offer the simplest option to influence the two-photon state, Torres and coworkers introduced the concept of shaping the pump beam in order to modify the generated two-photon state, in particular allowing the generation of a maximally entangled state in, e.g., three dimensions [38, 39].

OAM is related to azimuthal angle via a Fourier transform, i.e. a continuous (although periodic) variable. In turn, the periodic nature of the angular variable results in the angular momentum being described as a discrete, but unbounded series. A beam with a well-defined OAM value requires a uniform angular intensity distribution. Restricting the angular distribution of a light mode instead introduces OAM sidebands [40]. This highlights the fact that it is impossible to precisely know the azimuthal angle within a light mode and its OAM at the same time–suggesting an angular version of the EPR paradox. States that demonstrate EPR-type correlations are a subclass of entangled states, hence a measurement of EPR correlations therefore proves that the system is entangled.

In its original form, the EPR paradox related to nonlocal measurements of position and momentum. Einstein assumed that an experimental system could be found which exhibits nonlocal correlations in both position and momentum of two separated particles. A measurement of either position or momentum on one of these would then determine the position or momentum of the other, a localised measurement that should not 'disturb' the remote partner particle. Since the decision on whether to measure position or momentum can be delayed until the particles are already separated it seems that each particle has a well-defined position and momentum–a fact forbidden by the uncertainty principle. Finding such a system would suggest that quantum mechanics is either an incomplete theory or that objects are linked, i.e. entangled.

Measuring OAM can be achieved with a forked diffraction grating, and measuring the angular position by transmission through a radial slit. Again, both of these optical elements can be implemented using SLMs. Simply speaking, measuring the OAM of one photon allows to infer, with near certainty, the OAM of the other, which will have the opposite value. Similarly, measuring the angular position of one photon gives near certain knowledge of the angular position of the other, which will have the same value. This ability to show strong correlation in either of the two Fourier-related basis sets is completely analogous to the original formulation of the paradox and can therefore be considered as a demonstration of EPR for angles [34]. Measurements confirmed correlations of an order of magnitude stronger than those allowed by the uncertainty principle for independent (nonentangled) particles. Unlike demonstrations of Bell-type inequalities, which apply to discrete state spaces of two or more dimensions, EPR correlations test an extended range of OAM states in conjunction with the continuous state space of angular position.

Aside from fundamental tests of quantum mechanics, OAM also presents an advantage for practical applications such as communication. Encoding in higher-dimensional systems clearly increases the bandwidth, or the flux of information. However, transmission of OAM information, at least currently, relies on free-space transmission rather than fibres, thus limiting the use of OAM for long distance communication applications. At least in principle, however, it has been shown that higher-dimensional entangled states are more resistant to noise [7, 41].

12.7 Spiral bandwidth and dimensionality

The decomposition in Eq. (12.1) is often called the Schmidt decomposition of the quantum system. The Schmidt number, K, is defined as $\mathrm{K} = 1/Tr(\rho_{\mathrm{A}}^2) = 1/Tr(\rho_{\mathrm{B}}^2)$, where ρ_{A} and ρ_{B} are the reduced density matrices obtained by tracing $\rho_{2\,\mathrm{photon}}$ over the states associated with B and A respectively [36]. The Schmidt number characterises the effective dimensionality of the Hilbert space. For polarisation measurements, the Schmidt number is 2 (Fig. 12.6). The OAM Hilbert space is theoretically unbounded, that is, in writing the Schmidt decomposition as in Fig. 12.1, $-\infty < \ell < \infty$. However, in experiments where a

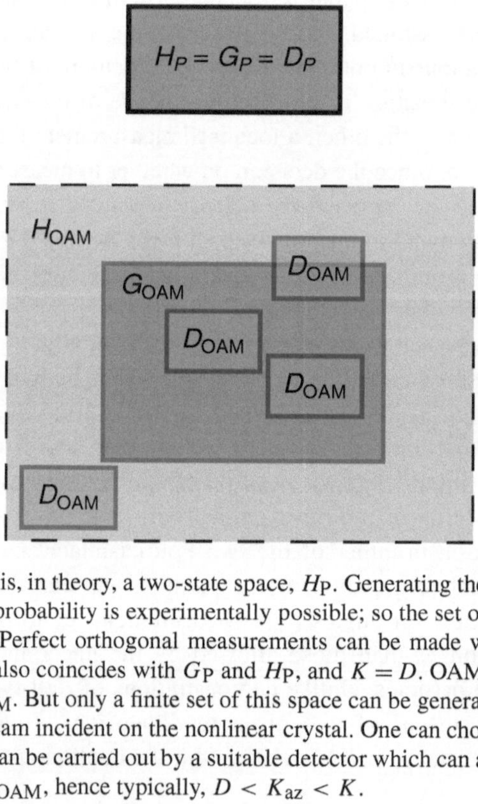

Figure 12.6 Polarisation is, in theory, a two-state space, H_{P}. Generating the two possible orthogonal polarisations with equal probability is experimentally possible; so the set of generated polarisations, G_{P}, coincides with H_{P}. Perfect orthogonal measurements can be made with polarisers, hence the measurement space M_{P} also coincides with G_{P} and H_{P}, and $K = D$. OAM is, in theory, an infinite-dimensional space, H_{OAM}. But only a finite set of this space can be generated, G_{OAM}, owing to the finite size of the pump beam incident on the nonlinear crystal. One can choose a subspace of H_{OAM} in which measurements can be carried out by a suitable detector which can access D_{OAM}. In practice D_{OAM} is smaller than G_{OAM}, hence typically, $D < K_{\mathrm{az}} < K$.

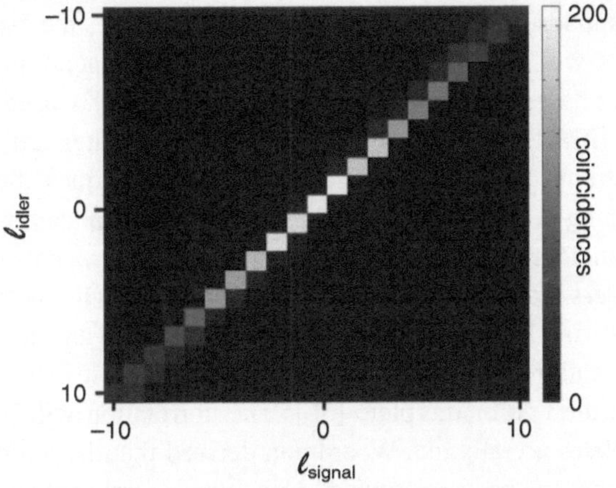

Figure 12.7 The number of OAM states which can be measured is very much dependent on the experiment (i.e. pump waist, signal and idler waist, phase-matching conditions of the crystal). The coincidences between signal and idler photons in OAM state $|\ell_{\text{signal}}\rangle$ and $|\ell_{\text{idler}}\rangle$ decreases as $|\ell|$ becomes larger, largely because of the finite size of the pump beam impinging on the crystal. Coincidences are low off the diagonal, as imposed by conservation of OAM.

finite-sized pump impinges on the nonlinear crystal, only a limited spectrum of ℓ modes can be generated. The width of this spectrum has been given the name *spiral bandwidth* and has been studied extensively [42–44]. The ℓ spectrum (i.e. the relative weight of each ℓ mode, $|c_\ell|$) can be calculated from the geometry of the experiment and has been shown to depend on the phase-matching conditions of the down-conversion crystal, width of the pump beam, and width of the modes being measured [42]. The spiral bandwidth is an indication of the amount of OAM entanglement present in the system. Being the modes associated with OAM only, the spiral bandwidth is naturally smaller than Schmidt number [36]. In the literature, spiral bandwidth is described by just the azimuthal Schmidt number, $K_{\text{az}} = 1/\sum |c_\ell|^2$, which is also smaller than K [45].

Measuring the spiral bandwidth of a system is easily done with holograms and single mode fibres: one only has to measure coincidences when the signal and idler photons are projected into modes ℓ and $-\ell$ respectively. It is a different matter to distinguish between the generated spiral bandwidth (the range of ℓ which actually comes out of the crystal) and the detection bandwidth (the range of ℓ which the hologram-fibre combination can actually measure). The former can be directly associated with K_{az}, and the latter is associated with the Shannon dimensionality, D, or the dimensionality of the *measured* entanglement [36, 46]. The concept of a dimensionality associated with a 'detector' becomes intuitive when one realises, for example, that the use of holograms and single mode fibres as an OAM mode projectors automatically sets only a range of p's or radial mode distributions which can be detected (Fig. 12.7).

The Shannon dimensionality was measured by Woerdman and coworkers [35, 36] by using phase plates. A phase plate is a transparent glass disc wherein the optical thickness at each point corresponds to the desired phase. For example, the optical thickness can be made to increases with azimuthal angle if one wants an incoming beam to be transformed to a helically phased beam. Such spiral phase elements are used in exactly the same way as the forked diffraction gratings, albeit spiral phase plates are zero-order devices. Although placing extreme demands upon manufacturing techniques, these plates have been successfully made using lithographic methods. Because phase plates are transmissive, they are more optically efficient than the equivalent diffractive components (particularly important when one considers that the coincidence count rates scale with the square of the optical efficiency). For measuring Shannon dimensionality, Woerdman and coworkers used single-mode fibres in combination with an angular phase plate – a phase plate that gives a π phase shift with respect to the rest of the plate [36]. The correlation will be highest when the two angular phase plates are aligned. Woerdman derived that the number of modes is the inverse of the area under the peak-normalised coincidence curve obtained when the relative orientation of the two angular phase plates is varied [35, 36]. This is consistent with the sinusoidal coincidence curves obtained when polarisers are rotated, because polarisation is indeed a two-state space.

Given the mentioned limitations for holograms and phase plates, one has to look beyond these devices for measuring the true spectrum of the entangled OAM state. As an example, van Exter and coworkers have measured the spiral spectrum by using an image rotator (consisting of three mirrors which can rotate one of the down-converted fields with respect to the other) and bucket photon counters to measure angle correlations. They then exploited the Fourier relationship between angle and OAM in order to derive K_{az}, which characterises the width of the spiral spectrum [45, 47]. They also showed that spiral bandwidth depends on phase-matching. Depending on the crystal type, the phase matching can be tuned by changing the temperature of the crystal or by changing the angular orientation of the crystal. They showed that the spiral spectrum can be flattened by changing the phase mismatch [45]. The spiral spectrum can also be tailored for specific applications, such as generating maximally entangled OAM states with a suitable engineering of the pump beam [38, 39].

12.8 Conclusions

The orbital angular momentum of light is now established as a useful degree of freedom, with many different research groups able to measure nonlocal OAM correlations. By restricting measurements to a bidimensional subspace, such that an analogy with the Poincaré sphere for polarisation is maintained, many of the experiments originally performed with polarisation can be repeated using holograms or phase plates rather than polarisers. Moreover, if SLMs are used, any point on the OAM Poincaré sphere is easily realisable – it is just a matter of programming the mode displayed on the SLM. This is one useful advantage over polarisation. In the case of polarisation, in fact, any arbitrary point on the Poincaré sphere corresponds to an elliptical polarisation which is not straightforwardly measurable with just one optical component.

Moreover, unlike polarisation, OAM has both an unbounded state space and a Fourier variable, namely angular position. In these regards, OAM and angle are similar to linear momentum and position and, consequently, it is possible to perform analogous experiments for angular and linear measurements. One advantage of the OAM basis is that OAM spatial modes provide a conceptual analogy with tests of quantum mechanics and system protocols that are already well understood in polarisation and linear position-momentum bases.

From a practical perspective, the OAM basis represents a convenient orthonormal basis set that can be used for encoding additional bits of information onto a single photon and consequently adds capacity to any communication or processing system. To date, most single photon measurements result only to two possible outcomes. This can change if the higher-dimensional OAM state space can be exploited–a clear challenge for the future is to realise an experiment where the high dimensionality of the OAM space results in a multi-state outcome at the level of single photons.

References

[1] L. Allen, M. W. Beijersbergen, R. J. C. Spreeuw and J. P. Woerdman. Orbital angular momentum of light and the transformation of Laguerre–Gaussian laser modes. *Phys. Rev. A*, **45**:8185–9, 1992.

[2] H. He, M. Friese, N. R. Heckenberg, and H. Rubinzstein-Dunlop. Direct observation of transfer of transfer of angular momentum to absorptive particles from a laser beam with a phase singularity. *Phys. Rev. Lett.*, **75**:826–9, 1995.

[3] M. Friese, J. Enger, H. Rubinzstein-Dunlop and N. R. Heckenberg. Optical angular momentum transfer to trapped absorbing particles. *Phys. Rev. A*, **54**:1593–6, 1996.

[4] J. T. Barreiro, T. C. Wei, and P. G. Kwiat. Beating the channel capacity limit for linear photonic superdense coding. *Nat. Phys.*, **4**(4):282–6, 2008.

[5] H. Bechmann-Pasquinucci and W. Tittel. Quantum cryptography using larger alphabets. *Phys. Rev. A*, **61**(6):62308, 2000.

[6] S. P. Walborn, D. S. Lemelle, M. P. Almeida and P. H. S. Ribeiro. Quantum key distribution with higher-order alphabets using spatially encoded qudits. *Phys. Rev. Lett.*, **96**(9):90501, 2006.

[7] D. Kaszlikowski, P. Gnaciński, M.Żukowski, W. Miklaszewski and A. Zeilinger. Violations of local realism by two entangled *n*-dimensional systems are stronger than for two qubits. *Phys. Rev. Lett.*, **85**(21):4418–21, 2000.

[8] D. Gottesman. Fault-tolerant quantum computation with higher-dimensional systems. *In Quantum Computing and Quantum Communications*. Berlin: Springer-Verlag, pages 302–13, 1999.

[9] B. P. Lanyon, M. Barbieri, M. P. Almeida *et al.* Simplifying quantum logic using higher-dimensional Hilbert spaces. *Nat. Phys.*, **5**(2):134–40, 2008.

[10] K. Dholakia, N. Simpson, M. J. Padgett and L. Allen. Second-harmonic generation and the orbital angular momentum of light. *Phys. Rev. A*, **54**(5):3742–5, 1996.

[11] L. C. Dávila Romero, D. L. Andrews and M. Babiker. A quantum electrodynamics framework for the nonlinear optics of twisted beams. *J. Opt. B*, **4**:S66–72, 2002.

[12] D. Burnham and D. Weinberg. Observation of simultaneity in parametric production of optical photon pairs. *Phys. Rev. Lett.*, **25**:84, 1970.

[13] C. K. Hong and L. Mandel. Theory of parametric frequency down conversion of light. *Phys. Rev. A*, **31**(4):2409–18, 1985.

[14] S. J. Freedman and J. F. Clauser. Experimental test of local hidden-variable theories. *Phys. Rev. Lett.*, **28**:938–41, 1972.

[15] A. Aspect, Dalibard and G. Roger. Experimental test of Bell's inequalities using time-varying analyzers. *Phys. Rev. Lett.*, **49**:1804–7, 1982.

[16] A. Aspect, P. Grangier and G. Roger. Experimental tests of realistic local theories via Bell's theorem. *Phys. Rev. Lett.*, **47**:460–3, 1981.

[17] J. Bell. *Speakable and Unspeakable in Quantum Mechanics*. Cambridge: Cambridge University Press, 1987.

[18] V. Bazhenov, M. V. Vasnetsov and M. S. Soskin. Laser beams with screw dislocations in their wavefronts. *JETP Lett.*, **52**:429–31, 1990.

[19] N. R. Heckenberg, R. McDuff, C. P. Smith, H. Rubinsztein-Dunlop and M. J. Wegener. Laser beams with phase singularities. *Opt. Quant. Electron.*, **24**(9):951–62, 1992.

[20] A. Mair, A. Vaziri, G. Weihs and A. Zeilinger. Entanglement of the orbital angular momentum states of photons. *Nature*, **412**:313–6, 2001.

[21] S. Franke-Arnold, S. Barnett, M. Padgett and L. Allen. Two-photon entanglement of orbital angular momentum states. *Phys. Rev. A*, **65**:033823, 2002.

[22] C. H. Monken, P. H. S. Ribeiro and S. Pádua. Transfer of angular spectrum and image formation in spontaneous parametric down-conversion. *Phys. Rev. A*, **57**(4):3123–6, 1998.

[23] S. P. Walborn, A. N. De Oliveira, R. S. Thebaldi, and C. H. Monken. Entanglement and conservation of orbital angular momentum in spontaneous parametric down-conversion. *Phys. Rev. A*, **69**(2):23811, 2004.

[24] H. Everett III. 'Relative state' formulation of quantum mechanics. *Rev. Mod. Phys.*, **29**(3):454–62, 1957.

[25] N. K. Langford, R. B. Dalton, M. D. Harvey *et al.* Measuring entangled qutrits and their use for quantum bit commitment. *Phys. Rev. Lett.*, **93**(5):53601, 2004.

[26] B. Jack, J. Leach, H. Ritsch, S. M. Barnett, M. J. Padgett and S. Franke-Arnold. Precise quantum tomography of photon pairs with entangled orbital angular momentum. *New J. Phys.*, **11**:103024, 2009.

[27] M. W. Beijersbergen, L. Allen, H. Van der Veen and J. P. Woerdman. Astigmatic laser mode converters and transfer of orbital angular momentum. *Opt. Commun.*, **96**(1–3):123–32, 1993.

[28] M. J. Padgett and J. Courtial. Poincaré-sphere equivalent for light beams containing orbital angular momentum. *Opt. Lett.*, **24**(7):430–2, 1999.

[29] L. Allen, J. Courtial and M. J. Padgett. Matrix formulation for the propagation of light beams with orbital and spin angular momenta. *Phys. Rev. E*, **60**:7497–503, 1999.

[30] D. Kawase, Y. Miyamoto, M. Takeda, K. Sasaki and S. Takeuchi. Observing quantum correlation of photons in Laguerre–Gauss modes using the Gouy phase. *Phys. Rev. Lett.*, **101**(5):50501, 2008.

[31] J. Romero, J. Leach, B. Jack *et al.* Entangled optical vortex links. *Phys. Rev. Lett.*, **106**(10):100407, 2011.

[32] A. Vaziri, G. Weihs and A. Zeilinger. Experimental two-photon, three-dimensional entanglement for quantum communication. *Phys. Rev. Lett.*, **89**(24):240401, 2002.

[33] A. Dada, J. Leach, G. Buller, M. J. Padgett and E. Andersson. Experimental high-dimensional two-photon entanglement and violations of generalized Bell inequalities. *Nature Phys.*, **7**:677–80, 2011.
[34] J. Leach, B. Jack, J. Romero *et al.* Quantum correlations in optical angle-orbital angular momentum variables. *Science*, **329**:662–5, 2010.
[35] J. B. Pors, A. Aiello, S. S. R. Oemrawsingh, M. P. van Exter, E. R. Eliel, and J. P. Woerdman. Angular phase-plate analyzers for measuring the dimensionality of multimode fields. *Phys. Rev. A*, **77**(3):033845, 2008.
[36] J. B. Pors, S. S. R. Oemrawsingh, A. Aiello *et al.* Shannon dimensionality of quantum channels and its application to photon entanglement. *Phys. Rev. Lett.*, **101**(12):120502, 2008.
[37] D. Collins, N. Gisin, N. Linden, S. Massar and S. Popescu. Bell inequalities for arbitrarily high-dimensional systems. *Phys. Rev. Lett.*, **88**(4):40404, 2002.
[38] G. Molina-Terriza, J. P. Torres and L. Torner. Management of the angular momentum of light: preparation of photons in multidimensional vector states of angular momentum. *Phys. Rev. Lett.*, **88**(1):13601, 2001.
[39] J. P. Torres, Y. Deyanova, L. Torner and G. Molina-Terriza. Preparation of engineered two-photon entangled states for multidimensional quantum information. *Phys. Rev. A*, **67**:052313, 2003.
[40] S. Franke-Arnold, S. M. Barnett, E. Yao, J. Leach, J. Courtial and M. J. Padgett. Uncertainty principle for angular position and angular momentum. *New J. Phys.*, **6**:103, 2004.
[41] D. L. Andrews. The effect of scattering on single photon transmission of optical angular momentum. *J. Opt.*, **13**:064003, 2011.
[42] J. P. Torres, A. Alexandrescu and L. Torner. Quantum spiral bandwidth of entangled two-photon states. *Phys. Rev. A*, **68**(5):050301, 2003.
[43] F. M. Miatto, A. M. Yao and S. M. Barnett. Full characterization of the quantum spiral bandwidth of entangled biphotons. *Phys. Rev. A*, **83**(3):033816, 2011.
[44] A. M. Yao. Spectral decomposition of entangled photons with an arbitrary pump. *New J. Phys*, **13**(5):053048, 2011.
[45] H. Di Lorenzo Pires, H. C. B. Florijn and M. P. van Exter. Measurement of the spiral spectrum of entangled two-photon states. *Phys. Rev. Lett.*, **104**(2):20505, 2010.
[46] C. E. Shannon and W. Weaver. *The Mathematical Theory of Communication*. New York: University of Illinois Press, 1949.
[47] A. K. Jha, B. Jack, E. Yao et al. Fourier relationship between the angle and angular momentum of entangled photons. *Phys. Rev. A*, **78**:043810, 2008.

13

Measurement of light's orbital angular momentum

M. P. J. LAVERY, J. COURTIAL AND M. J. PADGETT

13.1 Introduction

The orbital angular momentum (OAM) carried by light is widely seen as an extremely useful optical characteristic, with applications in many areas of optics. It was Allen *et al.* [1] who recognised that a helically phased light beam with a phase cross-section of $\exp(i\ell\phi)$ carries an OAM, with a value of $\ell\hbar$ per photon. Such a light beam contains an optical vortex line of ℓ on its axis. One issue that is yet to be completely resolved is the development of a simple and 100% efficient method for the measurement of OAM.

A better known case of optical angular momentum is spin angular momentum (SAM). SAM is associated with the polarisation state of the light; the spin angular momentum in a left and right circularly polarised beam is $\sigma\hbar = \pm 1$, per photon, respectively [2]. The SAM can be easily determined through the use of a polarising beam splitter, where a $\pi/4$ waveplate converts circular polarised light into a p- or s-polarised state which is then transmitted or reflected to give one of two outputs, as shown in Fig. 13.1(a).

OAM arises from the amplitude cross-section of the beam and is therefore independent of the spin angular momentum. One key characteristic of beams carrying OAM is that whereas SAM has only two orthogonal states, the OAM is described by an unbounded state space, i.e. ℓ (as in $\exp(i\ell\phi)$) can take any integer value [3]. Each state described by an integer value of ℓ is orthogonal to all of the others, which presents a distinct advantage for the encoding of information over polarisation alone, as polarisation is limited to two orthogonal states. However, this advantage presents a greater challenge in the methods required to make a measurement, as the use of N modes means there is in N-dimensional state space to be considered. The perfect detection of all N potential states is required to finally resolve this issue of measuring OAM (Fig. 13.1 (b)).

In this chapter we will review the current methods which have been investigated for the measurement of OAM. These experimental techniques have included measuring the torque applied to microscopic particles [4], counting the spiral interference fringes [5] and using optical transformations [6], among many others. However, all the current attempts have

The Angular Momentum of Light, ed. D. L. Andrews and M. Babiker. Published by Cambridge University Press.

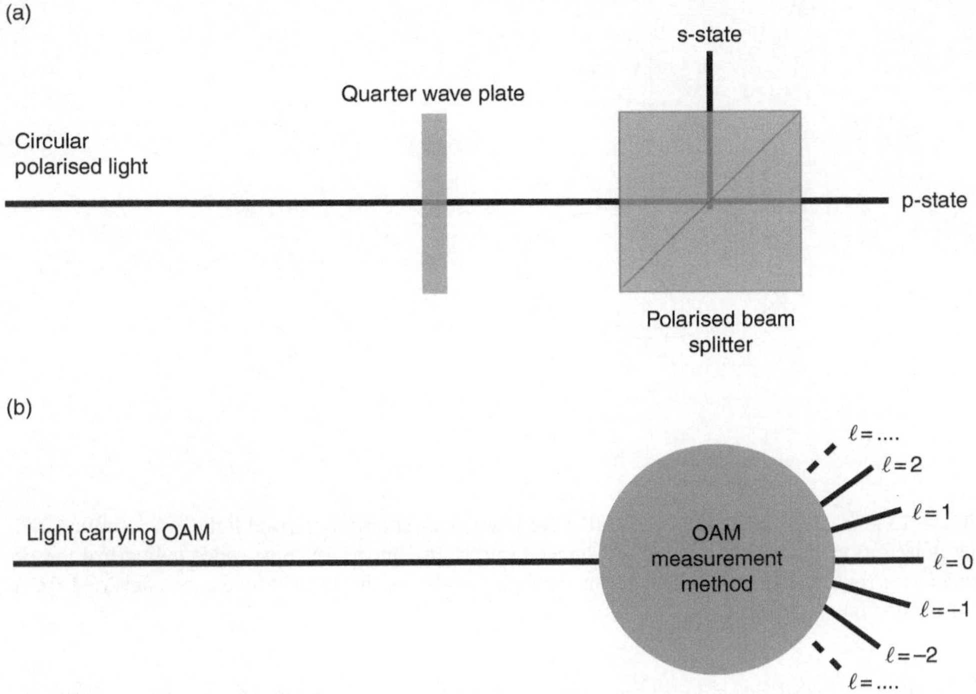

Figure 13.1 (a) A quarter-wave plate converts circularly polarised light into a linear state, the orientation of which is determined by the handedness of the original polarisation. A polarising beam splitter transmits p-polarised light and reflects s-polarised light, giving two outputs, allowing for the determination of the polarisation state. (b) A helically phased beam carries an OAM with a value of $\ell\hbar$ per photon. The variable ℓ, in principle, can take any integer value, offering a large range of potential outputs.

their strengths and weaknesses, but have yet to offer a simple solution as efficient as the polarising beam splitter is for measuring SAM.

13.2 Spinning trapped particles

A light beam carrying orbital angular momentum has an azimuthal component to the Poynting vector of the electro-magnetic field [7]. The first studies into the rotational movement of objects by light beams were considered theoretically by Poynting in 1909 [8], in which he suggested that circularly polarised light could contain angular momentum. In 1936, Beth carried out an experiment in which a quartz wave plate suspended on a length of quartz fibres was shown to rotate when circularly polarised light was passed through the wave plate centred on the fibre (Fig. 13.2) [2]. This experiment led Allen *et al.* to propose a similar experiment where the OAM present within Laguerre–Gaussian (LG) laser modes would cause the rotation of suspended cylindrical lenses [7], shown in Fig. 13.2 (b). However, a rotation of this type has yet to be experimentally observed.

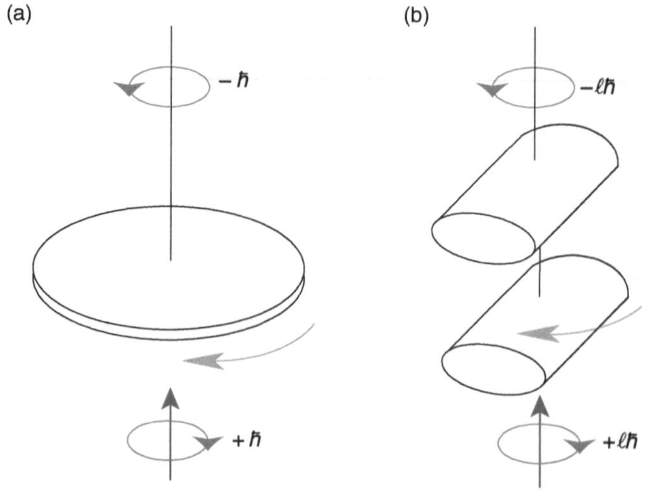

Figure 13.2 (a) Torque applied to a half-wave plate as circularly polarised light is transmitted and the polarisation state changes from left-handed into right-handed. (b) Suspended cylindrical lenses undergo a rotation when a helical beam through passes, as the OAM values are converted from left-handed into right-handed.

In 1995, He *et al.* considered the transfer of orbital angular momentum to trapped microscopic absorbing particles within optical tweezers (Fig. 13.3) [4]. Optical tweezers trap particles through the use of a force proportional to the intensity gradient in the focused light beam: dielectric particles are attracted to the centre of the region of high field strength, which then allows the trapping and movement of micro-scale particles [9]. When trapping using a helically phased beam, with a beam waist equivalent to the size of the particle, the centre of attraction corresponds to the vortex location of the beam. The early studies into the torque applied by beams carrying OAM used black micron-sized CuO particles, which were seen to spin when trapped using a plane polarised LG beam. In later studies, the torque due to OAM was combined with that due to the spin angular momentum of circular polarised light [10, 11].

A dielectric particle will partially absorb the light incident on it, hence absorbing any angular momentum in either the form of SAM or OAM, resulting in a torque on the particle. In beams with a helical phase structure the torque applied acts in addition to the torque applied from circular polarisation. When simply considering the paraxial limit, the interaction with the particle is affected by both the polarisation and helicity of the beam, causing a torque

$$\Gamma = \frac{P_{\text{abs}}}{\lambda}\,(\ell + \sigma_z)\,, \qquad (13.1)$$

where P_{abs} is the power absorbed by the dielectric bead, λ is frequency of the light and σ_z is the handedness of the polarised light. The application of torque to a particle is not restricted to the case of many incident photons, as every photon in the beam carries an

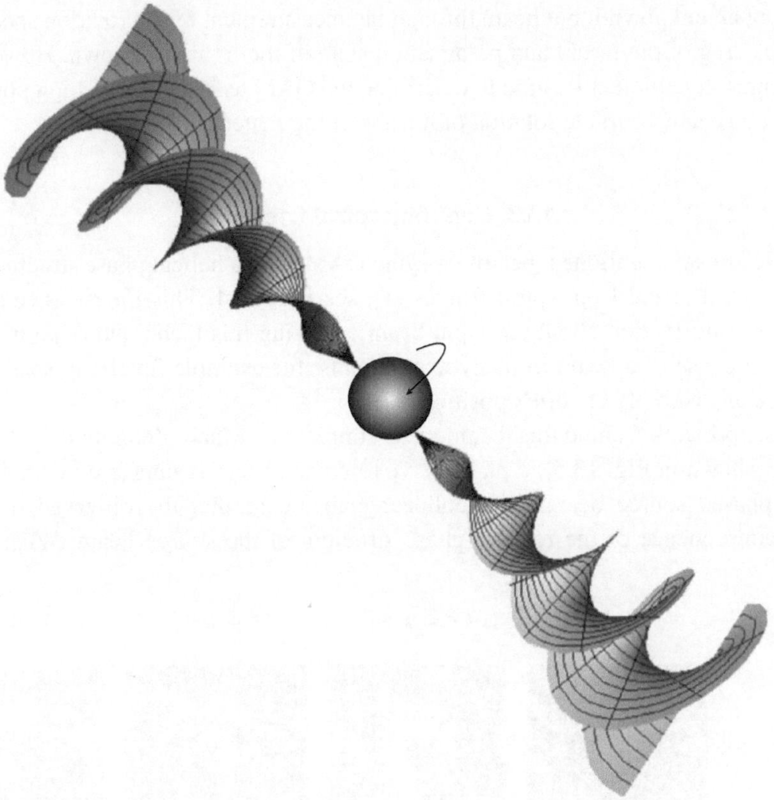

Figure 13.3 Focused helically phased beam trapping a dielectric bead. A partially absorbing bead absorbs the OAM, resulting in torque.

OAM of $\ell\hbar$. The total torque applied to a particle, within the paraxial approximation, can be determined as the photon flux multiplied by the angular momentum of the helical beam. In the case of LG beams, the total angular momentum per photon is $(\ell + \sigma_z)\,\hbar$. It was shown by Simpson *et al.* that the torque due to OAM of $\ell = 1$ and that due to a polarisation of $\sigma = -1$ can cancel each other [11].

Particles trapped in optical tweezers are usually in a viscous material, usually water, which will cause a rotational drag resulting in slower spinning. In the simplest case, this drag is proportional to the angular velocity of the particle. A measurement of the optical torque can be made when an equilibrium of the rational drag and applied torque is reached. This measurement can be captured using a high-speed digital cameras or a photo diode at the edge of the trapped particle measuring a flashing as the particle spins. From this rotation and knowledge of the viscous medium, the torque is calculated and can then be used in Eq. (13.1) to calculate ℓ.

After the initial experiment by He *et al.*, beams carrying OAM have become useful for optical spanners [11] and many other useful tasks. This method can be used to determine the OAM of an unknown input beam through the measurement of the rotation speed, when the viscous drag of the liquid and polarisation state of the light are known. However, this is not an optically efficient method to determine the OAM as it requires a high power laser source to cause a measurable rotation of the absorbing material.

13.3 Counting spiral fringes

As we have already mentioned, beams carrying OAM have a helical phase structure, which interference can reveal from spiral fringes [5], see Fig. 13.4. This interference approach yields phase information about the input beam, allowing this technique to be used in the study of the phase of a beam in many experiments, for example involving speckle fields [12] and beams with offset vortex positions [13].

The basic principle behind this technique is simple: in a Mach–Zehnder or similar interferometer (shown in Fig. 13.5), a plane-wave reference beam is interfered with a colinear helically phased source beam. This colinear path means that the observed fringes are purely a consequence of the relative phase structure of the source beam. When using a

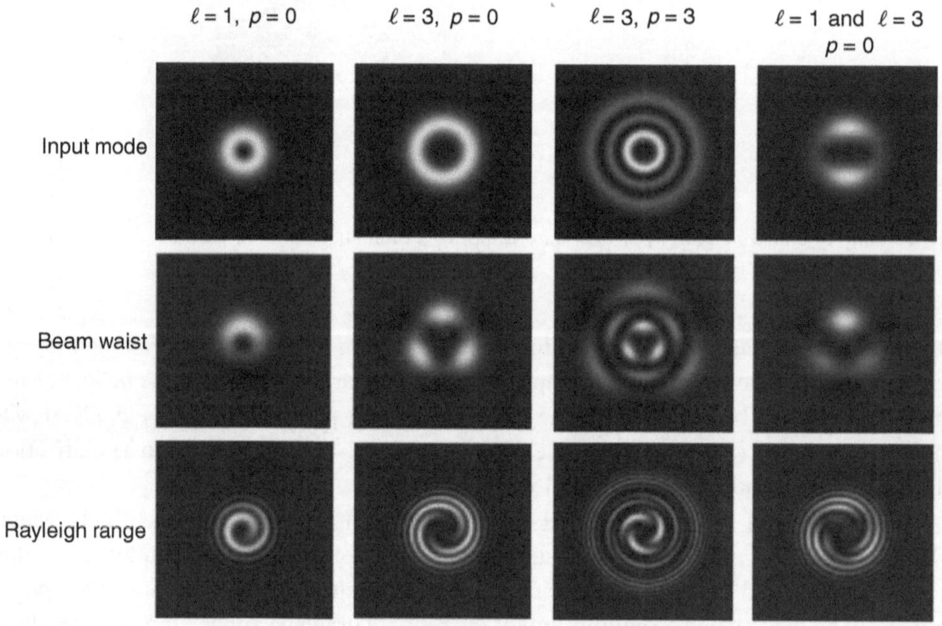

Figure 13.4 Interference of beam (LG beams, examples of beams carrying OAM where ℓ is the azimuthal component and p is the radial component), with a uniform plane-wave in the plane of the beam waist and at the Rayleigh range. These are shown for several input modes which contain helically phased modes with no radial component p in columns one and two. Column three shows the effect of a mode containing radial lobes and column four shows a superposition of helically phased beams.

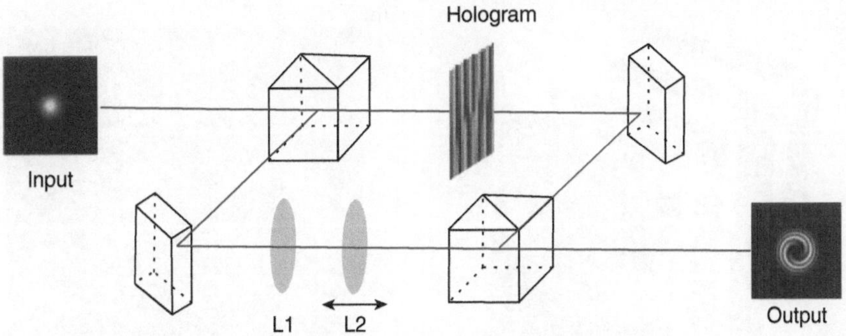

Figure 13.5 Schematic of a Mach–Zehnder type interferometer in which a beam containing OAM is generated in one arm of the interferometer through the use of a hologram. The fringe pattern is changed from a petal pattern to a spiral fringe pattern by changing the distance between lens L1 and L2, as this alteration changes the curvature of the phase front in the reference arm.

helically phased input, with a phase structure of $\exp(i\ell\phi)$, each fringe corresponds to one step through 2π and by simply counting the number of these fringes the value ℓ can be determined, as seen in Fig. 13.4. The pattern observed is dependent on the curvature of the phase front: when the curvature is matched, petal patterns are observed, as seen at the beam waist; and when there is a difference in curvature, spiral fringes are observed. In the case of Laguerre–Gaussian laser modes, where the modes contain both a helical phase term dependent on ℓ and a radial term dependent on p, the radial components are seen as radial nodes in the interferogram, the number of these is equal to $p + 1$, which is seen in column three of Fig. 13.4. Superpositions of Laguerre–Gaussian laser modes result in interleaved spiral fringes, seen in column four of Fig. 13.4.

Interferometric methods allow the OAM content to be easily determined. These are very useful techniques and have many applications, but due to the requirement to generate many interference images, they are not suitable for cases involving small numbers of detected photons.

The methods of interference can be applied to more complex structures by generating a detailed phase map of the source. The areas of darkness within a interference pattern can lead to ambiguity about the phase value at this point, as this darkness may be a result of a lack of intensity as opposed to destructive interference. Changing the phase of the plane-wave reference gives further information about these areas.

Each pixel in an image can be considered as an individual intensity measurement. An intensity reading is made for every pixel for each step in phase, usually 16 or 32 steps. When these steps are taken over a range 2π, the intensity of each point will change sinusoidally, the start point of this sinusoid represents the phase of that particular pixel.

Computationally this can be carried out very quickly, allowing for example the tracking of vortex positions in speckle fields, thereby enabling studies into the topology of vortex lines in these fields [14] (see Fig. 13.6).

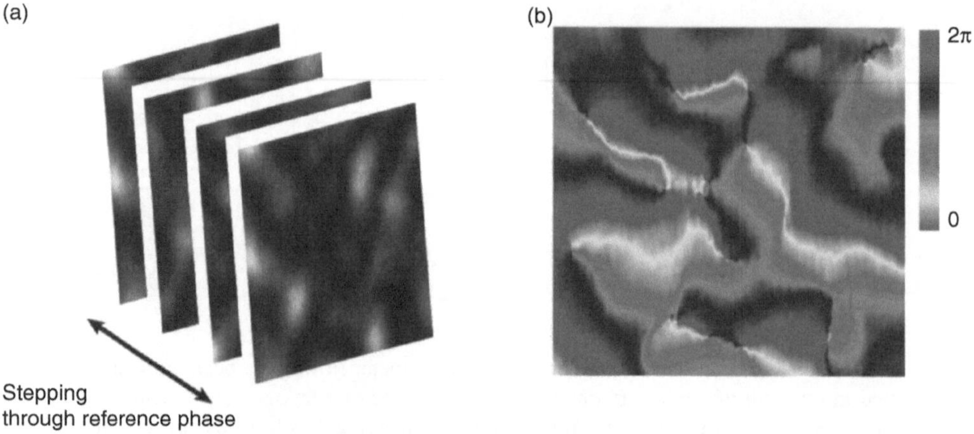

Figure 13.6 (a) A stack of intensity images where the phase of the reference beam changes in known steps of $2\pi/N$, allowing the phase of the intensity field to be determined. (b) Representation of the measured phase profile of a speckle field. Any phase change of 2π around a closed loop contains a vortex with topological charge $\ell = \pm 1$.

13.4 Diffraction effects from apertures

The requirement for a suitable reference beam places strict limitations on the areas where two-beam interference techniques can be used. For many cases where one would like to make a measurement of OAM, this is troublesome and in some cases impossible, for example in the detection of sources from astronomical objects.

One technique that does not require a reference beam is that of diffraction from apertures [15, 16]. A very elegant example of this was investigated by Hickmann *et al.*, where a helically phased beam incident on a triangular aperture resulted in a triangular lattice of constructive interference spots, the number and orientation of which was related to the ℓ value of the input (see Fig. 13.7)[15]. This method can be modelled by considering the Fraunhofer diffraction pattern from the aperture, i.e. the squared modulus of the two-dimensional Fourier transform of the amplitude in the transverse plane immediately behind the aperture.

This problem is not too distant from Fraunhofer diffraction being studied by undergraduates around the world, but had never been considered for beams containing a phase singularity. The observed interference effects arise from the light diffracted at the each side of the aperture, and in the case of helically phased beams there is a change in phase along the edges of the triangle which is directly related to the ℓ value of the input beam. This detailed phase structure at the edge causes an interference pattern which is a function of ℓ, where an increase in the mode index results in a new layer of spots within the interference pattern, as can be seen in Fig. 13.7. The triangular pattern is rotated with respect to the aperture. This rotation occurs due to the helical nature of the input mode and the direction of this rotation is reversed when the sign of ℓ is changed.

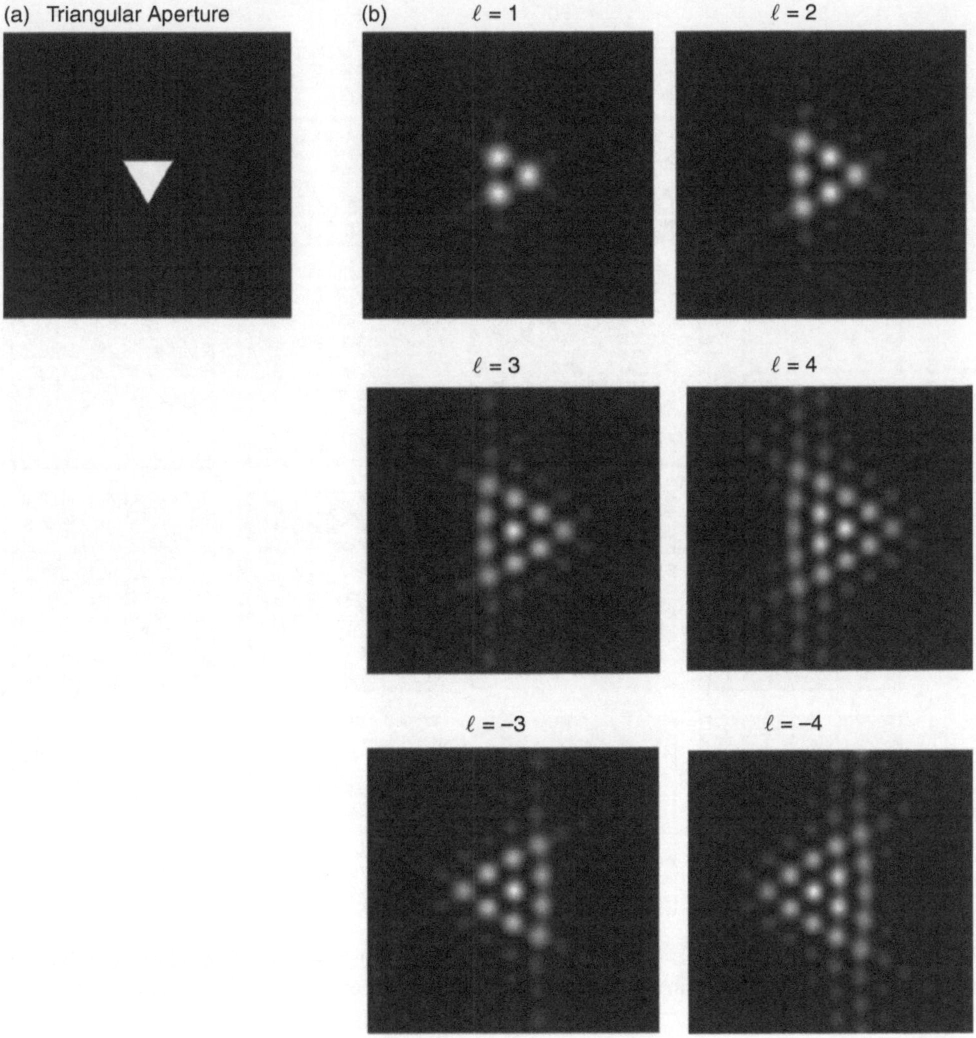

Figure 13.7 (a) Triangular aperture. (b) The far-field diffraction pattern generated by LG beams clipped by the aperture for various values of ℓ.

The use of a single aperture presents an elegant method to measure the OAM value of an input mode. This approach is not limited to the use of triangular apertures and has recently been shown to work with a hexagonal aperture, generating lines of constant intensity, the number of which correspond to the ℓ mode index of the input mode [17]. This method has also been used to measure fractional vortex charges [18], proving its effectiveness as a measurement technique.

An alternative aperture-based approach was independently developed by Berkhout and Beijersbergen, using a circle of pinholes to generate diffraction patterns where the pattern

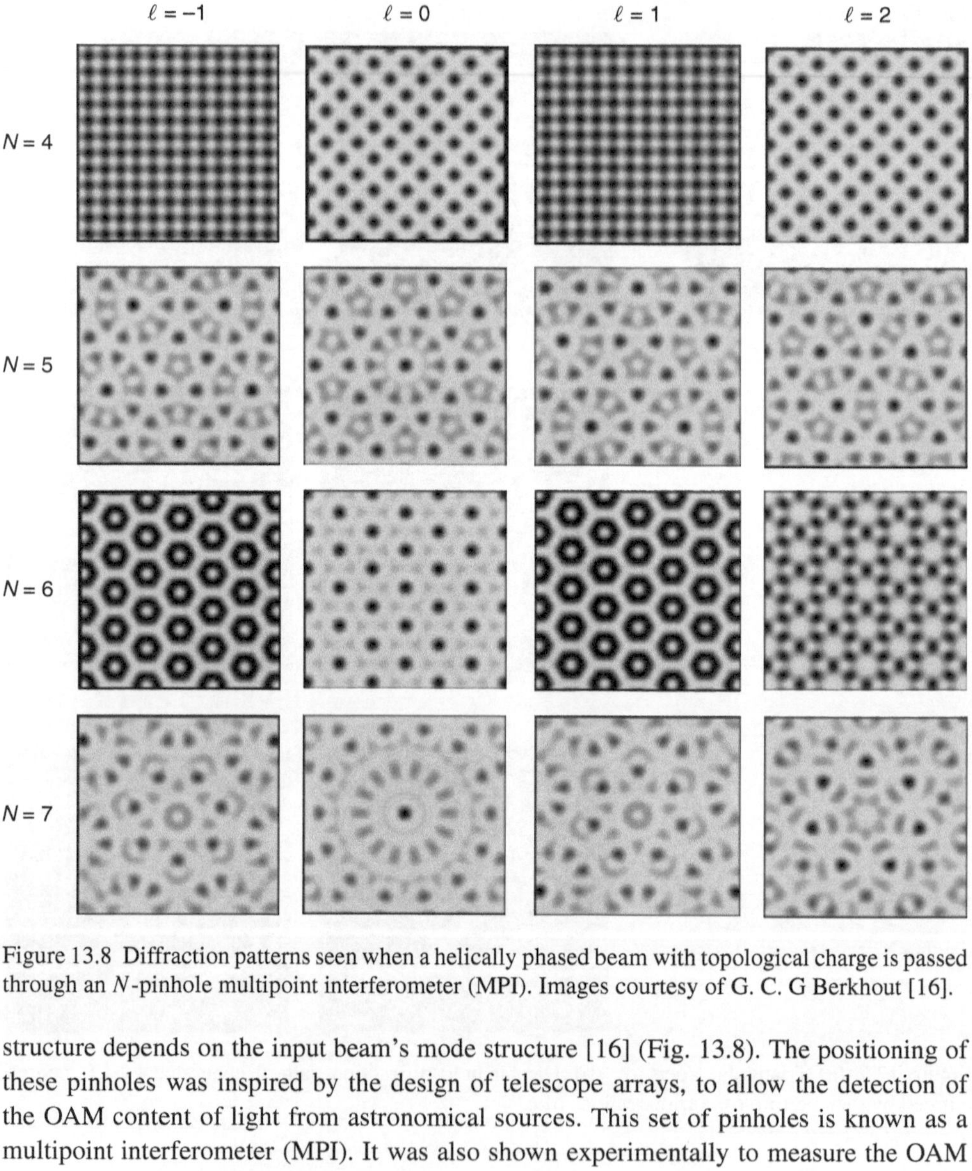

Figure 13.8 Diffraction patterns seen when a helically phased beam with topological charge is passed through an N-pinhole multipoint interferometer (MPI). Images courtesy of G. C. G Berkhout [16].

structure depends on the input beam's mode structure [16] (Fig. 13.8). The positioning of these pinholes was inspired by the design of telescope arrays, to allow the detection of the OAM content of light from astronomical sources. This set of pinholes is known as a multipoint interferometer (MPI). It was also shown experimentally to measure the OAM state of partially coherent light [16].

The use of a set of pinholes illuminated with a beam with the vortex at the centre results in a sampling of the phase at fixed locations around the beam axis. Each pinhole can be considered to be a point light source. This again can be analysed by considering the Fraunhofer limit, yielding the far-field intensity pattern,

$$I_\ell^N \propto \left| \sum_{n=0}^{N-1} \exp(-\mathrm{i}\ell\alpha_n) \times \exp\left(\mathrm{i}\frac{ka}{z}(x\cos\alpha_n + y\sin\alpha_n)\right) \right|^2, \tag{13.2}$$

where I_ℓ^N is the far-field intensity of the MPI, N is the number of pinholes, k is the wave number, a is the radius of the circle of pinholes and $\alpha_n = 2\pi n/N$ is the azimuthal coordinate of the nth point.

The ability of this approach to detect OAM is very dependent on the number of pinholes used, which can been seen in Fig. 13.8. This dependence can be understood in terms of the phase sampling that occurs at each of the pinholes. In the case of two or three pinholes, any phase difference between the pinhole positions in a helically phased beam could also simply be due to a suitably inclined plane-wave. This method is therefore only truly effective when four or more pinholes are arranged around the centre of the input mode's beam axis. In the specific case of four pinholes the MPI acts as distinguish between odd and even OAM states. The orientation of the diffraction patterns changes when inverting the mode index.

Either the Hickmann *et al.* [15] or the Berkhout and Beijersbergen [16] aperture-based methods are a very attractive option when a simple solution for the measurement of the OAM content of a specific source is required. However, these techniques are based on pattern recognition for the determination of the mode index of the light and this is difficult to automate. These methods also require the use of a large number of photons in the same state to make a measurement of OAM and this makes them inherently inefficient.

13.5 Diffractive holographic filters

In 1992, the generation of beams carrying OAM was first achieved through the use of a pair of cylindrical lenses [1]. Later Beijersbergen *et al.* developed helical phase plates [19] to do the same task. Prior to Allen and coworkers' recognition of the OAM within helically phased beams it was shown by Bazhenov *et al.* and Heckenberg *et al.* that computer-generated holograms could be used to generate beams containing phase singularities [20, 21].

Holograms have now become a standard tool for the generation of beams carrying OAM. In 1990, Bazhenov *et al.* took an approach based on a diffraction grating to produce multiple orders with the desired phase structure [20]. These holograms can be understood as the phase of the product of the vortex term $\exp(i\ell\phi)$ and a phase-ramp term $\exp(ik\sin(\alpha)x)$, filtered ('blazed') such that points whose phase is in the upper half of all possible phase values are coloured black, the remainder are coloured white. The effect of such a blazing is to distribute the phase into all possible diffraction orders. The ℓ-forked pattern was inspired by the interference pattern between a beam containing a phase singularity and a tilted plane wave, producing this distinctive pattern [20].

In 1992, Heckenberg *et al.* focused on the concept of a spiral zone plate as the basis for the holograms they investigated [21]. The spiral zone plates consisted of binary concentric circles with a singularity in the centre, acting in exactly the same way as a Fresnel zone plate but with a dislocation in the centre. The number of interleaved circles defines the charge of the singularity generated. These zone plates can be understood as the phase of the product of a vortex term $\exp(i\ell\phi)$ and a lens term $\exp(ikr^2/(2f))$, blazed just

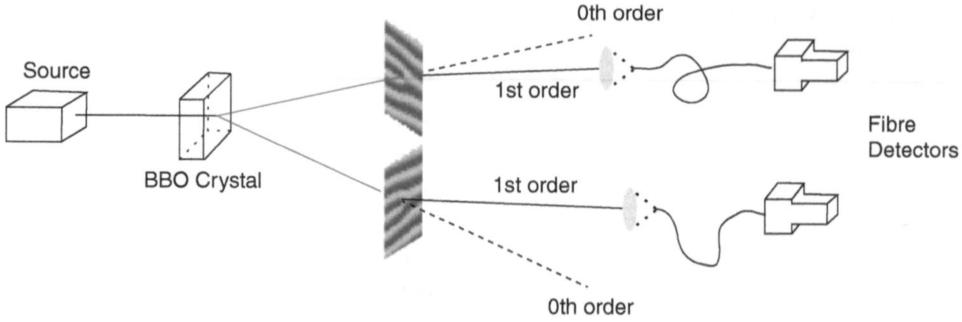

Figure 13.9 Schematic overview of the setup that is used to couple entangled photons from a BBO down-conversion crystal carrying a specific OAM content. A photon with the correct ℓ value becomes a Gaussian mode in the first order, which then gives a reading at the fibre detector.

like Bazhenov *et al.*'s phase patterns. These gratings have proven effective in generating singularities, but they suffer from a lack of separation of the different diffraction orders.

In Bazhenov *et al.*'s approach [20], different diffraction orders travel in different directions, and can therefore be separated through Fourier filtering. The method, using blazed holograms, has become one of the most common methods to generate vortex beams. More recently the use of digital diffractive elements, in the form of spatial light modulators (SLMs), has allowed the interactive generation of beams containing singularities.

Mair *et al.* showed that this generating method can be reversed, coupling a particular diffraction order into a single-mode fibre [22]. In this method the hologram, in combination with the fibre's entrance aperture, becomes a filter for OAM. It was used by Mair *et al.* to demonstrate the entanglement of the OAM of pairs of photons produced from a non-linear crystal. This approach is currently extensively used for the detection of the OAM at the single-photon level quantum optics experiments (see Fig. 13.9).

The concept was extended by Molina-Terriza *et al.* to couple multiple values of ℓ into multiple diffraction orders, allowing a larger bandwidth of detection [23]. However, this method requires many photons in the same state, as light can only be coupled into the fibre when the correct hologram is displayed. This restriction means that for an unknown helical beam, many holograms must be cycled through to test for the correct ℓ value.

13.6 More complex holograms

OAM as a basis for high-bandwidth free-space communications has become an area of interest, with many researchers studying the stability of vortices through atmospheric turbulence. It is also seen a possible method to communicate with space-based receivers [24, 25]. Gibson *et al.* developed a free-space communicator based on two interleaved ℓ-forked holograms generating a 3×3 grid of diffraction orders, each of which tests for a different ℓ of an incoming beam [26]. The system was shown to measure the OAM content for eight ℓ modes over a range of 15 metres, and was also shown to measure superpositions of these eight modes.

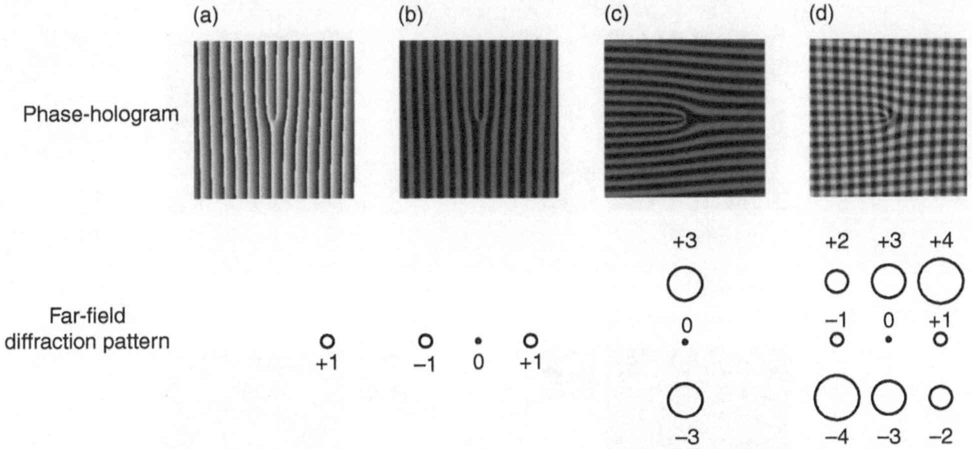

Figure 13.10 (a) A blazed ℓ-forked hologram, which when illuminated with a Gaussian beam will generate an $\ell = 1$ beam in the first order. (b) The blazed hologram can be further optimised to give an equal power weighing to the first and negative first orders. (c) A rotation of $\pi/2$ radians gives the same rotation in the position of the diffraction orders, as shown the charge can be increased to a higher ℓ. (d) The modulo-2π addition of the patterns shown in (b) and (c) gives the desired pattern of nine orders, generating a range of ℓ-valued beams.

This process of combining two diffractive phase elements, where each diffractive element corresponds to a different ℓ value, was outlined by Khonina *et al.* in 2000 [27]. The work by Gibson *et al.* [26] is similar to that of Mair *et al.* [25], where an ℓ-forked hologram converts an $\exp(i\ell\phi)$ beam into a beam with a planar phase front. A hologram can be optimised to generate specific diffraction orders, the phase handedness is inverted for negative orders. Gibson *et al.* made use of this effect by optimising the hologram to produce the -1, 0 and 1 orders, see Fig. 13.10, giving $\pm\ell$ in the first and negative first orders, respectively, in the horizontal direction (13.10(b)). When a beam carrying OAM is incident, this allows the detection of the corresponding ℓ mode. The same optimisation process can also be achieved in the vertical direction, where the direction of the phase ramp is simply rotated by 90°. To allow the detection of many modes simultaneously, these two patterns were superimposed, resulting in each horizontal order having a number of vertical diffraction orders, such that the ℓ value of the orders are added. For example, if the positive first order of each hologram is $\ell = 1$, then the vertical orders above and below the first horizontal order will be $\ell = 2$ and $\ell = 0$.

Using combinations of horizontal and vertical forks produces a grid of potential ℓ readings. When an input beam of the form $\exp(i\ell\phi)$ is incident on this pattern, a spot will appear in the centre of the diffraction order, for which the ℓ values of the beam and the diffraction order add up to zero. It should be noted that the central grid location contains the original beam. Gibson *et al.* used this light for system alignment [26].

For any input mode the power of the input is spread evenly across all of the nine diffracted orders, hence the efficiency of the method scales with the reciprocal of the

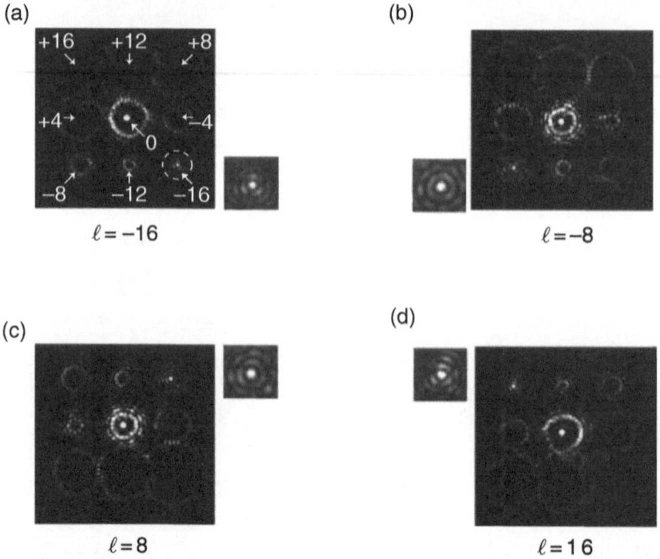

Figure 13.11 The far field observed for the ℓ modes ±16, and ±8. Images courtesy of G. Gibson [26].

number of modes being detected. A superposition of ℓ modes can be determined as two or more bright central spots, the relative intensity of which determines the mode weighting in the superposition (see Fig. 13.11). The system was tested for several cases of beam misalignment and use of a selective aperture, resulting in an observed OAM spectrum [28, 29].

Gibson *et al.* showed the detection of eight modes over a range of 15 metres, as at this range the atmosphere has only a small effect on the mode quality of the transmitted beams. Beams carrying a vortex charge of $\ell = \pm 1$ are quite resilient to atmospheric distortion as a single-charge singularity cannot simply disappear. However, this is not the case for higher charges. Beams with an azimuthal index of $|\ell| > 1$ are highly sensitive to aberrations, which cause the collapse of higher-order singularities [30]. In optical communications reliant on higher-order vortex beams, these atmospheric aberrations pose considerable problems.

This interleaved hologram approach offers the ability to easily measure the OAM of many states simultaneously, and is relatively simple in its experimental implementation. However, the loss of a large proportion of the input light into other diffractive beams inherently makes this quite an inefficient detection method unsuitable for single-photon measurements. In classic terms, beams carrying OAM have the ability to increase the potential bandwidth of free-space communications but the effect of turbulence on the beams carrying the information currently limits the distance at which OAM could be transmitted in this way.

13.7 The rotational Doppler effect

The measurement of the Doppler effect is used to determine the speed at which a light source is travelling. Garetz and Arnold [31] demonstrated that circularly polarised light experiences a frequency shift of $\Delta\omega = \sigma\Omega$, where σ determines the handedness of the circular polarisation, when passed through a half-wave plate rotating at an angular frequency Ω.

Nienhuis [32] predicted that spinning a cylindrical lens in a mode converter, at a frequency of $\Omega/2$, will result in a frequency shift in the beam of $\ell\Omega$. In 1998, Courtial *et al.* proposed an experimental method whereby spinning a beam carrying OAM at high speed will result in a Doppler shift in the observed frequency of a microwave source [33]. Specifically, rotating a beam with an azimuthal profile $\exp(i\ell\phi)$ at an angular frequency Ω will result in a frequency shift of $\Omega\ell$.

Courtial *et al.* combined their earlier works with that of Garetz and Arnold [31]. This time, the beam (phase, intensity and polarisation) was rotated by a Dove prism in combination with a half-wave plate (see Fig. 13.12). As the Dove prism is spun, this will cause a spinning of the input beam where the transmitted intensity and phase rotate at twice the angular velocity of the prism. As discussed previously, a half-wave plate was used to rotate the polarisation vectors again at twice the angular velocity, resulting in the entire beam being rotated without any rotation of the source required. This effect results directly from the total angular momentum carried, defined as $J = \hbar(\ell + \sigma)$, giving a total frequency shift of $\Delta\omega = \Omega J$. When the difference in frequency was measured, it was found to closely match that predicted between clockwise and counterclockwise spin, of $\omega_{\text{cw}} - \omega_{\text{ccw}} = 2\Omega J = 2\Omega(\ell + \sigma)$.

The observed Doppler shift can be further understood through the consideration of the field distribution of the circularly polarised LG beams. An $(\ell+\sigma)$-fold rotational symmetry is seen in the field distribution, see Fig. 13.13, and for every 2π rotation of the beam the electric field undergoes a change of $(\ell + \sigma)$ cycles. The angular momentum will then, as expected, cause a change in the frequency of the spinning electric field.

A well-defined change in frequency suggests a methods to accurately make a measurement of the OAM and SAM, as the change in frequency is well defined for a particular combination of OAM and SAM. The experimental procedure requires the high-speed

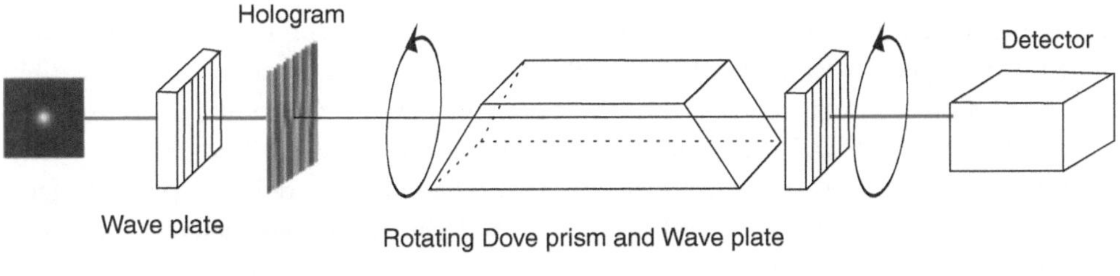

Figure 13.12 Experimental setup for the rotation of a beam using a Dove prism.

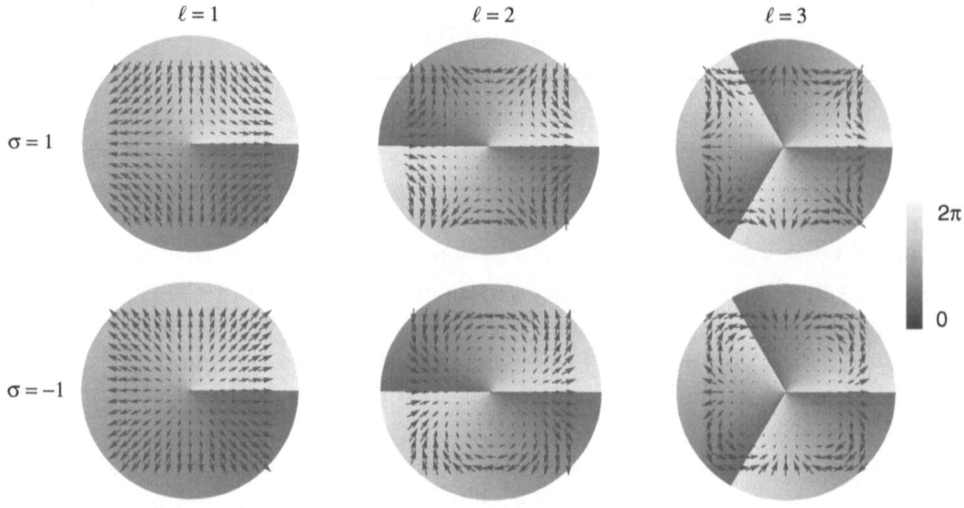

Figure 13.13 Vector plots of the transverse electric field.

rotation of optical components, which is challenging and not suitable for all circumstances. This effect is also much harder to observe at optical frequencies compared to microwave frequencies.

13.8 A Dove prism interferometer

A method to sort OAM at the single photon-level was outlined by Leach *et al.* based on a Mach–Zehnder interferometer containing a Dove prism in each arm [34], see Fig. 13.14. This allowed the sorting of beams with a particular ℓ value into one of the output arms of the interferometer without any loss of the helical structure of the beams. The process is also theoretically 100% efficient. It was proposed that a cascaded set of such interferometers would allow the detection of many OAM states, but requires $N-1$ of them to do so, which is technically very challenging.

The key component in the interferometer is a Dove prism, which flips the transverse cross section of any input beam [35]. An interferometer of this type is required have complete constructive interference in one output port, A, and complete destructive interference in the other output port, B, for a particular OAM mode. When the two Dove prisms are rotated with respect to each other by an angle of $\pi/2$, the transmitted beam is rotated by an angle of π and the relative phase difference between the two arms of the interferometer is $\Delta\psi = \ell\pi$ (Fig. 13.15). By appropriately adjusting the path length of the interferometer, constructive interference will occur in one of the two output ports for even ℓ-valued states and in the opposite port for odd ℓ-valued states. The light that is constructively interfered still carries OAM, allowing this light to be passed into a second interferometer to be further sorted.

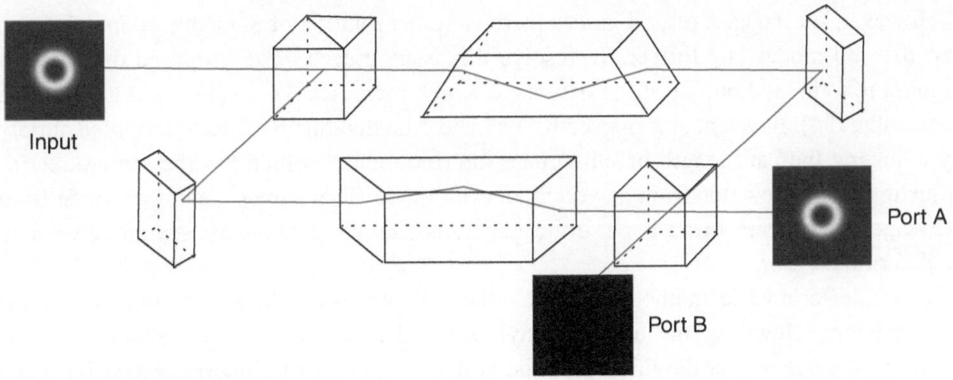

Figure 13.14 An experimental setup for a Dove prism interferometer. The layout is the same as a standard Mach–Zehnder interferometer, except a Dove prism is placed in each arm resulting in a rotation of the beam in one arm, with respect to the other. This rotation will cause complete constructive interference in either port A or B depending on the input mode value.

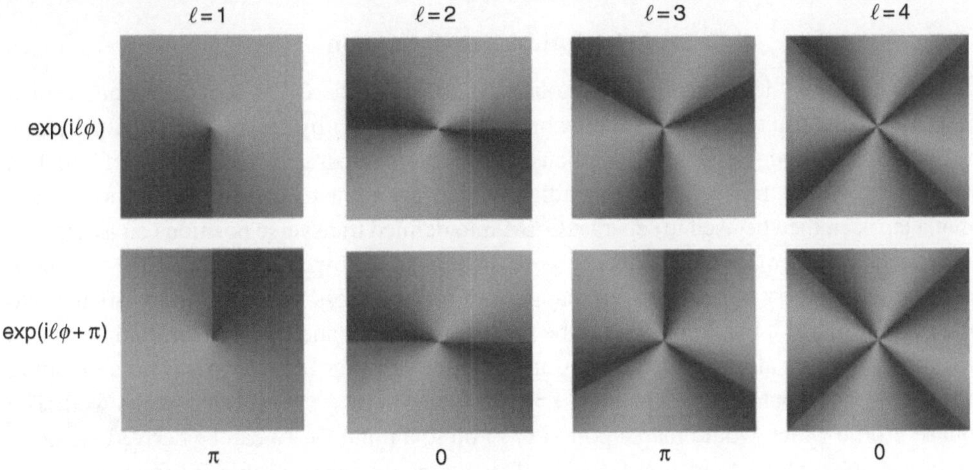

Figure 13.15 The phase profiles of several helical beams are shown in row one with no rotation and in row two with a rotation of π. It can be seen that even modes will constructively interfere as the profiles are identical, and that odd modes will destructively interfere as the phase profiles are the inverse of each other.

The interferometer is not limited to sorting odd from even, as altering the angular rotation between each of the Dove prisms will change the modes that are sorted into the output ports A and B. When the two Dove prisms are rotated by an angle of $\alpha/2$ with respect to each other, the beams transmitted through the two arms of the interferometer will be rotated with respect to each other by an angle of α. For $\alpha = \pi/2$ the phase difference between the two beams is $\alpha/2 = \pi/4$, which means that modes with mode index $\ell = 0, \pm 4, \pm 8, \ldots$ go into port A and beams with a mode index of $\ell = 0, \pm 2, \pm 6, \pm 10, \ldots$ go into port B.

However, for the case of odd modes there is no net rotation of the Dove prisms that can sort the odd modes any further. To resolve this issue, Leach *et al.* proposed the use of a hologram at the odd output of the first stage which increases the ℓ value by $\Delta\ell = 1$, to an even value [34]. It was later proposed by Wei and Xue that this issue can be tackled simply by adjusting the path length in subsequent interferometers, which has the same effect as adjusting the mode value from an odd to an even mode. This allows the output mode from the stage-one interferometer to be further separated, allowing, in theory, any mode value to be determined [36].

This interferometric method presents a theoretically very efficient method for OAM measurement. However, the use of many cascaded interferometers is technically challenging. The use of just the first stage, the odd-from-even sorter, interferometer has been applied to demonstrate amplitude damping [37] and in the development of a linear optical CNOT gate for OAM [38]. The interferometric approach offers many advantages where the non-destructive sorting of OAM is required, such as a logic gate within quantum computing.

13.9 Optical transformation

An approach was developed by Berkhout *et al.* whereby the OAM states are transformed into linear direction states [6]. The technique was inspired by the fact that a tilted plane wave passing through a lens will be focused into a spot, the transverse position of which is determined by the transverse phase gradient of that plane wave. This transformation means that a lens can then be used to separate OAM into defined transverse positions on a detector.

Optical transformations of this type were first discussed by Bryngdahl in 1974 who used a hologram to carry out the desired mapping of a beam cross-section in the xy-plane into the Fourier plane of a hologram described by coordinates u and v [39]. This method places restrictions on the allowed transformations. For the purposes of Berkhout *et al.* a mapping $(x, y) \mapsto (u, v)$, where $v = a\arctan(y/x)$ and $u = -a\ln(\sqrt{x^2+y^2}/b)$, works well. The phase profile which would map a point (x, y) on to a point (u, v) can be derived from the stationary-phase approximation [39–41]. This yields the phase profile for the first optical element (hologram (i) in Fig. 13.16) of the form

$$\phi_1(x, y) = \frac{2\pi a}{\lambda f}\left[y\arctan\left(\frac{y}{x}\right) - x\ln\left(\frac{\sqrt{x^2+y^2}}{b}\right) + x\right], \tag{13.3}$$

where λ is the wavelength of the incoming beam and f is the focal length of the Fourier lens (Fig. 13.16(a)). There are two parameters, a and b, which determine the scaling and position of the transformation in the Fourier plane, respectively. Berkhout *et al.* chose a scaling which would fill the output aperture in the u direction meaning $a = d/2\pi$, where d is the width of the aperture.

At the Fourier plane of the first element the transformation is complete. As described by Hossack *et al.* [40] the phase in the beam can then be restored using the stationary phase.

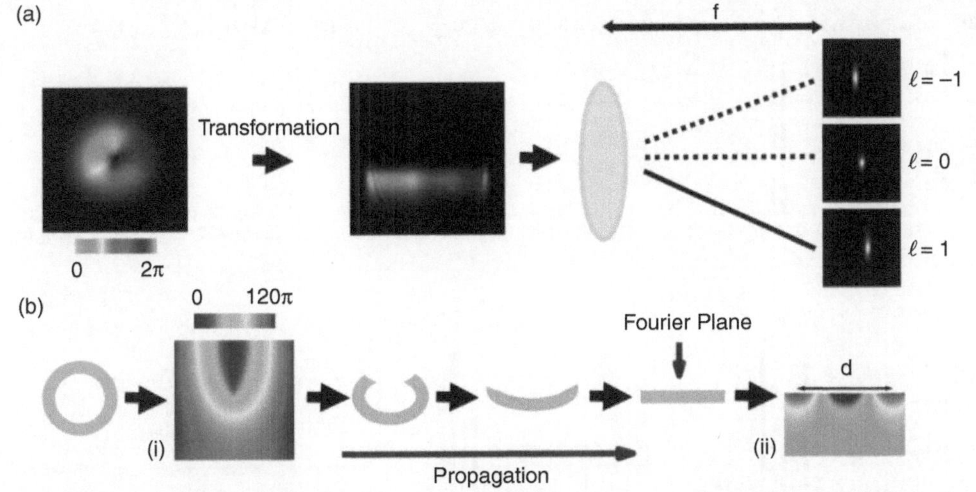

Figure 13.16 Transformation of beams with a helical-phase into a linear phase gradient. (a) The cross-section of a pure OAM mode is reshaped such that the azimuthal angle is transformed into a Cartesian coordinate, here the horizontal coordinate. Intensity is shown as brightness; phase is represented by shade; the cross-sections are drawn for $\ell = 1$. The resulting beam cross-section has a uniform phase gradient in the horizontal direction which is directly proportional to ℓ. When passed through a lens it gets focused to a position whose horizontal position depends on this phase gradient, and therefore ℓ. This is shown here for different values of ℓ. (b) Representation of the spatially varying phase added by the phase holograms used in our experiment. Hologram (i) initiates the Cartesian-to-polar geometric transformation, which is complete by the time light has reached the hologram's Fourier plane. Hologram (ii) corrects for the differences in optical path length between different parts of the beam.

The phase profile of the second optical element is thus the opposite of the phase component of this complex amplitude giving, specifically, a component that adds a phase

$$\phi_2(u, v) = -\frac{2\pi ab}{\lambda f} \exp\left(-\frac{u}{a}\right) \cos\left(\frac{v}{a}\right) \tag{13.4}$$

to the beam makes the phase at the transformed position that of the original position.

This method is not limited to the detection of single ℓ values and can be used in the detection of superposition also. When a superposition of OAM modes is transformed it becomes a combination of transverse phase gradients which are then focused into multiple locations in the detector plane. It was shown experimentally by Lavery *et al.* that this technique can be used to measure detailed superpositions [42] using the example of inclined and tilted LG modes (Fig. 13.17).

The optical-transformation method of OAM detection has several useful properties. It can work at the level of single photons, it can detect complex superpositions of modes,

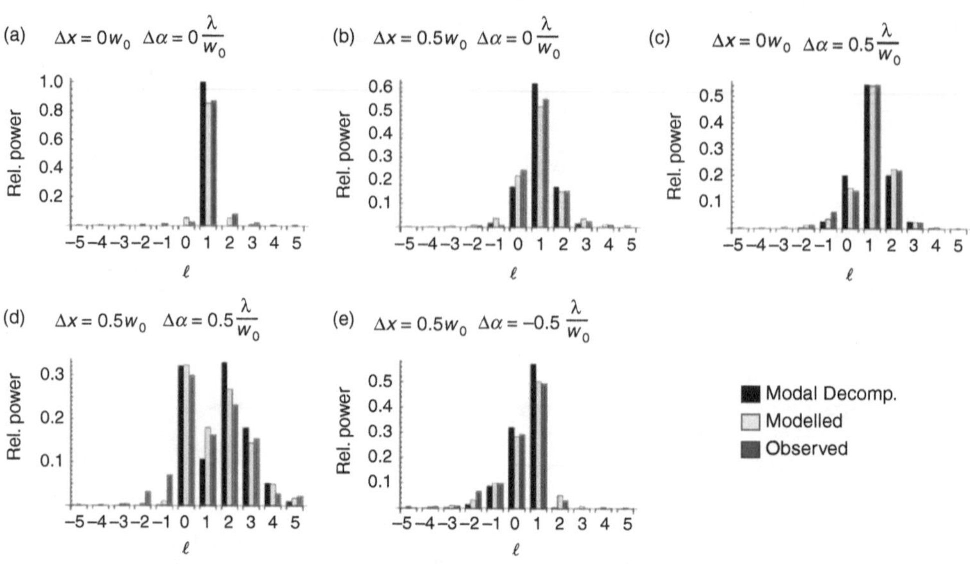

Figure 13.17 OAM spectrum of superposition beams created by the deliberate misalignment of a pure LG beam with $\ell = 1$: (a) aligned measurement axis and beam axis; (*b*) lateral movement of $x_0 = 0.5w_0$; (*c*) tilt angle $\alpha = 0.5w_0/\lambda$; (*d*) combination of a lateral movement of $x_0 = 0.5w_0$ and a tilt of $\alpha = 0.5\lambda/w_0$; (*e*) tilt $\alpha = -0.5\lambda/w_0$. This OAM spectrum shows good agreement with that expected from theory [28].

and it is efficient. However, it suffers from cross-talk due to a small overlap of the spots in the detection plane, the width of which correspond to the diffraction limit of the optical system.

13.10 Summary

Even though the issues surrounding the measurement of OAM have not been completely resolved, many of the methods outlined in this chapter have proved to be very useful for the task. An ideal solution is the development of the equivalent of a polarising beam splitter for OAM, giving 100% detection efficiency at the single-photon level. The Dove-prism interferometer method outlined by Leach *et al.* is the only method that can reach this target theoretically. However, to measure OAM over a wide mode range using this approach is technically very challenging [34]. Spiral fringe interference patterns [5] and a hologram selector approach [22] are the most tried and tested methods, but each have their weaknesses. The optical transformation approach [6] allows for the measurement of OAM at the single-photon level, without the requirement for many photons in the same state, and it is relatively simple to align as compared to the Leach *et al.* approach, but suffers from a small amount of channel overlap. We expect more methods to appear in the future and these may finally resolve this outstanding issue in the field of OAM.

References

[1] L. Allen, M. Beijersbergen, R. Spreeuw and J. Woerdman, Orbital angular-momentum of light and the transformation of laguerre-gaussian laser modes, *Physical Review A*, **45**(11),8185–9, 1992.

[2] R. Beth, Mechanical detection and measurement of the angular momentum of light, *Physical Review*, **50**, **2**, 115–25, 1936.

[3] S. Franke-Arnold, L. Allen and M. Padgett, Advances in optical angular momentum, *Laser & Photonics Reviews*, **2**(4), 299–313, 2008.

[4] H. He, M. Friese and N. Heckenberg, Direct observation of transfer of angular momentum to absorptive particles from a laser beam with a phase singularity, *Physical Review Letters*, **75**(5), 826–9, 1995.

[5] M. Padgett, J. Arlt, N. Simpson and L. Allen, An experiment to observe the intensity and phase structure of Laguerre–Gaussian laser modes, *American Journal Of Physics*, **64**(1), 77–82, 1996.

[6] G. C. G. Berkhout, M. P. J. Lavery, J. Courtial, M. W. Beijersbergen and M. J. Padgett, Efficient sorting of orbital angular momentum states of light, *Physical Review Letters*, **105**(15), 153601, 2010.

[7] L. Allen and M. Padgett, Optical tweezers and optical spanners with Laguerre–Gaussian modes, *Journal of Modern Optics*, **43**(12), 2485–91, 1996.

[8] J. Poynting, The wave motion of a revolving shaft, and a suggestion as to the angular momentum in a beam of circularly polarised light, *Proceedings Of The Royal Society Of London Series A*, **82**, 557, 560–7, 1909.

[9] A. Ashkin, J. Dziedzic, J. Bjorkholm and S. Chu, Observation of a single-beam gradient force optical trap for dielectric particles, *Optics Letters*, **11**(5), 288–90, 1986.

[10] M. Friese, J. Enger and H. Rubinsztein-Dunlop, Optical angular-momentum transfer to trapped absorbing particles, *Physical Review A*, **54**(2), 1593–6, 1996.

[11] N. Simpson, L. Allen and M. Padgett, Optical tweezers and optical spanners with Laguerre–Gaussian modes, *Journal Of Modern Optics*, **43**(12), 2485–91, 1996.

[12] K. O'Holleran, M. Padgett and M. Dennis, Topology of optical vortex lines formed by the interference of three, four, and five plane waves, *Optics Express*, **14**(7), 3039–44, 2006.

[13] M. Soskin, V. Gorshkov, M. Vasnetsov, J. Malos and N. Heckenberg, Topological charge and angular momentum of light beams carrying optical vortices, *Physical Review A*, **56**(5), 4064–75, 1997.

[14] J. Leach, M. Dennis, J. Courtial and M. Padgett, Laser beams: knotted threads of darkness, *Nature*, **432**(7014), 165–5, 2004.

[15] J. M. Hickmann, E. J. S. Fonseca, W. C. Soares and S. Chavez-Cerda, Unveiling a truncated optical lattice associated with a triangular aperture using light's orbital angular momentum, *Physical Review Letters*, **105**(5), 053904, 2010.

[16] G. C. G. Berkhout and M. W. Beijersbergen, Method for probing the orbital angular momentum of optical vortices in electromagnetic waves from astronomical objects, *Physical Review Letters*, **101**(10), 100801, 2008.

[17] Y. Liu, Measuring the orbital angular momentum of elliptical vortex beams by using a slit hexagon aperture, *Optics Communications*, **284**(10–11), 2424–9, 2011.

[18] A. Mourka, J. Baumgartl and C. Shanor, Visualization of the birth of an optical vortex using diffraction from a triangular aperture, *Optics Express*, **19**(7), 5760–71, 2011.

[19] M. Beijersbergen, L. Allen, H. Vanderveen and J. Woerdman, Astigmatic laser mode converters and transfer of orbital angular-momentum, *Optics Communications*, **96**(1–3), 123–32, 1993.

[20] V. Bazhenov, M. Soskin and M. Vasnetsov, Screw dislocations in light wave-fronts, *Journal of Modern Optics*, **39**(5), 985–90, 1992.

[21] N. Heckenberg, R. McDuff, C. Smith and A. White, Generation of optical-phase singularities by computer-generated holograms, *Optics Letters*, **17**(3), 221–3, 1992.

[22] A. Mair, A. Vaziri, G. Weihs and A. Zeilinger, Entanglement of the orbital angular momentum states of photons, *Nature*, **412**(6844), 313–6, 2001.

[23] G. Molina-Terriza, L. Rebane, J P. Torres, L. Torner and S. Carrasco, Probing canonical geometrical objects by digital spiral imaging, *Journal of the European Optical Society–Rapid Publications*, **2**, 07014, 2007.

[24] C. Paterson, Atmospheric turbulence and orbital angular momentum of single photons for optical communication, *Physical Review Letters*, **94**, 153901, 2005.

[25] G. A. Tyler and R. W. Boyd, Influence of atmospheric turbulence on the propagation of quantum states of light carrying orbital angular momentum, *Optics Letters*, **34**(2), 142–4, 2009.

[26] G. Gibson, J. Courtial, M. Padgett *et al.*, Free-space information transfer using light beams carrying orbital angular momentum, *Optics Express*, **12**(22), 5448–56, 2004.

[27] S. Khonina, V. Kotlyar and R. Skidanov, Gauss–Laguerre modes with different indices in prescribed diffraction orders of a diffractive phase element, *Optics Communications*, **175**, 301–8, 2000.

[28] M. Vasnetsov, V. Pas'ko and M. Soskin, Analysis of orbital angular momentum of a misaligned optical beam, *New Journal of Physics*, **7**(42), 2005.

[29] S. Franke-Arnold, S. M. Barnett, E. Yao, J. Leach, J. Courtial and M. Padgett, Uncertainty principle for angular position and angular momentum, *New Journal of Physics*, **6**, 103, 2004.

[30] J. Courtial and M. Padgett, Performance of a cylindrical lens mode converter for producing Laguerre–Gaussian laser modes, *Optics Communications*, **159**(1–3), 13–18, 1999.

[31] B. Garetz and S. Arnold, Variable frequency-shifting of circularly polarized laser-radiation via a rotating half-wave retardation plate, *Optics Communications*, **31**(1), 1–3, 1979.

[32] G. Nienhuis, Doppler effect induced by rotating lenses, *Optics Communications*, **132**, 8–14, 1996.

[33] J. Courtial, K. Dholakia, D. Robertson and L. Allen, Measurement of the rotational frequency shift imparted to a rotating light beam possessing orbital angular momentum, *Physical Review Letters*, **80**, 15, 1998.

[34] J. Leach, M. Padgett, S. Barnett, S. Franke-Arnold and J. Courtial, Measuring the orbital angular momentum of a single photon, *Physical Review Letters*, **88**(25), 257901, 2002.

[35] M. Born and E. Wolf, *Principles of Optics: Electromagnetic Theory of Propagation, Interference and Diffraction of Light*, 3rd edn. London: Pergamon Press, 1965.

[36] X. Xue, H. Wei and A. Kirk, Beam analysis by fractional Fourier transform, *Optics Letters*, **26**(22), 1746–8, 2001.

[37] A. Dudley, M. Nock, T. Konrad, F. S. Roux and A. Forbes, Amplitude damping of Laguerre–Gaussian modes, *Optics Express*, **18**(22), 22 789–95, 2010.

[38] L.-P. Deng, H. Wang and K. Wang, Quantum CNOT gates with orbital angular momentum and polarization of single-photon quantum logic, *Journal of Modern Optics B*, **24**(9), 2517–20, 2007.
[39] O. Bryngdahl, Geometrical transformations in optics, *Journal of the Optical Society of America*, **64**(8), 1092–9, 1974.
[40] W. Hossack, A. Darling and A. Dahdour, Coordinate transformations with multiple computer-generated optical-elements, *Journal of Modern Optics*, **34**(9), 1235–50, 1987.
[41] Y. Saito, S. Komatsu and H. Ohzu, Scale and rotation invariant real-time optical correlator using computer generated hologram, *Optics Communications*, **47**(1), 8–11, 1983.
[42] M. P. J. Lavery, G. C. G. Berkhout, J. Courtial and M. J. Padgett, Measurement of the light orbital angular momentum spectrum using an optical geometric transformation, *Journal of Optics*, **13**(6), 064006, 2011.

14

Efficient generation of optical twisters using helico-conical beams

V. R. DARIA, D. PALIMA AND J. GLÜCKSTAD

14.1 Introduction

Helical phase fronts are commonly associated with Laguerre–Gaussian (LG) beams and optical vortices. These beams are characterized by an input field normally expressed as $\exp(-il\varphi)$, where l is an integer called the topological charge, which describes the magnitude and handedness of the helical phase profile. In quantum theory, the topological charge relates to a quantized orbital angular momentum (OAM) of $l\hbar$ per photon [1–2]. When projected in the far field, the azimuthal components of the phase create destructive interference giving rise to a dark centre surrounded by a high-intensity ring of light via constructive interference. As the topological charge is increased, the steeper phase gradients deflect light farther off-axis, thereby enlarging the light ring surrounding the dark centre.

When focused using a lens, the propagation of an LG beam along the optical axis follows a conical ray until it reaches a minimum ring radius at the focus and then conically increases after the focus. A high concentration of photons is maintained at the outskirts of the conical beam forming a ring at the transverse plane. Increasing the topological charge (and consequently the OAM) disperses the distribution of photons around a larger ring. Aside from circularly symmetric beams, higher-order LG beams bring about unique laser transverse modes, which display unique three-dimensional (3D) light patterns. Hence, the interplay of helical phase fronts results in far field beam profiles that can be uniquely utilized in optical tweezers [3–7] or as means of engineering the point spread function in laser microscopy [8].

In this chapter, we describe a specific family of beams, so called helico-conical (HC) beams [9], which project unique three-dimensional intensity distributions in the far field based on helical wavefronts characterized not only by the topological charge, but also by a K-parameter (see Eq. (14.1)). Unlike Laguerre–Gaussian modes, which only have helical phase profiles, HC beams exhibit both helical phase and amplitude distributions. When focused, the direction of linear scaling of HC beams along the radial direction impart differences in the pitch of the helical amplitude profile along the axial direction in the focal

The Angular Momentum of Light, ed. D. L. Andrews and M. Babiker. Published by Cambridge University Press.

volume. Here, we characterize their differences by plotting the three-dimensional intensity distributions of HC beams. An HC beam with $K = 1$ is encoded with an apodized helical phase front that starts with a null phase at a predefined outer radius and linearly scales towards the phase singularity at the centre of the beam. Such a beam projects a helical amplitude profile with a pitch that is smaller than the axial point spread function (PSF) of the focusing optics. The photon distribution is skewed but the highest concentration of photons is still concentrated within the same plane, resulting in a dark core with a radius that is related to the topological charge. On the other hand, an HC beam with $K = 0$ has an opposite radial phase scaling compared to HC beam with $K = 1$. At the focus, the helical pitch of the amplitude profile is greater than the axial point (PSF). The skewed distribution of photons results in lesser interference that normally brings about the dark core in LG beams. Thus, an HC beam with $K = 0$ can be best described as an optical twister, where a high concentration of photons at the focus is maintained even when the topological charge is increased. Owing to the high photon concentration along a spiral path, optical twisters can efficiently impose a spiral motion on particles trapped along the spiralling optical orbit.

In Section 14.2, we characterize HC beams by plotting their 3D intensity distributions. We also plot the photon density as a function of topological charge for different K-parameters. In Section 14.3, we show how we can form longer optical twister chains by linking multiple HC beams. In Section 14.4, we describe the optical setup used to verify how HC beams induce helical motion on microparticles and then discuss the experimental results in Section 14.5.

14.2 Far-field projection of helico-conical beams

The interplay of helical phase profiles at an input plane can project doughnut beams and other unique beam profiles in the far field. Figure 14.1 shows the basic optical setup to project unique 3D intensity distributions from beams having helical phase fronts. A helical phase is introduced to the incident beam by a phase-only spatial light modulator positioned at the input plane located at the back-focal plane of a Fourier transform lens. HC beams have an input field of the form

$$e(\rho, \varphi) = A(\rho)\exp\left[-il\varphi\left(K - \frac{\rho}{\rho_o}\right)\right], \tag{14.1}$$

where the K−parameter is a constant that takes a value of either 1 or 0 and (ρ, φ) are polar coordinates at the input plane. For $K = 1$, the radial phase scaling factor, $K - \rho\rho_o$, linearly varies from null at the outskirts, $\rho = \rho_0$, to unity at the centre, indicating a phase singularity at the optical axis. For $K = 0$, the radial scaling of the phase is reversed, where the optical axis has null phase and no phase singularity. We introduce a truncating aperture $A(\rho)$ with radius ρ_o to distinguish a Gaussian beam profile with that of optical vortices with a top-hat input beam profile [11]. Nonetheless, optical vortices and LG beams exhibit similar helical phase profiles and doughnut-shaped beams at the output. We neglect the effect of rings due to the convolution of Bessel functions brought about by the top-hat input profile.

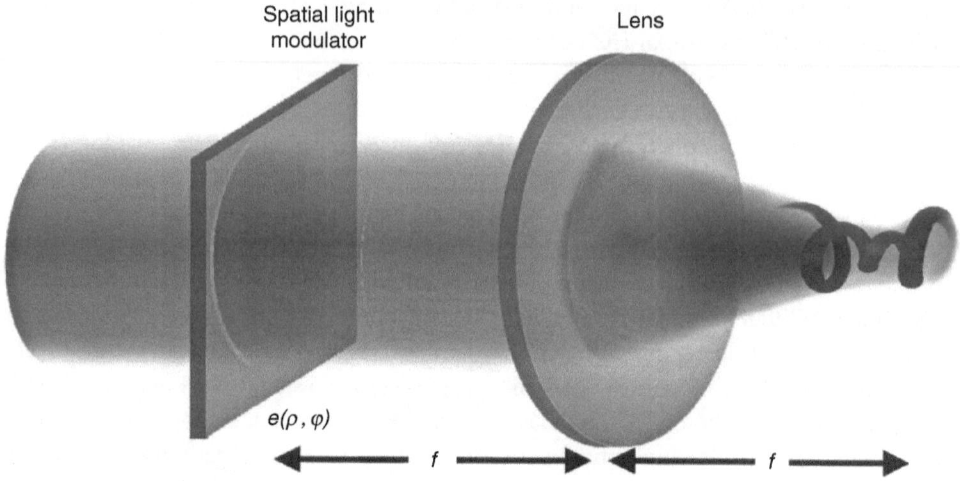

Figure 14.1 Far-field projection of unique 3D intensity distributions from beams having helical phase fronts.

After the lens, the beam tapers towards its smallest radius at the focus and enlarges after the focus. Hence, the beam has a region of highest concentration of photons at the focus. We have previously described the properties of the phase profile in Eq. (14.1) as a non-separable multiplication between the topological charge l and a conical phase of unitary scale [9]. The incorporation of a conical phase varies the helical phase linearly, generating a distinctly different beam as compared to LG and Bessel beams. Hence, the intensity distribution after the lens transformation is difficult to derive analytically due to the fact that the phase at the input has an inseparable dependence between ρ and φ.

To visualize the 3D intensity distribution at the output, we numerically evaluate the intensity distribution via the Fresnel diffraction integral [12]. To compare with a standard focusing without any phase encoding, we set $l = 0$ to yield an intensity distribution that can be characterized as a typical Airy function for a specified numerical aperture of NA = 0.2 and incident laser wavelength, $\lambda = 800$ nm. Figure 14.2(a) shows the 3D projection visualized via the Drishti (v1.1) Volume Exploration and Presentation Tool, a software developed by the Australian National University Supercomputer Facility [13]. The 3D projections are derived from 150 planes of intensity distributions calculated via the Fresnel diffraction integral. For comparison, we also visualize 3D projections of the intensity distribution of optical vortices beams for topological charges $l = 10$ (Fig. 14.2(b)) and $l = 20$ (Fig. 14.2(c)). Optical vortices are produced from Eq. (14.1) by setting $K = 0$ and $\rho = -\rho_o$ in the phase term. As the topological charge is increased, the light in an optical vortex has a lesser photon density since it is distributed to a ring with a larger radius.

The input phase pattern given by Eq. (14.1) and the corresponding 3D projections of the generated HC beams are shown in Fig. 14.3. The phase values are scaled from

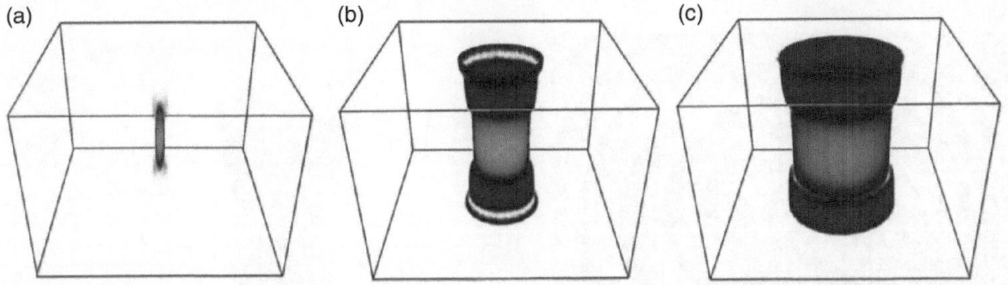

Figure 14.2 Three-dimensional projections of the focal region's intensity distribution for (a) a top-hat beam profile with no phase modulation and optical vortices with topological charge (b) $l = 10$ and (c) $l = 20$.

0 to 2π, corresponding to greyscale values from black to white. For $K = 1$ and $l = 10$, an asymmetric doughnut beam is projected at the focal region of the lens as shown in Fig. 14.3(b). The inset in Fig. 14.3(b) shows the top view of the beam featuring a circularly asymmetric intensity distribution. From this view, the asymmetry of the intensity distribution gives us an idea that such an HC beam follows a helical intensity distribution along the optical axis. The 3D intensity distribution shows that the highest photon concentration follows a skewed distribution along the optical axis. The helical dark bands in the 3D plot can be associated to the minimum in the axial point spread function (PSF), which takes the form of a sinc function, $u^{-1}\sin u$. Since the majority of the photons are still within the same plane, the interference of the helical phase brings about a similar type of doughnut beam as for optical vortices but with a slightly skewed axial projection. A larger ring is formed for $K = 1$ and $l = 20$ as shown in Fig. 14.3(d). Note that the helical pitch profile of the intensity distribution for both $l = 10$ and $l = 20$ is the same, indicating that the helicity in the amplitude profile is not affected by the helicity in the phase profile.

With $K = 0$, the beam's region of highest photon density spirals and tapers towards a focal point, which is both axially and transversely shifted as compared to the nominal focus of the lens. Fig. 14.3(c) shows the optical twister for $l = 10$. Compared to an optical vortex or HC beam with $K = 1$ of similar topological charge, the radius at the focal point of the optical twister is significantly smaller. This is due to the fact that the helical pitch of the amplitude profile is larger than the axial PSF of the lens. Fig. 14.3(d) shows an optical twister for $l = 20$, which shows a more pronounced twist in the intensity distribution. The insets in Fig. 14.3(b) and 14.3(d) shows the top view of the intensity distribution, which illustrates an asymmetric curved beam with one end closer to the optical axis, while the other end is dispersed outwards towards an increasing radius. As with the HC beam described earlier, the helical pitch of the amplitude profile is not affected by the topological charge or helicity of the phase profile. However, a negative topological charge will axially shift the focal point of the optical twister in the opposite direction and, hence, be located before the nominal focus of the lens. The beam will also spiral in the opposite direction.

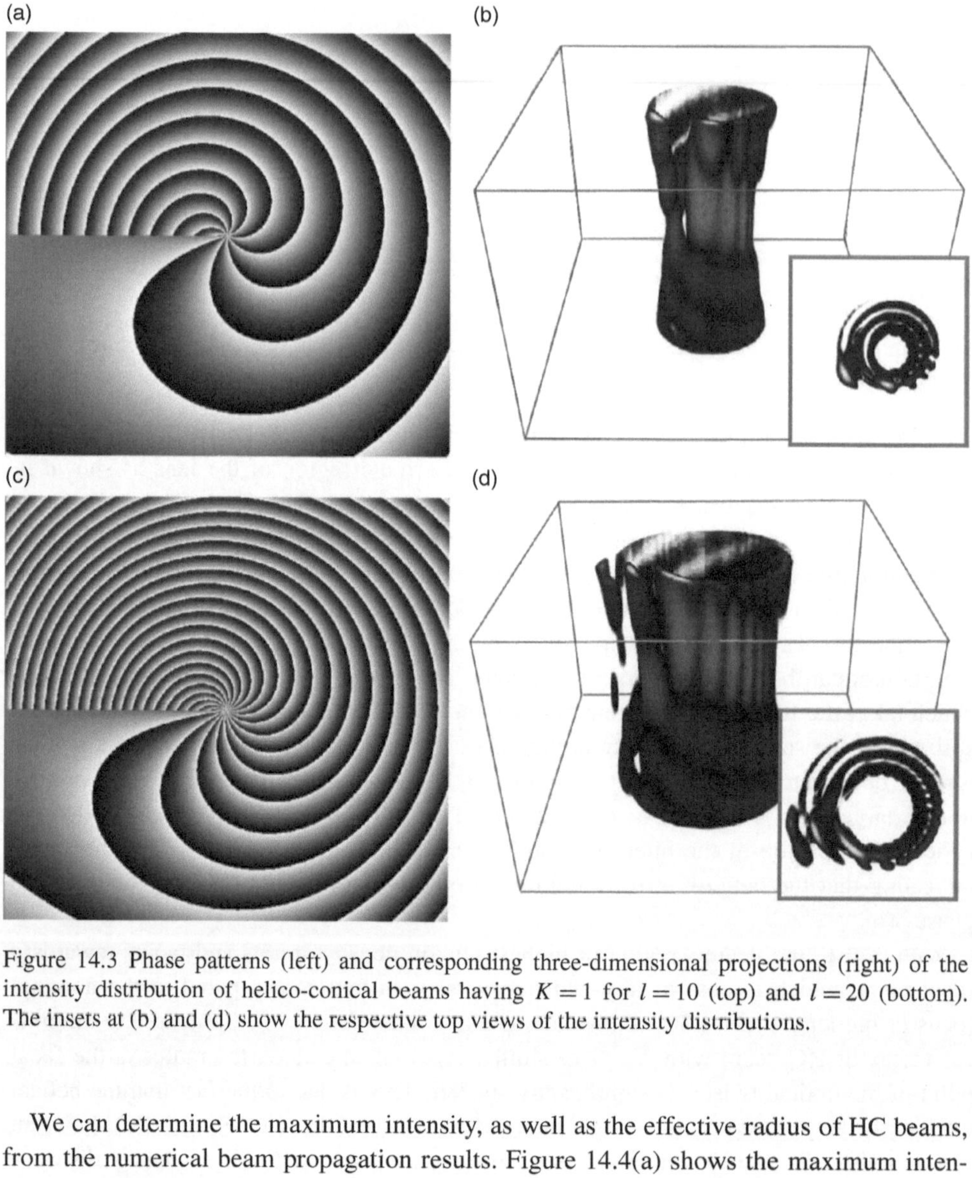

Figure 14.3 Phase patterns (left) and corresponding three-dimensional projections (right) of the intensity distribution of helico-conical beams having $K = 1$ for $l = 10$ (top) and $l = 20$ (bottom). The insets at (b) and (d) show the respective top views of the intensity distributions.

We can determine the maximum intensity, as well as the effective radius of HC beams, from the numerical beam propagation results. Figure 14.4(a) shows the maximum intensities as a function of topological charge for optical vortices and HC beams ($K = 0$ and $K = 1$). As the topological charge is increased, optical twisters – HC beams with $K = 0$ – exhibit higher intensities compared to optical vortices and HC beams with $K = 1$. Figure 14.4(b) shows the radial positions of highest intensity for both LG and HC beams. The rate at which the radius of optical vortices and HC beam ($K = 1$) increases with topological charge is 6× higher than that of optical twisters (HC beams with $K = 0$). These indicate that optical twisters are able to concentrate the majority of the photons within a small region around the optical axis even at higher topological charge.

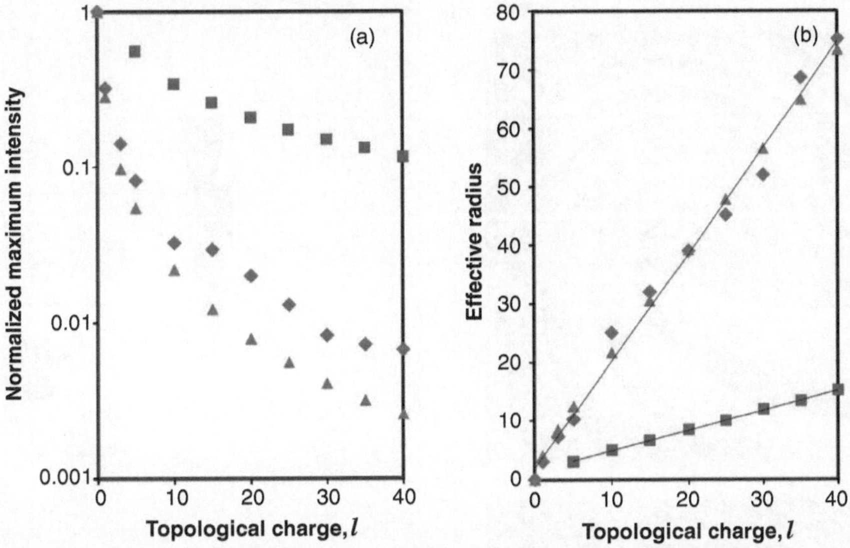

Figure 14.4 (a) Normalized maximum intensity distribution and (b) effective radius of the optical vortices (▲) and helico-conical beam with $K = 0$ (■) and $K = 1$ (♦).

14.3 Projecting multiple optical twisters and twister-chains

Multiple optical twisters (or HC beams with $K = 0$) can be generated by splitting a single incident beam into multiple spots based on the principle of field superposition [14–16]. The 3D positions of optical twisters can be set using a combination of lens and prism functions and multiplying with the field given by Eq. (14.1). Hence, each optical twister can be independently associated to a particular focal spot position in 3D. In Figure 14.5, we have shown that only a single-cycle optical twister is rendered within the region of maximum photon concentration, which is located at the focal region. By superposing various spot positions, the effective input field for generating multiple optical twisters is thus given by

$$e(\rho, \varphi) = A(\rho) \sum_{n=0}^{N} \exp\left[\mathrm{i}\frac{2\pi}{\lambda f}\left(\rho r_n \cos(\theta_n - \phi) - \rho^2 z_n\right)\right] \exp\left[-il\varphi\left(K - \frac{\rho}{\rho_o}\right)\right], \tag{14.2}$$

where λ is the wavelength of the incident laser, f is the focal length of the Fourier lens, while $r_n^2 = x_n^2 + y_n^2$, $\theta_n = \tan^{-1}(y_n/x_n)$ and z_n are cylindrical coordinates at the focal volume referring to the position of the nth optical twister. Figure 14.6 shows intensity distributions of multiple optical twisters visualized in 3D. Figure 14.6(a) shows two ($N = 2$) twisters positioned at $(x_n, y_n, z_n) = (-a, 0, -a)$ and $(+a, 0, +a)$, where a is an arbitrary position. Figure 14.6(b) also shows two ($N = 2$) twisters aligned along the optical axis forming a chain with $(x_n, y_n, z_n) = (0, 0, +a)$ and $(0, 0, -a)$.

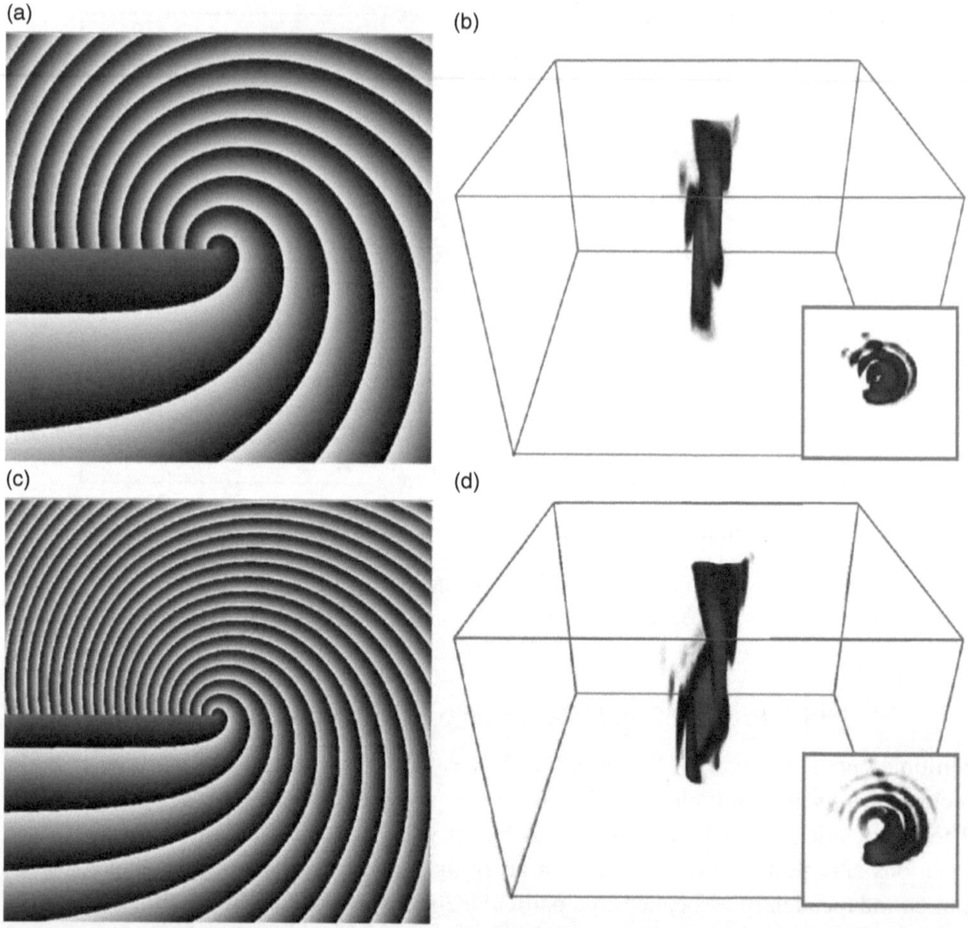

Figure 14.5 Phase patterns (left) and three-dimensional projections (right) of the intensity distribution of helico-conical beams having $K = 0$ for $l = 10$ (top) and $l = 20$ (bottom). The insets at (b) and (d) show the respective top views of the intensity distributions.

14.4 Optical setup

The optical setup for projecting HC beams in a microscope is schematically illustrated in Fig. 14.7. Computer-generated Fourier holograms, displayed as phase modulation on a spatial light modulator modulator (SLM, Hamamatsu Photonics), are read out by an expanded beam from a Ti:S laser (Spectra Physics 3900s, pumped by a Nd:YVO4 Spectra Physics Millenia 5W). The near infrared (NIR) readout beam ($\lambda \sim 800\,\text{nm}$) is expanded via L1 and L2 to illuminate the full area of the SLM. The second lens (L2) of the beam expander is reused as an optical Fourier transform lens for the modulated beams reflecting off the SLM. Thus, we are able to project the desired optical vortices and spiral beams at the Fourier plane of L2 while minimizing the tilt angle between the incident and reflected

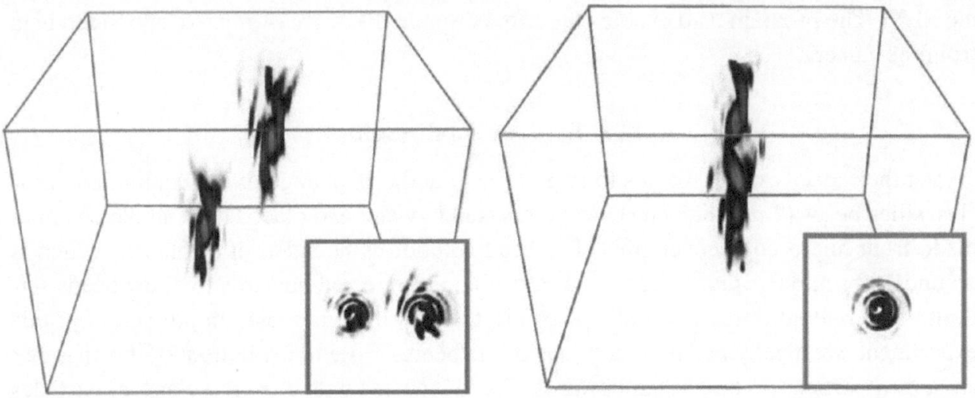

Figure 14.6 Multiple optical twisters: (a) two distinct optical twisters in 3D space; (b) two optical twisters forming a longer chain. Insets show the top view.

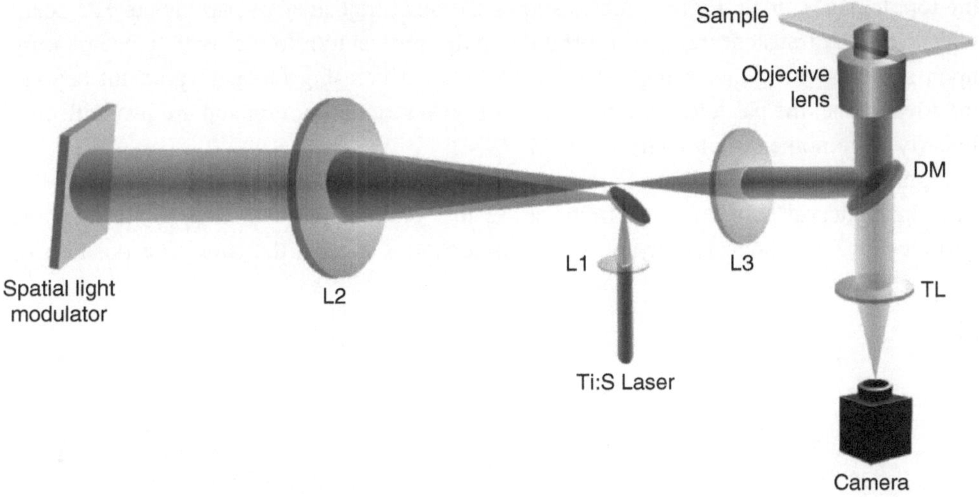

Figure 14.7 Optical setup for projecting HC beams at the focus of a microscope.

beams on the SLM. A minimal tilt angle ensures that the holographically generated beams are produced with a high degree of purity.

The generated beams are demagnified and relayed to a sample positioned on a microscope stage, through the microscope's fluorescence port, using a $4f$ configuration of lens L3 and the microscope objective lens (Leica). The dichroic mirror (DM) reflects the NIR beam from L3 to the sample through the microscope objective and also transmits visible light for illuminating and imaging the sample onto a CCD camera. A program developed in LabVIEW (National Instruments) provides a graphical user interface for controlling the generated beams via video control signals (800 × 600) that encode phase holograms on the SLM. The program also enables the acquisition and storage of microscope video data from the camera.

14.5 Experimental results

To test the capacity of HC beams to impart linear and orbital angular momentum on particles, silica beads (2 μm diameter) were dispersed in water and placed in a sample chamber made from microscope cover slips. The bead solution is bounded in a channel, which is around 100 μm deep. Our setup uses beams that are directed upwards to allow beads settling at the bottom of the cover slip to be ejected upwards via radiation pressure. For this experiment, we axially position the focus of the beams close to the bottom of the chamber and allow particles to be pushed upwards. For HC beams with $K = 1$, we expect particles to be attracted to the region of highest intensity and pushed upwards towards the top of the chamber. Since the helical pitch of the amplitude profile is less than the axial PSF of the focusing lens, such beams will form a skewed doughnut beam whose radius is defined by the topological charge. Figure 14.8(a) shows a typical trajectory of particles in HC beam with $K = 1$. A transient transfer of orbital angular momentum to the particle occurs only upon reaching the plane of highest intensity. With decreasing intensity gradient beyond the focal plane, the particles will no longer follow a circular motion and are just deflected linearly with momentary transfer of OAM.

Figure 14.9 shows time lapse sequence of particles affected by an HC beam with $K = 1$. The time interval between each frame is 32 ms. The sequence of images shows three particles going through the beam at various locations around the ring. The position of

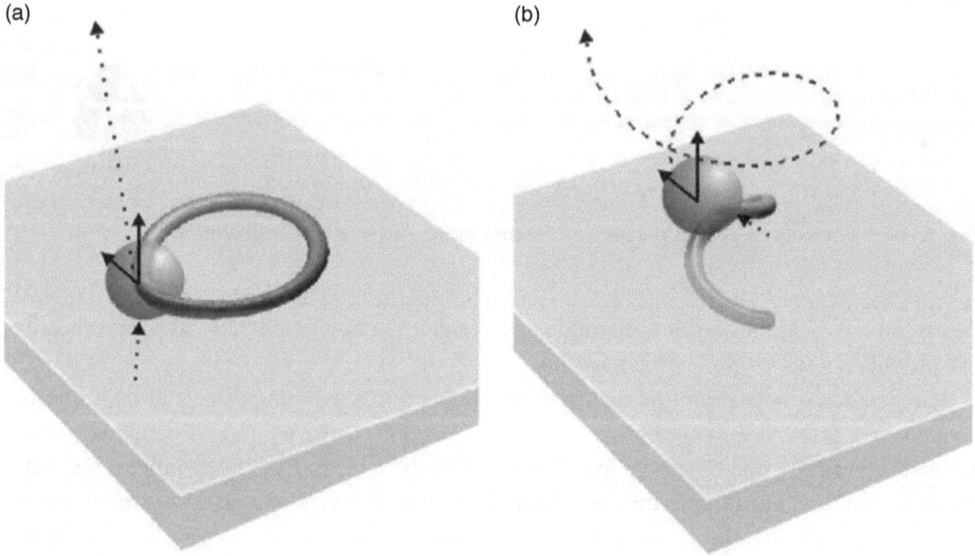

Figure 14.8 Trajectory of a particle illuminated using HC beams with (a) $K = 1$ and (b) $K = 0$.

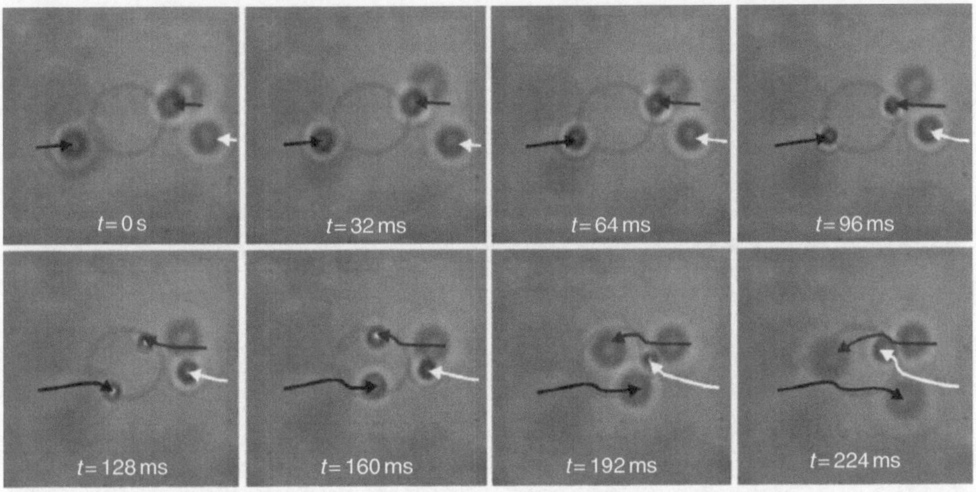

Figure 14.9 Time lapse images of particles in an HC beam with $K = 1$. The beam is indicated with a circle, while the arrows denote the trajectory of three particles.

the HC beam is indicated with a ring at the centre while the arrows denote the motion of the particles. The imaging camera is focused at the same plane as the focal plane of the HC beam. The particles appear to be out-of-focus below and above the focal plane. As the particles move up by radiation pressure, they pass through the ring and are instantaneously deflected.

Next, we show how an HC beam with $K = 0$ or an optical twister affect the motion of particles. Since the intensity of the beam follows a spiral profile, we can expect a dielectric particle approaching a spiral beam to be trapped along the region of highest photon density via gradient forces. The beam follows a spiral profile and particles are expected to be trapped along the orbit and follow a spiral motion with an axial component of the momentum towards the direction of the beam propagation. Figure 14.8(b) shows the trajectory of a particle in an optical twister.

Figure 14.10 shows time lapse images of multiple microparticles trapped and guided along a spiral orbit when they approach an optical twister with $l = 20$. The time interval between each frame is 8 ms. While both experiments ($K = 1$ and $K = 0$) are sampled using the same frame rate, we show a more temporally resolved sequence for $K = 0$ since optical twisters induce faster motion on the particles owing to their higher intensities. With maximum photon density at the focus, majority of the momentum transfer occurs at this region sending the particle into a spiral motion. The respective trajectories of the particles are indicated by the arrows. Compared to the HC beam with $K = 1$, all particles rotate around the optical twister and follow a circular path as viewed from the top. Note that, although the particles approach the twister at various entry points around the twister, all particles exit at the same angle. Such behavior indicates that majority of the particles go through the same track around the twister.

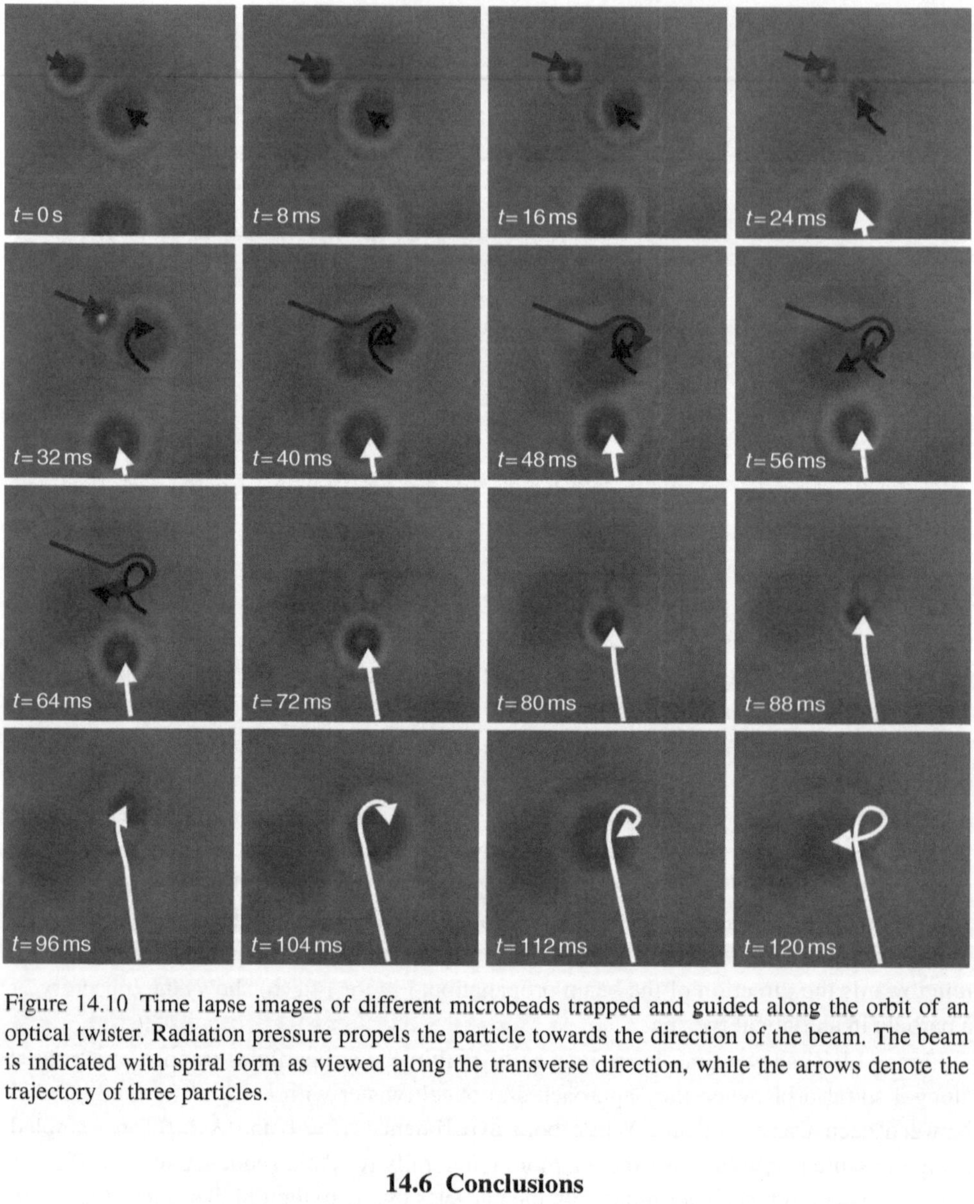

Figure 14.10 Time lapse images of different microbeads trapped and guided along the orbit of an optical twister. Radiation pressure propels the particle towards the direction of the beam. The beam is indicated with spiral form as viewed along the transverse direction, while the arrows denote the trajectory of three particles.

14.6 Conclusions

Helico-conical beams have spiral phase patterns that yield circularly asymmetric intensity distributions along the transverse direction. Such asymmetry affects the axial intensity distribution of such beams. An HC beam produces a beam with spiral patterns in both amplitude and phase profiles as it propagates in free space. An HC beam with $K = 1$ has a helical pitch along the optical axis that is smaller than the axial point spread function of the focusing optics. Such beam forms are like a skewed doughnut beam with radius dictated by the topological charge or helicity of the phase profile. Compared to an HC beam with

$K = 1$, the 3D beam structure of $K = 0$ HC beams displays an intensity distribution with a single cycle spiral or with a helical pitch that is equivalent to the axial PSF of the focusing optics. Such a beam is best described as an optical twister. In contrast to optical vortices, optical twisters can have a scalable helical phase profile or equivalently its orbital angular momentum while keeping its radius minimal. We have shown that optical twisters maintain a high concentration of photons at the focus even as the topological charge is increased. Finally, our experiments have shown the effective transfer of both orbital angular and linear momentum to induce spiral motion on particles trapped along the optical twister's orbit.

Acknowledgements

This collaborative work has been made possible via the Short Term Scientific Mission of the COST Action MP0604 and the Australian Academy of Science. V. R. Daria has been funded partially by the Australian National University – Office of the Vice-Chancellor. D. Palima and J. Glückstad acknowledge financial support from the Danish Technical Scientific Research Council (FTP). We also thank P. J. Rodrigo for discussions and assistance with the experiments.

References

[1] L. Allen, M. W. Beijersbergen, R. J. C. Spreeuw, and J. P Woerdman, Orbital angular momentum of light and the transformation of Laguerre–Gaussian laser modes. *Phys. Rev. A* **45**, 8185–89 (1992).

[2] L. Allen, M. J. Padgett, and M. Babiker, The orbital angular momentum of light, *Prog. Opt.* **39**, 291–372 (1999).

[3] H. He, M. E. J. Friese, N. R. Heckenberg, and H. Rubinsztein-Dunlop, Direct observation of transfer of angular momentum to absorptive particles from a laser beam with a phase singularity, *Phys. Rev. Lett.* **75**, 826–9 (1995).

[4] H. He, N. R. Heckenberg, and H. Rubinsztein-Dunlop, Optical particle trapping with higher-order doughnut beams produced using high efficiency computer generated holograms, *J. Mod. Opt.* **42**, 217 (1995).

[5] M. E. J. Friese, H. Rubinsztein-Dunlop, J. Enger, and N. R. Heckenberg, Optical angular momentum transfer to trapped absorbing particles, *Phys. Rev. A* **54**, 1593–6 (1996).

[6] V. Garces-Chavez, K. Volke-Sepulveda, S. Chavez-Cerda, W. Sibbett, and K. Dholakia, Transfer of orbital angular momentum to an optically trapped low-index particle, *Phys. Rev. A* **66**, 063402 (2002).

[7] N. B. Simpson, K. Dholakia, L. Allen, and M. J. Padgett, The mechanical equivalence of the spin and orbital angular momentum of light: an optical spanner, *Opt. Lett.* **22**, 52–54 (1997).

[8] S. R. P. Pavani and R. Piestun, Three dimensional tracking of fluorescent microparticles using a photon-limited double-helix response system, *Opt. Express* **16**, 22048–257 (2008).

[9] C. A. Alonzo, P. J. Rodrigo, and J. Glückstad, Helico-conical optical beams: a product of helical and conical phase fronts, *Opt. Express* **13**, 49–60 (2005).

[10] V.R. Daria, D. Palima, and J. Glückstad, Optical twists in phase and amplitude, *Opt. Express* **19**, 476–81 (2011).
[11] Z. S. Sacks, D. Rozas, and G. A. Swartzlander, Holographic formation of optical-vortex filaments, *J. Opt. Soc. Am. B* **15**, 2226–34 (1998).
[12] J. Goodman, *Introduction to Fourier Optics* (New York: McGraw-Hill, 1996).
[13] Ajay Limaye, *Drishti – Volume Exploration and Presentation Tool*, Poster presentation, Vis 2006, Baltimore, OH.
[14] E. R. Dufresne and D. G. Grier, Optical tweezer arrays and optical substrates created with diffractive optical elements, *Rev. Sci. Instrum.* **69**, 1974–17 (1998).
[15] M. Reicherter, T. Haist, E. U. Wagemann, and H. J. Tiziani, Optical particle trapping with computer-generated holograms written on a liquid-crystal display, *Opt. Lett.* **24**, 608–10 (1999).
[16] J.E. Curtis, B. Koss, and D. G. Grier, Dynamic holographic optical tweezers, *Opt. Commun.* **207**, 169–75 (2002).

15

Self-similar modes of coherent diffusion with orbital angular momentum

O. FIRSTENBERG, M. SHUKER, R. PUGATCH AND N. DAVIDSON

When a paraxial beam of light propagates in free space, it undergoes *diffraction*, i.e. the dispersion of the different spatial frequencies comprising its transverse shape. Although diffraction changes the transverse intensity pattern – for example, Gaussian profiles are broadened and images are distorted – no dissipation occurs, and the process can in principle be perfectly reversed. Such reversal is manifested in an optical imaging arrangement. In contrast, the process of *diffusion* is dissipative, and any initial field distribution undergoes an irreversible blurring when diffusing, due to the decay of the high spatial frequencies.

In Fourier space, where $\mathbf{k} = (k_x, k_y)$ are the transverse spatial frequencies of a field, the two-dimensional Laplacian is given by $\nabla_\perp^2 \to -k^2$, and the resulting equations of paraxial diffraction and diffusion, respectively, are

$$\left(\frac{\partial}{\partial z} + \mathrm{i}\frac{\lambda}{4\pi}k^2\right) E(\mathbf{k}, z) = 0, \quad \left(\frac{\partial}{\partial t} + Dk^2\right) E(\mathbf{k}, t) = 0. \tag{15.1}$$

The diffraction constant $\mathrm{i}\lambda/(4\pi)$ is imaginary (λ being the optical wavelength), while the diffusion constant D is real. In writing the diffraction equation in the paraxial approximation, $E(\mathbf{k}, z)$ is assumed to be a stationary envelope of the electric field that is slowly varying with respect to the propagation distance, z. As opposed to the "regular" diffusion of non-negative real-valued fields, such as the concentration of some physical quantity, here we describe the coherent diffusion of a complex-valued field. In the language of optics, we are interested in the diffusion of the electric field rather than the diffusion of its intensity.

In this chapter, we aim to investigate the evolution of fields with orbital angular momentum (OAM) that experience diffusion, in analogy to the evolution of light beams carrying OAM and diffracting in free space. We focus on the self-similar modes of diffusion and present slow-light and stored-light experiments in which they were realized and their dynamics was studied [1, 2]. Similarly to optical modes, the diffusion modes are only partially self-similar; their shape is preserved up to scaling and normalization, while their phase pattern curves or flattens. A self-similar contraction is also attainable, in analogy to a focusing OAM beam.

The Angular Momentum of Light, ed. D. L. Andrews and M. Babiker. Published by Cambridge University Press.

15.1 Coherent diffusion

The term *coherent diffusion* describes the diffusion of a multi-component field, either a complex-valued scalar field, containing both an amplitude and a phase, or a spinor field. Perhaps the most famous manifestation of coherent diffusion is that of the nuclear-magnetization vector field in NMR experiments [3]. While the spatial distribution of the magnetization, $\mathbf{M}(r)$, evolves locally according to the Maxwell–Bloch equations, the motion of the particles in the sample results in its diffusion. This separation of "internal" and "external" dynamics was first delineated by Torrey [4]. Coherent diffusion is also naturally found in various electronic spin ensembles, such as spin-polarized alkalis in vapor [5–7] and electronic or exciton spins in metals [8] and semiconductors [9–11].

In many cases, the vectorized equations can be reduced into a single diffusion equation of a scalar field, in the form of

$$\frac{\partial}{\partial t}\psi(\mathbf{r},t) = D\nabla^2\psi(\mathbf{r},t) - \gamma\psi(\mathbf{r},t), \tag{15.2}$$

where $\gamma = 1/T_2$ is an homogenous decay rate. The field is complex-valued ($\psi = M_x + \mathrm{i}M_y$) when the dynamics of the spin coherence is considered (transverse spin) [4], or real-valued ($\psi = M_z$) for spin-population dynamics (longitudinal spin) [12]. In fact, Eq. (15.2) is found in other disciplines as well, such as for the vorticity field in viscous fluids, where OAM is important in the context of cyclonic vortices [13].

Recently, coherent diffusion has been shown to be of major importance in the dynamics of slow and stored light in hot atomic vapor. An alkali-metal atom in a vapor maintains quantum coherence in its electronic ground state for as long as a few milliseconds [14] and even up to a second [15]. In the most simple case, this coherence would be a single complex scalar, having an amplitude and a phase, which oscillates with a known frequency (a phasor). When immersed in a dense buffer-gas, the alkalis undergo diffusion due to frequent velocity-changing collisions with the buffer-gas particles. For certain buffer gases, these collisions preserve the internal-state coherence of the alkalis, resulting in the actual diffusion of the spatial distribution of coherence [1]. Strictly speaking, diffusion is experienced by all slowly varying terms in the atomic density matrix [16], which are either real (populations) or complex (coherences, or dipoles).

With recently developed techniques for coherent light storage [17], an arbitrary three-dimensional light field, $E(\mathbf{r})$, can be imprinted onto an alkali vapor [7], linearly mapping the complex field envelope onto the atomic coherence field, $\psi(\mathbf{r})$. In the process, two laser beams–a weak pulsed "signal" and a strong auxiliary "control"–are tuned to resonance with two atomic transitions sharing the same excited state. The signal pulse traversing the medium, $E(\mathbf{r} - \mathbf{v}_g t)$, experiences very weak absorption due to electromagnetically induced transparency [18], and is accompanied by a traveling atomic coherence wave, $\psi(\mathbf{r} - \mathbf{v}_g t)$. The coupled light–matter excitation, known as a dark-state polariton [19], travels in the medium with a group-velocity, $|\mathbf{v}_g|$, smaller by orders of magnitude than the speed of light in vacuum. Storage is executed by shutting down the control field during the polariton motion, forcing a complete transfer of the signal field onto the atomic coherence and stopping the polariton. The dynamics of the complex field ψ during storage is governed by the motion of the atoms in space and by homogenous decoherence mechanisms, e.g. spin-exchange and spin-destruction collisions. Atomic diffusion, as occurring in a buffer-gas environment, thus results in a dynamics in the form of (15.2), with γ being the decoherence rate and D being the diffusion coefficient of the alkali within the buffer gas. Finally, the signal light field can be regenerated from the atomic coherence field $\psi(\mathbf{r})$ by reopening the control, and it subsequently leaves the cell and is measured.

Equation (15.2) does not include the dynamics resulting from possible spatial inhomogeneity of the system, such as magnetic-field gradients. For example, quadrupole magnetic fields have been shown to increase or decrease the OAM of stored light in both hot and cold atomic systems [20, 21]. In this chapter, we will only refer to uniform media, which do not introduce explicit spatial dependency in the dynamics.

In hot vapor, the diffusion of the atoms affects the light pulse also during its slow propagation in the medium, via motional spectral-broadening. From the viewpoint of the dark-state polaritons, which are a mutual excitation of light and matter, it is the "matter part" contribution to the dispersion relation that facilitates the effect of the atomic motion [16]. As a result, the polaritons spatially diffuse while slowly traversing the medium [22]. With respect to the delay time, the polariton field obeys the coherent diffusion equation (15.2), with additional optical diffraction coming from its "light part." Recent studies of this system have explored collisional narrowing [23–25], spectral fringes and coherent recurrence [6, 26, 27], transverse momentum diffusion [28, 29], slow-light diffraction manipulation [22], and spatial field diffusion [1, 7].

15.2 Diffusion of orbital angular momentum

The OAM of a light beam, parallel to its propagation direction $\hat{z}$, can be obtained from a cross-section in the transverse plane x-y at any given z. A field of the from $F_0(r)\exp(il\phi)$, where (r,ϕ) are the polar coordinates in the x-y plane, carries OAM of $l\hbar$ per photon [30]. The phase twist $\exp(il\phi)$, and thus the OAM, are maintained throughout the propagation. Now suppose that this field is put through diffusion. We expect, since diffusion is a continuous process and similarly to diffraction, that the phase singularity will not unwind and the topological charge will be maintained. Therefore, diffusion should also conserve OAM.

To show this for a single phase singularity, we utilize the Green function of the two-dimensional diffusion equation,

$$E(\mathbf{r},t) = \int \mathrm{d}x'\mathrm{d}y' E\left(\mathbf{r}',t=0\right) G\left(\mathbf{r}-\mathbf{r}',t\right), \text{ where } G(\mathbf{r},t) = \frac{\mathrm{e}^{-|\mathbf{r}|^2/(4Dt)}}{4\pi Dt}. \quad (15.3)$$

For the initial condition $E(\mathbf{r},t=0) = F_0(r)\exp(il\phi)$, we replace $\left|\mathbf{r}-\mathbf{r}'\right|^2 = r^2 + r'^2 - 2rr'\cos(\phi'-\phi)$, and use the modified Bessel function of order $|l|$,

$$I_{|l|}\left(\frac{r'r}{2Dt}\right) = \int_0^{2\pi} \mathrm{d}\phi' \mathrm{e}^{il(\phi'-\phi)} \mathrm{e}^{\frac{r'r}{2Dt}\cos(\phi'-\phi)}, \quad (15.4)$$

to find

$$E(\mathbf{r},t) = F(r,t)\mathrm{e}^{\mathrm{i}l\phi}, \tag{15.5}$$

where the radial cross-section is given by

$$F(r,t) = \frac{1}{2Dt}\mathrm{e}^{-\frac{r^2}{4Dt}} \int_0^\infty \mathrm{d}r' F_0(r') r' \mathrm{e}^{-\frac{r'^2}{4Dt}} I_{|l|}\left(\frac{r'r}{2Dt}\right). \tag{15.6}$$

Indeed while the radial pattern evolves with time, the phase twist in Eq. (15.5) is constant, and the OAM "per photon" is conserved.

A well-known property of the diffusion equation without homogenous decay ($\gamma = 0$) is that the two-dimensional integral $\int \mathrm{d}x\mathrm{d}y E(r,t)$ is always conserved. In the case of a vortex, it is zero at all times. Nevertheless, the L^2 norm of field, which stands for the total optical power in electromagnetics,

$$P(t) \equiv \int \mathrm{d}x\mathrm{d}y\, |E(\mathbf{r},t)|^2 = 2\pi \int_0^\infty \mathrm{d}r\, |F(r,t)|^2\, r, \tag{15.7}$$

is not maintained, and it decreases during diffusion. As opposed to optical diffraction, in which the total power is conserved for all beams, the rate of decay due to diffusion depends on the OAM, through the Bessel function $I_{|l|}(x)$, and on the initial radial cross-section. A detailed analysis of the power decay during diffusion is presented in the next section.

As a first experimental example, we consider the diffusion during a storage-of-light experiment, with a Laguerre–Gaussian (LG) mode of radial index $p = 0$ and orbital index $l = 1$ at its waist plane [31],

$$E_{0,1}^{\mathrm{LG}}(z=0) = F_0(r; w_0)\mathrm{e}^{\mathrm{i}\phi} \quad \text{with} \quad F_0(r; w_0) = \sqrt{\frac{P_0}{\pi}\frac{2r}{w_0^2}}\mathrm{e}^{-\frac{r^2}{w_0^2}}. \tag{15.8}$$

Here, $P_0 = P(t=0)$ is the total power, and we use the waist radius w_0 to explicitly parametrize the annular radial cross-section F_0. The experimental setup and procedure is described in Fig. 15.1. It is effectively assumed that the internal state of each individual atom does not change as a result of diffusion, and so it "carries" the stored complex amplitude, $E(\mathbf{r}, t=0)$, within its electronic quantum coherence. After storage, the ensemble average of the different atoms arriving to the same small macroscopic volume determines the retrieved signal field [6, 7]. The atomic diffusive motion can be described as a continuous diffusion process of the density matrix field, given by Eq. (15.2).

The diffusion of the vortex (15.8) with initial radius $w_0 = 670$ μm and diffusion constant $D = 11$ cm^2/s, after 50 μs of slowing and 30 μs of storage, is presented in Fig. 15.2 (left, upper panel) [1]. For the sake of comparison, the diffusion of a *flat-phase* Gaussian beam with a darkened center is also presented in the figure (lower panel). The latter was created by passing a Gaussian beam ($w_0 = 670$ μm) around a circular beam-stop of radius 300 μm and imaging it onto the cell. Evidently, while the central hole in the flat-phase pattern is completely filled, the hole of the vortex remains dark, implying that the phase vortex

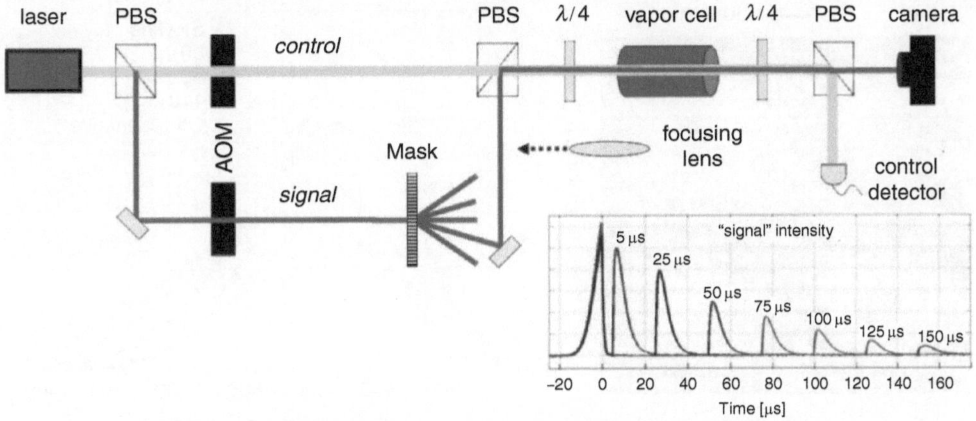

Figure 15.1 The experimental setup for light-storage experiments with vortex patterns. The experiments utilize the phenomenon of electromagnetically-induced transparency (EIT) within the $D1$ transition of ^{87}Rb [19]. A diode laser, locked to the $F = 2 \rightarrow F' = 1$ transition, is divided into two channels of perpendicular polarization, the "signal" and the "control", using a polarizing beam-splitter (PBS). The two beams pass through acousto-optic modulators (AOM), controlling both the frequency and the temporal shape of the pulses. The control is spatially shaped as a wide Gaussian beam ($w_{\text{control}} = 2.2$ mm), while the signal passes through an off-axis holographic binary mask with a fork dislocation. A diffraction order with the desired vortex mode is chosen, and the signal is reunited with the control and sent into the cell. The small size of the signal insures that it experiences a nearly constant control intensity, and also guarantees that diffusion is dominant over other decoherence mechanisms. A quarter wave-plate before the cell ($\lambda/4$) creates the two perpendicular circular polarizations suitable for EIT between two Zeeman sublevels. During light storage, the quantum coherence between these sublevels holds the stored information. A decoherence rate on the order of 10 000/s, due to spin-exchange collisions and magnetic-field inhomogeneity, causes uniform decay of the stored pattern and limits the storage time. The 5 cm long vapor cell contains isotopically pure ^{87}Rb, which diffuses through 10 Torr of Neon buffer gas. At a temperature of about 65°C, the Rb density is 4×10^{11}/cc. The beams are separated after the cell using polarization optics, and the signal is imaged onto a CCD camera. A typical experimental sequence starts by sending a short (30 μs) signal pulse into the cell while the control is on. The signal is slowed down to a group velocity of the order of km/s (delay of tens of μs). When the signal pulse is in the medium, the control is turned off, and the signal is mapped onto the atomic coherence in similarly to a hologram. After some storage time, during which diffusion of the atoms occur, the control is turned back on and the signal is retrieved. The graph shows traces of the signal intensity after the cell in several experiments (with a photodetector replacing the camera). In each experiment, half of the signal pulse is allowed to leave the cell before storage ($t \leq 0$), and the second half is stored, and retrieved after a certain duration. The homogenous decay during storage is also evident.

was preserved (a direct evidence for the preservation of OAM is given in the next section). From a microscopic point of view, the stability of the dark center is the result of destructive interference; two atoms that diffuse to the center from opposite sides of the vortex have the opposite phase, and their coherence destructively interfere. In other words, the phases of all atoms diffusing to the center sum up to zero.

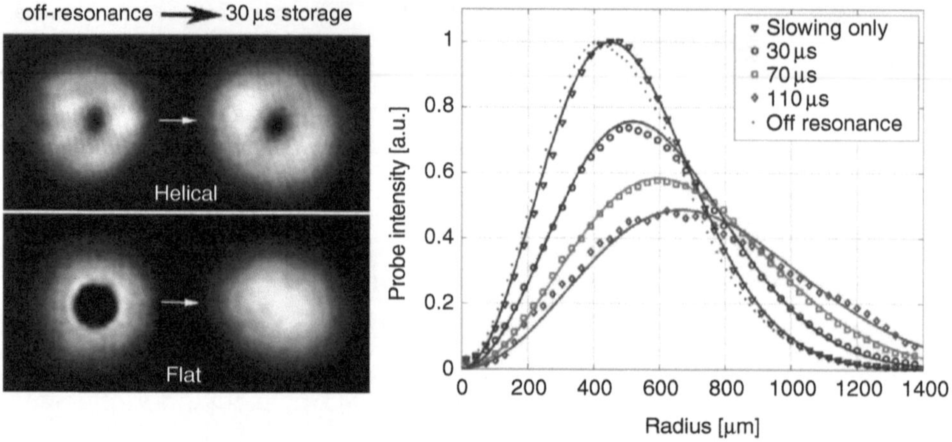

Figure 15.2 Left: Diffusion of a vortex beam (top) and a flat-phase beam (bottom). The left column shows the initial shape before any diffusion (off-resonance), and the right column shows both beams after storage of 30 μs. Right: Averaged radial cross-sections of the restored vortex after 30 μs, 70 μs, and 110 μs of storage. It is evident that the center remains completely dark in the vortex case, even for the longest storage duration, while completely filling with light in the flat-phase case. The theoretical prediction (solid lines) agrees with the experimental results (symbols). Note that the total intensity of the stored light also decays uniformly due to other mechanisms, and the data presented here are normalized to ease the comparison. Adapted from [1].

Figure 15.2 (right) presents the radial cross-sections of the retrieved light for different storage times. The theoretical curves are calculated from (15.6) and (15.8), and using (9.6.3) and (11.4.29) in [32],

$$F(r, t; w_0) = \frac{1}{s^2}\sqrt{\frac{P_0}{\pi}}\frac{2r}{s^2 w_0^2} e^{-\frac{r^2}{s^2 w_0^2}} = \frac{1}{s^2} F_0\left(r; s(t)\, w_0\right), \tag{15.9}$$

where s is the waist stretching factor,

$$s(t) = \sqrt{1 + \frac{4Dt}{w_0^2}}. \tag{15.10}$$

Indeed, the theoretical predictions exactly capture the effect of the diffusion, and the dark center stays dark at least up to 110 μs, in clear contrast to the flat-phase beam.

Most importantly, Fig. 15.2 (right) and expression (15.9) indicate that at any given time, the radial cross-section is merely a stretched version of the original cross-section, with the radial scaling given by $s(t)$. This self-similar evolution, which resembles the diffraction expansion of vortex beams in free-space, is the subject of the following section.

15.3 Laguerre–Gaussian modes of diffusion

We are interested in identifying the set of functions with definite orbital angular momentum, of the form $E(\mathbf{r}, t) = F(r, t)\mathrm{e}^{\mathrm{i}l\phi}$, which evolve self-similarly while diffusing, retaining their shape $|F(r, t)|$ up to scaling and normalization. The property of self-similarity is generally associated with the long-time behavior of dynamic processes [33]. In most dissipative systems, the similarity solutions decay with a characteristic rate, indicating the asymptotic evolution of a given initial condition [34, 35]. All the same, in free-space optics, which is ideally non-dissipative, and if the propagation distance is given the role of time, the similarity solutions often emerge, with a familiar example being the family of Laguerre–Gaussian beams [36]. These beams are broadly referred to as OAM *modes* of optical diffraction, even though they are not "stationary" as Bessel beams [37] and not eigenmodes of the underlying Hamiltonian. In analogy to optics, the self-similar solutions of the coherent diffusion equation should provide the family of OAM modes of diffusion. These modes can then be utilized for the construction and analysis of the stable evolution in coherent media, and for the description of the OAM dynamics of light-matter polaritons.

Consider first $\partial E/\partial z = -\mathrm{i}\lambda/(4\pi)\nabla_\perp^2 E$ the time-independent paraxial diffraction, for the slowly varying envelope, E, of a light field with a wavelength λ. Two different sets of polynomial-Gaussian solutions are known for this equation, namely the standard and the elegant modes [31]. The more familiar "standard" modes, e.g. the Hermite–Gaussian (HG) or Laguerre–Gaussian (LG), form complete sets of modes that are self-similar under diffraction. Their transverse intensity distribution, $I\,(x, y; z)$, is maintained along the propagation direction z, normalized, and scaled by the beam radius $w(z)$. In contrast, the transverse shape of the "elegant" solutions is generally not maintained, and originally they were investigated due to their elegant mathematical form [38, 39].

The elegant LG solution, $E_{p,l}^{\mathrm{LG}} \equiv E_{p,l}^{\text{elegant LG}}(\mathbf{r}; w_0)$, with w_0 being the radius at the waist plane $z = 0$, is written in terms of the associated Laguerre polynomials $L_p^{|l|}$ of indices p and l as

$$E_{p,l}^{\mathrm{LG}} = \sqrt{\frac{P_0}{\pi}} N_p^{|l|} \frac{(-1)^{p+|l|}}{w_0} \left[\mathrm{i}\frac{\pi w_0^2}{\lambda q(z)} \right]^{|l|/2+p+1} \tilde{r}^l L_p^{|l|}\left(\tilde{r}^2\right) \mathrm{e}^{-\tilde{r}^2} \mathrm{e}^{\mathrm{i}l\phi}. \qquad (15.11)$$

Here, $q(z; w_0) = \mathrm{i}z_{\mathrm{R}} + z$ is the complex distance, where $z_{\mathrm{R}} = \pi w_0^2/\lambda$ is the Rayleigh length. The normalization constant $N_p^{|l|}$ depends only on the radial and orbital indices[1] and is chosen such that $\int |E_{p,l}^{\mathrm{LG}}|^2 \mathrm{d}x\mathrm{d}y = P_0$. The scaled radius $\tilde{r}$, appearing in both the polynomial and the Gaussian terms, is a complex number depending on z and w_0, and given by

[1] The normalization N_p^l is given by

$$\frac{1}{N_p^l} = (l + p)! \sqrt{\frac{1}{2^{l+1} p!^3} \sum_{k,j \le p} \binom{p}{j}\binom{p}{k} \frac{(l + j + k)!}{(j + l)!(l + k)!} \left(-\frac{1}{2}\right)^{j+k-1}}.$$

$$\tilde{r} = \sqrt{\frac{\mathrm{i}\pi}{\lambda q(z;\, w_0)}}\, r. \tag{15.12}$$

The beam radius, $w(z)$, and the radius of curvature of the phase fronts, $R(z)$, are obtained from $q(z;\, w_0)^{-1} = R(z)^{-1} - (\mathrm{i}\lambda/\pi)w(z)^{-2}$. For the corresponding *standard* mode, the complex argument of the polynomial ($\tilde{r}$) is replaced by a real argument $\sqrt{2}r/w(z)$,

$$E_{p,l}^{\text{standard LG}}(\mathbf{r};\, w_0) \propto \left(\frac{\sqrt{2}r}{w(z)}\right)^{|l|} L_p^{|l|}\left(\frac{2r^2}{w(z)^2}\right) \mathrm{e}^{-\tilde{r}^2} \mathrm{e}^{\mathrm{i}l\phi}. \tag{15.13}$$

Both the elegant and the standard modes carry orbital angular momentum of $l\hbar$ per photon [40, 41].

To describe the effect of diffusion on the elegant modes, we utilize the fact that the z dependence in (15.11) comes only through the complex distance $q(z)$, so that

$$\left(\frac{\partial E}{\partial w_0}\right)_z = \left(\frac{\partial q}{\partial w_0}\right)_z \left(\frac{\partial E}{\partial q}\right)_{w_0} + \left(\frac{\partial E}{\partial w_0}\right)_q = \mathrm{i}\frac{2\pi}{\lambda} w_0 \left(\frac{\partial E}{\partial z}\right)_{w_0} + \left(\frac{\partial E}{\partial w_0}\right)_q . \tag{15.14}$$

Here we have used $(\partial q/\partial z)_{w_0} = 1$ and $(\partial q/\partial w_0)_z = \mathrm{i}2\pi/\lambda$. Since the elegant solutions satisfy the paraxial diffraction equation, $(\partial E/\partial z)_{w_0} = -\mathrm{i}\lambda/(4\pi)(\nabla_\perp^2 E)_{w_0}$, one arrives at

$$(\nabla_\perp^2 E)_{w_0} = \frac{2}{w_0}\left[\left(\frac{\partial E}{\partial w_0}\right)_z - \left(\frac{\partial E}{\partial w_0}\right)_q\right], \tag{15.15}$$

or for $E_{p,l}^{\mathrm{LG}}$ from (15.11),

$$\nabla_\perp^2 E_{p,l}^{\mathrm{LG}} = \frac{2}{w_0}\left(\frac{\partial E_{p,l}^{\mathrm{LG}}}{\partial w_0}\right)_z - \frac{2}{w_0^2}(N+1)\left(\frac{\partial E_{p,l}^{\mathrm{LG}}}{\partial w_0}\right)_q, \tag{15.16}$$

where $N = 2p + |l|$ is the total mode order. Therefore, if a two-dimensional diffusion takes place at a certain x-y plane, where z is held constant, the first term in the right-hand side of (15.16) accounts for a change in the waist radius, and the second term for a uniform decay of the field. Equation (15.2) is therefore solved by

$$\psi_{p,l}^{\mathrm{LG},z}(x, y, t) = \mathrm{e}^{-\gamma t} s(t)^{-(N+1)} E_{p,l}^{\mathrm{LG}}(\mathbf{r};\, w_0 s(t)), \tag{15.17}$$

where the diffusion coefficient again enters only through the stretching factor $[s(t) = (1 + 4Dt/w_0^2)^{1/2}]$. We thus find that the effect of diffusion on all the elegant LG solutions is the stretching of the waist radius, and consequently they all evolve self-similarly at their waist plane while diffusing. For example, the intensity and phase patterns at the waist plane of the $E_{2,1}^{\mathrm{LG}}$ mode are shown in Fig. 15.3 before and after diffusion. The field after diffusion was calculated by numerically solving the diffusion equation, and the numerical solution agrees with Eq. (15.17). It is evident that the phase pattern at the waist of the elegant mode

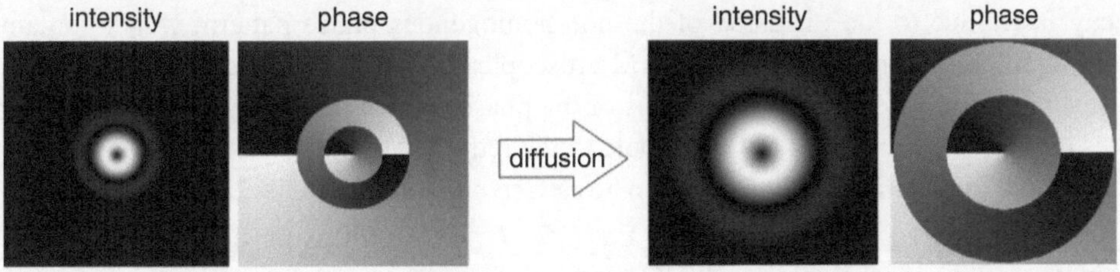

Figure 15.3 The self-similar evolution of both the amplitude and phase patterns of $E^{\mathrm{LG}}_{2,1}$, calculated numerically by directly solving the two-dimensional diffusion equation.

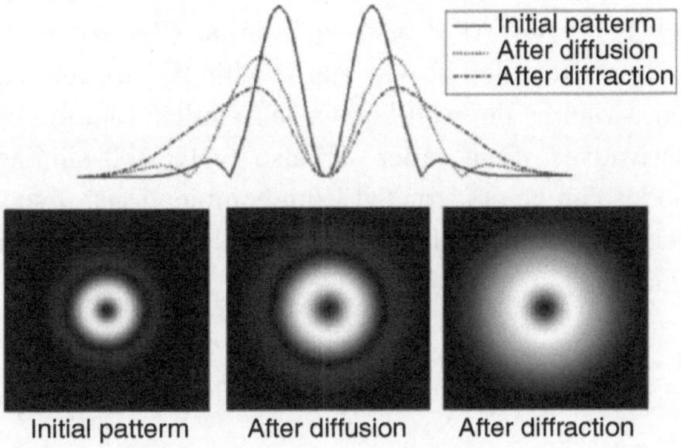

Figure 15.4 The self-similar evolution of the elegant mode $E^{\mathrm{LG}}_{1,2}$ under diffusion in comparison with the evolution under diffraction (top: Cartesian cross-sections of $|E(x, y)|$; bottom: field intensity $|E(x, y)|^2$).

expands self-similarly. Figure 15.4 compares the calculated evolution of the mode $E^{\mathrm{LG}}_{1,2}$ under diffusion and diffraction, demonstrating the preservation of the shape for diffusion and its alternation for diffraction.

As for the relations between w_0, z, and t, it is worth observing that if diffusion is addressed as an imaginary-time evolution of diffraction, it is readily seen from the definition of the complex distance $q(z; w_0)$ that exchanging the real evolution in z for an imaginary one corresponds to a real increase in the waist radius. Also note that diffusion leads to expansion only at the vicinity of the beam's waist, while far from the waist plane, contraction occurs. This contraction is addressed in the next section.

The total power (15.7) is not preserved during diffusion even when $\gamma = 0$, following a mode-dependent algebraic decay,

$$P(t) = \mathrm{e}^{-2\gamma t} s\,(t)^{-2(N+1)}\, P(0). \qquad (15.18)$$

The algebraic decay affects even the bare Gaussian mode ($N = 0$), because it is the field, rather than the intensity, that is diffusing. Thus, despite the uniformity of the phase in the waist plane of a Gaussian beam, and while $\int E(x, y)\mathrm{d}x\mathrm{d}y$ is conserved as expected from real-valued diffusion, the integral $\int \mathrm{d}x\mathrm{d}y|E(x, y)|^2$ decays. Higher-order modes ($N > 0$)

decay faster due to the diffusion of the non-homogenous phase pattern, which contains larger gradients for higher N [34, 42]. Microscopically speaking, the coherence of atoms diffusing through the spatial oscillations of the phase pattern in higher-order modes interferes destructively and does not contribute to the retrieved light.

A similar mathematical derivation can be carried out with the elegant HG solution, $E^{\mathrm{HG}}_{n,m}$, of Cartesian orders n and m, yielding Eq. (15.17) for the diffusing field $\psi^{\mathrm{HG},z}_{n,m}(x,y,t)$, with $N=n+m$. At their waist, all HG modes $\psi^{\mathrm{HG},z=0}_{n_1,n_2}$ are real-valued, and they identify with the expanding similarity solutions of real-valued diffusion [34], occurring, e.g., for the vorticity field of a viscous fluid [13, 35].

In general, any spatial derivative of a self-similar solution of a linear partial differential equation ($\psi(\mathbf{r},t)=h(t)f[\mathbf{r}/w(t)]$) is also self-similar ($\partial\psi/\partial x=h(t)F[\mathbf{r}/w(t)]/w(t)$, where $F=\partial f/\partial x$). For example, if one starts with the expanding two-dimensional Gaussian distribution, which is the well-known self-similar solution of real non-negative diffusion, then its derivatives of any order will also evolve self-similarly. Indeed, all elegant HG and LG modes can be constructed from combinations of such derivatives, with the derivative order corresponding to the total mode order N [43]. Furthermore, the unified Hermite–Laguerre–Gaussian modes [44],

$$\left(\cos\theta\frac{\partial}{\partial x}+\mathrm{i}\sin\theta\frac{\partial}{\partial y}\right)^n\left(\cos\theta\frac{\partial}{\partial x}-\mathrm{i}\sin\theta\frac{\partial}{\partial y}\right)^m\left(\frac{\mathrm{e}^{-\tilde{r}^2}}{q(z)}\right), \tag{15.19}$$

yield generalized self-similar modes for diffusion, with $\theta=0$ being the elegant HG solutions (indices n and m), and $\theta=\pi/4$ the elegant LG solution (indices $p=\min(n,m)$ and $l=n-m$). Note that *integrals* of the fundamental Gaussian are also self-similar and may be employed in the expansion of solutions with infinite energy.

We conclude from (15.17) that any linear sum of modes sharing the same waist and the same total order N evolves self-similarly. In contrast, a general complex "image" (in the x-y plane), formed from the superposition of OAM modes of various orders, will not be maintained. The elegant LG modes do not form an orthonormal basis, and a linear decomposition with elegant modes $E^{\mathrm{LG}}_{p,l}$ requires the biorthogonal set [45]

$$\bar{u}^p_l=\tilde{r}^{|l|}L^{|l|}_p\left(\tilde{r}^2\right)\mathrm{e}^{-\tilde{r}^2}\mathrm{e}^{\mathrm{i}l\phi}. \tag{15.20}$$

Thus, an arbitrary complex field evolves in time according to

$$E(x,y,t)=\sum_{p,l}a^l_p\psi^{\mathrm{LG}}_{p,l}(x,y,t)=e^{-\gamma t}\sum_{p,l}a^l_p s(t)^{-(2p+l+1)}E^{\mathrm{LG}}_{p,l}(\mathbf{r};w_0 s(t)), \tag{15.21}$$

where $a^l_p=\int\mathrm{d}x\mathrm{d}y\bar{u}^p_l E(x,y,t=0)$. The image undergoes diffusion blurring, and asymptotically the lowest-order solution prevails [7].

15.4 Experiments with vortex modes

15.4.1 Expansion

Diffusion experiments with OAM modes were carried by storing light beams onto the ground-state coherence of diffusing atoms in a vapor cell [2]; see caption of Fig. 15.1 for more experimental details. The diffusion of the LG modes $E_{0,0}^{\mathrm{LG}}$, $E_{0,1}^{\mathrm{LG}}$, and $E_{0,2}^{\mathrm{LG}}$ with the cell located at the beam waist ($z = 0$) is presented in Fig. 15.5. The retrieved images, proportional to $|\psi(x, y, t)|^2$, are shown for storage durations of $t = 2$, 30, and 60 μs (left). Evidently, the intensity distribution becomes wider while maintaining the shape throughout the diffusion process. The radially weighted cross-sections of each mode at different storage times (right) coincide when scaled by $s(t)^{1/2}$ and normalized accordingly, clearly demonstrating the self-similarity.

To verify that the OAM is maintained, a complementary test was carried out on the retrieved beam after diffusion. By passing the beam through a binary grating mask with a fork-dislocation, diffraction orders were generated with a phase function $\mathrm{e}(\mathrm{i}M\phi)$ added to the M-th diffraction order [46]. As shown in Fig. 15.6, the retrieved $E_{0,0}^{\mathrm{LG}}$ mode then produced a central Gaussian ($l = 0$) with two side-band vortices ($l = \pm 1$), while the retrieved $E_{0,1}^{\mathrm{LG}}$ produces a central vortex ($l = 1$), a Gaussian ($l = 0$) on the right, and a higher-order vortex ($l = 2$) on the left. This result confirms the preservation of the helicity and hence of the OAM through diffusion.

The waist radius after diffusion is found by fitting the retrieved images to the expected mode pattern (15.11) at $z = 0$. The waist-radii squared grow linearly with the storage duration, as shown in Fig. 15.7, with the mutual linear slope being $4D$. The diffusion coefficient

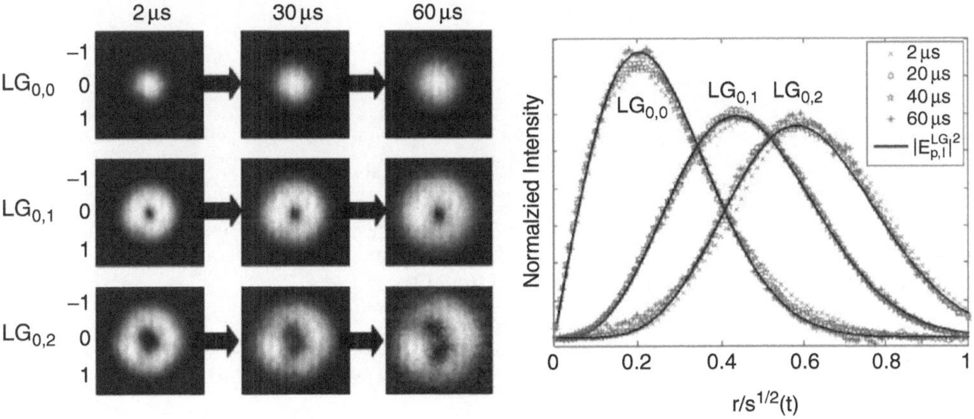

Figure 15.5 Left: Diffusion of the complex field of atomic coherence in a storage-of-light experiment, with Laguerre–Gaussian modes of radial index $p = 0$ and orbital indices (top to bottom) $l = 0$, 1, and 2. All images are 1.6 × 1.6 mm. Right: Self-similarity–the radially weighted cross-sections at different times are congruent when plotted versus the scaled coordinate $s(t)^{-1/2}r$. The solid lines are the analytic forms of the elegant LG modes. Adapted from [2].

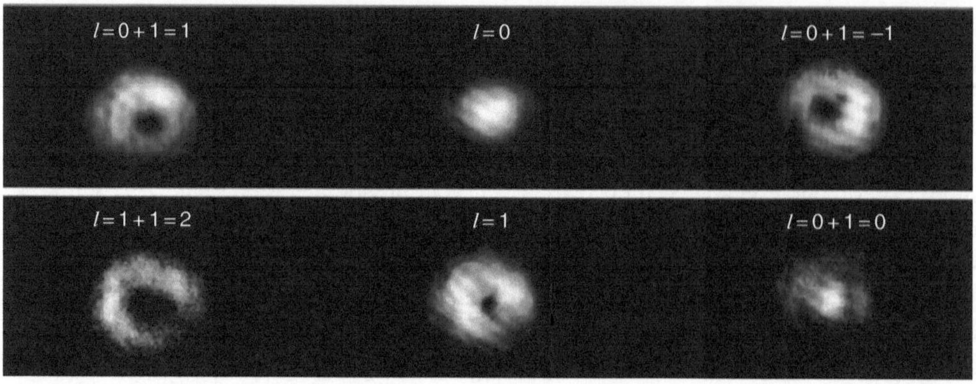

Figure 15.6 Diffraction orders −1, 0, +1 of the retrieved images after passing through a grating mask with a fork dislocation. The retrieved Gaussian mode (top) produces two $l = \pm 1$ vortex modes, while the retrieved vortex $l = +1$ (bottom) produces the $l = 0$ and $l = +2$ modes. Adapted from [2].

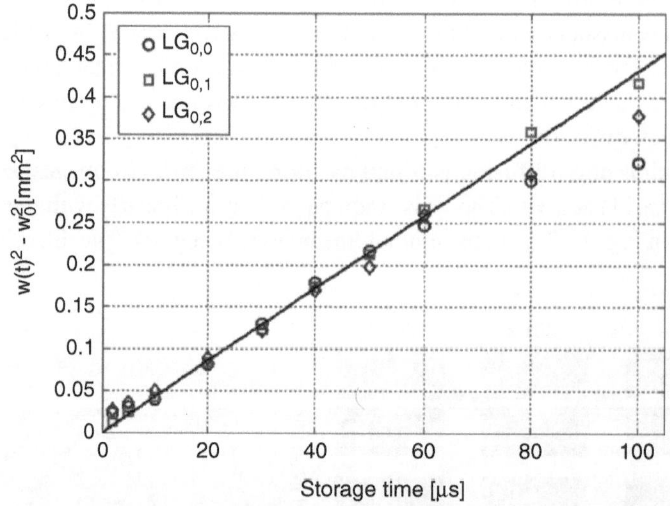

Figure 15.7 The linear growth of the mode area (w^2) versus diffusion time. The line is $w^2 - w_0^2 = w_0^2[s(t)^2 - 1] = 4Dt$, with $D = 10.8$ cm^2/s. Both D and w_0 are fit parameters ($w_0 = 0.4 \sim 0.5$ mm, varying for the different modes). Adapted from [2].

and initial waist radii obtained from these fits were used to determine $s(t)$ and to scale the cross-sections in Fig. 15.5.

The remaining optical power after diffusion is measured by integrating over the intensity of the retrieved images. In order to find the homogenous decay γ, the data are first divided by $s(t)^{-2(N+1)}$ to remove the algebraic decay and then plotted on a semilog scale (inset in Fig. 15.8). The exponential decay expected from the intrinsic homogenous decay is the dashed line in Fig. 15.8 (and the linear fit in the inset). The OAM modes exhibit

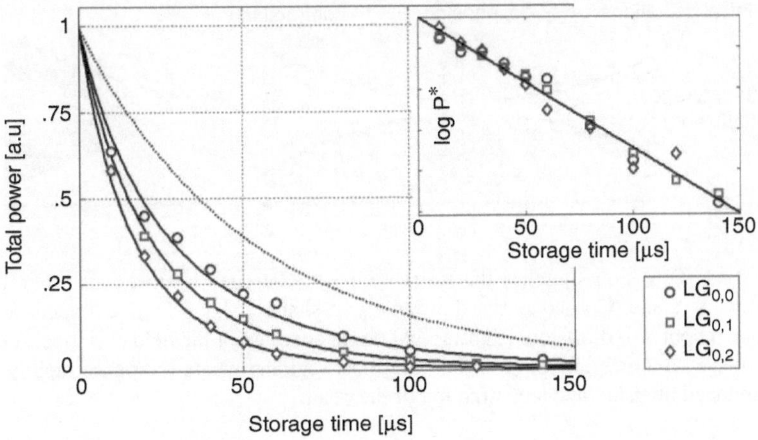

Figure 15.8 Decay of the total power in the retrieved images $P(t) = \int dxdy|\psi(t)|^2$. In the inset, $P^*(t) = P(t)s(t)^{2(N+1)}$ (with $N = 0, 1, 2$) compensates for the algebraic decay, with $s(t)$ known from the fit in Fig. 15.7. All data points collapse onto a single straight line in the semilog scale, yielding the homogenous decay rate $2\gamma = (68\ \mu s)^{-1}$. In the main graph, the dashed line is $e^{-2\gamma t}$ and the solid lines are $e^{-2\gamma t}s(t)^{-2(N+1)}$, demonstrating the faster decay of the higher-order modes. Adapted from [2].

a significant algebraic decay on top of the homogenous decay, with the higher-order modes decaying faster. The solid lines predicted from (15.17) agree perfectly with the data.

An imperfection of the experimental realization is noticeable for the $E^{LG}_{0,2}$ mode in Fig. 15.5 after 60 μs of storage. Evidently, the $l = 2$ vortex was unstable during the diffusion and broke down into two vortices, probably of $l = 1$. Higher-order vortices decaying into lower-order ones have been observed also in optics, quantum fluids, and Bose–Einstein condensates [47]. Several mechanisms may account for this splitting [13], though most likely it is the small fraction of lower-order components existing in the original beam created by the binary grating [46]. A similar OAM bifurcation due to imperfect generation procedure in *free-space optics* was observed by Soskin *et al.* [48]. It is worth noticing, however, that despite the splitting, the diffusion process still conserves the shape of the radial cross-section of the original $l = 2$ vortex (Fig. 15.5, the $LG_{0,2}$ curve).

15.4.2 Contraction

According to Eq. (15.17), diffusion results in an effective stretching of the waist-radius parameter. For Gaussian beams, such an expansion of the waist radius results in an increase of the transverse size at $|z| < z_R$ and in a decrease at $|z| > z_R$, as illustrated in Fig. 15.9. Therefore, if the diffusion is examined *at the waist* (*cell A* in the figure), an expansion of the transverse pattern is observed. In those experiments, the beam is focused to the middle of the cell, and the succeeding optics image this focus onto the camera. However, if the

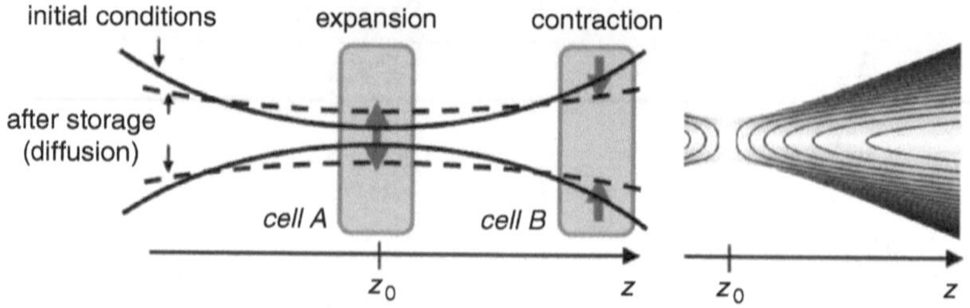

Figure 15.9 Left: Diffusion stretches the waist radius of Gaussian beams, even when it does not occur at the waist plane. Consequently, the transverse shape at $|z| > z_R$ contracts–a surprising outcome when it comes to diffusion. Right: Equal phase contours of the field envelope. A destructive interference between atoms diffusing in the outermost regions, where the phase rapidly oscillates, results in a reduced intensity and a contraction of the beam.

camera is to image a plane *far from the waist*, behind the Rayleigh distance, the same diffusion process will be captured as a contraction.

The contraction versus storage time, $C(t) = w(t)/w(0)$, is given by

$$C(t)^2 = \frac{s(t)^4 + \rho_0^2}{s(t)^2(1 + \rho_0^2)}, \tag{15.22}$$

where $\rho_0 = z/z_R$ specifies the initial distance from the waist. As the waist radius increases during the diffusion, the Rayleigh distance increases and eventually crosses the observed plane (which z coordinate is constant). At this time, the maximal contraction $C_{\min}^2 = 2\rho_0/(1 + \rho_0^2)$ is obtained, and thenceforth the beam expands indefinitely.

Since the diffusion (during storage) and the free-space diffraction (before and after the cell) are linear and commute, the evolution described above will also occur if the cell is moved away from the waist plane, as illustrated by *cell B* in Fig. 15.9. Indeed, (15.17) and (15.22) hold for all z, regardless of where the diffusion "takes place." We thus find, perhaps counterintuitively, that the initial effect of diffusion occurring at $|z| > z_R$ is the *contraction of the diffusing field.* Such contraction is possible only for complex-valued diffusion, and locally it is the consequence of a destructive interference in regions of the beam where the phase pattern rapidly oscillates (Fig. 15.9, right).

As opposed to the "standard" optical Gaussian modes, the "elegant" Gaussian modes of diffusion are generally not self-similar under optical diffraction, and the shape of the beam depends on z/z_R. Hence, even though the diffusion occurs and observed at a constant z, the aforementioned increase of z_R changes the transverse shape during diffusion. This means that at planes other than the waist plane, the expansion or contraction are generally not self-similar. To achieve self-similarity for all z, a mode needs to be simultaneously standard and elegant, i.e. self-similar to both diffusion and diffraction. The vortex modes ($p = 0$) with any OAM belong to this special subset, since $L_{p=0}^{l} = 1$, and the resulting polynomial structure is homogeneous. In this subset of modes, the complex scaling of the polynomial only introduces a uniform phase factor. Therefore, the vortex modes evolve self-similarly

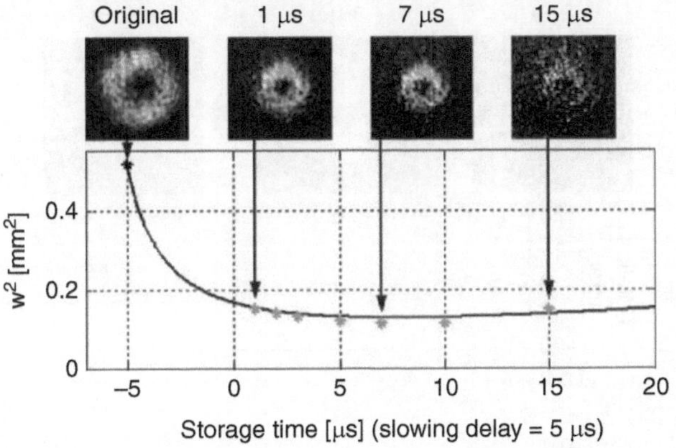

Figure 15.10 The self-similar contraction and subsequent expansion of the $LG_{0,1}$ mode upon diffusion. Much contraction occurs also during the slow propagation (not during storage), when the light, in the form of a light–matter polariton, propagates slowly in the cell. The black line is calculated from Eq. (15.22). The destructive interference at the outer region of the mode decreases the overall retrieved power much faster than in the expansion experiments, resulting in a poor signal-to-noise ratio already after 15 μs of storage. Adapted from [2].

in all three-dimensions, appearing as either an expansion or a contraction when observed at a definite x-y plane.

In experiments, the diffusion far from the waist was demonstrated by introducing a focusing lens into the signal channel in front of the cell (Fig. 15.1) [2]. The diffusion of the vortex mode $E^{\mathrm{LG}}_{0,1}$, focused at a distance $8z_{\mathrm{R}}$ before the cell ($\rho = 8$), is shown in Fig. 15.10. The maximum contraction ratio at this distance from the waist plane is $C^2_{\min} \approx 1/4$.

15.5 Fractional and varying OAM

Recently formulated elegant LG beams of fractional orders carry non-integer OAM [41]. Following (15.21), any mode of fractional order can be decomposed into a spectrum of integer-order OAM modes. During free-space diffraction, the different OAM modes acquire different phases, but still maintain their relative contribution to the mean (fractional) OAM. However if put through diffusion, the fractional-order modes do not preserve their OAM and obviously do not evolve self-similarly, since each of the different constituent integer-order modes decays in a different rate. In practice, a general superposition of OAM modes will be maintained during storage only if the diffusion could be neglected, as was predicted by Dutton *et al.* [49] and demonstrated for cold atomic ensembles [50–52].

Nevertheless, there are in principle superpositions of LG modes with non-integer mean OAM invariant under diffusion. These superposition span a spectrum of orbital indices l, while the total order of all constituent modes ($N = 2p + |l|$) is the same. For example, any sum of the vortices $E^{\mathrm{LG}}_{0,l}$ and $E^{\mathrm{LG}}_{0,-l}$,

$$E = C_l E^{\mathrm{LG}}_{0,l} + C_{-l} E^{\mathrm{LG}}_{0,-l}, \tag{15.23}$$

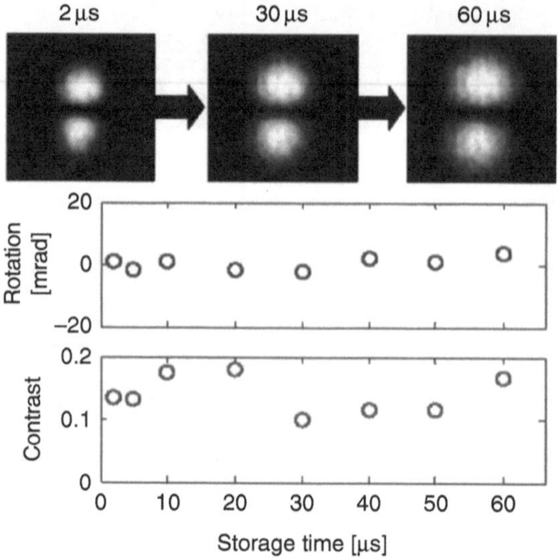

Figure 15.11 Diffusion of the HG mode $E^{HG}_{0,1} = (E^{LG}_{0,1} + E^{LG}_{0,-1})/\sqrt{2}$. The mode expands self-similarly (top), since both LG modes decay in the same rate and acquire the same phase. The preservation of the superposition of the LG modes is quantified by the "rotation" angle of the lobes pattern and the "contrast" of the pattern (bottom). The latter is calculated from the power difference between the two lobes.

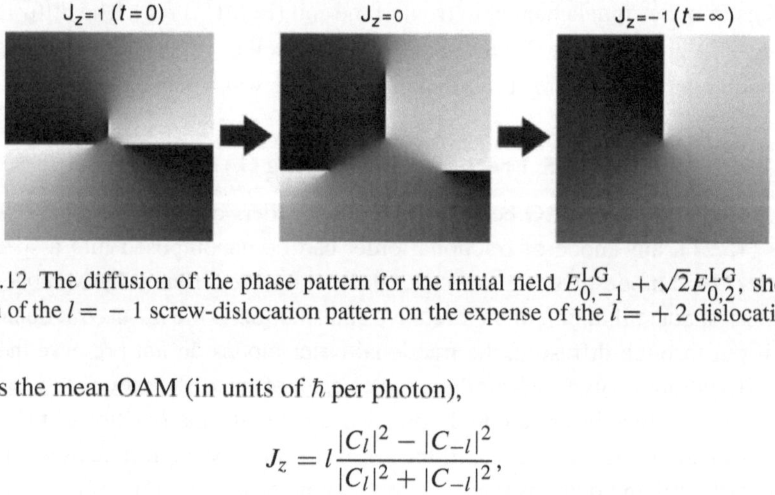

Figure 15.12 The diffusion of the phase pattern for the initial field $E^{LG}_{0,-1} + \sqrt{2}E^{LG}_{0,2}$, showing the expansion of the $l = -1$ screw-dislocation pattern on the expense of the $l = +2$ dislocation.

maintains the mean OAM (in units of $\hbar$ per photon),

$$J_z = l\frac{|C_l|^2 - |C_{-l}|^2}{|C_l|^2 + |C_{-l}|^2}, \tag{15.24}$$

since the rate of the diffusion-induced decay is independent of the sign of l. In (15.24), J_z can take any value between $-l$ and l. A trivial experimental example is presented for $l = 1$ in Fig. 15.11, with $C_1 = C_{-1}$ and $J_z = 0$. The resultant mode is an HG mode, and it is indeed maintained when diffusing.

Alternatively, one can utilize the diffusion-induced decay to design modes with the desired initial and final mean OAM. An example is given in Fig. 15.12, where the diffusion of the phase pattern is shown for the initial field $E = E^{LG}_{0,-1} + \sqrt{2}E^{LG}_{0,2}$. This field carries initial mean OAM of $J_z = (-1+2\cdot 2)/(1+2) = 1$, which gradually decreases, since $E^{LG}_{0,2}$ decays faster than $E^{LG}_{0,-1}$. Asymptotically, the lowest order $E^{LG}_{0,-1}$ prevails, with OAM of $J_z = -1$.

15.6 Non-diffusing modes with OAM

In optics, a unique family of OAM-carrying beams are the so-called non-diffracting or Bessel beams, formulated by Durnin in 1987 [37]. The electric field of a Bessel beam carrying an OAM of $l\hbar$ per photon is given by

$$\varepsilon(x, y, z, t) = e^{ik_z z - \omega t} E^{\text{Bessel}}_l(x, y), \quad \text{with } E^{\text{Bessel}}_l(x, y) = J_{|l|}(k_r r)\, e^{il\phi}, \qquad (15.25)$$

where $k_r^2 = (\omega/c)^2 - k_z^2$. A Bessel beam preserves its shape and phase pattern, $E^{\text{Bessel}}_l(x, y)$, without expanding or contracting, throughout its propagation. Occupying only a thin ring in $k_\perp$-space (of radius k_r), the field does not suffer from spatial dispersion and thus only acquires a uniform phase when diffracting. The intricate structure of the beam, such as its OAM, is encoded in the intensity and phase distribution *along the ring* and is therefore not altered.

Following the same arguments, it follows immediately from Eq. (15.1) that the Bessel beams are also *non-diffusing*. Having a single ($|\mathbf{k}_\perp| = k_r$) transverse frequency, diffusion will preserve the structure along the $k_\perp$-space ring and consequently the structure in real space. Strictly, this result can be derived from Eq. (15.6). While diffraction introduces a uniform phase variation, diffusion acts to decrease the total power of the mode, even when $\gamma = 0$, according to

$$E^{\text{Bessel}}_l(x, y, t) = e^{-Dk_r^2 t} E^{\text{Bessel}}_l(x, y, 0). \qquad (15.26)$$

As outlined in the summary below, the Bessel beams, being simultaneously non-diffracting and non-diffusing, are expected to propagate unchanged in a slow-light medium, where they experience diffraction and diffusion at the same time.

15.7 Summary

Coherent diffusion, like regular diffusion, is a dissipative process, wiping out the high spatial frequencies of the diffusing field and leading to spatial-information loss. However similarly to its "imaginary" counterpart – paraxial diffraction – coherent diffusion in two dimensions exhibits intricate coherent dynamics, via constructive and destructive interference effects, and sustains a set of expanding and contracting self-similar modes. Specifically, it is possible to construct a complete basis of self-similar modes with definite integer orbital angular momentum "per photon," which is maintained throughout the diffusion. The "power" of each mode, that is, its L^2 norm, is not maintained and decays algebraically with time, with the lowest-order mode prevailing.

The discussion in this chapter focused on the diffusion of coherent fields during storage-of-light, but diffusion also takes place during the slow propagation of a light pulse that is not stored. The physical diffusion of the atoms leads to the diffusion of the strongly coupled resonant light-field. Diffusion and diffraction then act together on the light field, and modes that are self-similar to both (simultaneously standard and elegant) will maintain their shape. Still, any light beam carrying a definite OAM will conserve the OAM, since neither diffraction nor diffusion are able to change it.

Acknowledgements

The author would like to acknowledge a fruitful collaboration with Amiram Ron, Donald R. Fredkin, Paz London, and Dimitry Yankelev.

References

[1] R. Pugatch, M. Shuker, O. Firstenberg, A. Ron, and N. Davidson. Topological stability of stored optical vortices. *Phys. Rev. Lett.*, **98**(20):203601, 2007.
[2] O. Firstenberg, P. London, D. Yankelev, R. Pugatch, M. Shuker, and N. Davidson. Self-similar modes of coherent diffusion. *Phys. Rev. Lett.*, **105**:183602, 2010.
[3] D. S. Grebenkov. Nmr survey of reflected Brownian motion. *Rev. Mod. Phys.*, **79**(3):1077–137, Aug 2007.
[4] H. C. Torrey. Bloch equations with diffusion terms. *Phys. Rev.*, **104**(3):563–5, Nov 1956.
[5] J. Skalla, G. Wbkerle, M. Mehring, and A. Pines. Optical magnetic resonance imaging of Rb vapor in low magnetic fields. *Phys. Lett. A*, **226**:69–74, 1997.
[6] Y. Xiao, I. Novikova, D. F. Phillips, and R. L. Walsworth. Diffusion-induced ramsey narrowing. *Phys. Rev. Lett.*, **96**(4):043601, 2006.
[7] M. Shuker, O. Firstenberg, R. Pugatch, A. Ron, and N. Davidson. Storing images in warm atomic vapor. *Phys. Rev. Lett.*, **100**(22):223601, 2008.
[8] J. I. Kaplan. Application of the diffusion-modified bloch equation to electron spin resonance in ordinary and ferromagnetic metals. *Phys. Rev.*, **115**(3):575–7, 1959.
[9] J. M. Kikkawa and D. D. Awschalom. Lateral drag of spin coherence in gallium arsenide. *Nature*, 397:139–41, 1999.
[10] Igor Zutic, Jaroslav Fabian, and S. Das Sarma. Spintronics: fundamentals and applications. *Rev. Mod. Phys.*, **76**(2):323–410, 2004.
[11] S. P. Dash, S. Sharma, R. S. Patel, M. P. de Jong, and R. Jansen. Electrical creation of spin polarization in silicon at room temperature. *Nature*, **462**:491–4, 2009.
[12] D. Giel, G. Hinz, D. Nettels, and A. Weis. Diffusion of CS atoms in Ne buffer gas measured by optical magnetic resonance tomography. *Opt. Express*, **6**(13):251–6, 2000.
[13] R. C. Kloosterzielt and G. J. F. Van Heijst. An experimental study of unstable barotropic vortices in a rotating fluid. *J. Fluid Mech.*, **223**:1–24, 1991.
[14] M. Shuker, O. Firstenberg, Y. Sagi, A. Ben-kish, N. Davidson, and A. Ron. Ramsey-like measurement of the decoherence rate between Zeeman sublevels. *Phys. Rev. A*, **78**(6):063818, 2008.

[15] M. V. Balabas, T. Karaulanov, M. P. Ledbetter, and D. Budker. Polarized alkali-metal vapor with minute-long transverse spin-relaxation time. *Phys. Rev. Lett.*, **105**(7):070801, 2010.

[16] O. Firstenberg, M. Shuker, R. Pugatch, D. R. Fredkin, N. Davidson, and A. Ron. Theory of thermal motion in electromagnetically induced transparency: effects of diffusion, Doppler broadening, and Dicke and Ramsey narrowing. *Phys. Rev. A*, **77**(4):043830, 2008.

[17] D. F. Phillips, A. Fleischhauer, A. Mair, R. L. Walsworth, and M. D. Lukin. Storage of light in atomic vapor. *Phys. Rev. Lett.*, **86**(5):783, 2001.

[18] A. Kasapi, M. Jain, G. Y. Yin, and S. E. Harris. Electromagnetically induced transparency: propagation dynamics. *Phys. Rev. Lett.*, **74**:2447, 1995.

[19] M. Fleischhauer and M. D. Lukin. Dark-state polaritons in electromagnetically induced transparency. *Phys. Rev. Lett.*, **84**(22):5094–7, 2000.

[20] D. Akamatsu and M. Kozuma. Coherent transfer of orbital angular momentum from an atomic system to a light field. *Phys. Rev. A*, **67**(2):023803, 2003.

[21] R. Inoue, T. Yonehara, Y. Miyamoto, M. Koashi, and M. Kozuma. Measuring qutrit-qutrit entanglement of orbital angular momentum states of an atomic ensemble and a photon. *Phys. Rev. Lett.*, **103**(11):110503, 2009.

[22] O. Firstenberg, P. London, M. Shuker, A. Ron, and N. Davidson. Elimination, reversal and directional bias of optical diffraction. *Nature Physics*, **5**:665–8, 2009.

[23] C. Bolkart, D. Rostohar, and M. Weitz. Dark resonances with variable Doppler sensitivity. *Phys. Rev. A*, **71**:043816, 2005.

[24] M. Shuker, O. Firstenberg, R. Pugatch, A. Ben-Kish, A. Ron, and N. Davidson. Angular dependence of Dicke-narrowed electromagnetically induced transparency resonances. *Phys. Rev. A*, **76**(2):023813, 2007.

[25] O. Firstenberg, M. Shuker, A. Ben-Kish, D. R. Fredkin, N. Davidson, and A. Ron. Theory of Dicke narrowing in coherent population trapping. *Phys. Rev. A*, **76**:013818, 2007.

[26] A. S. Zibrov and A. B. Matsko. Optical Ramsey fringes induced by Zeeman coherence. *Phys. Rev. A*, **65**(1):013814, 2001.

[27] R. Pugatch, O. Firstenberg, M. Shuker, and N. Davidson. Universal spectra of coherent atoms in a recurrent random walk. *Phys. Rev. Lett.*, **102**(15):150602, 2009.

[28] L. Zhao, T. Wang, Y. Xiao, and S. F. Yelin. Image storage in hot vapors. *Phys. Rev. A*, **77**(4):041802, 2008.

[29] P. K. Vudyasetu, R. M. Camacho, and J. C. Howell. Storage and retrieval of multimode transverse images in hot atomic rubidium vapor. *Phys. Rev. Lett.*, **100**(12):123903, 2008.

[30] L. Allen, S. M. Barnett, and M. J. Padgett. *Orbital Angular Momentum*. Institute of Optics Publishing, 2003.

[31] A. E. Siegman. *Lasers*. Sausalito, CA: University Science Books, 1986.

[32] M. Abramowitz and I. A. Stegun, editors. *Handbook of Mathematical Functions with Formulas, Graphs, and Mathematical Tables*. New York: Dover, 1972.

[33] G. I. Barenblatt. *Scaling, Self-Similarity, and Intermediate Asymptotics*. New York: Cambridge University Press, 1996.

[34] R. C. Kloosterzielt. On the large-time asymptotics of the diffusion equation on infinite domains. *J. Eng. Math.*, **24**:213–36, 1990.

[35] M. P. Satijn, M. G. van Buren, H. J. H. Clercx, and G. J. F. van Heijst. Vortex models based on similarity solutions of the two-dimensional diffusion equation. *Phys. of Fluids*, **16**(11):3997–4011, 2004.

[36] A. E. Siegman and E. A. Sziklas. Mode calculations in unstable resonators with flowing saturable gain. 1: Hermite–Gaussian expansion. *Appl. Opt.*, **13**(12):2775–91, 1974.

[37] J. Durnin, J. J. Miceli, and J. H. Eberly. Diffraction-free beams. *Phys. Rev. Lett.*, **58**(15):1499–501, 1987.

[38] A. E. Siegman. Hermite–gaussian functions of complex argument as optical-beam eigenfunctions. *J. Opt. Soc. Am.*, **63**(9):1093–4, 1973.

[39] A. Wunsche. Generalized Gaussian beam solutions of paraxial optics and their connection to a hidden symmetry. *J. Opt. Soc. Am. A*, **6**:1320–9, 1989.

[40] L. Allen, M. W. Beijersbergen, R. J. C. Spreeuw, and J. P. Woerdman. Orbital angular momentum of light and the transformation of Laguerre–Gaussian laser modes. *Phys. Rev. A*, **45**(11):8185–9, 1992.

[41] J. C. Gutierrez-Vega. Fractionalization of optical beams: II. elegant Laguerre–Gaussian modes. *Optics Express*, **15**:6300–13, 2007.

[42] T. Wang, L. Zhao, L. Jiang, and S. F. Yelin. Diffusion-induced decoherence of stored optical vortices. *Phys. Rev. A*, **77**(4):043815, 2008.

[43] E. Zauderer. Complex argument Hermite–Gaussian and Laguerre–Gaussian beams. *J. Opt. Soc. Am.* ***A***, 3:465–9, 1986.

[44] D. Deng and Q. Guo. Elegant Hermite–Laguerre–Gaussian beams. *Optics Letters*, **33**:1225–7, 2008.

[45] A. Kostenbauder, Y. Sun, and A. E. Siegman. Eigenmode expansions using biorthogonal functions: complex-valued Hermite–Gaussians. *J. Opt. Soc. Am. A*, **14**:1780–90, 1997.

[46] I. V. Basistiy, V. Yu. Bazhenov, M. S. Soskin, and M. V. Vasnetsov. Optics of light beams with screw dislocations. *Opt. Comm.*, **103**:422–8, 1993.

[47] C. Ryu, M. F. Andersen, P. Cladé, V. Natarajan, K. Helmerson, and W. D. Phillips. Observation of persistent flow of a Bose–Einstein condensate in a toroidal trap. *Phys. Rev. Lett.*, **99**(26):260401, 2007.

[48] M. S. Soskin, V. N. Gorshkov, M. V. Vasnetsov, J. T. Malos, and N. R. Heckenberg. Topological charge and angular momentum of light beams carrying optical vortices. *Phys. Rev. A*, **56**(5):4064–75, 1997.

[49] Z. Dutton and J. Ruostekoski. Transfer and storage of vortex states in light and matter waves. *Phys. Rev. Lett.*, **93**(19):193602, 2004.

[50] S. Barreiro, J. W. R. Tabosa, J. P. Torres, Y. Deyanova, and L. Torner. Four-wave mixing of light beams with engineered orbital angular momentumin cold cesium atoms. *Opt. Lett.*, **29**(13):1515–7, 2004.

[51] M. F. Andersen, C. Ryu, P. Cladé *et al.* Quantized rotation of atoms from photons with orbital angular momentum. *Phys. Rev. Lett.*, **97**(17):170406, 2006.

[52] D. Moretti, D. Felinto, and J. W. R. Tabosa. Collapses and revivals of stored orbital angular momentum of light in a cold-atom ensemble. *Phys. Rev. A*, **79**(2):023825, 2009.

16

Quantum entanglement of orbital angular momentum

M. P. VAN EXTER, E. R. ELIEL AND J. P. WOERDMAN

16.1 Introduction

In this chapter, we will be concerned with entangled states of light. Historically, quantum states of light, i.e., photons, have played a dominant role in the experimental study of quantum entanglement. For instance, the first Bell tests and the first realizations of quantum teleportation were performed using two-dimensional polarization-entangled two-photon states [3]. From an experimental point of view these states are readily prepared and manipulated using conventional polarization optics.

Recently, however, both theoretical and experimental efforts have shifted towards entanglement in *more than two modes*, which is often referred to as high-dimensional entanglement. This is much richer than two-dimensional entanglement, as is, for instance, reflected in stronger violations of various measures of nonlocality which reveal the complexity of high-dimensional Hilbert spaces [4–7]. Also, from an applications perspective, it promises a larger channel capacity [8, 9] and improved security for quantum communication [10, 11].

There are several ways to implement high-dimensional two-photon entanglement. One can exploit the temporal [12], frequency [13], or spatial [14–18] degrees of freedom of the optical field. A promising way to implement high-dimensional two-photon entanglement is to use the orbital-angular-momentum (OAM) degree of freedom of light. Light beams that carry OAM have a helical wavefront that spirals around the optical axis during propagation [19]. The winding number m, which characterizes the OAM eigenmodes, can adopt any integer value between $-\infty$ and ∞; since all coaxial OAM modes are orthogonal they can thus span a very high-dimensional Hilbert space. This high-dimensional OAM entanglement of two photons is the topic of the present chapter.

The simplest way to produce entangled photon pairs is through the process of spontaneous parametric down-conversion [1]. In this nonlinear process, a 'blue' pump photon is spontaneously converted into two 'red' photons, while conserving energy, momentum, and angular momentum [2]. We focus on the latter aspect, and, more specifically, on the

The Angular Momentum of Light, ed. D. L. Andrews and M. Babiker. Published by Cambridge University Press.

angular transverse momentum, and configure our system such that it produces a very large number of entangled OAM modes [20–22]. The two photons of a pair have undetermined but opposite OAM. The number of entangled modes is typically of the order of 60.

In the literature one finds various experimental techniques to sort the modes of an OAM superposition state by their mode number, some of which have been applied to the quantum regime [23–25]. However, to assess entanglement this generally does not suffice, as one has to measure observables in various unbiased state bases. In the first experiments on OAM entanglement, which probed a three-dimensional space, this issue was tackled by cleverly displacing the OAM analyzers [26, 27]. This approach was improved by either using holographic [27] or interferometric [28] means, providing the possibility to perform full state tomography. However, the scalability of these methods to systems of dimensionality larger than three has not been demonstrated, and may not be realistic. Recently, OAM and its conjugate variable, the transverse angular position, have been shown to exhibit correlations that are a signature of entanglement [29]. Also in the domain of continuous-variable states, quadrature entanglement has been demonstrated between low-order OAM modes [30, 31]. Obviously, there are many approaches possible, and the challenge is to find one that is easily implementable, reliable, and yields as much information as possible.

Our approach to high-dimensional OAM entanglement has been, as discussed in the following sections, to use only two suitably designed rotating state analyzers. As we will show, this easily allows to extract some genuine features of high-dimensional OAM entanglement. In a second approach that we have applied, we instead rotate (part of) the entangled state. Using two-photon interference and large (= position-insensitive) detectors, we could thus perform a full OAM Schmidt decomposition of the generated state [22].

This chapter is organized as follows. The theoretical Section 16.2 describes the Schmidt decomposition of the two-photon field and the importance of OAM in the rotationally symmetric non-critical phase-matching geometry. Section 16.3 discusses several important issues in OAM entanglement in more details. Section 16.4, entitled "Experimental analysis of OAM states," forms the experimental heart of the paper. It introduces various experimental tools to modify and detect specific OAM states. It also contains a description of two sets of experiments that we performed to analyze the OAM entanglement. The first set of experiments uses *single-mode detectors* in combination with suitably designed state analyzer. The second set of experiments uses *large (bucket) detectors* in combination with image rotation and two-photon interference. The final Section 16.5 contains a concluding discussion of the obtained results and some open questions.

16.2 Theory of OAM entanglement

We generate quantum-entangled photon pairs with the nonlinear optical process of spontaneous parametric down-conversion (SPDC), where single pump photons occasionally split into two photons of lower energy, traditionally called the signal and idler photon (see Fig. 16.1). We operate in the quasi-monochromatic limit, combining a narrow-band cw pump laser at frequency ω_p with narrow-band spectral filters at the degenerate frequency

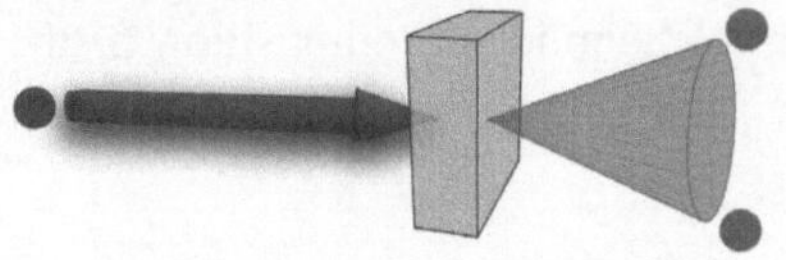

Figure 16.1 Quantum-entangled photon pairs (red) are generated by splitting single pump photons (blue) via spontaneous parametric down-conversion (SPDC) in a nonlinear optical crystal.

$\omega_1 = \omega_2 = \omega_p/2$. The two-photon state generated under these conditions has the general form [32]

$$|\Psi\rangle = \int\int A(\mathbf{q}_1, \mathbf{q}_2)\hat{a}^\dagger(\mathbf{q}_1)\hat{a}^\dagger(\mathbf{q}_2)|0\rangle, \tag{16.1}$$

where $\hat{a}^\dagger(\mathbf{q}_i)$ is the creation operator of a plane wave with transverse momentum $\mathbf{q}_i$ and i is the photon label. The generated two-photon state is pure when the pump field is spatially coherent and the spectral filtering is sufficiently narrow to remove any spatial-spectral correlations. The reduced one-photon operator $\rho^{(1)}$, obtained by tracing $|\Psi\rangle\langle\Psi|$ over all possible states of the second photon, is typically mixed, being an incoherent (and weighted) superposition over all available Schmidt states.

The general momentum representation of the two-photon amplitude is [33]

$$A(\mathbf{q}_1, \mathbf{q}_2) \propto \mathcal{E}_p(\mathbf{q}_1 + \mathbf{q}_2) \times \text{sinc}\left(\tfrac{1}{2}\Delta k_z L\right), \tag{16.2}$$

where $\mathcal{E}_p(\mathbf{q}_1 + \mathbf{q}_2)$ is the transverse momentum profile of the pump beam ($\mathbf{q}_p = \mathbf{q}_1 + \mathbf{q}_2$). The function sinc $\left(\frac{1}{2}\Delta k_z L\right)$ quantifies the influence of phase-matching, where L is the crystal length and where the projected wave-vector mismatch $\Delta k_z = \Delta k_z(\mathbf{q}_1, \mathbf{q}_2)$ is still a function of $\mathbf{q}_1$ and $\mathbf{q}_2$. We make the general expression more specific by inserting a (rotationally symmetric) Gaussian pump profile. We have also assumed non-critical phase matching, to remove any linear dependence of Δk_z on $\mathbf{q}_i$, and neglect the small dependence of Δk_z on $|\mathbf{q}_1 + \mathbf{q}_2|^2$, a term that scales with the difference between the refractive indices at the pump and SPDC wavelength. This allows us to rewrite Eq. (16.2) as [33]

$$A(\mathbf{q}_1, \mathbf{q}_2) \propto \exp\left(-|\mathbf{q}_1 + \mathbf{q}_2|^2/\sigma^2\right) \times \text{sinc}\left(b^2|\mathbf{q}_1 - \mathbf{q}_2|^2 + \varphi\right), \tag{16.3}$$

where $\sigma = 2/w_p$, for a Gaussian pump profile $E_p(\rho) \propto \exp(-|\rho|^2/w_p^2)$, and $b^2 = L/(4k_p)$, where $k_p = n\omega_p/c$ is the momentum of a pump photon in the crystal [34]. Equation (16.3) also applies to critically phase-matched type-I SPDC, where the crystal cut angle differs from 0° or 90° and where the two generated photons have the same polarization. It is generally not valid for the critically phase-matched type-II geometry, where transverse beam walk-off plays an important role and extra terms proportional to $\mathbf{q}_i$ show up in the argument of the sinc-function. Even in this geometry, a modified version of the presented description can still be used by limiting the detected SPDC light to two small angular regions of approximate uniform intensity and by redefining $\mathbf{q}_i$ with respect to these two beam lines.

Before we perform a formal Schmidt decomposition of the two-photon amplitude of Eq. (16.3), we note that the rotational symmetry of this function yields the intermediate result [33]

$$A(\mathbf{q}_1, \mathbf{q}_2) = \sum_{l=-\infty}^{\infty} \sqrt{P_l} F_l(q_1, q_2)/(2\pi) \exp[\mathrm{i}l(\theta_1 - \theta_2)], \tag{16.4}$$

where q_i and θ_i are the polar coordinates of $\mathbf{q}_i$ and where the radial amplitude is normalized as $\int\int |F_l(q_1, q_2)|^2 q_1 q_2 \mathrm{d}q_1 \mathrm{d}q_2 = 1$. Some nomenclature: the probability $P_l \equiv |\lambda_l|^2$ is called the *spiral weight*; its full distribution $\{P_l\}$ is called the *spiral spectrum*; and the width of this distribution is called the *spiral bandwidth* [20].

As a side remark, we note that Eq. (16.4) is exact whereas a related equation for the two-photon state

$$|\Psi\rangle = \sum_{l=-\infty}^{\infty} \sqrt{P_l} |l\rangle \otimes |-l\rangle \tag{16.5}$$

is not exact, but merely a convenient short-hand notation that singles out the azimuthal dependence of the two-photon state. This equation hides the fact that the outcome of most experiments on spatial entanglement will also depend on the radial dependence of the field, via projection on the (radial dependence of the) detection system.

The complete Schmidt decomposition of the two-photon state still requires a modal decomposition of each of the functions $F_l(q_1, q_2)$. Each of these "diagonalizations by basis transformation" yields a complete set of Schmidt modes, with radial mode label p, for a fixed OAM mode number l. After these final transformations, we obtain the two-dimensional Schmidt form

$$A(\mathbf{q}_1, \mathbf{q}_2) = \sum_{l=-\infty}^{\infty} \sum_{p=0}^{\infty} \lambda_{l,p} u_{l,p}(\mathbf{q}_1) u_{-l,p}(\mathbf{q}_2), \tag{16.6}$$

where p is the radial mode number, l is the azimuthal mode number l, and where OAM conservation has been accounted for via $l_1 + l_2 = 0$. The radial mode number $p = \{0, 1, 2, \ldots\}$ counts the number of zero crossings in the radial profile of the Schmidt eigenmodes $u_{l,p}(\mathbf{q}) = u_{l,p}(q) \exp(\mathrm{i}l\theta)$, outside the origin. The azimuthal mode number $l = \{-\infty, \ldots, -2, -1, 0, 1, 2, \ldots, \infty\}$ is associated with an OAM of $l\hbar$ per photon. The azimuthal dependence of each Schmidt mode has its common $\exp(\mathrm{i}l\theta)$ form. The radial shape, however, depends on the collinear phase mismatch φ; it resembles the well known Laguerre–Gaussian (l, p) profiles only for $\varphi \approx 0$ (or even better for $\varphi \approx -1$) but deviates strongly from these profiles at $|\varphi| \gg 1$.

To obtain analytic solutions, Law and Eberly replace the sinc-type phase-matching function at $\varphi = 0$ by its Gaussian approximation, using $\mathrm{sinc}(b^2|\mathbf{q}_1 - \mathbf{q}_2|^2) \approx \exp(-\mathbf{b}^2|\mathbf{q}_1 - \mathbf{q}_2|^2)$. This Gaussian approximation of the two-photon field has an exact Schmidt decomposition with true Hermite–Gaussian or Laguerre–Gaussian eigenmodes and modal weights

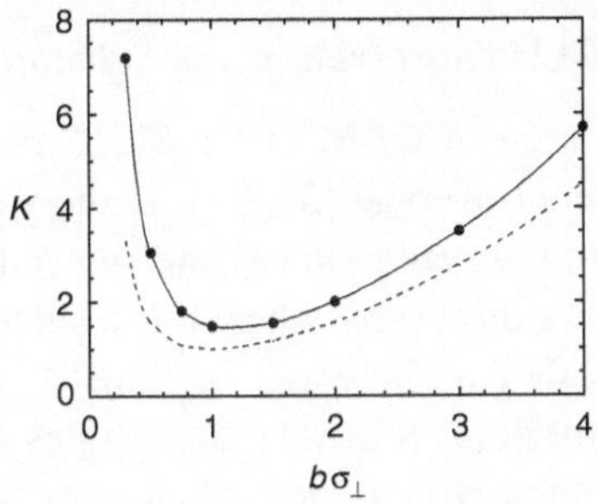

Figure 16.2 The Schmidt number K of the (full two-dimensional) spatial entanglement depends on the product $b\sigma = \sqrt{L/(2z_p)}$. The dashed curve results shows the analytic approximation of Eq. (16.8). The solid curve is obtained by numerical simulations (printed with permission from [33]).

$$P_{l,p} \equiv |\lambda_{l,p}|^2 = C_1 \exp\left(-(2p + |l|)/C_2\right), \tag{16.7}$$

where C_1 and C_2 are two constants. The Schmidt number of this decomposition, i.e., the effective number of modes that participates in the modal decomposition, is

$$K_{2D} \equiv \frac{\left(\sum_{p,l} P_{l,p}\right)^2}{\sum_{p,l} P_{l,p}^2} = \frac{(1 + b^2\sigma^2)^2}{(2b\sigma)^2}. \tag{16.8}$$

Mosts experiment operate in the weak-focusing limit (large w_p), where $b\sigma \ll 1$ and $K_{2D} \approx 1/(2b\sigma)^2$. Rewritten in a different form, that is more useful in experiments, $K_{2D} \approx z_p/(2L)$, where $z_p = \frac{1}{2}k_p w_p^2$ is the Rayleigh range of the pump beam inside the crystal.

Figure 16.2 shows the dependence of the two-dimensional Schmidt number K_{2D} on the dimensionless parameter $b\sigma = \sqrt{L/(2z_p)}$, which can also be written as $b\sigma = w_c/w_p$, where $w_c \equiv \sqrt{L/k_p}$ is the effective transverse width of the photon pair inside the crystal [35]. The dashed curve depicts the analytical approximation of Eq. (16.8). The solid curve is the result of a numerical calculation, performed by Law and Eberly [33], on the exact Eq. (16.3). The exact K values depend in a similar way on the product $b\sigma$, but are typically 20–60% larger than the approximate results obtained by Gaussian expansion.

16.3 Issues in OAM entanglement

The previous section contained a compact description of the spatial entanglement generated in a typical non-critically phase-matched SPDC geometry. In this section, we will highlight several issues that were only briefly discussed above and that have raised considerable interest in the literature. The discussion, which is divided into three subsections, concentrates on the limitations of the simple description and possible short-cuts and further understanding.

16.3.1 OAM conservation and Schmidt modes

The literature on the conservation of OAM in the SPDC process, specified by the conservation law $l_1 + l_2 = l_p$, is controversial [36, 37], as the validity of OAM conservation depends on the geometry. The conservation law is undisputed in the collinear non-critically phase-matched geometry that we discussed, where it is a natural consequence of rotational symmetry. The conservation law can, however, be violated in non-collinear geometries, when the signal and idler photons are emitted at large angles with respect to the pump and when their OAM is defined with respect to these tilted axes. A simple geometric argument indicates that, even in these geometries, OAM is conserved up to values of $l_{\max}\Theta \leq 1$ at an opening angle Θ. We will stick to the collinear geometry.

The spatial entanglement generated in the non-critically phase-matched collinear geometry is always of a special form. The conservation of OAM allows us to label each Schmidt mode by a fixed azimuthal quantum number l and an accompanying radial quantum number p, and write its profile as $u_{l,p}(\mathbf{q}) = u_{l,p}(q) \times \exp(il\theta)$. Despite the similarity with the Laguerre–Gaussian (LG) modes, which have the same quantum numbers and the same azimuthal dependence, the radial profile of the Schmidt modes can be very different from that of the LG modes. The shape of this radial profile depends critically on the on-axis phase mismatch φ. The Schmidt modes resemble the LG modes at $\varphi = 0$, but are very different at $|\varphi| \gg 0$. At $\varphi \ll 0$, where the SPDC ring has opened, even the fundamental $l = p = 0$ Schmidt mode is ring-shaped and has a wide region of zero intensity around the central axis.

16.3.2 Dimensionalities ($K_{2\mathrm{D}}$ vs K_{az}) and optical etendue N

The Schmidt number can be easily estimated by comparing the one-photon coherence of the SPDC emission with that of more standard optical sources. At sufficiently large $K_{2\mathrm{D}} \gg 1$, the Schmidt number of the entangled state reduces to the classical concept of the normalized optical etendue $N \equiv A\Omega/\lambda^2$, where A is the effective area of the source and Ω is its effective (solid) opening angle [35, 38]. The optical etendue quantifies the effective number of transverse modes contained in a partially coherent beam. The approximate sign in the expression $K_{2\mathrm{D}} \approx N$ refers to the way in which we define the effective area A and opening angle Ω of a source with a smooth radiation profile.

The Schmidt number $K_{2\mathrm{D}}$, often simply denoted as K, is the dimensionality of the full two-dimensional spatial entanglement. Many experiments, however, concentrate on the azimuthal part of this entanglement, as only this part has a nice generic form. The dimensionality of the azimuthal entanglement is defined as $K_{\mathrm{az}} = 1/\sum_l P_l^2$, where $P_l = \sum_{p=0}^{\infty} P_{l,p}$ and we assumed $P_l = 1$. For high-dimensional entanglement ($K_{2\mathrm{D}} \gg 1$) and $\varphi \approx 0$, the relation between both forms of entanglement is $K_{\mathrm{az}} \approx 2\sqrt{K_{2\mathrm{D}}}$, as the effective range of the l labels is typically $\approx 4\times$ as large as that of the p labels (see discussion below Eq. (16.6)). For the Gaussian approximation of the two-photon field, this statement can easily be checked with the help of Eq. (16.7). It is unclear whether this approximate relation also applies at $|\varphi| \ll 0$.

16.3.3 Generated vs detectable entanglement

The generated entanglement can only be used efficiently if the detection system is sufficiently large and angle insensitive to detect all Schmidt modes with a comparable (preferably high) quantum efficiency, i.e., if one uses so-called bucket detectors. For most other detectors the finite optical etendue of the detection systems tends to limit the dimensionality of the detected two-photon field. Under these limiting conditions, one is tempted to include the spatial limitation of the detection system in the two-field and perform a Schmidt decomposition of the *detectable* instead of the *generated* two-photon field; this has been done in [38]. Alternatively, one can define the dimensionality of the detected entanglement on the basis of the detector geometry alone. This concept is discussed in a series of papers on the detection of azimuthal entanglement through coincidence measurement with two single-mode detectors equipped with rotatable phase plates [39, 40].

16.4 Experimental analysis of OAM states

16.4.1 Experimental tools for OAM measurement

The decomposition of optical beams into their OAM states is quite challenging as it requires detailed information on the full optical field. Optical interference with a suitably designed reference beam would seem an obvious choice to access the optical phase. However, the conversion of part of the original beam to a suitable reference beam tends to be very inefficient, and the analysis of the resulting interference patterns is cumbersome. An attractive alternative is to interfere the original beam with a rotated version thereof in a modified Mach–Zehnder interferometer, as described in Section 16.4.4.

The phase-sensitivity required for OAM analysis can also be obtained by using mode-selective detectors, i.e., detectors that are only sensitive to light with one particular l value, or certain combinations thereof. These detectors generally acquire their mode selectivity from a single-mode fiber that transmits only the fundamental TEM_{00}-like $l = 0$ (and $p = 0$) mode. Mode selectivity for other OAM states is obtained by positioning a programmable or fixed OAM mode converter in front of the fiber such that the desired OAM state (or combination thereof) is converted into the $l = 0$ fiber mode before mode projection and detection. To be more precise, the detector count rate after mode projection is

$$R \propto \int \Psi^{*}_{\mathrm{det}}(\mathbf{x})\Psi_{\mathrm{in}}(\mathbf{x})\mathrm{d}\mathbf{x}, \tag{16.9}$$

where $\Psi_{\mathrm{in}}(\mathbf{x})$ is the incident field and $\Psi_{\mathrm{det}}(\mathbf{x})$ is the field profile of the detection mode. For the commonly used azimuthal phase plates, the field profile of the detector, $\Psi_{\mathrm{det}}(\mathbf{x}) = \exp[\mathrm{i}f(\theta)]\Psi_{\mathrm{fiber}}(r)$, factorizes in a pure azimuthal phase factor originating from the phase plate and a pure radial amplitude originating from the $l = 0$ fiber mode. It is a matter of choice, whether the mentioned phase factor is included in the effective detection mode or whether its action is interpreted as a mode conversion of the incident light prior to projection. Note that the count rate R involves a two-dimensional integration. Hence, the outcome of OAM analysis by mode projection will necessarily depend on the radial overlap between the source and the detection mode, a dependence that is not always fully appreciated.

Two types of mode converters are in use: so-called phase plates, with a fixed phase pattern, and spatial light modulators, liquid crystal devices that can be programmed to produce any phase-shift pattern. The flexibility of SLMs is an obvious advantage but it is somewhat offset by their limited reflection/transmission efficiency. OAM analysis with SLMs has been demonstrated in several experiments, among others the combined OAM analysis of the two photons out of a quantum-entangled pair [41].

OAM analysis by mode projection is convenient but inefficient, as the projection removes all available information and power from the other modes. A mode sorter, which simply sorts the available power over different OAM-dependent output channels, doesn't have this disadvantage. The first mode sorter was based on a Mach–Zehnder interferometer with image rotation. This device can easily sort the odd and even l modes over two output ports and can be extended to multiple output ports by cascading interferometers [24]. A different design, based on a complete reshaping of the optical field with two SLMs, has recently been demonstrated to sort all OAM modes over a continuum of output ports [25, 42].

16.4.2 Production of phase plates

In this section we will discuss the production of various types of phase plates. There are two types: those with a continuously varying phase profile, and binary phase plates. For the latter the phase takes on only two values, associated with a digital 0 and 1 and preferably π radians apart. We have made binary phase plates with various designs: those with a single elevated segment, multi-segment plates, and a truly two-dimensional design.

Figure 16.3 shows the design (left) and production technique (right) of our spiral phase plates. These are transparent plates with a uniform refractive index n and a smooth thickness variation of the form $h(\theta) = h_0 + h_s\theta/(2\pi)$, which varies linear from $\theta \in \langle 0, 2\pi \rangle$ and suddenly jumps at $\theta = 0 = 2\pi$. When such a spiral phase plate is inserted in the waist of a Gaussian beam it imprints a "vortex" with a "charge" equal to $\tilde{l} = h_s(n - n_0)/\lambda_0$, where n_0 is the refractive index of the embedding medium and λ_0 is the vacuum optical wavelength, thereby generating an output beam that carries on average $\tilde{l}\hbar$ OAM per photon. For the plate described below $\tilde{l} = 3.48 \pm 0.02$ at $\lambda = 813$ nm. Non-integer spiral phase plates allow one to probe a large dimensional phase space [44].

The right-hand side of Fig. 16.3 shows the manufacturing technique. We diamond-machine a mold that, naturally, is the negative of the plate that we wish to produce. The shape-accuracy of this mold is impressive: the interferometrically measured step height of 5.07 μm is very close to the designed value of 5.0 μm, the relation between the local height h and the azimuthal angle deviates only 15 nm (rms) from the ideal linear relation, the step discontinuity is limited to about 6°, and the diameter of the central anomaly is only ≈ 50 μm, very small compared to the overall diameter of the device. Subsequently, a

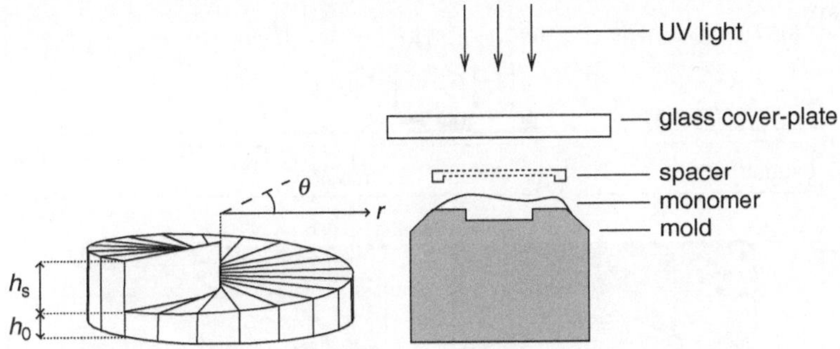

Figure 16.3 Sketch of (a) spiral phase plate and (b) its production method. The top surface spirals upwards from a height h_0 to $h_0 + h_s$. The plate adds a helical phase factor $\exp i\tilde{l}\theta$ to the transmitted optical field. These plates are produced by pressing an accurate mold into UV-curable monomer (printed with permission from [43]).

50 μm thick ring-shaped spacer is place on top of the mold, filled with curable monomer and topped with a cover glass. Curing under UV light converts the monomer to a polymer layer that remains stuck to the cover-glass after removal from the mold (see [43] for further details). One factor that makes this molding approach attractive is that it allows for the production of a couple of identical copies of a certain spiral phase plate design.

The performance of spiral phase plates for OAM conversion is superior to that of the related dislocation (fork) hologram, obtained by interference of a spiral wave front with a tilted plane wave. First of all, because the conversion efficiency of the former can be close to 100% for AR-coated devices whereas that of dislocation holograms is limited to a theoretical maximum of $(2/\pi)^2 \rightarrow 38\%$, but also because the former doesn't deflect the beam whereas the latter does.

For the production of the binary phase plates we use a polymer-based photolithography process. Figure 16.4 sketches the photolithographic production, which starts with the deposition of a uniform layer of the negative photoresist SU-8 onto the substrate by spin coating. This is more challenging than it would seem because the product of refractive index and layer thickness has to be just right and the required layer thickness is unusually large for conventional spin coating. After baking, the phase mask is imprinted by soft-contact lithography, the layers are processed and the left-over removed, to leave a patterned layer with a well-defined thickness (see [45] for details). This method is well suited when the phase pattern has limited spatial complexity.

For complex patterns the best results are obtained with laser-beam lithography in combination with reactive-ion etching. It allows for the fabrication of structures with sub-micron resolution.

16.4.3 OAM analysis with mode-selective detectors

The fabricated phase plates have been used for the OAM analysis of both classical light and quantum-entangled light. The experiments involving classical light were aimed at checking

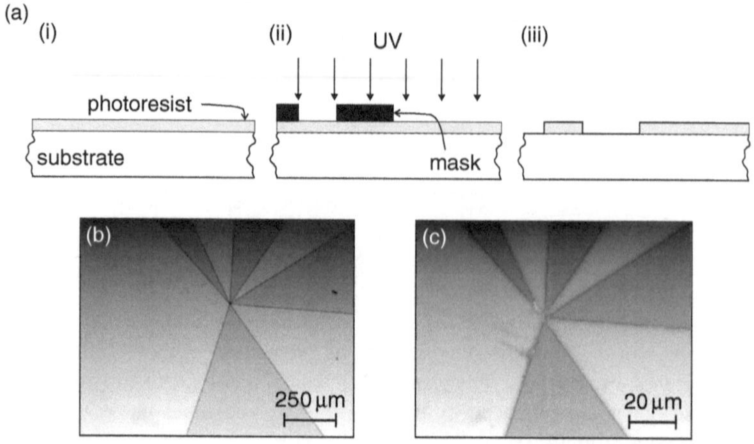

Figure 16.4 Photolithography of polymer-based phase plate. (a) Schematic of the fabrication: (i) a coating of diluted SU-8 is spun onto a glass substrate tailored to form a homogeneous layer of thickness $h = 707 \pm 10$ nm; (ii) a mask of the phase-plate profile is brought into soft contact and the stack is treated photolithographically; after which (iii) the sample is baked and the unexposed regions are lifted. (b) and (c) Microscopic images of a multi-segment phase plate made with this technique at two magnifications.

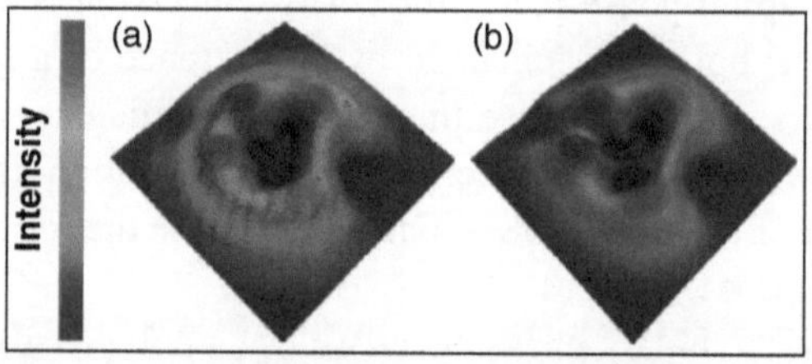

Figure 16.5 3-D representation of intensity profile of the far-field diffraction pattern of a $\tilde{l} = 3.5$ spiral phase plate for a Gaussian input beam: (a) experimental result; (b) theoretical result calculated from diffraction.

the quality of the plates. The imprinted phase profile can be accurately assessed by illuminating the plate with a fundamental Gaussian beam and observing its far-field diffraction pattern. Figure 16.5 shows the measured and calculated diffraction pattern behind the spiral phase plate of Fig. 16.3. In both cases, the pattern is ring-like, with three slightly separated regions of very low intensity inside. Each of these regions contains a phase vortex. An additional vortex lies outside the high-intensity ring. Diffraction calculations show that the far-field position of the latter vortex provides a direct measure of the fractional part of the beam's vortex charge; in the present case this fractional part equals 0.5 resulting in a total "vorticity" of $\tilde{l} = 3.5$ for this phase plate.

Figure 16.6 shows our generic setup for the OAM analysis of quantum-entangled light with phase plates. Quantum-entangled photon pairs, generated via collinear spontaneous

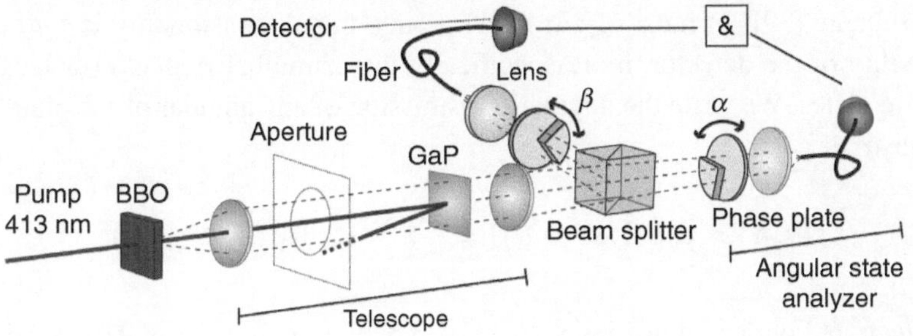

Figure 16.6 Our generic setup for the OAM analysis of quantum-entangled light with phase plates (printed with permission from [39]).

parametric down conversion (SPDC) in a nonlinear crystal are separated at a beam splitter and detected in coincidence by two photon counters positioned behind a combination of a rotatable phase plate and single-mode fiber. An optical low-pass GaP wafer removes the pump beam, narrow-band interference filters provide further spectral selection of the SPDC light. We measure the count rate R_{cc} of the two-photon coincidence detection as a function of the angular settings α and β of the two phase plates. This allows us to demonstrate the presence of OAM entanglement and determine the dimensionality of the detected entanglement.

Some technical details about this setup: as non-linear crystals we have used either BBO or periodically poled KTP; the SPDC efficiency of PP-KTP is about 25× larger than that of BBO [46]. We typically use 100–200 mW of pump power from a Kr-ion laser at an optical wavelength of either 406.7 or 413.1 nm, yielding frequency-degenerate SPDC light at either 813 or 826 nm. For our first experiments we used photon pairs generated by non-collinear SPDC, where the signal and idler beam are emitted at small (and opposite) angles with respect to the pump. This geometry provides for easy spatial separation of the signal and idler beams and facilitate suppression of the pump beam. A disadvantage of the non-collinear geometry is that the spatial emission is concentrated in ring-like structures, which frustrate convenient detection; it yields either elliptical beams or circular beams with a strongly reduced number of OAM modes [38]. Later experiments used the collinear geometry sketched in Fig. 16.6. This geometry allows for easy collection of all the SPDC light and thus profits maximally from the high spatial dimensionality of the source.

The OAM analysis of quantum-entangled light with phase plates is based on measurements of the coincidence count rate $R_{cc}(\alpha, \beta)$ as a function of the orientation angles α and β of the phase plates positioned in the signal and idler beam, respectively (see Fig. 16.6). When the centers of both plates are properly aligned with the axis of the pump beam (a difficult and time-consuming task), the coincidence rate $R_{cc}(\alpha, \beta)$ is observed to depend on the angle difference $\alpha - \beta$ only. This observation alone proofs the non-local nature of the photons' OAM and hence the spatial entanglement of the photon pair [47].

In order to quantify the number of azimuthal modes generated and detected in our experiment we have introduced the Shannon dimensionality D of the combined generation and detection scheme [39]. In most of our experiments, this dimensionality is limited by the dimensionality of the detector, more specifically the azimuthal richness (or lack thereof) of our phase plates. We write the angular transmission of any angular phase plate, oriented under angle α as

$$t(\theta;\alpha) = \exp\left[\mathrm{i}f(\theta-\alpha)\right] = \sum_{m=-\infty}^{\infty} \lambda_m \exp\left[\mathrm{i}m(\theta-\alpha)\right], \tag{16.10}$$

and introduce its Fourier intensities $\gamma_m \equiv |\lambda_m|^2$, where $\sum \gamma_m = 1$ as $|t(\theta,\alpha)| = 1$ for all angles. We define the Shannon dimensionality of the detector as

$$D \equiv 1/\sum_{m=-\infty}^{\infty} \gamma_m^2 \,. \tag{16.11}$$

Note the analogy with the earlier definition of the dimensionality or Schmidt number K of the generated state (see Eq. (16.8) and Section 16.3.2). The shape of the coincidence curve $R_{cc}(\alpha-\beta)$ contains complete information on the Fourier intensities γ_m [40]. For the assumed wide-band OAM source, the Shannon dimensionality D can, however, already be determined by comparing the area $\int R_{cc}(\alpha-\beta)d(\alpha-\beta)$ under the coincidence curve with the area of a rectangular block with a width 2π and height equal to the maximum R_{cc}; the Shannon dimensionality D equals the ratio of the latter over the former area [40]. For this approach to be applicable, the two-photon state must be pure up to the point of detection.

Figure 16.7 shows the measured coincidence rate for three different experiments, using three different sets of identical phase plates in signal and idler beams. Figure 16.7(a) shows the coincidence count rate measured with two half-sector plates oriented at angles α and β. The solid curve is a parabolic fit through the experimental points (circles). A Fourier decomposition of this parabolic shape yields non-zero OAM intensities $\gamma_m = (2/\pi)^2(1/m)$ only for odd m and a (Shannon) dimensionality of $D=3$. The squares and triangles, fitted by the dashed curves, are measured at reduced apertures. This reduces the number of generated OAM modes, limiting the dominant modes to $|l|=1$, modifies the coincidence curve into a sine function and reduces the dimensionality to $D=2$.

Figure 16.7(b) shows the results obtained with a quarter-sector plate and open aperture. The measured circles yield an experimental dimensionality of 5.8, which is close to the maximum value of $D=6$ obtainable with single-sector phase plates.

Figure 16.7(c), and its expanded view (d), show the coincidence count rate measured with two four-sector phase plates with the indicated design. The sketched plate has a design dimensionality of $D=23.0$, which is the optimum for $N=4$. The experimental value of the dimensionality we obtain from the data equals $D=16.5$. There are two causes for this difference: first, the experiment is very sensitive to even the slightest misalignment of the phase plates; the discrepancy between theory and experiment at the bottom of the curve is a direct result hereof. Second, in our derivation of Eq. (16.11) we assumed the source to have a flat spiral spectrum. With an increasing number of sectors on the phase plates this assumption breaks down causing the peak around $\alpha=\beta$ to be less sharp, thus reducing

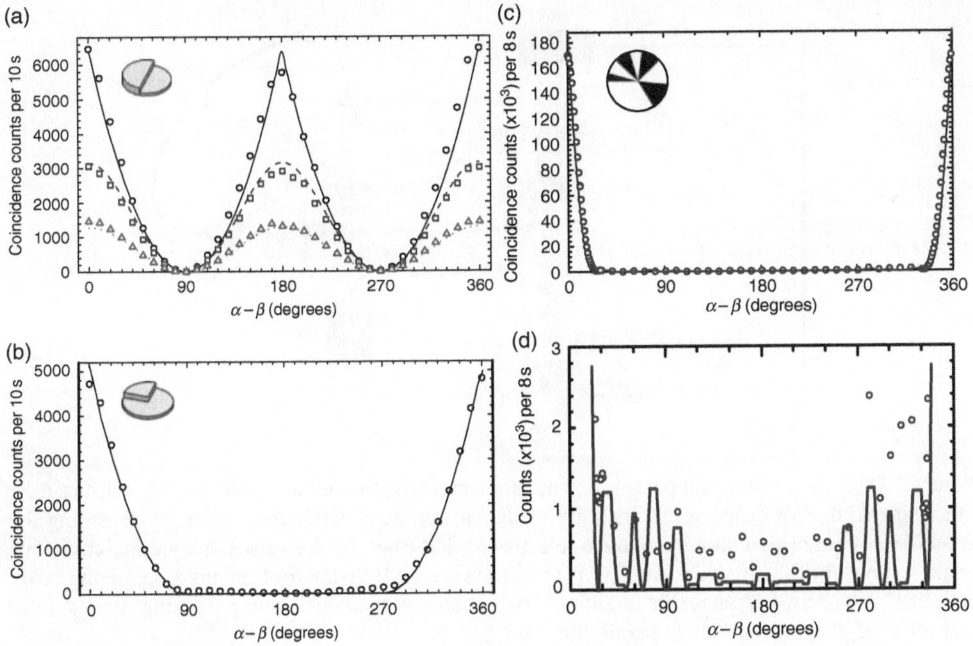

Figure 16.7 Coincidence count rate versus the relative orientation of two analyzing phase plates, for three different designs. The points denote experimental data, the curves are theoretical predictions. (a) *Half-sector plate*. The parabolic fringe (circles) measured at open aperture corresponds to a dimensionality $D = 3$. Truncating the number of modes, by closing the aperture, gradually reduces the parabola into a sine of dimensionality $D = 2$ (squares and triangles). (b) *Quarter-sector plate*. The experimental dimensionality $D = 5.8$, based on the measurements (circles), is close to the theoretical $D = 6$ associated with the piece-wise parabolic curve. (c) and detail (d) *Four-sector plate* designed for maximum dimensionality. The circles represent experimental date, the solid curve is based on theory. The theoretically predicted dimensionality equals $D = 23.0$ Experimentally, we obtain $D = 16.5$ (printed with permission from [39, 45]).

the experimental dimensionality. For N-sector plates with even larger N (up to $N = 14$), numerical simulations show that the optimum D increases approximately linear with N at a slope of 5.9, a behavior that is hitherto not understood [45]. As a general trend, the best designs have the lowest symmetry.

16.4.4 OAM analysis with bucket detectors

It is virtually impossible to perform an accurate unbiased OAM analysis with mode-selective detectors, as the outcome of such an analysis depends crucially on the modal overlap between the radial profiles of the detector modes and the source. A complete unbiased OAM analysis hence requires the use of large (position-insensitive) bucket detectors instead of mode-selective detectors. This subsection describes an experimental implementation of this idea [22]. Using two-photon interference in combination with image rotation,

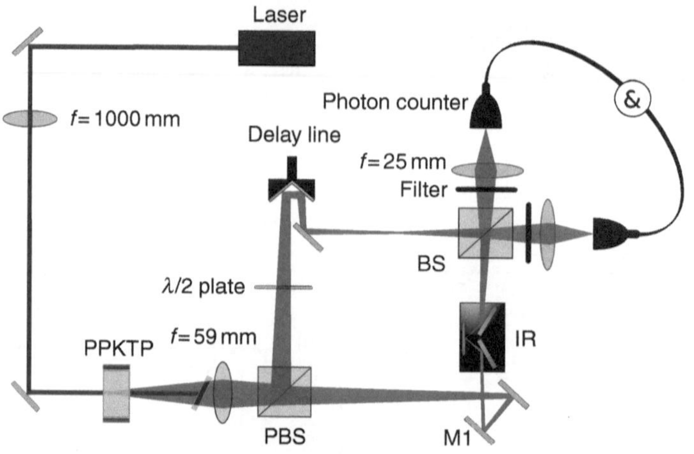

Figure 16.8 Experimental setup used to generated entangled two-photon pairs and to measure their OAM spectrum. We measure the visibility of the two-photon interference $V(\theta)$, by recording the coincidence rate of two photon counters (electronics indicated by &) versus delay time, at various angle settings θ of the image rotator (IR). The original type II geometry contains a polarizing beam splitter (PBS) and a $\lambda/2$ plate; the modified type I geometry contains a non-polarizing beam splitter and no wave plate. (Printed with permission from [22].)

we were able to obtain a full OAM analysis of the generated entanglement and demonstrate its tunability with crystal length L and collinear mismatch phase φ.

Figure (16.8) shows the experimental setup. Entangled photon pairs are generated by collinear SPDC in a periodically poled KTP crystal that is pumped with a weakly focused laser beam ($\lambda_p = 413$ nm, $w_p = 150$ μm). After removal of the pump light, with an AR-coated GaP wafer, all SPDC light is collected with a $f = 59$ mm lens, send through a Mach–Zehnder interferometer and collected by two large bucket detectors. Spectral filters (2 nm @ 826 nm) in front of these detectors are used to select photons close to frequency degeneracy. Fast coincidence logic is used to measured the two-photon coincidence count rate as a function of the path length difference in the interferometer, the so-called HOM (Hong, Ou, and Mandel) dip [48]. We measure the visibility of this two-photon interference $V(\theta)$ for various angles θ of the image rotator (IR) and normalize the measured visibility by the value obtained at $\theta = 0°$ (0.80 for 2 mm type-II PPKTP). Such normalization is allowed as the visibility is limited by an understandable amount of spatial-spectral correlations that doesn't affect further analysis. The image rotator, which contains three metallic mirrors, has the convenient property that it only rotates the image but leaves the optical polarization practically intact [21].

We will first describe the OAM analysis of light generated by a 2 mm thick PPKTP crystal tuned for optimum phase matching ($\varphi = 0$). Figure (16.9) shows the measured HOM dip at $\theta = 0°$. The temporal width of this dip is linked to the spectral width of the filters and irrelevant for our analysis. We are only interested in the visibility of the HOM dip.

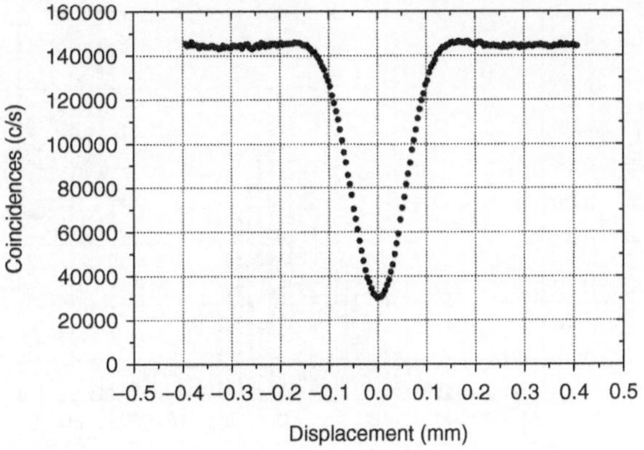

Figure 16.9 The coincidence rate observed behind the interferometer exhibits a so-called HOM-dip as a function of mirror displacement.

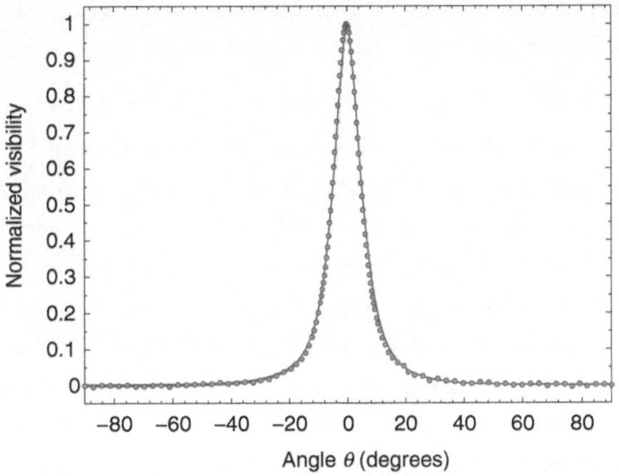

Figure 16.10 Normalized visibility of the measured HOM-dip as a function of the image rotation angle θ. The theoretical fit (solid curve) practically coincides with the measurements (circles).

Figure 16.10 shows the normalized visibility of similar HOM dips as a function of the image rotation angle θ. A Fourier analysis of this function

$$V(\theta) = \sum_{l=-\infty}^{l=\infty} P_l \cos 2l\theta, \tag{16.12}$$

yields the full spiral spectrum $\{P_l\}$. A crucial assumption for the derivation of this equation is the requirement of mirror symmetry $P_{-l} = P_l$. As a cute aside, we note that Eq. (16.12) predicts a recurrence of the two-photon interference already at a rotation angle of $\theta = 180°$ instead of the obvious $\theta = 360°$. This early recurrence, which demonstrates the two-photon character of the interference, has been observed experimentally and is an accurate check of the alignment of the image rotator.

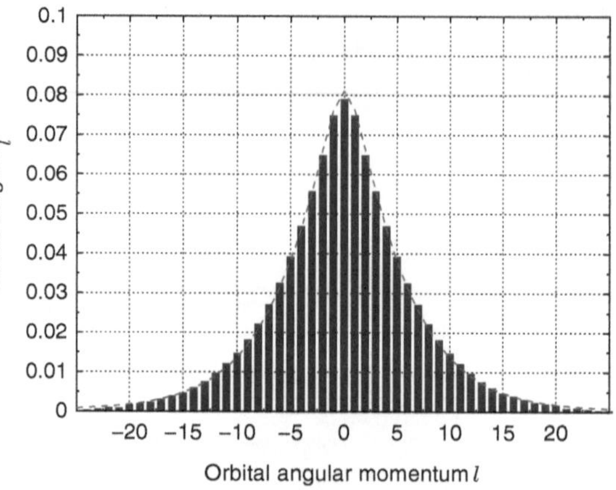

Figure 16.11 OAM modal weight P_l versus OAM l as measured (bars) and calculated (dashed curve) for a 2 mm thick PPKTP crystal operating at $\varphi = 0$ and pumped by a $w_p = 150$ μm pump beam. The agreement is perfect and corresponds to a spiral bandwidth of $K_{az} = 21.4 \pm 0.5$ (printed with permission from [22]).

Figure 16.11 shows the results of a Fourier analysis of the date in Fig. 16.10. The agreement between experiment (bars) and theory (dashed curve) is almost perfect. The experiment yields an azimuthal Schmidt number of $K_{\rm az}^{\rm ex} = 21.4 \pm 0.5$, while the theoretical value is $K_{\rm az}^{\rm th} = 21.6$ at our normalized pump waist $\overline{w}_p \equiv w_p/\sqrt{\lambda_p L} = 5.2$ (the Gaussian approximation predicts a substantially lower value of $K_{\rm az}^{\rm Gauss} \approx \sqrt{2\pi}\,\overline{w}_p \approx 13.0$). Figure (16.11) demonstrates that we completely understand the generation of OAM entanglement!

Next, we consider the OAM analysis of SPDC light at imperfect phase matching ($\varphi \neq 0$). For this experiment we use a 5 mm thick PPKTP crystal, for which the collinear mismatch phase is known to tune as $\varphi = 1.04 \times (T - T_0)$, where $T_0 \approx 60°$ K is the crystal temperature for perfect phase matching. At each setting of φ, we again measure the visibility of the two-photon interference versus the image rotation angle and perform a Fourier analysis to obtain the OAM modal decomposition $\{P_l\}$. Figure 16.12 shows the obtained spiral spectra for four values of φ. At $\varphi = 0$, we now measure an azimuthal Schmidt number of $K_{\rm az}^{\rm ex} = 13.8 \pm 0.5$, while the theoretical $K_{\rm az}^{\rm th} = 13.9$, in agreement with the expected scaling $K_{\rm az} \propto 1/\sqrt{L}$. At a positive detuning of $\varphi = 1.0$, where the rings close and the coincidence rate decreases, the measured azimuthal entanglement has decreased somewhat to $K_{\rm az}^{\rm ex} = 12.0$. For negative detuning, where the rings open up and the coincidence rate increases, the azimuthal Schmidt number increases dramatically. At $\varphi = -2.1$ we already measure $K_{\rm az}^{\rm ex} = 21.4$, while this increases to $K_{\rm az}^{\rm ex} = 34.2$ at $\varphi = -6.2$. Within the considered OAM range of $|l| \leq 10$, the probability distribution is now so uniform that there is virtually no need for entanglement concentration [49]. This is important because most of the potential applications of quantum entanglement work best for maximally entangled states. Phase mismatch therefore not only extends the range of useful l modes, allowing larger alphabets, but can also increase the conversion efficiency.

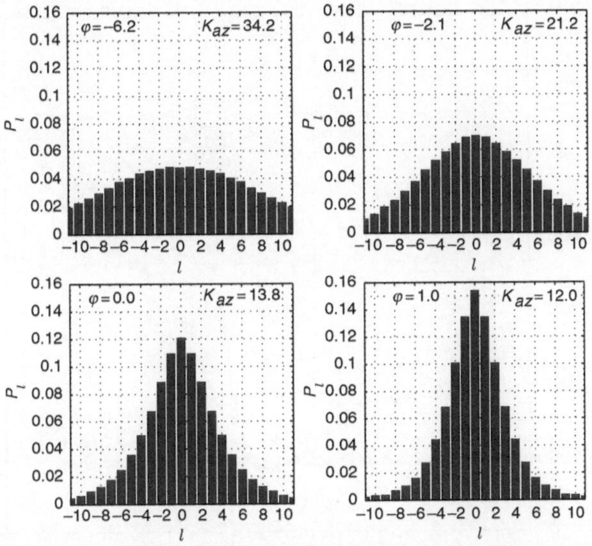

Figure 16.12 The effect of the phase mismatch on the spiral spectrum $\{P_l\}$ as demonstrated for a 5 mm long PPKTP crystal pumped by a $w_p = 160\,\mu$m beam. For negative φ, where the SPDC rings are open and the pair generation is more efficient, the spectrum is flatter and the Schmidt number K_{az} is larger than at $\varphi = 0$ (printed with permission from [22]).

The collinear mismatch phase φ will obviously also affect the full two-dimensional Schmidt number $K_{2\text{D}}$ of the generated state. Our experimental method to determine $K_{2\text{D}}$ is based on its close relation with the normalized optical etendue $N \equiv A\Omega/\lambda^2$. Figure 16.13 shows the experimental results, obtained by measuring the effective area A and (solid) opening angle Ω of the SPDC source, at various values of φ. The circles are the measurements. The solid/dashed curves are theoretical predictions, without/with a correction for the finite detection area, based on a fitted value of $b\sigma = 0.077$. The qualitative and quantitative agreement is almost perfect; even the peculiar wiggles in the curves, associated with the (dis)appearance of SPDC rings, are reproduce experimentally.

The four insets in Fig. 16.13 show far-field intensity profiles at four different φ values. At $\varphi = 0$ the center of the fundamental SPDC ring coincides with the pump beam. At $\varphi < 0$ this ring opens up, which results in an approximately twofold increase in the pair flux and a similar (or even somewhat larger) increase in the two-dimensional Schmidt number $K_{2\text{D}}$, here simply denoted as K. At $\varphi > 0$ the Schmidt number first decreases somewhat, but then increases steadily exhibiting wiggles when consecutive SPDC rings disappear from view. At $\varphi \approx 1$ the central peak resembles a Gaussian and K reaches its minimum values. For $\varphi > 1$, the fundamental ring disappears and weaker, secondary, peaks of the sinc-function result in a wider far-field and an increasing value of K.

A big disadvantage of positive detuning ($\varphi > 0$) is that the down-conversion efficiency drops dramatically due to lack of phase matching at any emission angle. More precisely, the

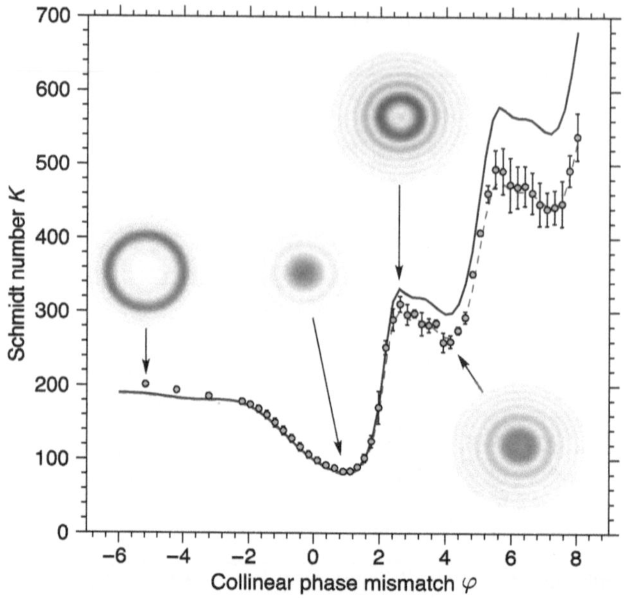

Figure 16.13 The Schmidt number K varies with the collinear phase mismatch φ. Experimental results for the measured etendue $N \approx K_{2D}$ for a 5 mm long PPKTP crystal pumped by a $w_p = 160\,\mu m$ beam are indicated as circles. These circles fall slightly below the theoretical prediction for infinite detection area (solid line), but agree well with the theoretical prediction that includes the finite size of the detection area (dashed curve). The four ring-shaped patterns show the measured angular emission profiles; the strong fundamental ring is open for $\varphi < 0$, shrinks to a central lobe around $\varphi \approx 0$, and disappear at $\varphi > 0$ leaving only weaker higher-order rings (printed with permission from [35]).

angle-integrated photon flux $F(\varphi)$ scales as $F(\varphi) \propto 1 + (2/\pi)[\varphi \mathrm{sinc}^2(\varphi) - \mathrm{Si}(2\varphi)]$, where $\mathrm{Si}(x) \equiv \int_0^x \sin(x')\,dx'$ is the sine integral function, yielding for instance $F(0) = 1.0$, $F(4) = 0.09$, and $F(8) = 0.04$.

We close this experimental section with two remarks on alternative experiments aimed at obtaining the OAM contents of a source. First, we note that the two-photon interference experiment with image rotation, was originally performed on only a small fraction of the SPDC light generated from a non-collinear SPDC source in the form of a BBO crystal [21, 50]. Although this experiment yielded convincing results and azimuthal Schmidt numbers up to $K_{az} \approx 8$, the measured spiral spectrum was limited by the size of the apertures that were used to select the interfering beams and contained practically no information on the (much higher) dimensionality of the source. Second, the OAM contents of partially coherent light can be measured with a similar scheme as the one presented here. By injecting the light into a Mach–Zehnder interferometer with image rotator, and measuring the visibility of the interfering optical fields at various rotation angles, one can again apply a Fourier decomposition to obtain the OAM spectrum of the injected light. This technique has been applied in [51].

16.5 Concluding discussion and challenges

In this chapter we have discussed the production and detection of OAM-entangled photon pairs generated via SPDC with a rotationally symmetric pump beam. We started out by presenting general expressions that relate the (modal distribution of the) entanglement to the angular size of the pump beam and the phase matching function and by highlighting intriguing aspects thereof. In the experimental section, we discussed two sets of experiments that use either (i) mode transformation and projection on single-mode detectors, or (ii) interference and detection with large mode-insensitive detectors. We introduced experimental tools for OAM analysis, demonstrated the presence of OAM entanglement and determined a lower limit for its dimensionality based on the detectable entanglement. In the second set of experiments, we performed an unbiased measurement of the spiral spectrum and showed how the spiral bandwidth can be increased by operating at a collinear phase mismatch $\varphi < 0$. We now end by pointing out some remaining challenges.

The biggest challenge in OAM-entanglement is probably the question how to properly deal with the radial dependence of the optical fields. For the generated two-photon field, the mathematical approach is clear and described in Eqs. (16.2)–(16.6). However, practically all experiments performed up to now use a combination of mode transformations and projections. As these projections are critically sensitive to the radial profile of the field, they unavoidably introduce a detection bias for some OAM states over others. Furthermore, the limited size of mirrors and lenses introduces unwanted apertures in any setup, potentially reducing the dimensionality of the available entanglement and modifying the associated Schmidt modes. These issues have to be considered in any OAM-based system.

A more general challenge in spatial entanglement is the potential to actually benefit from the mentioned radial variations and use both transverse degrees of freedom. From a theoretical point of view, the two-dimensional (l, p) Schmidt modes are the best basis to operate in. In practice, the discrete (top-hat or δ-like) states used in pixel entanglement could be an easy alternative [15]. Although this basis cannot access the full two-dimensional entanglement, on account of modal overlap, the part that it can access might well be more than enough in practice.

As a third challenge we mention the desirability to transport spatial entanglement over large distances. This requires an image-preserving optical transmission channel that also maintains the optical phase front over the full image. The transmission in free space is dominantly limited by optical turbulence [52, 53]. The transmission in optical fibers is limited by the optical dispersion between the guided transverse modes [54].

Finally, we make a comparison between the use of spatial entanglement versus temporal (= time-frequency or time-bin) entanglement. Both forms of entanglement exist in continuous variables and are thus high-dimensional in nature. The dimensionality of temporal entanglement can potentially be much higher than that of spatial entanglement. However, a high dimensionality $K \approx \Delta\omega_{\mathrm{SPDC}}/\Delta\omega_p$, where $\Delta\omega_{\mathrm{SPDC}}$ and ω_p are the SPDC and pump bandwidth, can only be obtained at a long measurement times $t > 1/\Delta\omega_p$.

High-dimensional spatial entanglement doesn't have this disadvantage and can thus be operated at much higher speeds as it uses many parallel channels, instead of a single time-frequency channel. In the (near) future, the combination of both forms of entanglement could allow one to benefit from the best of both worlds.

References

[1] P. G. Kwiat, K. Mattle, H. Weinfurter, A. Zeilinger, A. M. Sergienko, and Y. H. Shih. New high-intensity source of polarization-entangled photon pairs. *Phys. Rev. Lett.*, **75**(24):4337–41, 1995.

[2] S. P. Walborn, S. Pádua, and C. H. Monken. Hyperentanglement-assisted Bell-state analysis. *Phys. Rev. A*, **68**(4):042313, 2003.

[3] D. Bouwmeester, A. K. Ekert, and A. Zeilinger. *The Physics of Quantum Information*. New York: Springer, 2000.

[4] D. Kaszlikowski, P. Gnaciński, M. Żukowski, W. Miklaszewski, and A. Zeilinger. Violations of local realism by two entangled n-dimensional systems are stronger than for two qubits. *Phys. Rev. Lett.*, **85**(21):4418–21, 2000.

[5] D. Collins, N. Gisin, N. Linden, S. Massar, and S. Popescu. Bell inequalities for arbitrarily high-dimensional systems. *Phys. Rev. Lett.*, **88**(4):040404, 2002.

[6] M. Barbieri, F. De Martini, P. Mataloni, G. Vallone, and A. Cabello. Enhancing the violation of the Einstein-Podolsky-Rosen local realism by quantum hyperentanglement. *Phys. Rev. Lett.*, **97**(14):140407, 2006.

[7] C.-M. Li, K. Chen, A. Reingruber, Y.-N. Chen, and J.-W. Pan. Verifying genuine high-order entanglement. *Phys. Rev. Lett.*, **105**(21):210504, 2010.

[8] J. T. Barreiro, T.-C. Wei, and P. G. Kwiat. Beating the channel capacity limit for linear photonic superdense coding. *Nature Phys.*, **4**(4):282–6, 2008.

[9] C. Brukner, T. Paterek, and M. Zukowski. Quantum communication complexity protocols based on higher-dimensional entangled systems. *Int. J. Quantum Inf.*, **1**(4):519, 2003.

[10] N. J. Cerf, M. Bourennane, A. Karlsson, and N. Gisin. Security of quantum key distribution using d-level systems. *Phys. Rev. Lett.*, **88**(12):127902, 2002.

[11] T. Durt, D. Kaszlikowski, J. L. Chen, and L. C. Kwek. Security of quantum key distributions with entangled qudits. *Phys. Rev. A*, **69**(3):032313, 2004.

[12] R. T. Thew, A. Acín, H. Zbinden, and N. Gisin. Bell-type test of energy-time entangled qutrits. *Phys. Rev. Lett.*, **93**(1):010503, 2004.

[13] L. Olislager, J. Cussey, A.T. Nguyen, P. Emplit, S. Massar, J.-M. Merolla, and K. Phan Huy. Frequency-bin entangled photons. *Phys. Rev. A*, **82**(1):013804, 2010.

[14] L. Neves, G. Lima, J. G. Aguirre Gómez, C. H. Monken, C. Saavedra, and S. P.ádua. Generation of entangled states of qudits using twin photons. *Phys. Rev. Lett.*, **94**(10):100501, 2005.

[15] M. N. O'Sullivan-Hale, I. Ali Khan, R. W. Boyd, and J. C. Howell. Pixel entanglement: experimental realization of optically entangled $d = 3$ and $d = 6$ qudits. *Phys. Rev. Lett.*, **94**(22):220501, 2005.

[16] T. Yarnall, A. F. Abouraddy, B. E. A. Saleh, and M. C. Teich. Experimental violation of Bell's inequality in spatial-parity space. *Phys. Rev. Lett.*, **99**(17):170408, 2007.

[17] G. Molina-Terriza, J. P. Torres, and L. Torner. Twisted photons. *Nature Phys.*, **3**(5):305–10, 2007.

[18] S. P. Walborn, C. H. Monken, S. Padua, and P. H. Souto Ribeiro. Spatial correlations in parametric down-conversion. *Phys. Rep.*, **495**(4–5):87–139, 2010.

[19] L. Allen, M. W. Beijersbergen, R. J. C. Spreeuw, and J. P. Woerdman. Orbital angular momentum of light and the transformation of Laguerre-Gaussian laser modes. *Phys. Rev. A*, **45**(11):8185–9, 1992.

[20] J. P. Torres, A. Alexandrescu, and L. Torner. Quantum spiral bandwidth of entangled two-photon states. *Phys. Rev. A*, **68**(5):050301, 2003.

[21] M. P. van Exter, P. S. K. Lee, S. Doesburg, and J. P. Woerdman. Mode counting in high-dimensional orbital angular momentum entanglement. *Opt. Exp.*, **15**(10):6431–8, 2007.

[22] H. Di Lorenzo Pires, H. C. B. Florijn, and M. P. van Exter. Measurement of the spiral spectrum of entangled two-photon states. *Phys. Rev. Lett.*, **104**:020505, 2010.

[23] S. N. Khonina, V. V. Kotlyar, R. V. Skidanov, V. A. Soifer, P. Laakkonen, and J. Turunen. Gauss–Laguerre modes with different indices in prescribed diffraction orders of a diffractive phase element. *Opt. Comm.*, **175**(4–6):301–8, 2000.

[24] J. Leach, M. J. Padgett, S. M. Barnett, S. Franke-Arnold, and J. Courtial. Measuring the orbital angular momentum of a single photon. *Phys. Rev. Lett.*, **88**(25):257901, 2002.

[25] G. C. G. Berkhout, M. P. J. Lavery, J. Courtial, M. W. Beijersbergen, and M. J. Padgett. Efficient sorting of orbital angular momentum states of light. *Phys. Rev. Lett.*, **105**(15):153601, 2010.

[26] A. Mair, A. Vaziri, G. Weihs, and A. Zeilinger. Entanglement of the orbital angular momentum states of photons. *Nature*, **412**(6844):313–6, 2001.

[27] N. K. Langford, R. B. Dalton, M. D. Harvey *et al.* Measuring entangled qutrits and their use for quantum bit commitment. *Phys. Rev. Lett.*, **93**(5):053601, 2004.

[28] R. T. Thew, K. Nemoto, A. G. White, and W. J. Munro. Qudit quantum-state tomography. *Phys. Rev. A*, **66**(1):012303, 2002.

[29] J. Leach, B. Jack, J. Romero *et al.* Quantum correlations in optical angle-orbital angular momentum variables. *Science*, **329**(5992):662–5, 2010.

[30] A. M. Marino, V. Boyer, R. C. Pooser, P. D. Lett, K. Lemons, and K. M. Jones. Delocalized correlations in twin light beams with orbital angular momentum. *Phys. Rev. Lett.*, **101**(9):093602, 2008.

[31] M. Lassen, G. Leuchs, and U. L. Andersen. Continuous variable entanglement and squeezing of orbital angular momentum states. *Phys. Rev. Lett.*, **102**(16):163602, 2009.

[32] L. Mandel and E. Wolf. *Optical Coherence and Quantum Optics*. Cambridge: Cambridge University Press, 1995.

[33] C. K. Law and J. H. Eberly. Analysis and interpretation of high transverse entanglement in optical parametric down conversion. *Phys. Rev. Lett.*, **92**(12):127903, 2004.

[34] C. H. Monken, P. H. S. Ribeiro, and S. Padua. Transfer of angular spectrum and image formation in spontaneous parametric down-conversion. *Phys. Rev. A*, **57**(4):3123–36, 1998.

[35] H. Di Lorenzo Pires, C. H. Monken, and M. P. van Exter. Direct measurement of transverse-mode entanglement in two-photon states. *Phys. Rev. A*, **80**(2) 2009.

[36] H. H. Arnaut and G. A. Barbosa. Orbital and intrinsic angular momentum of single photons and entangled pairs of photons generated by parametric down-conversion. *Phys. Rev. Lett.*, **85**(2):286–9, 2000.

[37] H. H. Arnaut and G. A. Barbosa. Comment on "Orbital and intrinsic angular momentum of single photons and entangled pairs of photons generated by parametric down-conversion"–Arnaut and Barbosa reply. *Phys. Rev. Lett.*, **86**(22):5209, 2001.

[38] M. P. van Exter, A. Aiello, S. S. R. Oemrawsingh, G. Nienhuis, and J. P. Woerdman. Effect of spatial filtering on the Schmidt decomposition of entangled photons. *Phys. Rev. A*, **74**:012309, 2006.

[39] J. B. Pors, S. S. R. Oemrawsingh, A. Aiello *et al.* Shannon dimensionality of quantum channels and its application to photon entanglement. *Phys. Rev. Lett.*, **101**(12):120502, 2008.

[40] J. B. Pors, A. Aiello, S. S. R. Oemrawsingh, M. P. van Exter, E. R. Eliel, and J. P. Woerdman. Angular phase-plate analyzers for measuring the dimensionality of multimode fields. *Phys. Rev. A*, **77**:033845, 2008.

[41] E. Yao, S. Franke-Arnold, J. Courtial, M. J. Padgett, and S. M. Barnett. Observation of quantum entanglement using spatial light modulators. *Opt. Expr.*, **14**(26):13089–94, 2006.

[42] M. P. J. Lavery, G. C. G. Berkhout, J. Courtial, and M. J. Padgett. Measurement of the light orbital angular momentum spectrum using an optical geometric transformation. *J. Opt.*, **13**:064006, 2011.

[43] S. S. R. Oemrawsingh, J. A. W. van Houwelingen, E. R. Eliel *et al.* Production and characterization of spiral phase plates for optical wavelengths. *Appl. Opt.*, **43**(3):688–94, 2004.

[44] S. S. R Oemrawsingh, J. A. de Jong, X. Ma *et al.* High-dimensional mode analyzers for spatial quantum entanglement. *Phys. Rev. A*, **73**(3), 2006.

[45] J. B. Pors, F. Miatto, G. W. 't Hooft, E. R. Eliel, and J. P. Woerdman. High-dimensional entanglement with orbital-angular-momentum states of light. *J. Opt.*, **13**:064008, 2011.

[46] W. H. Peeters and M. P. van Exter. Optical characterization of periodically-poled $KTiOPO_4$. *Opt. Expr.*, **16**(10):7344–60, 2008.

[47] S. S. R. Oemrawsingh, X. Ma, D. Voigt *et al.* Experimental demonstration of fractional orbital angular momentum entanglement of two photons. *Phys. Rev. Lett.*, **95**(24), 2005.

[48] C. K. Hong, Z. Y. Ou, and L. Mandel. Measurement of subpicosecond time intervals between two phtons by interference. *Phys. Rev. Lett.*, **59**:2044, 1987.

[49] C. H. Bennett, H. J. Bernstein, S. Popescu, and B. Schumacher. Concentrating partial entanglement by local operations. *Phys. Rev. A*, **53**, 2046, 1996.

[50] W. H. Peeters, E. J. K. Verstegen, and M. P. van Exter. Orbital angular momentum anaylsis of high-dimensional entanglement. *Phys. Rev. A*, **76**:042302, 2007.

[51] H. Di Lorenzo Pires, J. Woudenberg, and M. P. van Exter. Measurement of the orbital angular momentum spectrum of partially coherent beams. *Opt. Lett.*, **35**(6): 889–91, 2010.

[52] B. J. Smith and M. G. Raymer. Two-photon wave mechanics. *Phys. Rev. A*, **74**(6):062104, 2006.

[53] J. B. Pors, C. H. Monken, E. R. Eliel, and J. P. Woerdman. Transport of orbital-angular-momentum entanglement through a turbulent atmosphere. *Opt. Expr.*, **19**(7):6671–83, 2011.

[54] W. Löffler, T. G. Euser, E. R. Eliel, M. Scharrer, P. St. J. Russell, and J. P. Woerdman. Fiber transport of spatially entangled photons. *Phys. Rev. Lett.*, **106**:240505, 2011.

Index